E-Mobility in Electrical Energy Systems for Sustainability

Lakshmi D.
VIT Bhopal University, India

Neelu Nagpal
Maharaja Agrasen Institute of Technology, India

Neelam Kassarwani
Maharaja Agrasen Institute of Technology, India

Vishnu Varthanan G.
VIT Bhopal University, India

Pierluigi Siano
University of Salerno, Italy

A volume in the Advances in Mechatronics and Mechanical Engineering (AMME) Book Series

IGI Global
PUBLISHER of TIMELY KNOWLEDGE

Published in the United States of America by
IGI Global
Engineering Science Reference (an imprint of IGI Global)
701 E. Chocolate Avenue
Hershey PA, USA 17033
Tel: 717-533-8845
Fax: 717-533-8661
E-mail: cust@igi-global.com
Web site: http://www.igi-global.com

Library of Congress Cataloging-in-Publication Data

CIP Data in progress

This book is published in the IGI Global book series Advances in Mechatronics and Mechanical Engineering (AMME) (ISSN: 2328-8205; eISSN: 2328-823X)

British Cataloguing in Publication Data
A Cataloguing in Publication record for this book is available from the British Library.

For electronic access to this publication, please contact: eresources@igi-global.com.

Advances in Mechatronics and Mechanical Engineering (AMME) Book Series

J. Paulo Davim
University of Aveiro, Portugal

ISSN:2328-8205
EISSN:2328-823X

Mission

With its aid in the creation of smartphones, cars, medical imaging devices, and manufacturing tools, the mechatronics engineering field is in high demand. Mechatronics aims to combine the principles of mechanical, computer, and electrical engineering together to bridge the gap of communication between the different disciplines.

The **Advances in Mechatronics and Mechanical Engineering (AMME) Book Series** provides innovative research and practical developments in the field of mechatronics and mechanical engineering. This series covers a wide variety of application areas in electrical engineering, mechanical engineering, computer and software engineering; essential for academics, practitioners, researchers, and industry leaders.

Coverage

- Computer-Based Manufacturing
- Intelligent Sensing
- Mechanisms and machines
- Intelligent Navigation
- Medical Robotics
- Sustainable and green manufacturing
- Tribology and surface engineering
- Bioengineering Materials
- Autonomous Systems
- Micro and nanomechanics

IGI Global is currently accepting manuscripts for publication within this series. To submit a proposal for a volume in this series, please contact our Acquisition Editors at Acquisitions@igi-global.com or visit: http://www.igi-global.com/publish/.

The Advances in Mechatronics and Mechanical Engineering (AMME) Book Series (ISSN 2328-8205) is published by IGI Global, 701 E. Chocolate Avenue, Hershey, PA 17033-1240, USA, www.igi-global.com. This series is composed of titles available for purchase individually; each title is edited to be contextually exclusive from any other title within the series. For pricing and ordering information please visit http://www.igi-global.com/book-series/advances-mechatronics-mechanical-engineering/73808. Postmaster: Send all address changes to above address.

Titles in this Series

For a list of additional titles in this series, please visit:
www.igi-global.com/book-series/advances-mechatronics-mechanical-engineering/73808

Harnessing Digital Innovation for Air Transportation
Ayse Asli Yilmaz (Atilim University, Trkey)
Engineering Science Reference • © 2024 • 300pp • H/C (ISBN: 9798369307328) • US $275.00

Studies on Single and Double Actuator Based DC Attraction Type Levitation Systems Optimization Techniques
Rupam Bhaduri (Dayananda Sagar University, ndia)
Engineering Science Reference • © 2023 • 231pp • H/C (ISBN: 9781668473887) • US $225.00

Advances in MEMS and Microfluidic Systems
Rajeev Kumar Singh (Department of Mechanical Engineering, Amity University, Noida, India) Rakesh Kumar Phanden (Department of Mechanical Engineering, Amity University, Noida, India) Basant Singh Sikarwar (Department of Mechanical Engineering, Amity University, Noida, India) and J. Paulo Davim (Department of Mechanical Engineering, University of Aveiro, Portugal)
Engineering Science Reference • © 2023 • 331pp • H/C (ISBN: 9781668469521) • US $240.00

Structural Mathematical Modeling Applications in Technological Machines and Transportation Vehicles
Andrey Eliseev (Irkutsk State Transport University, Russia & Irkutsk National Research Technical University, Russia)
Engineering Science Reference • © 2023 • 288pp • H/C (ISBN: 9781668472378) • US $225.00

Handbook of Research on Emerging Designs and Applications for Microwave and Millimeter Wave Circuits
Jamal Zbitou (LABTIC, ENSA of Tangier, University of Abdelmalek Essaadi, Morocco) Mostafa Hefnawi (Royal Military College of Canada, Canada) Fouad Aytouna (LABTIC, ENSA of Tetouan, University of Abdelmalek Essaadi, Morocco) and Ahmed El Oualkadi (LabTIC, ENSA of Tangier, University of Abdelmalek Essaadi, Morocco)
Engineering Science Reference • © 2023 • 542pp • H/C (ISBN: 9781668459553) • US $325.00

Trends, Paradigms, and Advances in Mechatronics Engineering
Mohamed Arezki Mellal (M'Hamed Bougara University, Algeria)
Engineering Science Reference • © 2023 • 299pp • H/C (ISBN: 9781668458877) • US $225.00

Handbook of Research on Aspects and Applications of Incompressible and Compressible Aerodynamics
Sathish K. Kumar (Nehru Institute of Engineering and Technology, India) and Naren Shankar Radhakrishnan (Vel Tech Rangarajan Dr. Sagunthala R&D Institute of Science and Technology, India)
Engineering Science Reference • © 2022 • 429pp • H/C (ISBN: 9781668442302) • US $335.00

IGI Global
PUBLISHER of TIMELY KNOWLEDGE

701 East Chocolate Avenue, Hershey, PA 17033, USA
Tel: 717-533-8845 x100 • Fax: 717-533-8661
E-Mail: cust@igi-global.com • www.igi-global.com

Table of Contents

Detailed Table of Contents

This study presents a systematic review of the studies and a comprehensive assessment of the ongoing implementation of e-mobility, evaluating social, monetary, policy, and sustainability factors associated with the circular economy (CE) ecosystems framework. It also discusses the current difficulties and proposes methods to address them to improve urban quality of life and the sustainability of electro-mobility. The objective of this study is to discover the particular explanation as to why industrialized countries are able to efficiently introduce cutting-edge electric fuel automobiles (e.g., electric vehicles) while underdeveloped economies cannot find it extremely difficult to implement these vehicles, particularly in terms of infrastructure and downstream activities and how these issues can be addressed effectively.

This study comprehensively evaluates the role of architectural planning in developing sustainable electric mobility infrastructure. As transport emissions rise with motorization, designing low-carbon alternatives requires nuanced, interdisciplinary solutions. Through analysis of urban form, charging access, street layouts, and public transit integration influence on electric vehicle adoption, the research assesses retrofitting challenges and opportunities afforded by forward-thinking design. Conceptualizing mobility ecosystems as sociotechnical systems, the chapter explores dimensions of accessibility, integration, and adaptability in architectural approaches. Sustainability, scalability, and flexibility are theorized as principles fundamental to resilient infrastructure planning. Empirical studies on smart grid technologies' potential to optimize renewable energy charging management are also critically appraised. Finally, analyses of how inclusive, equitable planning supports social equity and policy goals qualitatively examine best practices.

The journey of electric vehicles (EVs) involves myriad challenges and opportunities amid their widespread adoption. EVs, powered by rechargeable batteries and electric motors, are transforming the automotive industry, offering environmental benefits like zero tailpipe emissions, improved air quality, and reduced noise. With instant torque and acceleration, EVs provide a responsive driving experience. The realm of e-mobility holds a promising future with dynamic challenges, opportunities, and emerging trends. Challenges include establishing robust charging infrastructure, addressing range anxiety, ensuring ethical supply chains, and advancing battery technology. Integration with autonomous and connected technologies, supportive policies, and creating a more sustainable transportation landscape are also vital. This chapter explores technical obstacles and opportunities in the industry's pursuit of greater electrification of energy conversion systems in mobile platforms, often referred to as e-mobility.

As transportation, communications, and energy increasingly overlap, new business models may emerge, necessitating new tools and expertise. Researchers predict these will center on three key aspects: external connectivity, internal information and communication technology (ICT) architecture, and human-machine interaction (HMI). In this chapter, the authors thoroughly examine these themes, bringing together insights from requirements analysis and prototype creation to share with the reader. This work also outlines the skills and techniques that will be essential for capitalizing on the many opportunities presented by the imminent market disruption. Data processing, data mining, and data sharing "vehicles" will play a crucial role in the future of our mobility partners. New HMI and driver-assistance apps require a stable and reliable runtime environment, which must be provided by the vehicles of the future. This study further argues that the electric car is an opportunity in and of itself for the future of product lifecycle management and the new Industry 4.0.

Power distribution networks (PDNs) are most susceptible to extreme events, such as cyberattacks, hurricanes, floods, and wildfires, in a power system. Enhancing the resilience to withstand and recover after disruptive events is crucial for planners and operators of PDNs. The resilience-oriented planning and operation can enhance the resilience of PDNs. Furthermore, the significant EV integration, in recent years, provides a potential for improving resilience by restoration in PDNs. Considering the numerous studies regarding deployment of EVs for load restoration after disruptive events, this chapter provides the idea of resilience and the vulnerability of different components of PDNs due to various HILP (high impact, low probability) events. Furthermore, EV application to enhance the resilience of PDNs is reviewed with a detailed discussion of the methods, objectives, and solution strategies. In addition, a qualitative assessment of the studied literature is provided to enhance the comprehension of state-of-the-art resilience improvement methods of EV-enabled PDNs.

The transportation industry is undergoing a rapid transition from fossil fuels to green energies. The emerging concept of mobility includes vehicles that not only provide environmentally friendly transportation, but also serve as storage for excess energy from renewable sources. The e-mobility infrastructure is becoming more complicated, which increases the danger of security threats. A valuable target for cyberattacks, the grid is one of the most significant vital infrastructures. Data exchange between EVs and smart grids is technically necessary. Smart grids must overcome these significant obstacles in order to guarantee security and privacy for eMobile charging. Blockchain when integrated with smart grid, as a possible solution, can solve these problems with its inherent trust building technology and makes it possible for the smart grid to manage demand response and engage in effective and reliable energy trading. The aim of this chapter is to describe the potential privacy risks of an ongoing connection between electric vehicles (EVs) and the smart grid.

Power distribution networks are swiftly transforming into active distribution networks (ADNs) due to increasing penetration of renewables and electric vehicles. This raises concerns among distribution system operators regarding the power quality issues. Frequent over/under voltages in ADNs are some of the major issues. The advanced devices and control techniques are to be used to prevent system voltage violations to enable efficient and reliable operation. To enhance the insight and comprehension, a precise description regarding the state-of-the-art devices used for voltage control in ADNs, their models and characteristics are provided in this chapter. Various advanced control strategies used for voltage control under different operation scenarios to highly uncertain future power networks are systematically presented, along with their merits and limitations. Finally, a voltage control approach based on dual-stage model predictive control framework is presented and tested on modified 33-bus distribution network to study the voltage control capabilities of PVs and EV charging stations.

The ever-increasing energy demand and global environmental crisis have acted like a growing push towards the adoption of electric vehicles (EVs). However, integration of EVs brings forth several challenges concerning the efficiency and reliability of power distribution networks (PDNs). These challenges must be addressed through advanced technologies and smart grid solutions. Distribution network reconfiguration (DNR) has emerged as a promising technique to enhance the reliability and efficiency of PDNs, while supporting the integration of EVs. This chapter provides a comprehensive analysis on DNR method and its application in improving the performance of smart grid enabled PDNs. The study incorporates an in-depth literature survey, highlighting the use of DNR to facilitate EV integration in PDNs. Furthermore, quantitative assessments of the literature are presented. These assessments showcase the effectiveness of DNR in achieving improved performance metrics, such as reducing losses, operational cost, enhancing voltage profile, and ensuring the reliability of EV penetrated PDNs.

Preface

The landscape of electrical energy systems is undergoing a transformative shift, propelled by the integration of electric vehicles (EVs) into power distribution networks. "E-Mobility in Electrical Energy Systems for Sustainability" explores the intricate interplay between smart grids, electric transportation ecosystems, and the multifaceted challenges and opportunities presented by this paradigm shift. Edited by Lakshmi D, Neelu Nagpal, Neelam Kassarwani, Vishnu Varthanan G, and Pierluigi Siano, this reference book provides a comprehensive analysis of the latest developments in the distribution grid, addressing key areas such as infrastructure frameworks, planning, operations, control strategies, and software applications.

NAVIGATING THE LANDSCAPE OF E-MOBILITY INTEGRATION

Electric mobility stands as a crucial pillar for sustainable transportation, driven by the utilization of EVs powered by renewable energy sources. However, the electrification of transportation introduces formidable challenges, particularly in the context of the distribution system's dependability and resilience. The widespread adoption of EVs significantly impacts the administration, monitoring, and control of power systems, necessitating comprehensive planning strategies, advanced technologies, and intelligent grid solutions.

KEY THEMES EXPLORED IN THIS BOOK

The book proposal focuses on four pivotal areas: the architectural reconfiguration of electric transportation ecosystems, restoration strategies in the face of disruptive events, power quality control, and the establishment of rules and regulations for the seamless integration of EVs into smart grid-enabled distribution systems. It delves into the intricate process of establishing and executing e-mobility frameworks, emphasizing robust cooperation among governmental entities, industry players, and business stakeholders.

TECHNOLOGICAL ADVANCEMENTS DRIVING E-MOBILITY

The proposal highlights technological advancements facilitating the functioning of smart grids, encompassing enhanced metering, computing, and communication. Topics covered include smart meters, sensors, Vehicle-to-Grid (V2G) systems, plug-in hybrid electric vehicles, batteries, and their charging mechanisms. The book caters to a diverse audience, including students, professionals in the energy and transportation sectors, and individuals seeking to expand their knowledge on this dynamic topic.

ORGANIZATION OF THE BOOK

Chapter 1: Sustainable Mobility Revolution: Best Practices and Strategies for Enhanced Energy Efficiency in E-Mobility

Authored by Sona S, Arnel Lobo, Mohammed Ali A H, Sanketh Naik, and Manjunatha Badiger from Sahyadri College of Engineering & Management, India, this chapter delves into the sustainable mobility revolution, emphasizing the need for enhanced energy efficiency in E-Mobility. The authors navigate through best practices, cutting-edge technologies, infrastructure development, and legislative frameworks that contribute to the overall efficiency of electric vehicles. Their exploration extends to advancements in battery science and the crucial role of smart grids in integrating electric vehicles seamlessly into existing transportation systems, ultimately fostering innovation and job creation in clean transportation-related industries.

Chapter 2: E-Mobility in Electrical Energy Systems for Sustainability

Authored by Yadav Rajnath from the Institute of Engineering and Rural Technology (IERT), Prayagraj (Allahabad), UP-211002, India, this chapter offers a transformative perspective on the integration of e-mobility into electrical energy systems. Rajnath explores various aspects, including reducing reliance on fossil fuels, addressing emissions, and promoting sustainable transportation solutions. The chapter ensembles electrification of transportation, energy storage integration, development of smart charging infrastructure, challenges to the electrical grid, synergy with renewable energy sources, environmental benefits, life cycle assessment, policy, incentives, and technological advancements, providing a comprehensive understanding of the interplay between e-mobility and electrical energy systems.

Chapter 3: E-Mobility as a Service and its Economic Impact: Regulations, Policies, Governance and Best Practices in E-Mobility Ecosystems

Authored by Leo John, Rojalini Patro, Padmakala S, and Madhumala RB from various academic institutions in India, this chapter conducts a systematic review and assessment of E-mobility as a service, evaluating social, monetary, policy, and sustainability factors within the circular economy ecosystem framework. The authors delve into challenges and propose methods to improve urban quality of life and sustainability in the electro-mobility space. The study aims to uncover why industrialized countries can efficiently introduce electric automobiles while underdeveloped economies face challenges, particularly in terms of infrastructure and downstream activities, and suggests effective solutions.

Chapter 4: Designing Architecture for Sustainable Electric Mobility - Ecosystems Integrating Urban Planning and Infrastructure Design

Authored by Ashish Makanadar from the Indian Institute of Management, Mumbai, India, this chapter provides a comprehensive evaluation of the role of architectural planning in developing sustainable electric mobility infrastructure. This study explores interdisciplinary solutions to address rising transport emissions, analyzing urban form, charging access, street layouts, and public transit integration. The chapter conceptualizes mobility ecosystems as sociotechnical systems, emphasizing sustainability, scalability, and flexibility as fundamental principles of resilient infrastructure planning. Smart grid technologies' potential to optimize renewable energy charging management is critically appraised, offering insights into inclusive, equitable planning that supports social equity and policy goals.

Chapter 5: Electrifying Transportation and Pioneering the Future of Sustainable Electricity Mobility

Authored by Anil Bharti and Neha Gupta from various academic institutions in India, this chapter explores the challenges and opportunities in the journey of electric vehicles (EVs). The authors highlight the transformative impact of EVs, addressing challenges like establishing robust charging infrastructure, range anxiety, ethical supply chains, and advancing battery technology. The chapter delves into the integration of EVs with autonomous and connected technologies, supportive policies, and creating a more sustainable transportation landscape. It provides insights into the technical obstacles and opportunities in the electrification of energy conversion systems within the dynamic realm of E-Mobility.

Chapter 6: Introduction to Smart Grids and e-Mobility Ecosystem

Authored by Leo John, Fairosebanu Abdul, and Jaffer Alzubi from various academic institutions in India and Jordan, this chapter examines the overlapping realms of transportation, communications, and energy. The authors predict emerging business models centered on external connectivity, internal Information, and Communication Technology (ICT) architecture, and human-machine interaction (HMI). The study thoroughly examines these themes, bringing insights from requirements analysis and prototype creation, outlining skills and techniques essential for capitalizing on market disruption. The chapter argues that electric cars represent an opportunity for the future of product lifecycle management and Industry 4.0.

Chapter 7: Study of Power Distribution System Resilience in the Presence of e-Mobility Ecosystem

Authored by Vandana Kumari and Sanjib Ganguly from IIT Guwahati, India, this chapter delves into the concept of resilience and the vulnerability of different components of Power distribution networks (PDNs) due to various HILP (high impact, low probability) events. The authors discuss the potential deployment of Electric Vehicles (EVs) for load restoration following disruptive events, drawing upon existing studies in this area. Furthermore, EV application to enhance the resilience of PDNs is reviewed with a detailed discussion of the methods, objectives, and solution strategies presented. In addition, a qualitative assessment of the studied literature is provided to enhance the comprehension of state-of-the-art resilience improvement methods of EV-enabled PDNs. This study is significant as the resilient-oriented planning and operation can enhance the resilience of PDNs. Furthermore, the significant EV integration, in recent years, provides a potential for improving resilience by restoration in PDNs.

Chapter 8: An Overview of E-Mobility-Based Threats to Power Grid: Introduction to Smart Grids and Mobility Ecosystem

Authored by Neha Gupta, Sushant University and Nidhi Gupta from MSIT, India, this chapter addresses the transition from fossil fuels to green energies in the transportation industry. The authors explore the emerging concept of e-mobility and its increasing infrastructure complexity, leading to security threats. The chapter highlights the vulnerability of the grid to cyberattacks and the technical necessity of data exchange between electric vehicles (EVs) and smart grids. The potential privacy risks of ongoing connections between EVs and smart grids are described, with a focus on blockchain as a solution for ensuring security and privacy in eMobile charging.

Chapter 9: Voltage Control of EV Enabled Future Power Distribution Networks

Authored by Mudaser Dar and Sanjib Ganguly from the Indian Institute of Technology, Guwahati, India, this chapter discusses challenges and opportunities in EV development, focusing on charging infrastructure weaknesses and unstructured charging schedules. The authors review the integration of EV charging infrastructure with blockchain and AI, providing a secure, effective, and decentralized charging ecosystem. The chapter serves as a starting point for stakeholders and policymakers to determine prospective paths and manage charging infrastructure for EVs, aiming to offer consumers exceptional service quality and expand opportunities for researchers and financial investors.

Chapter 10: Optimal Network Reconfiguration for Efficient EV Integration in the Smart Grid Enabled Power Distribution Networks

Authored by Subhadarshini Panda and Sanjib Ganguly from the Indian Institute of Technology Guwahati, India, this chapter provides a comprehensive analysis of distribution network reconfiguration (DNR) and its application in improving the performance of smart grid-enabled power distribution networks (PDNs). The study highlights the challenges and opportunities associated with EV integration and presents a quantitative assessment of the literature, showcasing the effectiveness of DNR in improving performance metrics. The chapter emphasizes reducing losses, operational costs, enhancing voltage profiles, and ensuring the reliability of EV-penetrated PDNs through advanced technologies and smart grid solutions.

Chapter 11: Empowering Future of Smart Grids: Unveiling the Role of Electric Vehicles in V2G Integration for Sustainable Infrastructure

Authored by Aanal Raval, Arpita Maheriya, Shailesh Panchal, and Komal Borisagar from Gujarat Technological University, India, this chapter explores the mission of Vehicle-to-Grid (V2G) within the context of a smart grid. The authors delve into establishing bidirectional energy flow between EVs and the grid, facilitating demand response, balancing supply and demand, promoting renewable energy integration, and supporting grid resilience. The chapter acts as a reference and guide for the forthcoming technological development and commercialization of electric vehicles (EVs), providing insights and recommendations for the future of smart grid transportation.

Chapter 12: A Comprehensive Review of Electric Vehicles A Smart Choice for Environmental Sustainability and Energy Conservation

Authored by Sanketh Naik, Mohammed Ali A. H., Arnel Lobo, Sona S., and Manjunatha Badiger from Sahyadri College of Engineering & Management, India, this chapter provides a comprehensive review of electric vehicles (EVs) as a smart choice for environmental sustainability and energy conservation. The authors explore the ecological footprint, energy efficiency, challenges like range anxiety, and charging infrastructure. Beyond environmental benefits, the chapter discusses social and economic implications, including job creation, lowered healthcare costs, and improved air quality. Global policy initiatives and case studies highlight the transformative potential of EVs in reshaping transportation for environmental sustainability and global energy conservation.

Chapter 13: A Comprehensive Review on Artificial Intelligence and Blockchain Technologies for Electric Vehicle Charging Ecosystem

Authored by Poras Khetarpal and Mahesh Kumar from BVCOE, India, Lakshay and Neelu Nagpal from MAIT, India, and Pierluigi Siano from University of Salerano, Italy, this chapter provides a comprehensive review of artificial intelligence (AI) and blockchain technologies for the electric vehicle (EV) charging ecosystem. The study addresses challenges in EV development, focusing on charging infrastructure weaknesses and unstructured charging schedules. The chapter explores current AI and blockchain charging solutions for EVs and reviews the combination of both technologies to develop a secure, effective, and decentralized EV charging ecosystem. Stakeholders and policymakers can use this chapter as a starting point to determine prospective paths and manage charging infrastructure for EVs.

Chapter 14: Modeling and Simulation of Electric Vehicle for Performance Analysis BEV and HEV Electrical Vehicle Implementation using Simulink for e-mobility Ecos

Authored by Kutubuddin Kazi from BMIT; Solapur, India, this chapter focuses on the mathematical model of Battery Electric Vehicles (BEVs) created in the MATLAB/Simulink environment. The study evaluates BEV performance using standard drive cycles and employs fuzzy logic control for regenerative braking. AI techniques are used for performance analysis, including renewable energy sources, regenerative braking, and battery performance. The chapter emphasizes sensors, two-way communication systems using cloud computing, and wireless technology for monitoring overall performance within the e-mobility ecosystem.

Chapter 15: Technological Trends in Electric Passenger Car Segment in India and Medium-term Demand for Magnet-based Rare Earth Materials

Authored by Arpita Pandey and Rudrodip Majumdar from The Energy and Resources Institute (TERI), New Delhi, India, and National Institute of Advanced Studies, India, respectively, this chapter explores the technological trends in the electric passenger car segment in India. The authors discuss the green mobility aspirations of India and the substitution of conventional vehicles with electric motor-driven counterparts. The study estimates the magnet-based rare earth demand for realistic growth trajectories of electric four-wheelers up to 2030, emphasizing India's potential collaboration in research and development, technology transfer, and securing raw material supply for a better reach in the global automotive market.

CONCLUSION

As editors of "E-Mobility in Electrical Energy Systems for Sustainability," we take immense pride in presenting this comprehensive compilation that unravels the intricate tapestry of the e-mobility ecosystem. The diverse perspectives and expertise shared by esteemed authors contribute to a holistic understanding of the challenges, opportunities, and advancements in this dynamic field.

Throughout the chapters, the authors have meticulously explored various dimensions, from sustainable mobility practices to the integration of electric vehicles into smart grids, underscoring the significance of technological advancements, policy frameworks, and societal implications. Each chapter adds a valuable layer to the narrative, offering insights into critical aspects such as energy efficiency, economic impact, urban planning, and the role of electric vehicles in reshaping our future transportation landscape.

The collective wisdom shared by researchers, scholars, and industry experts provides a roadmap for navigating the complexities of integrating e-mobility into electrical energy systems. From the design of sustainable infrastructure to the challenges of power distribution system resilience, the chapters in this book reflect the interdisciplinary nature of this transformative field.

As we conclude this preface, we extend our gratitude to the contributing authors for their dedication and scholarly contributions, which enrich the academic discourse on e-mobility. This book aims not only to serve as a valuable resource for U.G. and P.G. students, research scholars, academicians, policymakers, and industry experts but also to inspire further research and dialogue in the unexplored realms of this evolving domain.

We hope that readers find this book insightful, informative, and instrumental in advancing their understanding of the integration of electric vehicles in the distribution grid. May it spark innovative ideas, foster collaboration, and contribute to the ongoing revolution in electrical energy systems for a sustainable and resilient future.

D. Lakshmi
VIT Bhopal University, India

Neelu Nagpal
Maharaja Agrasen Institute of Technology, Delhi, India

Neelam Kassarwani
Maharaja Agrasen Institute of Technology, Delhi, India

G. Vishnu Varthanan
VIT Bhopal University, India

Pierluigi Siano
University of Salerno, Italy

Chapter 1
Sustainable Mobility Revolution:
Best Practices and Strategies for Enhanced Energy Efficiency in E-Mobility

Sona S. S.
Sahyadri College of Engineering and Management, India

Arnel Ralf Lobo
Sahyadri College of Engineering and Management, India

Mohammed Hyder Ali A.H.
Sahyadri College of Engineering and Management, India

Sanketh C. Naik
Sahyadri College of Engineering and Management, India

Manjunatha Badiger
Sahyadri College of Engineering and Management, India

ABSTRACT

Navigating the sustainable mobility revolution effectively requires investigating best practices and approaches to improve e-mobility's energy efficiency. The demand for environmentally friendly transportation, including a thorough analysis of cutting-edge technologies, the creation of new infrastructure, and legislative frameworks designed to maximize energy use are explored in order to increase the overall efficiency of electric vehicles. Adapting to new technologies and developments in battery science promotes the creation of electric vehicles that are more capable and efficient. Widespread acceptance of electric vehicles is facilitated by infrastructure investment, such as smart grids, which enable the smooth integration of these vehicles into current transportation systems. Sustainable mobility revolution propels innovation and employment creation in clean transportation-related industries by opening up new markets and economic opportunities. Thus, transformative experience is essential in paving the way for a future in which transportation is more sustainable and kind to the environment.

DOI: 10.4018/979-8-3693-2611-4.ch001

1. INTRODUCTION

As the E-Mobility landscape experiences rapid expansion, driven by the surge in urbanization and the pressing need to address environmental concerns, a comprehensive analysis of the challenges becomes imperative. It delves into the Complexities of optimizing E-Mobility services, focusing on key factors such as user adoption, energy efficiency, and infrastructure development. It opens with a contextualization of the global energy landscape, underlining the critical role played by the transport sector, which accounts for nearly one-third of total energy consumption. The overwhelming reliance on fossil fuels, especially oil, underscores the urgency for a paradigm shift, making E-Mobility a focal point in the pursuit of sustainable development.

Sustainable mobility marks a transformative shift in our approach to transportation, prioritizing environmentally friendly, efficient, and inclusive solutions. The embrace of green transportation modes, such as improved public transit, cycling, and walking, not only diminishes carbon footprints but also fosters healthier urban living. The adoption of electric vehicles and alternative fuels further contributes to cleaner air and reduces our reliance on fossil fuels. Smart infrastructure, encompassing intelligent transportation systems and smart cities, optimizes both traffic flow and energy consumption. Car-sharing and ride-sharing initiatives promote resource-efficient transportation, complemented by supportive policies and regulations fostering the widespread adoption of sustainable practices.

The E-Mobility sector still faces immense obstacles in attaining sustainability and smooth integration, even with great progress. The Sustainable Mobility Revolution is emphasized in this discourse as a vital tactic for reducing environmental effect in the context of e-mobility. To create a truly sustainable transportation environment, it highlights the need for the integration of renewable energy sources. Acknowledging both the advancement of technology and the larger context of environmental sustainability, a comprehensive viewpoint is essential for successfully navigating the complexities of E-Mobility. Strategic planning that considers the interdependence of sustainable practices and the smooth integration of electric vehicles into current transportation frameworks is crucial in the pursuit of sustainability within E-Mobility. Gaining insight into the complex relationship between sustainable mobility and e-mobility enables stakeholders to create comprehensive plans that not only overcome challenges but also shape a more integrated and ecologically conscious transportation industry in the future.Efforts are directed towards understanding and mitigating the environmental impact, particularly in response to evolving climate regulations (Bibri, 2018).

Technological innovations play a crucial role in enhancing energy efficiency within the E-Mobility sector. A wide range of advancements, from cutting-edge battery technologies to the implementation of intelligent charging infrastructure, contribute to this transformative journey.

Recognizing the environmental footprint of transportation, a key focus is on promoting battery electric vehicles as a solution with the potential to revolutionize the industry. The multifaceted nature of the Sustainable Mobility Revolution is underscored, encompassing technological advancements, policy interventions, economic considerations, and societal implications. The integration of renewable energy sources into the fabric of E-Mobility is positioned as a linchpin for achieving genuine sustainability. The introduction elucidates the interconnectedness of the transport and energy sectors, emphasizing the potential for synergies and spillover effects in an integrated approach to energy transition (Hlela, 2022). Patience It serves as a comprehensive prelude, weaving together the intricate threads of global energy challenges, the urgency of sustainable mobility, and the promise held by E-Mobility. It establishes a robust foundation for the subsequent deep dive into best practices and strategies, setting the tone for a nuanced and insightful exploration of the Sustainable Mobility Revolution.

As the narrative progresses, the transitions into an exploration of the broader context of the Sustainable Mobility Revolution, touching on the significance of smart cities. The paper conducts a literature review to discern current research directions and the role of big data tools and technologies in different components of smart cities. The interplay between human-nature interactions, big data, and sustainable urban mobility emerges as a dynamic area ripe for exploration. Additionally It provides a thematic overview of the paper's structure, foreshadowing the subsequent sections that will delve into best practices, case studies, and the potential synergies between energy and transport transitions. The anticipation of a case study on traffic congestion data in the city of Mumbai adds a practical dimension to the exploration, offering a real-world lens through which to analyze the charging behavior of users and draw relevant conclusions regarding the Charging Pile Network (Picon, 2015).

Exploring innovative strategies to address environmental concerns and enhance sustainability within the broader framework of sustainable mobility, the focus is specifically on the E-Mobility sector. Key themes encompass the integration of renewable energy sources, harnessing technological advancements, and ensuring economic viability (Owusu & Asumadu-Sarkodie, 2016). The abstract underscores the critical role of electric vehicles in achieving energy efficiency goals and reducing carbon emissions within the sustainable mobility context. By examining successful practices and strategic frameworks, the aim is to provide actionable insights for stakeholders, policymakers, and industry leaders actively involved in shaping a sustainable and efficient future for both sustainable mobility and E-Mobility.

In conclusion, It unfolds to the contemporary landscape of sustainability-oriented changes in the transport sector, driven in part by increasingly stringent climate and environmental regulations. It delves into the role of technological innovations, with a particular focus on battery-electric vehicles (BEVs) as potential game-changers. BEVs are presented as not only vehicles but as boundary objects that reconcile economic success and environmental goals, thereby preserving established mobility patterns. the imperative of a Sustainable Mobility Revolution in the EMobility context. By integrating renewable energy sources, leveraging technological advancements, and aligning with evolving regulatory landscapes, the industry can overcome challenges and pave the way for a sustainable and seamlessly integrated future in transportation (Hassan et al., 2023).

2. CHALLENGES

The obstacles associated with e-mobility include a lack of adequate charging infrastructure, battery constraints, cost hurdles, gaps in customer awareness, and supply chain sustainability issues. In order to improve energy efficiency, it is recommended to give priority to incentives for advanced technologies, adopt intelligent charging solutions, carry out lifecycle evaluations, encourage cooperative research, and progress integrated multimodal transportation networks. A thorough strategy to tackle these issues can help the e-mobility industry move toward greater energy efficiency and sustainability.

3. LITERATURE SURVEY

Adib et al. (2019), the trend of rising electrification in mobile platforms, including a wide range of applications such as transportation, robotics, drones, and electric devices. Emphasizing the end-to-end energy systems, it highlights the shared technical demands despite varying power ratings. Given limited access to the power grid, the focus lies on achieving extended operation through ultrafast charging, charging during motion, and advancements in battery technologies. Addressing space, shape, and weight constraints, the paper stresses the need for compact and lightweight onboard energy technologies, including power electronic converters and magnetic components. Efficiency and cost-effectiveness are critical considerations, pushing for improved solutions compared to conventional designs. The paper delineates the technical challenges hindering the transition towards greater electrification in mobile platforms (E-Mobility) while reviewing recent advancements documented in literature.

Castillo-Calzadilla et al. (2022) have used MATLAB-Simulink simulations to study the effects of electric vehicle (EV) proliferation inside Positive Energy Districts (PEDs) in Europe.Focusing on a small district in northern Spain, five scenarios are analyzed, varying in building energy efficiency, street lighting, and EV numbers. Findings highlight that buildings' energy ratings significantly influence the annual energy balance. A PED comprising six interconnected buildings and 405 EVs achieves a positive energy balance only with high-rated buildings (A or B certification). In the most efficient scenario (A-rated buildings), the PED can sustain 695 EVs, covering nearly 9 million green kilometers. This achievement indicates a potential 71% reduction in carbon emissions solely from e-mobility, contributing significantly to decreasing the district and city's overall carbon footprint.

Tundys and Wisniewski (2023) provides a comprehensive analysis of the integration of electric mobility into smart urban frameworks, emphasizing its pivotal role in fostering sustainable development. The study extensively explores key challenges, such as the inadequacy of charging infrastructure, concerns related to energy crises, and technological limitations, which collectively hinder the widespread adoption of electric vehicles in urban environments.Central to these discussions is the role of electromobility in curbing urban pollution through the promotion of zeroemission vehicles, aligning with the overarching goal of smart cities towards environmental sustainability. Market analyses and projections, particularly focusing on the light commercial vehicle (LCV) segment, delineate current EV market shares and predict future growth trajectories considering technological advancements and policy support. Researchers contemplate the potential ramifications of an impending energy crisis on the developmental trajectory of electromobility, especially in urban contexts, probing its impact on EV expansion and the realization of zero-emission urban economies. Throughout this discourse, there's a recurring emphasis on pragmatic assessments, cautioning against overly optimistic expectations concerning EV market penetration and the attainment of zero-emission targets, factoring in economic, technological, and policy-related complexities.

Cascetta and Henke (2023) explores the historical development of transportation, distinguishing between revolutions and evolutions based on the speed of change and their consequential effects on political, social, and economic structure. It identifies six historical transport revolutions and subsequent evolutions, suggesting a seventh revolutionary phase driven by energy source innovations, connected and autonomous vehicles, and smart mobility services. Anticipating the magnitude of change is challenging due to the potential amplification of unforeseen consequences from past revolutions. The seventh phase's effects on environmental, social, and economic sustainability in freight and passenger transportation are unpredictable in the short to medium term. It details past transport revolutions, discussing elements and potential interactions of the ongoing seventh revolution, urging researchers to holistically analyze its effects to mitigate negative outcomes and maximize positive impacts. Future transport policies, particularly in urban areas, must account for the opportunities, risks, and uncertainties posed by this ongoing transport revolution.

Lamedica et al. (2022) began with the observation that in a collective effort to mitigate CO2 emissions, European governments are enacting economic policies to incentivize e-mobility. The growing prevalence of e-mobility is then highlighted as a direct outcome of these policy interventions. Furthermore, the study explores initiatives aimed at enhancing the quality of life for individuals with disabilities through the introduction of e-mobility services. This motivation is rooted in a commitment to inclusivity and accessibility, reflecting a broader societal emphasis on creating transportation solutions that cater to diverse needs and contribute to a more sustainable future.A feasibility assessment and cost-benefit analysis are included in the report that describes an e-mobility service project for individuals with disabilities that is supported by Sapienza University of Rome. This project aims to identify an optimal, sustainable e-mobility solution considering technical and environmental aspects. A methodology is proposed, encompassing route optimization, electric vehicle selection, and charging scheduling based on energy consumption, travel time, costs, and service quality. A 5-step methodology calculates energy-efficient routes, defines optimal charging schedules considering charging points along routes, and selects vehicle types based on economic assessments. Additionally, a software tool has been developed to automate this methodology for practical implementation.

Moyo et al. (2020)'s exploration of contemporary advancements in public transportation, researchers highlight global transformation fueled by increased investments in alternative energy sources like e-mobility. Nevertheless, despite its considerable potential benefits, the study identifies low levels of adoption and implementation of e-mobility, pointing to existing challenges in its widespread integration.. Conducted via bibliometric analysis using Web of Science, the study reviewed 578 articles on electric vehicles, with China, the European Union, and the United States leading in publications. Key insights suggest the crucial role of a robust charging infrastructure in enabling widespread adoption of electric vehicles in Africa. To facilitate the transition to emobility, the study recommends government intervention through the development of policies and incentives. The findings and implications highlighted in the literature review provide crucial insights to guide the deployment and integration of e-mobility within African contexts, emphasizing the significance of energy, public transportation, and infrastructure development. Keywords: e-mobility, electric vehicle, Africa, energy, public transportation.

Banik (n.d.) refers to their commitment to addressing environmental concerns and the impending threat of climate change, the Indian government and Niti Ayog have introduced a series of initiatives. These endeavors distinctly prioritize zero-emission technologies, with a significant emphasis on nurturing the e-mobility sector as a crucial step towards ensuring a sustainable future.To restore India's historical prominence and position on the global stage, there's a focus on bolstering the indigenous automobile sector, promising significant employment opportunities for the country's youth. This involves supporting initiatives for eco-friendly mobility solutions, including solar-powered electric vehicles (EVs). However, the efficacy of the system is hindered by challenges like inadequate charging infrastructure, power management, and technological innovation systems (TIS). This paper outlines the evolution, standards, features, economic aspects, and policies pertinent to EVs, emphasizing the crucial role of various factors— social, economic, and policy-related—in driving vehicle adoption. By systematically recognizing these factors, the paper aims to pave the way for a sustainable future, aligning with green initiatives and sustainable development goals.

Carteni et al. (2020) tackles the major environmental problem of mobility, highlighting the possibility of more environmentally friendly solutions in the future, especially with the help of technology like electric mobility that may have a positive influence on mobility habits.. It critically analyzes the strengths, weaknesses, and application domains of electric mobility within Italy. Presently, electric vehicles face challenges such as high production costs, limited range, and environmental impacts associated with vehicle production, usage, and recycling. However, they offer zero local emissions, making them beneficial for urban mobility, especially in densely populated areas. Moreover, in the realm of emerging mobility paradigms like Mobility as a Service (MaaS), electric mobility finds utility. This includes micro-mobility, shared mobility, urban bus fleets, and freight distribution, where higher user/operator willingness to pay could amplify its advantages, mitigating its drawbacks and enhancing its suitability in diverse mobility scenarios.

Shalender and Yadav (2018) examines the reasons behind the limited adoption of electric vehicles (EVs) in India and devises strategies to overcome these hurdles while considering diverse stakeholder perspectives. Combining in-depth literature review and a three-phase Delphi technique involving experts, the study identifies and categorizes four primary challenges specific to India's e-mobility landscape. It presents an integrated framework aimed at addressing these impediments and provides a comprehensive roadmap for the widespread incorporation of EVs across the country. What sets this research apart is its holistic approach, thoroughly analyzing barriers to EV adoption and offering a clear, actionable plan to mitigate these obstacles. By outlining steps for overcoming challenges, this study endeavors to encourage greater acceptance and utilization of electric vehicles, fostering a more sustainable future for transportation in India.

Papa and Lauwers (2015) delves into the impact of smart mobility on urban landscapes, questioning whether it serves as an opportunity or a threat to city innovation. The paper explores how advancements in smart mobility technologies, such as ride-sharing apps, autonomous vehicles, and IoT-connected transportation systems, influence the dynamics and functionality of urban spaces. It debates whether these innovations enhance urban life by improving transportation efficiency, reducing congestion, and enhancing accessibility, or if they pose challenges such as data privacy concerns, increased dependence on technology, and potential social disparities in access to smart mobility solutions. By examining both sides of the argument, the paper seeks to uncover the nuanced implications of smart mobility on cities and communities, shedding light on its potential benefits and drawbacks in shaping the urban environment.

Ashok et al. (2022) examines the control strategy architecture of battery management systems (BMS) designed for lithium-ionbatterypowered electric vehicles.The review also discusses modern technologies such as IoT and cloudbased BMS, and the challenges associated with designing efficient BMSs for electric vehicles. The paper concludes with recommendations to tackle these challenges and highlights future prospects for the development of BMSs.

Goel et al. (2021) is a review of the barriers and challenges of electric vehicle adoption in India, as well as the potential benefits of vehicle-to-grid optimization. It provides an overview of the current state of electric vehicles and their impact on reducing greenhouse gas emissions. The paper also discusses various government initiatives and subsidies aimed at promoting electric vehicle adoption in India. Additionally, the paper explores different optimization techniques and their potential to increase the efficiency of electric vehicles. Overall, the paper aims to provide a comprehensive understanding of the current state of electric vehicles in India and the challenges and opportunities associated with their adoption.

Rocco et al. (2023) explores an overview of urban mobility, revolutionizing the landscape through innovative smart parking systems, as outlined in this paper titled "An Overview of Urban Mobility."It explores the latest IoT technologies and proposes a smart parking architecture that can efficiently manage urban parking areas. The paper discusses the benefits of this innovative solution for both public authorities and private individuals, and also highlights potential challenges and limitations to implementing this system in urban areas. The paper concludes by emphasizing the importance of intelligent mobility in improving human interaction with the urban environment and its liveability.

Pochont and Sekhar (2023) is a comprehensive exploration of Vehicle Integrated Photovoltaics (VIPV) as a promising solution for sustainable transportation. It particularly focuses on its application in passenger vehicles within tropical climates like India. It begins by addressing the escalating demand for passenger vehicles and the imperative need to reduce carbon footprints amid rapid urbanization. The concept of VIPV, integrating photovoltaic systems within vehicles for solar energy generation, is discussed, acknowledging the initial challenges related to reliability affecting vehicle performance. The review delves into recent technological advancements, encompassing module integration methods, advancements in solar cell technologies, and strategies to optimize energy efficiency within vehicles. Furthermore, it includes a case study analyzing solar energy potential (E) across different driving routes in a tropical climate, highlighting the impact of factors such as module wattage, vehicle surface area, travel duration, and weather conditions on energy generation. Conclusively, the paper emphasizes the identified technical challenges and proposes innovative pathways for future research, underlining the crucial need for further exploration and development in VIPV for sustainable transportation solutions in tropical regions.

Rimal et al. (2022) centers on emerging electric and autonomous electric vehicles between 2009 and 2020. The objective is to identify knowledge gaps and provide insights into sustainability assessments. It underscores the limitations of traditional LCA methodologies in addressing broader economic and social impacts associated with new transportation technologies. Highlighting the evolution from LCA to Life Cycle Sustainability Assessment (LCSA), the study emphasizes the necessity of early-stage evaluations for emerging technologies to guide their development amidst challenges such as data scarcity and market uncertainties. Analyzing 138 LCA studies, the paper classifies and scrutinizes them based on sustainability indicators, life cycle approaches, phases, regions, and vehicle technology. It identifies five key knowledge gaps, including the lack of socioeconomic assessments, absence of integrated modeling approaches, and limited consideration of circular economy aspects, offering recommendations for future research to advance the field of sustainability assessment for emerging electric vehicles. Overall, this review provides a comprehensive overview, shedding light on critical gaps and suggesting pathways for further advancements in evaluating the sustainability of electric vehicle technologies.

Onat and Kucukvar (2022) explores the Internet of Vehicles (IoV), focusing on its integration with electric vehicles (EVs), power grids, renewable energy, and communication/computing infrastructures for smart grids and sustainable urban development. It addresses challenges in grid management and proposes a cloud-based EV charging framework to manage peak-hour demand at charging stations. Hierarchical models, including price incentives and electricity supply expansion, are presented and demonstrated to enhance system performance and customer service. Additionally, it discusses seven emerging trends in smart EVs, such as blockchain-based energy trading and artificial intelligence applications, while outlining research areas in IoV. The paper emphasizes the need for integrating communication, power systems, EV control, and computing technologies for grid stability and the evolution of large-scale EV charging networks in the IoV ecosystem.

Kesselring et al. (2021) introduced a discussion on the urgent need for transitioning to sustainable automobility, prompted by the inefficiencies and environmental unsustainability of the current automobile system. It explores various emerging concepts and technologies aiming for a post-car paradigm shift, leveraging digitalization and technological advancements to integrate different transport modes and infrastructures. Emphasizing the social and economic impacts, it addresses the implications of this transition on jobs, livelihoods, and societal structures. The paper includes insights from multiple articles within a special issue, covering topics such as car-sharing practices, automated driving, decarbonization efforts, shared mobility, e-mobility impacts, policy dynamics, road transport emissions, and the socio-technical effects of the COVID-19 pandemic on automobility in Europe. These contributions collectively envision pathways toward sustainable mobility amidst the challenges posed by the existing automobile system.

Agyekum (2023) conducts a survey on the topic of electric vehicles (EVs) in India. They analyze articles and research papers to gather in-depth information about factors, barriers, and the availability of EVs. They also use questionnaires and surveys to collect valuable insights from experts in the field. The research follows ethical considerations and limitations, including the use of secondary data collection methods due to the unavailability of appropriate data and time constraints. The questionnaire survey design includes demographic analysis and focuses on impactful factors affecting the poor adoption rate of EVs. Data analysis involves regression analysis and ANOVA analysis to measure the correlation between variables and assess their significance. The results show that most consumers are satisfied or dissatisfied with state and federal policies, and a significant number of EV users are aware of the economic and environmental benefits. The research philosophy used in this study aims to enhance the understanding of research workers. Overall, the literature survey provides comprehensive insights into the EV sector in India.

Ephraim Bonah Agyekum et al. highlights that the transportation sector heavily relies on internal combustion engines and fossil fuels, contributing significantly to emissions and environmental degradation. This study delves into the current state of electric vehicles (EVs) in Africa, scrutinizing the impediments hindering their widespread adoption. Challenges highlighted include the considerable upfront costs, inadequate infrastructure, unreliable power grids, deficient charging systems, and the prevalence of conventional used vehicles. Moreover, the paper assesses cleaner alternative fuels such as hydrogen, biogas, ethanol, methanol, and biodiesel, exploring their potential application across the African continent. In the latter section, the study employs a SWOT analysis to evaluate the strengths, weaknesses, opportunities, and threats within Africa's alternative fuel industry. Conversely, opportunities arise from declining renewable energy technology costs and a heightened global understanding of the adverse environmental effects of greenhouse gases (GHGs), potentially fostering the growth of alternative fuels in Africa's transportation sector

4. ELECTRIC VEHICLE TYPES AND CLASSIFICATIONS

Electric vehicles (EVs) come in a variety of forms and categories, each designed to serve particular functions and inclinations. Zero tailpipe emissions are produced by Battery Electric Vehicles (BEVs), which run exclusively on electricity stored in batteries (Liu et al., 2020). Despite not being made to be charged externally, hybrid electric vehicles, or HEVs, include both an internal combustion engine and an electric motor. Despite not being made to be charged externally, hybrid electric vehicles, or

HEVs, include both an internal combustion engine and an electric motor. Hydrogen fuel cells are used by Fuel Cell Electric Vehicles (FCEVs), such as the Hyundai Nexo and Toyota Mirai, to produce energy while only releasing water vapor. In addition, a variety of electrified vehicles that serve different purposes in urban commuting, leisure, and commercial applications . Figure 1 shows the classifications of Electric Vehicles.

Figure 1. Electric vehicles classification according to their engine technologies and settings

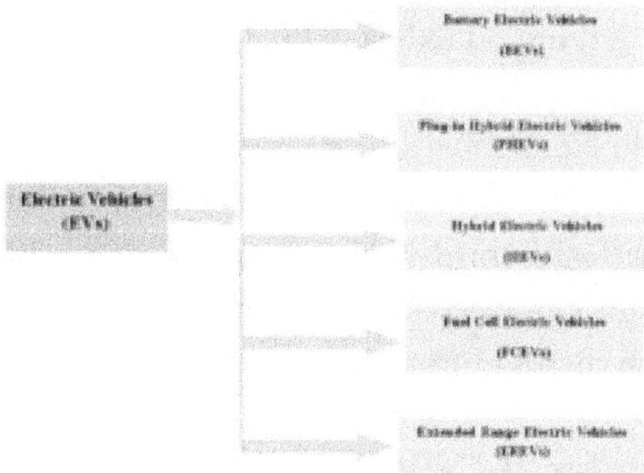

4.1 Battery Electric Vehicles (BEVs)

Battery Electric Vehicles (BEVs) are at the forefront of environmentally friendly transportation, exclusively utilizing electric batteries to eliminate emissions from internal combustion engines. Ongoing advancements in battery technology enhance the efficiency and range of BEVs, catering to consumers seeking longer-lasting and more sustainable mobility options. The development of a robust charging infrastructure is crucial for the widespread adoption of BEVs, addressing concerns about charging accessibility and range anxiety. BEVs contribute significantly to reducing dependence on fossil fuels, aligning with global efforts to combat climate change and air pollution. Their zero-emission profile appeals to environmentally conscious consumers, reflecting a commitment to reducing the overall carbon footprint. Government incentives, such as tax credits and rebates, play a pivotal role in making BEVs economically attractive, driving further adoption. As the automotive industry transitions, BEVs emerge as a key player in shaping a cleaner, more sustainable future for transportation.

4.2 Plug-In Hybrid Electric Vehicles (PHEVs)

Plug-in Hybrid Electric Vehicles (PHEVs) seamlessly combine internal combustion engines with electric batteries, offering users the flexibility of both electric-only and gasoline backup modes. This versatility caters to diverse driving needs, making PHEVs a practical choice for individuals transitioning to environmentally friendly transportation. PHEVs serve as a transitional alternative, allowing users to ease into electric mobility without sacrificing the longer range associated with conventional vehicles. Their dual-mode capability addresses concerns about charging infrastructure and range limitations, providing a convenient solution for various driving scenarios.

This adaptability makes PHEVs an attractive option for those not ready to fully commit to Battery Electric Vehicles (BEVs). In essence, PHEVs offer a balanced approach, satisfying the demands of users seeking both eco-friendly options and the practicality of extended driving ranges.

4.3 Hybrid Electric Vehicles (HEVs)

Hybrid Electric Vehicles (HEVs) optimize fuel efficiency by combining internal combustion engines with electric batteries, utilizing regenerative braking to recharge the batteries during deceleration. Unlike plug-in hybrids, HEVs cannot be externally charged, making them a practical choice for consumers seeking increased efficiency without relying on external charging infrastructure. This characteristic renders HEVs a convenient option for individuals who may not have regular access to charging stations. Serving as an initial step toward complete electrification, HEVs play a crucial role in raising awareness about hybrid technology and fostering environmentally friendly driving habits. Their existence encourages consumers to embrace the benefits of hybrid vehicles, acting as a bridge between traditional gasoline-powered cars and fully electric alternatives. The gradual transition facilitated by HEVs contributes to the overall promotion of sustainable transportation practices and the reduction of greenhouse gas emissions.

4.4 Fuel Cell Electric Vehicles (FCEVs)

Fuel Cell Electric Vehicles (FCEVs) present a green alternative by harnessing hydrogen fuel cells to produce power, emitting only water vapor as a byproduct. Despite challenges linked to hydrogen infrastructure, ongoing research and development initiatives aim to enhance the accessibility and efficiency of FCEVs. Positioned as a sustainable option, FCEVs represent a cutting-edge approach to zero-emission transportation, emphasizing the continual advancement of hydrogen fuel cell technology. The unique selling point of FCEVs lies in their ability to offer a clean driving experience with minimal environmental impact. Overcoming infrastructure obstacles is crucial for wider FCEV adoption, and ongoing efforts signal a commitment to addressing these challenges. FCEVs contribute to the vision of a more environmentally friendly future by promoting the use of hydrogen as a clean energy source in the automotive sector.

4.5 Extended Range Electric Vehicles (EREVs)

Extended Range Electric Vehicles (EREVs), akin to Plug-in Hybrid Electric Vehicles (PHEVs), emphasize electric driving by offering an extended electric-only range before seamlessly transitioning to internal combustion engines when required. This dual-mode operation prioritizes reduced emissions and heightened energy efficiency, showcasing the impact of advancements in battery technology on overall performance. EREVs play a pivotal role in the ongoing commitment of the automotive industry to furnish electric alternatives with extended ranges, aligning with diverse user preferences and accommodating various driving patterns. The extended electric-only range addresses concerns related to environmental impact and promotes sustainable mobility practices. The evolution of EREVs signifies a continuous effort to refine and expand electric vehicle technology, contributing to a more comprehensive and efficient transition toward electrified transportation. This versatile approach underscores the industry's responsiveness to user needs, offering a balance between electric driving and the flexibility of internal combustion engines for an optimal driving experience.

5. BATTERY TECHNOLOGIES

Battery technologies have undergone remarkable progress, powering a wide array of applications ranging from electric vehicles to portable electronics. Lithium-ion (Li-ion) batteries, characterized by their high energy density and rechargeable nature, dominate the market, propelling innovations in devices like smartphones and electric cars. Lithium Iron Phosphate (LiFePO4) batteries offer improved safety and longer cycle life, making them particularly suitable for certain applications. Nickel-Metal Hydride (NiMH) batteries find use in hybrid vehicles and various electronic devices, balancing cost and performance. Lead-Acid batteries, an established technology, persist in automotive and stationary applications due to their low cost and recyclability. Emerging technologies include Solid-State batteries, aiming to enhance safety and energy density, SodiumIon batteries with potential cost advantages, and Flow batteries designed for scalable energy storage. As the field evolves, ongoing research seeks to address limitations, enhance performance, and drive the next wave of advancements crucial for sustainable energy solutions and technological innovation. Figure2 shows the various critical applications of BMS in EV technology.

Figure 2. Various critical applications of BMS in EV technology

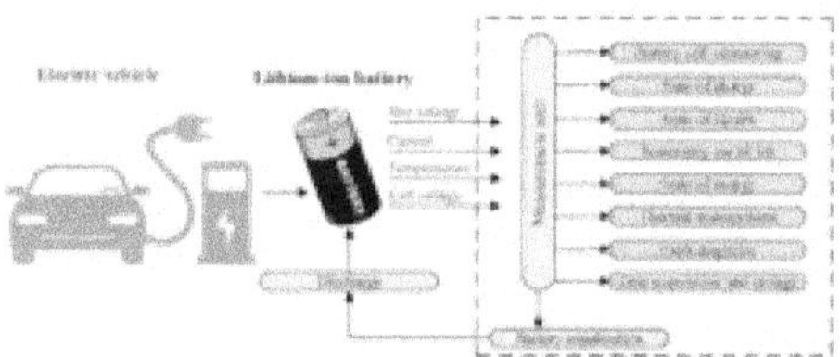

A number of essential technologies are needed for battery management in order to maximize battery systems' lifespan, safety, and performance. Advanced algorithms are used in State of Charge (SOC) and State of Health (SOH) estimation to determine the remaining capacity and general health of batteries, which helps with maintenance and usage decisions. Cell balancing prevents capacity imbalances by ensuring that charges are distributed equally throughout the cells in multi-cell packs. By preserving ideal operating temperatures, thermal management systems reduce the danger of overheating and thermal deterioration. Real-time performance data is provided by voltage and current monitoring, which makes it possible to identify abnormalities and guarantee safe operation. By quickly identifying problems, fault detection and diagnostics improve system reliability (Russell & Benner, 2010). One essential method for estimating remaining capacity is Coulomb counting. Overcurrent and overvoltage protection measures guard against harm, while communication protocols enable smooth connection with external systems. Batteries last longer because predictive maintenance algorithms use data to foresee problems and suggest prompt repairs. Together, these technologies provide the foundation for efficient battery management, which is essential for a variety of applications like renewable energy systems and electric vehicles. Future developments in these important technologies will have a significant impact on energy storage systems.

6. ENERGY MANAGEMENT TECHNIQUES

Techniques for managing energy include a wide range of approaches and tools intended to maximize energy production, distribution, and consumption. Demand response is a basic strategy that modifies energy use in response to outside variables such as grid circumstances or electricity costs. Smart grids increase total grid efficiency by integrating cutting-edge technologies for realtime monitoring and control. Batteries and other energy storage devices help maintain grid stability by storing excess energy during times of low demand and releasing it during times of peak demand. Sophisticated management strategies, such as grid balancing and predictive analytics, are necessary for the integration of renewable energy sources, such as solar and wind (Jones, 2017). One of the main goals of energy management is increasing energy efficiency through actions like implementing energy-efficient technologies and carrying out energy audits. During times of heavy demand, load shedding and peak shaving techniques help to reduce grid instability. Energy management capabilities are further enhanced by integration of electric vehicles, machine learning, artificial intelligence, and distributed energy resources (DERs). When combined, these methods provide an energy infrastructure that is more responsive, sustainable, and robust and meets the changing demands of contemporary civilization.

6.1 Regenerative Braking

Regenerative braking serves as a crucial energy management technique in Electric Vehicles (EVs) by recovering kinetic energy during braking maneuvers and converting it back into electrical energy. As the driver applies the brakes, the electric motor seamlessly transforms into a generator, efficiently capturing and storing the otherwise dissipated energy in the vehicle's battery. This innovative process not only enhances the overall efficiency of EVs but also significantly reduces energy wastage that typically occurs as heat during traditional braking systems. The stored energy can be subsequently utilized to power the vehicle, extending its driving range and improving energy conservation. Regenerative braking thus contributes to the sustainability of EVs, making them more energy-efficient and environmentally friendly.

6.2 Battery Thermal Management

Maintaining the ideal temperature is paramount for Electric Vehicle (EV) batteries' efficiency and longevity. Battery thermal management systems play a pivotal role in regulating temperatures by employing cooling or heating mechanisms to keep the battery within an optimal operating range. This meticulous temperature control is essential for maximizing energy efficiency and preserving the battery's lifespan. By preventing extremes in temperature, these systems mitigate the risks of accelerated degradation and enhance overall performance. Efficient thermal management translates into improved energy retention and release during various driving conditions. It also helps manage charging and discharging processes, minimizing stress on the battery cells. The result is a more sustainable and reliable electric vehicle, as optimal battery temperature management becomes a key factor in ensuring the long-term viability and effectiveness of electric mobility solutions.

6.3 Predictive Energy Management

Predictive energy management is a sophisticated approach that leverages advanced algorithms and real-time data in Electric Vehicles (EVs) to foresee driving conditions and strategically optimize energy consumption. By factoring in elements like route, traffic patterns, and weather conditions, the EV's on-board computer dynamically adjusts power delivery and energy consumption, thereby maximizing both efficiency and driving range. This predictive technique contributes to a more seamless driving experience, as the vehicle intelligently adapts to its surroundings. Ultimately, predictive energy management not only enhances the overall efficiency of EVs but also plays a vital role in extending their range, making electric mobility a more practical and user-friendly solution for diverse driving scenarios.

6.4 Smart Charging and Grid Integration

Smart charging systems empower Electric Vehicles (EVs) to engage intelligently with the electrical grid, optimizing charging based on various factors. This includes leveraging off-peak charging periods to capitalize on lower electricity rates, reducing costs for EV owners. Moreover, these systems dynamically adjust charging times in response to grid demand, contributing to more efficient and balanced energy distribution. Bidirectional energy flow facilitated by grid integration enables EVs to not only draw power but also feed excess stored energy back into the grid during peak demand. The synergy between smart charging and bidirectional energy flow promotes a more resilient and adaptable energy infrastructure, fostering a sustainable and interconnected ecosystem between electric vehicles and the broader electrical grid.

6.5 Power Electronics Optimization

Power electronics are integral in Electric Vehicles (EVs), responsible for converting and managing electrical power efficiently. Optimizing power electronics involves enhancing components such as inverters and converters to improve overall efficiency. This optimization aims to minimize energy losses during the conversion processes, ensuring a higher proportion of energy is effectively transferred to the electric motor. By reducing inefficiencies in power electronics, EVs can operate more resourcefully, thereby extending driving range and enhancing overall performance.

7. LIST OF EXAMPLES RELATED TO ELECTRIC VEHICLES

7.1 Tesla Model 3

The Tesla Model 3 is a legendary electric sedan that is well-known for its outstanding range and performance. It has fundamentally changed people's opinions about electric automobiles by demonstrating that they can deliver premium performance together with sustainability. The Model 3, with its innovative technology and elegant appearance, has come to represent Tesla's dedication to innovation in the automotive sector. Its enormous appeal is a reflection of the rising desire for stylish and efficient electric cars. Beyond only its stylish appearance, the Model 3 has had a significant impact on the transition to high-performance, environmentally friendly electric vehicles (Van Mierlo et al., 2021).

7.2 Nissan Leaf

One of the main contributors to the accessibility of electric transportation is the best-selling Nissan Leaf electric vehicle. Known for being reasonably priced and useful, the Leaf has improved with new features and a longer range. Its widespread success demonstrates how electric cars are becoming more and more popular (Cano et al., 2018). Nissan's dedication to ongoing enhancement is seen in the Leaf's role in mitigating environmental effects. The Nissan Leaf continues to be a major participant in the shift to environmentally friendly transportation thanks to expanding charging infrastructure.

7.3 Government Incentives

The broad adoption of electric cars (EVs) has been greatly aided by government incentives like those found in Norway. Customers find electric vehicles (EVs) more tempting since the government effectively lowers the financial barriers to ownership by offering tax cuts and toll exemptions. In addition to encouraging the initial purchase, these incentives also help to lower overall costs over the course of the vehicle's lifetime (Hagman et al., 2016). Norway is a prime example of how well-thought-out regulations may accelerate the shift to environmentally friendly transportation, and it also shows how governments can have a significant impact on how people move around in the future.

7.4 Hero Electric

Offering a range of environmentally friendly solutions, Hero Electric is a prominent player in the Indian electric scooter market. Their electric scooters, which are well-known for their inventiveness and affordability, make commuting greener and more effective, meeting India's growing need for environmentally friendly transportation (Aravamudhan & Raj, 2022). As electric two-wheelers become more and more popular, Hero Electric is essential in boosting electric mobility and lowering pollution in Indian cities.

8. CONCLUSION

The arrival of the Sustainable Mobility Revolution, which emphasizes improving e-mobility's energy efficiency, marks a revolutionary development in the transportation sector. The integration of optimal practices and strategic strategies is critical to this transformative path towards a more sustainable and environmentally friendly future. Reducing the environmental impact associated with traditional transportation systems requires utilizing the potential of electric vehicles and increasing renewable energy sources. The pursuit of increased energy efficiency requires implementing state-of-the-art technology like smart grids, creative battery solutions, and the construction of sturdy infrastructure. To achieve increased energy efficiency, cutting-edge technology like smart grids, creative battery solutions, and sturdy infrastructure are essential. In addition, it is critical for communities, businesses, and governments to work together to remove obstacles and hasten the transition to sustainable mobility.A comprehensive approach that includes focused research and development, public awareness efforts, and legislative incentives is necessary to create an atmosphere that encourages the adoption of electric vehicles.Governments, corporations, and communities must work together to eliminate obstacles and hasten the shift to sustainable mobility. To promote an environment that is favorable to e-mobility, a complete strategy including public awareness campaigns, focused research and development projects, and regulatory incentives is essential. It is necessary to place a high priority on accessibility, affordability, and diversity-embracing while implementing electric transportation solutions.Beyond merely advancing technology, the Sustainable Mobility Revolution represents a societal commitment to a greener, cleaner future. It is a shared commitment to change the course of transportation to one that is sustainable, energy-efficient, and mindful of the environment. Our collective efforts, prudent consumption, and persistent innovation are what are taking us to this revolutionary place. We can reshape the mobility landscape and provide the foundation for a future that is more sustainable and environmentally benign by guiding technology and societal norms in tandem.

REFERENCES

Adib, A., Afridi, K. K., Amirabadi, M., Fateh, F., Ferdowsi, M., Lehman, B., & Lewis, L. H. (2019). E-mobility—Advancements and challenges. *IEEE Access : Practical Innovations, Open Solutions*, *7*, 165226–165240. doi:10.1109/ACCESS.2019.2953021

Agyekum, E. B. (2023). Transportation in Africa under Paris Agreement 2° C goal—A review of electric vehicle potentials, cleaner alternative fuels for the sector, challenges, and opportunities. *Environmental Science and Pollution Research International*, 1–36.

Aravamudhan, V., & Raj, P. M. (2022). Impact Of Artificial Intelligence On Ev Industry And Perceptual Mapping. Academic Press.

Ashok, B., Kannan, C., Mason, B., Ashok, S. D., Indragandhi, V., Patel, D., Wagh, A. S., Jain, A., & Kavitha, C. (2022). Towards safer and smarter design for lithium-ion-battery-powered electric vehicles: A comprehensive review on control strategy architecture of battery management system. *Energies*, *15*(12), 4227. doi:10.3390/en15124227

Banik, A. (n.d.). *Comprehensive Investigation on India's Solar Powered E-Mobility Sector: Charging the Future*. Academic Press.

Bibri, S. E. (2018). A foundational framework for smart sustainable city development: Theoretical, disciplinary, and discursive dimensions and their synergies. *Sustainable Cities and Society*, *38*, 758–794. doi:10.1016/j.scs.2017.12.032

Cano, Z. P., Banham, D., Ye, S., Hintennach, A., Lu, J., Fowler, M., & Chen, Z. (2018). Batteries and fuel cells for emerging electric vehicle markets. *Nature Energy*, *3*(4), 279–289. doi:10.1038/s41560-018-0108-1

Carteni, Henke, Molitierno, & Errico. (2020). *Towards Emobility: Strengths and Weaknesses of Electric Vehicles.* Academic Press.

Cascetta & Henke. (2023). *The seventh transport revolution and the new challenges for sustainable mobility.* Academic Press.

Castillo-Calzadilla, T., Alonso-Vicario, A., Borges, C. E., & Martin, C. (2022). *E-Mobility in Positive Energy Districts Buildings*. Academic Press.

Goel, Sharma, & Rathore. (2021). A review on barrier and challenges of electric vehicle in India and vehicle to grid optimisation. *Transportation Engineering, 4*.

Hagman, J., Ritzén, S., Stier, J. J., & Susilo, Y. (2016). Total cost of ownership and its potential implications for battery electric vehicle diffusion. *Research in Transportation Business & Management*, *18*, 11–17. doi:10.1016/j.rtbm.2016.01.003

Hassan, Q., Algburi, S., Sameen, A. Z., Salman, H. M., & Jaszczur, M. (2023). Green hydrogen: A pathway to a sustainable energy future. *International Journal of Hydrogen Energy*.

Hlela, S. B. P. (2022). *Towards understanding geographic information competencies: the case of decision-making in environmental impact assessments in South Africa* [PhD diss.]. University of Pretoria.

Jones, L. E. (2017). *Renewable energy integration: practical management of variability, uncertainty, and flexibility in power grids*. Academic press.

Kesselring, S., Canzler, W., & Kaufmann, V. (2021). Sustainable Automobilities in the Mobile Risk Society. *Sustainability (Basel)*, *13*(10), 5648. doi:10.3390/su13105648

Lamedica, R., Maccioni, M., Ruvio, A., Carere, F., Mortelliti, N., Gatta, F. M., & Geri, A. (2022). Optimization of e-Mobility Service for Disabled People Using a Multistep Integrated Methodology. *Energies*, *15*(8), 2751. doi:10.3390/en15082751

Liu, X., Elgowainy, A., Vijayagopal, R., & Wang, M. (2020). Well-to-wheels analysis of zero-emission plug-in battery electric vehicle technology for medium-and heavy-duty trucks. *Environmental Science & Technology*, *55*(1), 538–546. doi:10.1021/acs.est.0c02931 PMID:33356189

Moyo, Onososen, Musonda, & Muzioreva. (2020). *Advancements in E-mobility: A bibliometric literature review on battery technology, charging infrastructure, and energy management*. Academic Press.

Onat, N. C., & Kucukvar, M. (2022). A systematic review on sustainability assessment of electric vehicles: Knowledge gaps and future perspectives. *Environmental Impact Assessment Review*, *97*, 106867. doi:10.1016/j.eiar.2022.106867

Owusu, P. A., & Asumadu-Sarkodie, S. (2016). A review of renewable energy sources, sustainability issues and climate change mitigation. *Cogent Engineering*, *3*(1), 1167990. doi:10.1080/23311916.2016.1167990

Papa & Lauwers. (2015). *Smart Mobility: Opportunity or Threat to Innovate Places and Cities.* Academic Press.

Picon, A. (2015). *Smart cities: a spatialised intelligence*. John Wiley & Sons. doi:10.1002/9781119075615

Pochont, N. R., & Sekhar Y, R. (2023). Recent trends in photovoltaic technologies for sustainable transportation in passenger vehicles–A review. *Renewable & Sustainable Energy Reviews*, *181*, 113317. doi:10.1016/j.rser.2023.113317

Rimal, B. P., Kong, C., Poudel, B., Wang, Y., & Shahi, P. (2022). Smart electric vehicle charging in the era of internet of vehicles, emerging trends, and open issues. *Energies*, *15*(5), 1908. doi:10.3390/en15051908

Rocco, G., Pipino, C., & Pagano, C. (2023). An Overview of Urban Mobility: Revolutionizing with Innovative Smart Parking Systems. *Sustainability (Basel)*, *15*(17), 13174. doi:10.3390/su151713174

Russell, B. D., & Benner, C. L. (2010). Intelligent systems for improved reliability and failure diagnosis in distribution systems. *IEEE Transactions on Smart Grid*, *1*(1), 48–56. doi:10.1109/TSG.2010.2044898

Shalender, K., & Yadav, R. K. (2018). Promoting e-mobility in India: Challenges, framework, and future roadmap. *Environment, Development and Sustainability*, *20*(6), 2587–2607. doi:10.1007/s10668-017-0006-x

Tundys & Wisniewski. (2023). *Smart Mobility for Smart Cities—Electromobility Solution Analysis and Development Directions Energies.* Academic Press.

Van Mierlo, J., Berecibar, M., El Baghdadi, M., De Cauwer, C., Messagie, M., Coosemans, T., Jacobs, V. A., & Hegazy, O. (2021). Beyond the state of the art of electric vehicles: A fact-based paper of the current and prospective electric vehicle technologies. *World Electric Vehicle Journal*, *12*(1), 20. doi:10.3390/wevj12010020

Chapter 2
E-Mobility in Electrical Energy Systems for Sustainability

Yadav KrishnaKumar Rajnath
Institute of Engineering and Rural Technology, Prayagraj, India

ABSTRACT

The integration of e-mobility into electrical energy systems stands as a transformative initiative with profound implications for sustainability in the transportation sector. This study explores the multifaceted aspects of e-mobility, emphasizing its role in reducing reliance on fossil fuels, decreasing emissions, and promoting cleaner and more sustainable transportation solutions. Key considerations include the electrification of transportation, integration of energy storage, development of smart charging infrastructure, challenges and enhancements to the electrical grid, synergy with renewable energy sources, environmental benefits, lifecycle assessment, policy and incentives, and ongoing technological advancements. By delving into these aspects, this research aims to provide a comprehensive understanding of the interplay between e-mobility and electrical energy systems, contributing valuable insights to the broader discourse on sustainable transportation.

1. INTRODUCTION

The surge in e-mobility is a direct response to the urgent environmental concerns associated with traditional fossil fuel-dependent transportation. Given that the transport sector significantly contributes to carbon emissions and air pollution, the transition to electric vehicles (EVs) emerges as a strategic measure to address these environmental challenges. This section provides a succinct overview of the historical background propelling the emergence of e-mobility as a transformative solution.

The imperative for implementing electric mobility (e-mobility) becomes increasingly apparent, given its pivotal role in the ongoing discourse on the future of transportation. The transport sector, responsible for over 25% of global energy-related greenhouse gas emissions, demands resolute action by governments worldwide to combat global warming and align with the objectives of the Paris Agreement (Bleicher, A., & Pehlken, A. (Eds.). (2020)). Additionally, the substantial contribution of the transportation sector

DOI: 10.4018/979-8-3693-2611-4.ch002

to local air pollutants results in seven million premature deaths annually. Acknowledged as a decisive solution, e-mobility is considered essential for reshaping mobility to align with sustainable development goals and decarbonization objectives (Ruggieri, R., et al (2021)).

Despite this recognition, ongoing debates revolve around the optimal policy framework for the integrated and sustainable adoption of electric mobility. This chapter aims to assist public policymakers at various levels—international, national, regional, and local—in accelerating the development of sustainable electric mobility. The objective is to formulate actionable recommendations for different policy levels while considering the suitability of various e-mobility models based on local circumstances.

Drawing insights from the Catalogue of Policy Measures (CPM) in the Global Roadmap of Action toward Sustainable Mobility (GRA), encompassing over 180 tested policy measures globally, this paper specifically delves into policy measure 116, focusing on "integrating new mobility solutions into existing transport." The explored associated measures include "establishing electric vehicle manufacturing mandates," "supporting vehicle connectivity and smart charging regulations," "utilizing public procurement to enhance vehicle electrification," "developing infrastructure for road transport electrification," "investing in railway electrification," "facilitating public discourse on new mobility solutions," "supporting research and development to optimize the life cycle of vehicle batteries," and "educating users about new sustainable solutions."The chapter is divided into three primary sections: the first introduces electric mobility as an integral component of sustainable mobility, the second delineates seven essential building blocks for successful public policy closely linked to measures in the CPM, and the last provides actionable policy recommendations for international, national, and local institutions to support the scaling up of sustainable electric mobility in the short and medium term.

A global analysis underscores the indispensability of sustainable mobility in achieving development and climate goals. The emphasis is on the notion that e-mobility transcends electric cars, emphasizing the need to contemplate all transport modes for feasibility. An advocated comprehensive approach includes the electrification of the existing fleet, greening of the grid, and promotion of modal shifts to realize the substantial benefits of electric mobility while contributing to sustainable mobility development (Sovacool, B. K., Kester, et al. (2019)). An examination of existing barriers reveals a diverse range, including policy and legislative hindrances, knowledge gaps, market and financial challenges, and technical obstacles. Addressing these barriers mandates a coordinated set of interventions at various levels—international, national, and local. Consequently, the document defines seven essential building blocks for successful public policy, laying the groundwork for policy recommendations across different decision-making levels. These building blocks are categorized into three action fields: vision, policy, and implementation, as illustrated in Figure 1.

1.1 Background and Rationale: Unveiling the Need for Transformation

The automotive sector has historically been dominated by internal combustion engines, a factor that has significantly contributed to environmental challenges such as air pollution and climate change. The increasing awareness of these environmental issues has instigated a shift towards sustainable alternatives, with electric mobility emerging as a prominent solution (Dijk, M., et al. (2013), Hoogma, R., et al. (2002)). This section explores the historical context and reasoning behind the ascent of e-mobility, underscoring the imperative need to transition from conventional to environmentally conscious transportation solutions.

Figure 1. The three action fields encompass seven essential building blocks

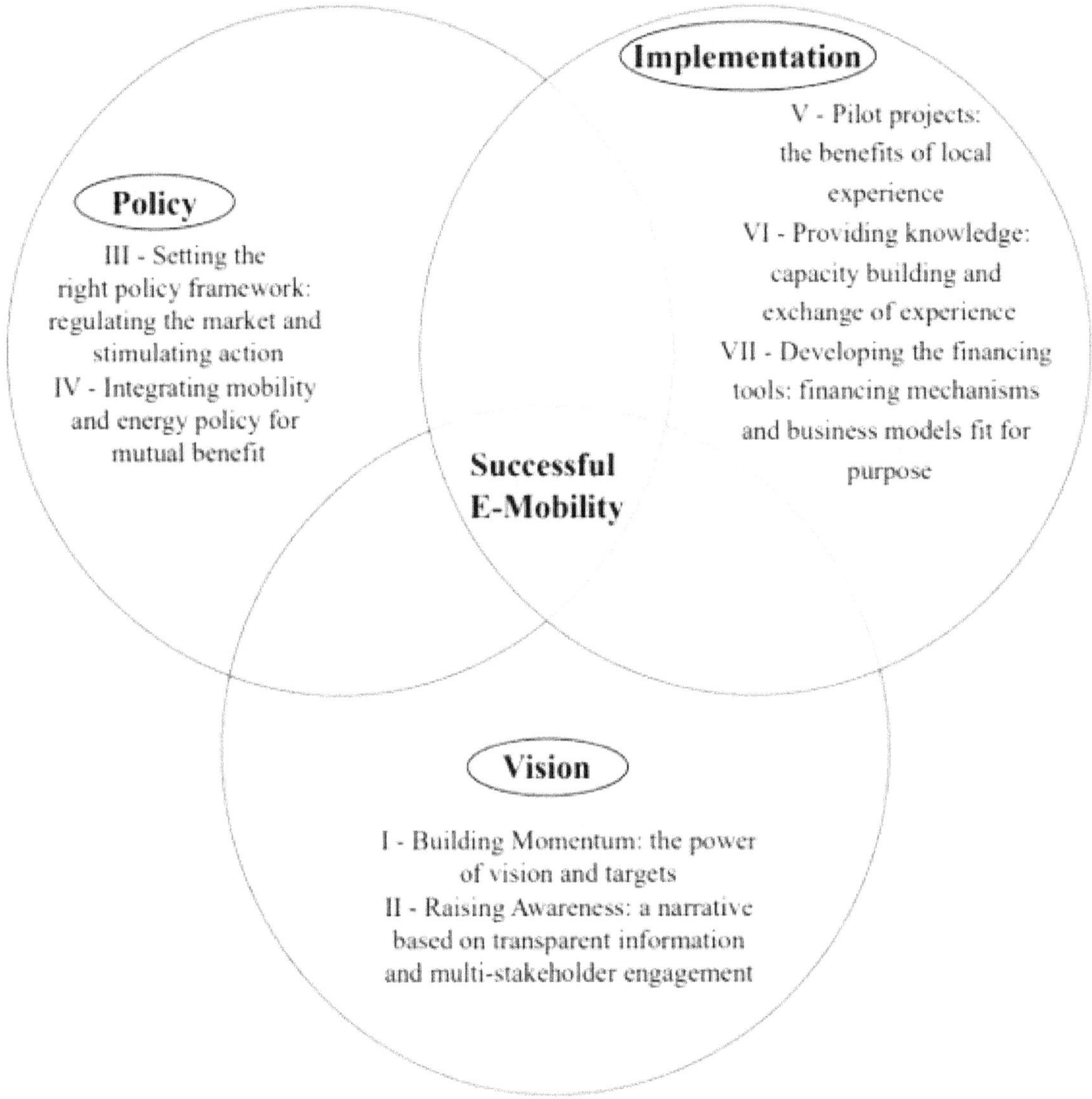

The inception of e-mobility can be linked to concerns regarding finite fossil fuel resources, unpredictable oil prices, and the adverse environmental effects associated with traditional vehicles (Kemp, R., et al.(1998)). The acknowledgment that the transport industry is a major contributor to greenhouse gas emissions has hastened the progress and acceptance of electric vehicles (EVs) (Berkeley, et al. (2018)). Grasping this historical context is essential for understanding the urgency and significance of incorporating e-mobility into electrical energy systems..

1.2 Objectives of the Study: Navigating the Complex Interplay

To comprehend the multifaceted relationship between e-mobility and electrical energy systems, a set of clear objectives guides this research. These objectives serve as a roadmap for the ensuing chapters, providing a structured approach to examining various dimensions of the integration. Key objectives include unraveling electrification trends, exploring energy storage integration, understanding the evolution of smart charging infrastructure, addressing challenges to the electrical grid, elucidating the synergy with renewable energy sources, evaluating environmental benefits, conducting lifecycle assessments, analyzing policy frameworks, and staying abreast of ongoing technological advancements.

Each objective is designed to contribute a distinct piece to the overarching puzzle, collectively forming a comprehensive understanding of the interdependence between e-mobility and sustainable electrical energy systems.

1.3 Significance of E-Mobility in Sustainability: Transformative Impacts

The transition towards e-mobility represents more than just a technological progression; it symbolizes a transformative approach towards sustainability in the realm of transportation. This section emphasizes the profound importance of incorporating electric vehicles into the broader energy landscape. Through a reduction in dependence on fossil fuels and the mitigation of emissions, e-mobility emerges as a pivotal element in the collective endeavor to attain global sustainability objectives (Mahiban, L. N., & Emimal, M. (2023)). This significance goes beyond environmental considerations, extending to economic and societal dimensions. E-mobility introduces opportunities for innovation, job creation, and heightened energy security (Biresselioglu, et al. (2018)). Furthermore, it aligns with the evolving preferences of consumers who seek cleaner and more responsible modes of transportation. Acknowledging and comprehending the holistic impact of e-mobility is essential for stakeholders across various sectors (Spickermann, A., et al. (2014)).

In essence, this introduction sets the groundwork for a thorough exploration of how e-mobility, intricately integrated into electrical energy systems, serves as a catalyst for a sustainable and resilient future. It encourages readers to embark on a journey through the upcoming chapters, each dedicated to unraveling a specific facet of this transformative integration.

2. ELECTRIFICATION OF TRANSPORTATION

The automotive sector's historical trajectory has long been dominated by internal combustion engines, contributing significantly to environmental challenges such as air pollution and climate change. Acknowledging these issues, there has been a substantial shift towards more sustainable alternatives, with electric mobility emerging as a pivotal contender (Schot, J., & Geels, F. W. (2008)). This section delves into the historical context and rationale behind the rise of e-mobility, emphasizing the critical necessity to transition from traditional to eco-conscious modes of transportation. The inception of e-mobility is rooted in concerns about finite fossil fuel resources, unpredictable oil prices, and the adverse environmental consequences associated with conventional vehicles (Nikitas, A., et al. (2017)). Understanding that the transportation sector is a major contributor to greenhouse gas emissions has expedited the innovation and adoption of electric vehicles (EVs). This historical context is essential for contextualizing the urgency and importance of seamlessly integrating e-mobility into contemporary electrical energy systems.

To decipher the intricate relationship between e-mobility and electrical energy systems, a set of explicit objectives guides this research. These objectives serve as a roadmap for the ensuing chapters, offering a methodical approach to dissecting various dimensions of the integration. They include unraveling electrification trends, exploring energy storage integration, comprehending the evolution of smart charging infrastructure, addressing challenges posed to the electrical grid, elucidating the synergy with renewable energy sources, evaluating environmental benefits, conducting lifecycle assessments, analyzing policy frameworks, and staying abreast of ongoing technological advancements.The transition towards e-mobility signifies more than just a technological advancement; it represents a transformative approach to sustainability in transportation (Hensher, D. A., et al. (2020)). This section emphasizes the profound significance of seamlessly integrating electric vehicles into the broader energy landscape, diminishing dependence on fossil fuels, and curbing emissions, aligning with global sustainability objectives.

This significance extends beyond environmental considerations to encompass economic and societal dimensions. E-mobility presents opportunities for innovation, job creation, and enhanced energy security (Biresselioglu, et al. (2018)). Furthermore, it aligns with evolving consumer preferences for cleaner and more responsible modes of transportation. Recognizing and appreciating the holistic impact of e-mobility is paramount for stakeholders across sectors. In essence, this introduction sets the stage for a nuanced exploration of how e-mobility, intricately woven into electrical energy systems, becomes a catalyst for a sustainable and resilient future, inviting readers to embark on a journey through various chapters, each dedicated to unraveling specific facets of this transformative integration.

2.1 Evolution of Electric Vehicles

The development of Electric Vehicles (EVs) represents a noteworthy journey marked by substantial advancements and transformative changes. From their inception to the present, electric vehicles have undergone significant evolution, reshaping the automotive industry and contributing to the global transition towards sustainable transportation. The advantages of EVs over internal combustion engine (ICE)-based vehicles are widely recognized, alongside acknowledging the challenges associated with their integration (Ghosh, A. (2020), Lebrouhi, B. E., et al. (2021)). The business landscape is adapting to the imminent shift towards EVs, driven by factors such as rising fuel prices and the necessity for change in the transportation sector. Essentially, the evolution of EVs has spurred technological progress and a notable shift towards sustainable transportation alternatives.

The history of Electric Vehicles can be traced back to their early conceptualization and experimental phases. In the late 19th century, pioneers laid the groundwork for electric propulsion systems. However, it wasn't until the 21st century that EVs gained substantial traction, fueled by concerns about environmental sustainability and the imperative to reduce reliance on traditional internal combustion engine vehicles. A pivotal moment in EV evolution was the introduction of the Tesla Roadster in 2008. This high-performance electric sports car not only demonstrated the potential of electric drivetrains but also challenged perceptions of EVs being limited to compact, low-range options. Tesla's innovative approach to battery technology set a new industry standard. Following this, there was a surge in electric vehicle development by traditional automakers and new entrants alike. Advances in battery technology, emphasizing energy density and charging capabilities, played a crucial role in extending EV range. Worldwide investments in charging infrastructure aimed to alleviate range anxiety and promote EV adoption (Saif, M.A., et al.. 2024)

The evolution of electric vehicles extends beyond cars alone. Electric buses, trucks, and two-wheelers have emerged as viable alternatives, contributing to a comprehensive electrification of the transportation sector (Muratori, M., et al. (2021)). Advances in autonomous driving technology have synergized with the electric vehicle revolution, paving the way for intelligent and sustainable mobility solutions. The ongoing evolution involves addressing challenges like charging infrastructure expansion, cost reduction, and sustainable material sourcing (Hasan, M. M., et al. 2023). Governments continue implementing policies to incentivize EV production and adoption, further accelerating their integration into mainstream transportation. In conclusion, the evolution of electric vehicles has been a dynamic and multifaceted process, driven by technological innovation, environmental consciousness, and a collective commitment to a sustainable future. As EVs continue to evolve, their widespread adoption holds the potential to revolutionize how we perceive and engage with transportation, ushering in an era of cleaner, greener, and more efficient mobility. A representation outlining the evolution of electric vehicles is presented in Table 1

Table 1. The evolution of electric vehicles

Period	Key Developments
Late 19th Century .	Invention of the electric vehicle (EV) with crude batteries
Early 20th Century	EVs gain popularity; Detroit Electric Car Company thrives.
Mid-20th Century	Decline due to the rise of internal combustion engine cars.
Late 20th Century	Brief resurgence with experimental EVs, like GM's EV1
Early 21st Century	Major advancements, Tesla Roadster introduced, EVs gain traction.

In tracing the historical trajectory of electric vehicles (EVs), recent years have borne witness to a remarkable surge in both development and global adoption. The narrative surrounding the marketing of EVs spans a spectrum of perspectives, encompassing environmental advantages and considerations of consumer perceptions. As we move forward, the historical narrative of EV development and marketing remains dynamic, shaped by a complex interplay of technological advancements, regulatory frameworks, and evolving consumer sentiments.

2.2 Types of Electric Vehicles

Electric vehicles (EVs) come in various types, each designed to cater to different preferences and purposes. These innovative transportation options utilize electric power as their primary source of energy, contributing to a more sustainable and eco-friendly mode of mobility.

- Battery Electric Vehicles (BEVs): BEVs are fully electric vehicles that rely solely on rechargeable batteries for power. These batteries store electricity, which is then used to propel the vehicle's electric motor. BEVs are known for producing zero emissions during operation, making them a popular choice for environmentally conscious individuals.

- Plug-in Hybrid Electric Vehicles (PHEVs): PHEVs combine an electric motor with an internal combustion engine and a battery that can be charged by plugging into an electric power source. These vehicles offer the flexibility of driving in electric-only mode for short distances while providing the extended range of a traditional combustion engine for longer trips.
- Hybrid Electric Vehicles (HEVs): HEVs feature both an electric motor and an internal combustion engine, but unlike PHEVs, they do not have a plug-in option. Instead, the electric motor is primarily charged through regenerative braking, capturing and storing energy during deceleration.
- Fuel Cell Electric Vehicles (FCEVs): FCEVs use hydrogen fuel cells to generate electricity, which powers an electric motor. These vehicles emit only water vapor and heat as byproducts, making them a promising option for achieving clean and sustainable transportation.
- Micro Electric Vehicles (MEVs): MEVs are compact electric vehicles designed for short-distance urban commuting. They are often smaller in size and feature lightweight construction, making them an efficient and maneuverable choice for navigating congested city streets.
- Electric Bikes and Scooters: As a more sustainable alternative to traditional two-wheelers, electric bikes and scooters are gaining popularity. These vehicles are equipped with electric motors powered by rechargeable batteries, providing an eco-friendly and energy-efficient mode of personal transportation.
- Autonomous Electric Vehicles: The integration of electric propulsion with autonomous technology is shaping the future of transportation. These vehicles have the capability to navigate and operate without human intervention, offering a glimpse into a potential revolution in the automotive industry.
- By understanding the diverse types of electric vehicles available, consumers can choose the option that best aligns with their needs, contributing to a cleaner and greener future of transportation.

2.3 Impact on Greenhouse Gas Emissions

The impact on greenhouse gas emissions is a critical aspect of discussions surrounding environmental sustainability and climate change. Greenhouse gases (GHGs), such as carbon dioxide (CO2), methane (CH4), and nitrous oxide (N2O), contribute to the greenhouse effect, trapping heat in the Earth's atmosphere and leading to global warming (Ming, T., et al. (2016)). Understanding and addressing the sources of these emissions are essential for mitigating the adverse effects of climate change. Several factors contribute to the impact on greenhouse gas emissions:

- The combustion of fossil fuels, including coal, oil, and natural gas, is a major contributor to the emission of greenhouse gases. Yoro and Daramola (2020) highlight that power plants, industrial facilities, and transportation are key sources of these emissions. Shifting towards renewable energy sources like solar, wind, and hydropower presents a viable solution to significantly reduce emissions associated with energy production.
- Changes in land use, particularly deforestation, contribute to the release of greenhouse gases. Trees, through photosynthesis, absorb CO2, and when forests are cleared, stored carbon is released. Sustainable land management practices, afforestation, and reforestation efforts can help counterbalance these emissions.

- Certain industrial processes, like cement production, release CO2 as a byproduct. Innovations in manufacturing technologies and the adoption of cleaner production methods can play a crucial role in reducing emissions from these sectors.
- Agricultural activities, including enteric fermentation in livestock, rice cultivation, and the use of synthetic fertilizers, contribute to methane and nitrous oxide emissions (Montes et al., 2013). Sustainable agricultural practices, such as precision farming and improved livestock management, can aid in minimizing these emissions.
- Landfills emit methane as organic waste decomposes in anaerobic conditions. Enhanced waste management practices, including increased recycling and composting, can effectively reduce methane emissions from landfills (Themelis & Ulloa, 2007).
- The transportation sector significantly contributes to greenhouse gas emissions through fossil fuel combustion in vehicles. Transitioning to electric vehicles, improving public transportation, and promoting alternative fuels emerge as effective strategies to curtail emissions in this sector.
- Government policies and regulations play a pivotal role in shaping the behavior of industries and individuals concerning greenhouse gas emissions (Prado-Lorenzo et al., 2009). The implementation of carbon pricing mechanisms, establishment of emission reduction targets, and promotion of clean energy policies are crucial for driving positive change.
- Advancements in technology, such as carbon capture and storage (CCS) and sustainable farming practices, present potential solutions to mitigate greenhouse gas emissions (Rahman et al., 2017). Continued research and development in clean energy technologies are imperative for achieving long-term sustainability goals.

3. ENERGY STORAGE INTEGRATION

Energy storage integration in electric vehicles (EVs) involves incorporating systems to enhance performance, efficiency, and functionality. The primary component is the traction battery, with lithium-ion batteries significantly improving energy density and range. Super capacitors complement batteries, offering quick bursts of power during acceleration and braking (Lemian, D., & Bode, F. 2022). Regenerative braking captures kinetic energy, recharging the battery and extending the vehicle's range. Plug-in Hybrid Electric Vehicles (PHEVs) combine an internal combustion engine, electric motor, and rechargeable battery for flexible energy use. Vehicle-to-Grid (V2G) technology allows EVs to contribute excess energy back to the grid, balancing demand and supply. Advanced Energy Management Systems optimize stored energy, considering driving conditions and user preferences. Innovations like high-power chargers and fast-charging networks reduce charging times, while solid-state batteries aim to overcome limitations in current technologies. Efficient thermal management systems are crucial for battery health. Continuous advancements in energy storage for EVs focus on improving performance, reducing costs, and addressing environmental concerns in production and disposal. Future innovations are anticipated in this evolving field.

3.1 Battery Technologies in E-Mobility

Battery technologies are pivotal in advancing E-Mobility, significantly influencing electric vehicle (EV) performance, range, charging times, and costs. Lithium-ion (Li-ion) batteries dominate the EV landscape, known for their high energy density and efficiency (Itani, K, & De Bernardinis, A. (2023), Mohammadi, F., & Saif, M. (2023).) . Solid-state batteries offer potential improvements in energy density and safety. Lithium-sulfur batteries, utilizing sulfur as the cathode material, present high theoretical energy density but face challenges in cycle life and performance maintenance. Graphene batteries leverage graphene materials for enhanced conductivity and performance, with potential for faster charging times. Nickel-Cobalt-Manganese (NCM) and Nickel-Cobalt-Aluminum (NCA) batteries provide high energy density but raise concerns about cobalt-related issues. Ongoing research explores post-lithium batteries, including sodium-ion, potassium-ion, and magnesium-ion, aiming for reduced reliance on scarce materials and environmental benefits. Flow batteries, known for stationary energy storage, offer scalability but face challenges in energy density for electric vehicles (Pokhriyal, A., et al., 2023). The dynamic landscape of battery technologies requires continuous research and development to address challenges and drive sustainable advancements in E-Mobility.

3.2 Role of Advanced Batteries

Advanced batteries play a pivotal role in diverse industries, driving innovation and sustainability. In electric vehicles (EVs), they serve as primary energy storage, enhancing range and performance. In renewable energy, batteries store energy from sources like solar and wind, ensuring a stable power supply. Grid energy storage balances demand, enhancing overall grid efficiency. For portable electronics, medical devices, and consumer electronics, advanced batteries offer high energy density and reliability. In aerospace, batteries power auxiliary units and electric propulsion in aircraft. Stationary energy storage systems enable time-shifting and backup power. Uninterruptible Power Supply (UPS) systems rely on batteries for continuous power, crucial for critical infrastructure. In military applications, batteries provide reliable power for communication devices and electric vehicles, emphasizing durability and high energy density. Across these sectors, advanced batteries play a multifaceted role in driving efficiency, resilience, and technological advancement.

3.3 Storage Solutions for Optimal Energy Use

Optimal energy use entails the efficient management of energy generation, storage, and consumption to meet demand while minimizing waste (Omer, A. M. (2008)). Diverse energy storage systems play crucial roles in ensuring a reliable and efficient energy landscape. Battery Energy Storage Systems (BESS), like lithium-ion batteries, are integral for electric vehicles, renewable energy, and grid storage. Pumped Hydro Storage offers large-scale energy storage with high capacity and a long cycle life. Compressed Air Energy Storage (CAES) provides scalable and cost-effective grid storage. Flywheel Energy Storage excels in grid stabilization with fast response times and high cycling capability. Thermal Energy Storage enhances heating and cooling efficiency in buildings. Hydrogen Energy Storage allows for long-duration storage and sector coupling. Superconducting Magnetic Energy Storage (SMES) contributes to grid stabilization with high power density. Advanced Lead-Acid Batteries offer cost-effective solutions for grid storage and uninterruptible power supplies. Vanadium Redox Flow Batteries (VRFB) provides scalability and long cycle life for grid storage. Smart Energy Management Systems optimize various storage technologies based on real-time demand, pricing, and grid conditions in residential and commercial applications, maximizing renewable energy use and minimizing costs.

4. SMART CHARGING INFRASTRUCTURE

Smart charging infrastructure optimizes the charging process for electric vehicles (EVs), enhancing efficiency, convenience, and sustainability while managing grid load. Key components include Vehicle-to-Grid (V2G) communication, smart meters for real-time data and dynamic pricing, and load balancing to avoid peak-hour grid congestion. Remote monitoring via mobile apps and cloud-based solutions enables user control and predictive analytics for optimizing performance. Payment solutions, billing integration, and user authentication ensure secure access. Grid integration allows EVs to provide ancillary services, while energy management systems and dynamic pricing models promote efficient charging (Barman, P., et al . 2023). Fleet management integration and scalability ensure a seamless and intelligent EV charging ecosystem. Smart charging infrastructure is vital for widespread EV adoption, addressing grid concerns and enhancing user experience.

4.1 Importance of Smart Charging

Smart charging plays a crucial role in the widespread adoption and successful integration of electric vehicles (EVs) into the broader energy ecosystem. It contributes to grid stability by managing electrical grid loads and preventing peaks in demand. Through load balancing, it avoids concentrated demand during peak hours, reducing the need for costly infrastructure upgrades. Smart charging optimizes EV charging based on renewable energy availability, promoting the use of solar or wind power. It enables cost optimization through time-of-use pricing, resulting in savings for both EV owners and utilities. By considering factors such as energy demand and user preferences, smart charging improves energy efficiency and reduces waste. Users benefit from remote monitoring and control, enhancing convenience and flexibility. Additionally, it helps businesses optimize charging schedules for electric vehicle fleets, reducing downtime and supporting grid stability. Smart charging also aligns EV charging with renewable energy, contributing to a lower carbon footprint. User-friendly features and regulatory compliance further enhance the overall EV ownership experience. In essence, smart charging is essential for efficient, convenient, and sustainable EV charging while ensuring the stability and reliability of the electrical grid in the face of growing EV numbers.

4.2 Technologies and Innovations

Smart charging infrastructure for electric vehicles (EVs) entails the integration of diverse technologies to bolster efficiency, grid management, and user experience. A pivotal innovation is Vehicle-to-Grid (V2G) Technology, enabling bidirectional communication between EVs and the grid, facilitating energy return to the grid. This aids grid balance by leveraging EV batteries for peak shaving, frequency regulation, and load balancing. Advanced Metering Infrastructure (AMI) with smart meters furnishes real-time energy consumption data, fostering dynamic pricing models for EV charging. This allows time-of-use pricing, incentivizing users to charge during off-peak hours for grid equilibrium.

Grid Connectivity and Communication, employing technologies like cellular networks and the Internet of Things (IoT), facilitates real-time communication among charging stations, vehicles, and central grid management systems (Das, H. S., et al. (2020)). This enables remote monitoring, data analytics, and dynamic control of charging infrastructure. Energy Management Systems (EMS) optimize energy use by coordinating charging schedules based on grid demand, renewable energy availability, and user preferences. This maximizes charging infrastructure efficiency and minimizes grid impact during peak demand.

Smart Charging Platforms and Mobile Apps provide real-time information on charging station availability, pricing, and progress. They empower users to remotely control and monitor charging sessions, enhancing the user experience and promoting off-peak charging. Blockchain Technology ensures secure and transparent transactions within the smart charging ecosystem, enhancing security and trust. Dynamic Pricing Models, like time-of-use pricing and demand-response mechanisms, encourage users to charge during low-demand periods, optimizing energy consumption (Kapassa, E., & Themistocleous, M. (2022)). Fleet Management Integration integrates with smart charging infrastructure, optimizing charging schedules for electric vehicle fleets, improving efficiency, reducing downtime, and ensuring timely charging. Predictive Analytics utilize historical and real-time data to predict charging demand, allowing proactive management of charging infrastructure. Bi-Directional Charging (V2H and V2B) goes beyond V2G, enabling EVs to supply power to homes or businesses and support grid services, enhancing energy resilience, providing backup power, and generating additional revenue streams. Wireless Charging Technology, utilizing electromagnetic fields, improves user convenience and seamlessly integrates into urban environments.

AI and Machine Learning analyze data from charging stations, grid conditions, and user behavior, optimizing charging schedules and predicting demand (Mazhar, T., et al. (2023)). This enhances the intelligence of charging infrastructure, improving adaptability and efficiency over time. The integration of these technologies in smart charging infrastructure is vital for creating a resilient, user-friendly, and grid-friendly ecosystem supporting the growing adoption of electric vehicles. Continued advancements and collaborations in these areas will contribute to a more sustainable and efficient transportation landscape.

4.3 Grid Connectivity and Load Management

The smart charging infrastructure for electric vehicles (EVs) relies on robust grid connectivity and load management to ensure grid stability, optimize energy use, and provide a seamless charging experience. Grid connectivity involves communication protocols like OCPP, IoT integration for remote monitoring, G2V communication for dynamic pricing, and smart meters for accurate data. Integration with grid management systems enables intelligent load balancing.

Load management strategies include load balancing to evenly distribute charging loads, demand response to adjust charging rates based on real-time demand, and peak shaving to minimize consumption during peak hours. Time-of-use pricing incentivizes off-peak charging, and predictive analytics anticipate charging demand for proactive load adjustments. User preferences and scheduling are considered, and vehicle-to-grid (V2G) integration allows bidirectional energy flow. Dynamic charging profiles adapt to real-time conditions, considering factors like grid congestion, energy prices, and user preferences. Overall, these systems are crucial for creating a reliable, efficient, and sustainable charging network as EV adoption continues to grow.

5. GRID INTEGRATION CHALLENGES

Grid integration confronts numerous challenges stemming from the unpredictability of renewable sources, grid stability concerns, and transmission limitations. The intermittent nature of solar and wind energy requires advanced forecasting and balancing mechanisms for stability. Issues like frequency control and voltage regulation become critical. Energy storage, vital for grid flexibility, faces hurdles in compatibility, efficiency, and cost (Tan, K. M., et al. (2021)). Cybersecurity risks add complexity, demanding robust measures to safeguard digitized grids. Regulatory gaps create uncertainties, hindering the seamless integration of diverse energy resources.

Demand response struggles with consumer engagement, requiring clear communication and incentives. Inadequate planning leads to grid congestion, while traditional systems prove inflexible in adapting to rapid energy shifts (Koirala, B. Pet al. (2016)). Cross-border integration faces challenges in regulatory misalignment and geopolitical factors. Market design must evolve to accommodate decentralized generation. Addressing aging infrastructure demands substantial upgrades, constrained by regulatory and financial limitations. A collaborative approach involving policymakers, utilities, and technology developers is crucial. Solutions hinge on innovation, regulatory reforms, infrastructure investment, and stakeholder engagement for a sustainable energy future.

5.1 Increased Electricity Demand

The widespread adoption of electric mobility (E-Mobility) presents both opportunities and challenges for electrical energy systems. On the positive side, the increased demand for electricity driven by electric vehicles (EVs) can open up additional revenue streams for utilities while optimizing the use of existing grid infrastructure. This surge in demand also provides an opportunity to integrate EV charging with renewable energy sources, such as solar and wind, promoting sustainability and reducing the carbon footprint of transportation. Moreover, the deployment of smart charging infrastructure offers grid flexibility by adapting charging times based on grid conditions, contributing to overall grid stability and efficiency. The implementation of Vehicle-to-Grid (V2G) technology enables EVs to provide valuable grid support services by returning excess energy during peak demand, enhancing grid resilience (Brown, M. A., & Soni, A. (2019)). Innovative load management strategies can optimize EV charging schedules, preventing grid congestion, reducing peak loads, and improving overall grid performance. Additionally, the energy storage capacity of EV batteries can be leveraged as distributed energy storage systems, storing excess energy during low-demand periods and releasing it during peak demand or when renewable energy generation is low. The necessity to accommodate increased electricity demand from EVs can stimulate investment in grid modernization, fostering advancements in monitoring, communication, and control systems, thereby presenting economic growth opportunities.

However, challenges accompany these opportunities. The existing grid infrastructure may require upgrades to handle the growing demand from a rising number of EVs, especially in areas with high EV concentrations. Managing the concentrated demand for electricity during peak hours, particularly for charging EVs, poses challenges that necessitate load management strategies to prevent overloads. The deployment of adequate and accessible charging infrastructure must keep pace with the growing number of electric vehicles to avoid charging bottlenecks and ensure user convenience.

Moreover, the intermittent nature of renewable energy sources and variability introduced by EV charging may challenge grid stability, necessitating advanced control mechanisms. Local distribution networks may face challenges in accommodating increased demand for EV charging, potentially leading to voltage fluctuations and overloads. The absence of clear and standardized regulations for EV charging, grid interaction, and incentives for sustainable practices creates uncertainties for investors and stakeholders.

Significant investments in grid infrastructure, charging stations, and supportive policies pose financial challenges for utilities and governments. The increased demand for electricity from EVs may impact the overall energy mix, requiring a delicate balance between clean energy sources and traditional power generation. Additionally, handling and securing sensitive data related to EV charging patterns, locations, and user information present challenges in terms of data security and privacy. High-density urban areas may experience charging grid congestion, necessitating effective urban planning and investments in charging infrastructure. To address these opportunities and challenges, a comprehensive and collaborative approach involving utilities, policymakers, technology developers, and other stakeholders is essential. The implementation of smart grid technologies, demand-response mechanisms, and the establishment of clear regulatory frameworks will play a crucial role in ensuring sustainable and resilient electrical energy systems.

5.2 Grid Stability Considerations

Grid stability considerations are crucial when integrating electric mobility into existing electrical energy systems. The increased adoption of electric vehicles (EVs) and the growing demand for electric charging infrastructure can impact the stability of the power grid (Hoarau, Q., & Perez, Y. (2018)). Charging electric vehicles (EVs) poses challenges to grid stability, such as increased peak loads and uneven distribution leading to congestion. Smart charging systems and load-balancing algorithms help manage these issues. The charging cycles, especially in bidirectional energy flow scenarios like Vehicle-to-Grid (V2G), impact grid frequency and voltage, necessitating advanced control systems. Grid infrastructure upgrades may be required for handling higher power demand, involving enhancements to transformers, distribution lines, and substations. Predictive analytics and machine learning assist in forecasting EV charging patterns, allowing utilities to plan and address potential stability issues. Integrating energy storage systems, like large-scale batteries, supports grid stability by storing excess energy and releasing it during peak demand. Considering EV charging in the context of overall grid resilience is crucial, assessing its impact on the grid's ability to withstand disruptions. Coordinating grid-interactive vehicles, including V2G capabilities, is critical for stability. Voltage regulation measures are essential in local networks with high concentrations of EV charging stations to address rapid voltage changes. Ensuring the cybersecurity of charging infrastructure and communication networks is vital for preventing disruptions. Establishing clear regulations and standards for electric mobility integration is essential to ensure coordinated and stable operations for grid operators, utilities, and charging station providers.

5.3 Smart Grid Technologies for E-Mobility

Smart grid technologies play a crucial role in integrating electric mobility into the broader energy infrastructure. These technologies enhance the efficiency, reliability, and sustainability of the electrical grid while supporting the growth of electric vehicle (EV) adoption. Smart grid technologies form a symbiotic relationship between electric mobility and the broader energy ecosystem, enhancing the sustainability and resilience of the energy infrastructure. Demand Response enables adjusting electricity consumption in response to grid signals, with smart grids optimizing EV charging based on grid conditions and demand fluctuations. Advanced Metering Infrastructure (AMI) with smart meters ensures accurate billing and real-time monitoring of EV charging. Grid-to-Vehicle (G2V) Communication and Vehicle-to-Grid (V2G) Technology allow bidirectional information flow, optimizing charging rates and providing grid support during peak demand. Smart Charging Stations enable real-time monitoring and contribute to grid stability. Predictive Analytics anticipates EV charging patterns for effective grid management. Intelligent Energy Management Systems coordinate EV charging, integrate renewables, and balance energy demand. Grid Automation, Cyber-Physical Systems (CPS), and Decentralized Energy Resources Integration enhance grid responsiveness, security, and sustainability. Dynamic Pricing adjusts electricity prices based on real-time conditions, encouraging grid-friendly charging behavior. As electric vehicle adoption grows, these technologies foster a resilient and sustainable energy infrastructure (Das, H. S., et al. (2020)).

6. SYNERGY WITH RENEWABLE ENERGY

The integration of electric mobility with renewable energy sources represents a pivotal advancement towards sustainable transportation. Aligning electric vehicle (EV) charging infrastructure with renewable sources such as solar, wind, and hydropower contributes to a cleaner and more eco-friendly transportation ecosystem (Soares, N., et al. (2018)). This section delves into the strategies and technologies that facilitate the seamless alignment of E-Mobility with renewable sources.

- Technological Integration: Explore how advancements in smart grids and energy storage technologies enable the harmonious integration of EV charging with renewable energy sources.
- Grid Management: Discuss the role of intelligent grid management systems in optimizing the utilization of renewable energy for electric vehicle charging, ensuring efficiency and reliability.
- Smart Charging Infrastructure: Highlight the development of smart charging stations equipped with capabilities to harness energy from renewable sources and their impact on reducing the carbon footprint of electric vehicles.

6.1 Distributed Energy Resources From Electric Vehicles

The emergence of electric vehicles not only revolutionizes personal transportation but also transforms them into distributed energy resources. This section explores the concept of leveraging electric vehicles as decentralized sources of energy and their potential to contribute to the overall energy grid.

- Vehicle-to-Grid (V2G) Technology: Investigate the applications of V2G technology, allowing electric vehicles to feed surplus energy back into the grid during peak demand, enhancing grid stability.
- Demand Response: Discuss how electric vehicles can actively participate in demand response programs, adjusting their charging patterns based on grid conditions and contributing to load management.
- Energy Storage: Explore the role of electric vehicle batteries as mobile energy storage units and their potential in balancing intermittent renewable energy sources.

6.2 Environmental Benefits of the Synergy

This section provides an in-depth analysis of the environmental benefits stemming from the synergy between electric mobility and renewable energy.

- Lifecycle Assessment: Conduct a comprehensive lifecycle assessment, evaluating the overall environmental impact of electric vehicles when charged using renewable energy sources (Nordelöf, A., et al. (2014)).
- Emission Reduction: Quantify the reduction in greenhouse gas emissions and pollutants achieved through the integration of renewable energy with electric mobility.
- Societal Implications: Explore how the environmental benefits positively impact communities, public health, and global efforts to combat climate change.

7. POLICY AND INCENTIVES

This part focuses on the critical role of policy frameworks and incentives in shaping the future of electric mobility and its synergy with renewable energy.

- Government Initiatives: Discuss existing policies supporting electric vehicle adoption and the integration of renewable energy, highlighting successful case studies from around the world.
- Incentive Programs: Explore incentive programs aimed at encouraging consumers, businesses, and utility providers to actively participate in the transition towards sustainable electric mobility.
- Regulatory Frameworks: Analyze the regulatory frameworks that govern the integration of electric mobility with renewable energy, addressing challenges and proposing solutions for a conducive environment.
- Global Collaborations: Highlight the importance of international collaborations in developing unified policies that foster the growth of electric mobility and renewable energy integration on a global scale.

8. TECHNOLOGICAL ADVANCEMENTS

Technological advancements in e-mobility are pivotal for fostering sustainability within electrical energy systems. The integration of electric vehicles (EVs) into these systems holds the potential to curtail greenhouse gas emissions, enhance energy efficiency, and pave the way for a more sustainable future. Continuous advancements in battery technology, emphasizing enhanced energy density and fast-charging capabilities, are extending the driving range and reducing charging times for electric vehicles (EVs). Solid-state batteries, a focus of ongoing research, promise improved safety and longevity. Bidirectional charging through Vehicle-to-Grid (V2G) technology not only supports grid stability but enables demand response, allowing EVs to adjust charging based on grid conditions. Inductive charging technologies provide user convenience by eliminating the need for physical cables. AI and machine learning in smart charging optimize schedules based on electricity prices, grid demand, and user preferences. Lightweight materials like carbon fiber and aluminum contribute to increased energy efficiency and prolonged battery life. Real-time IoT communication enables seamless transactions, remote monitoring, and predictive maintenance. Retired EV batteries finding a second life as stationary energy storage support grid stability, and regenerative braking systems increase overall energy efficiency by capturing kinetic energy during braking. These integrated technologies collectively contribute to a more sustainable and efficient electrical energy system, aligning with the goal of reducing carbon emissions and mitigating the environmental impact of transportation.

8.1 Continuous Improvements in Battery Technology

Continuous advancements in battery technology play a crucial role in advancing E-mobility and reinforcing electrical energy systems for sustainability. This comprehensive exploration within the realm of E-mobility, covering electric vehicles (EVs) and various electric transportation modes, emphasizes the pivotal role of efficient and sustainable battery systems. Researchers actively strive to enhance energy density, a critical parameter influencing the driving range of EVs. Progress in materials and chemistry, including advanced cathode and anode materials in lithium-ion batteries, contributes to achieving higher energy density (Tian, Y., et al. (2020), Zeng, X., et al. (2019)). Ongoing research in fast-charging technologies addresses the imperative to improve charging speed, crucial for the widespread adoption of electric vehicles. Innovations like solid-state batteries aim to overcome limitations in traditional lithium-ion batteries, promising faster and more efficient charging. The focus on enhancing the cycle life of batteries emerges as a critical aspect, impacting cost-effectiveness by reducing the frequency of battery replacements. Advances in anode and cathode materials, electrolyte formulations, and cell design contribute to batteries with longer life spans.

Safety remains paramount, with continuous research focused on developing batteries equipped with enhanced safety features, including advanced thermal management systems. Solid-state batteries, utilizing solid electrolytes, show promise in mitigating safety risks associated with flammable materials. Addressing cost concerns involves advances in manufacturing processes, economies of scale, and recycling technologies, with government incentives playing a pivotal role in making electric vehicles more affordable. A sustainability focus is evident in efforts to reduce the environmental impact of batteries (Peters, J. F., et al. (2017)). Researchers explore materials with lower environmental footprints, and recycling processes for end-of-life batteries are actively being developed. The commitment to sustainable battery development is underscored by efforts to minimize reliance on scarce or environmentally harmful materials, such as cobalt.

Integration with renewable energy sources and smart grid systems is a strategic approach to enhance overall sustainability in E-mobility. Optimizing charging patterns to align with renewable energy availability and incorporating vehicle-to-grid (V2G) technologies showcase the potential for electric vehicles to contribute to a greener energy ecosystem. This narrative highlights the need for continuous collaboration among researchers, manufacturers, and policymakers to drive advancements in battery technology. The ultimate goal is to cultivate a sustainable and efficient E-mobility ecosystem, contributing to a cleaner and greener future for transportation and energy systems.

8.2 Vehicle Efficiency Enhancements

Efficiency enhancements in E-mobility are a critical focus for maximizing the benefits of electric vehicles (EVs) and ensuring the sustainability of electrical energy systems, as detailed in this book chapter. The research emphasizes optimizing electric drive trains and power electronics, minimizing energy losses during energy conversion processes. Regenerative braking systems play a pivotal role by capturing kinetic energy during braking and converting it back into electrical energy, thus improving overall energy efficiency (Naseri, F., et al. (2016)). The chapter explores the use of lightweight materials like advanced alloys and carbon fiber composites to reduce vehicle weight, resulting in improved energy efficiency and extended driving ranges. The role of tire technology is highlighted, showcasing advancements in low-rolling-resistance tire designs and materials to reduce friction and enhance overall vehicle efficiency. Managing component temperatures, especially for batteries, motors, and power electronics, is discussed, emphasizing advanced thermal management systems for optimal efficiency and longevity. This topic also delves into smart energy management systems with predictive algorithms to optimize energy usage based on driving conditions and user behavior. Bidirectional energy flow through Vehicle-to-Grid (V2G) technology is explored as a means to contribute surplus energy back to the grid, enhancing grid stability and overall energy efficiency. The importance of efficient charging infrastructure, including high-power chargers and smart charging stations, is emphasized for minimizing charging times and maximizing energy transfer efficiency. Lastly, this addresses the significance of user education in promoting efficient driving practices, such as smooth acceleration, judicious use of climate control systems, and optimizing charging patterns. This multidimensional approach to enhancing vehicle efficiency underscores the evolution of E-mobility and its role in fostering a cleaner and more energy-efficient transportation future.

8.3 Innovations in Charging Infrastructure

Innovations in charging infrastructure are pivotal for advancing E-mobility and ensuring the sustainability of electrical energy systems. As electric vehicles (EVs) gain popularity, high-power charging stations, surpassing 150 kW, offer faster charging, addressing range anxiety and making long-distance travel more convenient. The development of ultra-fast charging stations exceeding 350 kW further reduces charging times, making EVs comparable to traditional refueling (Muzir, N. A. Q., et al. (2022)). Battery swapping stations provide a quick exchange of depleted batteries, eliminating lengthy charging stops and addressing infrastructure availability concerns. Wireless charging technology, including inductive charging and dynamic wireless charging roads, enhances user convenience and driving range. Smart charging infrastructure with communication capabilities optimizes charging based on grid conditions, contributing to energy efficiency and grid stability. Grid integration and demand response capabilities adjust charging rates based on renewable energy availability and overall energy demand. Solar-powered charging stations harness renewable energy, reducing reliance on the grid. Integrating energy storage systems at charging stations manages peak demand and provides a stable power supply. Universal charging standards like CCS and CHAdeMO ensure interoperability and user convenience. Charging hubs with multi-modal integration offer various charging options and flexibility for different EV models. User-friendly apps and payment systems simplify access to charging services, encouraging broader EV adoption. Continuous innovation in charging infrastructure is essential for overcoming adoption barriers, making electric transportation an attractive and viable choice for a diverse range of users.

9. FUTURE TRENDS AND PROSPECTS

The future of e-mobility within electrical energy systems is marked by promising trends driving sustainability in transportation. Advancements in battery technology, particularly the development of solid-state batteries, hold the potential to revolutionize electric vehicle (EV) performance. Smart charging infrastructure and Vehicle-to-Grid (V2G) technology play pivotal roles in optimizing energy usage and promoting grid stability. The integration of electric vehicles with renewable energy enhances the overall sustainability of the transportation ecosystem. Autonomous electric vehicles (AEVs) are expected to reshape transportation, optimizing energy efficiency and reducing congestion. Efforts to electrify various modes of transportation contribute comprehensively to carbon emission reduction. Innovations in sustainable materials for EV manufacturing and regulatory support are key components of the e-mobility landscape. Consumer education and international collaboration are crucial for fostering acceptance and global integration (Gilpin, R. G. (2016)). In conclusion, the future of e-mobility is characterized by a holistic and interconnected approach, combining technological innovation, renewable energy integration, supportive policies, and collaborative efforts to create a sustainable transportation ecosystem.

9.1 Anticipated Developments in E-Mobility

The evolution of e-mobility within electrical energy systems is positioned to significantly impact sustainability in transportation. Anticipated advancements include breakthroughs in battery technologies, such as higher energy density and faster charging, essential for enhancing the efficiency of electric vehicles (EVs). Integration with renewable energy, smart charging infrastructure, and vehicle electrification in public transportation are expected to contribute to cleaner and eco-friendly mobility. Regulatory initiatives and incentives, along with the electrification of last-mile transportation, are pivotal for widespread adoption. Collaboration among stakeholders is crucial for addressing infrastructure challenges and ensuring a seamless integration of e-mobility.

9.2 Emerging Technologies

In the evolving realm of e-mobility, advancements in diverse technologies are shaping a more sustainable future. Solid-state batteries and enhanced lithium-ion technologies drive increased energy density and safety in electric vehicles (EVs). High-power charging stations and wireless charging are pivotal for faster and more convenient EV charging (Feng, H., et al. (2020)). Energy storage integration explores concepts like Vehicle-to-Grid (V2G) systems and smart charging infrastructure, enabling EVs to contribute to grid stability and optimize charging times. Advanced power electronics, such as wide-bandgap semiconductors, improve efficiency in electric propulsion. Vehicle electrification encompasses Hybrid Electric Vehicles (HEVs), Plug-in Hybrid Electric Vehicles (PHEVs), and Fuel Cell Electric Vehicles (FCEVs), contributing to fuel efficiency and zero-emission transportation. Smart grid technologies and energy management systems integrate EV charging with intelligent grids for enhanced load management. Innovations in lightweight materials extend the range of EVs, while the integration of autonomous features improves efficiency and safety (Czerwinski, F. (2021)). Data analytics, connectivity, and Vehicle-to-Everything (V2X) communication optimize infrastructure and enhance traffic flow. Circular economy initiatives, including battery recycling, underscore the commitment to reducing environmental impact (Demestichas, K., & Daskalakis, E. (2020)). Continuous research and innovation across these domains are imperative for ensuring the widespread adoption of sustainable e-mobility solutions.

9.3 Implications for Sustainable Transportation

Sustainable transportation, characterized by the use of modes with lower environmental impact, holds far-reaching implications for various facets of society, the economy, and the environment. The environmental advantages are evident in the diminished emissions resulting from the adoption of sustainable modes such as electric vehicles (EVs), bicycles, and public transit. This not only reduces greenhouse gas emissions and air pollution but also diminishes dependence on fossil fuels, promoting enhanced air quality and contributing to the global battle against climate change. Furthermore, the endorsement of alternatives like walking, biking, and public transportation in sustainable transportation practices leads to decreased pressure on land and resources, playing a role in the conservation of natural resources (Schiller, P. L., & Kenworthy, J., 2017).

From an economic perspective, sustainable transportation yields substantial cost savings through decreased fuel consumption, lower maintenance expenditures, and a reduction in healthcare costs associated with air pollution and traffic accidents (World Health Organization, 2012). Moreover, the transition to sustainable transportation, exemplified by the production of electric vehicles and the development of public transit and cycling infrastructure, generates employment opportunities, contributing to job creation. Socially and health-wise, sustainable transportation brings positive implications. Modes like walking and cycling not only minimize environmental impact but also promote increased physical activity, mitigating sedentary lifestyles and related health issues. Public transportation, as a vital element of sustainable mobility, ensures fair access to employment, education, and services, particularly benefiting those without the means or choice to own a private vehicle.

In the domain of urban planning and design, sustainable transportation fosters smart growth by encouraging compact, mixed-use development, thereby reducing urban sprawl and endorsing efficient land use (Furlan, R., & Al-Mohannadi, A., 2020). Technological innovation plays a pivotal role in sustainable transportation, with ongoing advancements in electric vehicle technology, battery development, and charging infrastructure driving widespread adoption. Additionally, the integration of technology for traffic management, real-time information dissemination, and efficient public transportation systems contributes to the overall sustainability of mobility.

Policy implications are crucial in steering societies toward sustainable transportation. Governments can enact incentives and regulations such as tax incentives for EVs, congestion pricing, and stricter emission standards to promote sustainable transportation practices. Public investment in infrastructure, including public transportation, cycling lanes, and pedestrian-friendly urban planning, serves as a tangible commitment to encouraging sustainable modes of transportation. On a global scale, international collaboration is imperative to address cross-border challenges inherent in transportation, such as emissions from the aviation and shipping industries. Culturally, a shift towards sustainable transportation necessitates changing mindsets, a process that can be fostered through public awareness campaigns and education, emphasizing the importance of embracing environmentally conscious transportation choices.

10. CONCLUSION

In conclusion, the integration of Electric Mobility (E-Mobility) into electrical energy systems holds significant promise for sustainability, but it also presents a set of challenges that need to be addressed comprehensively. The transition to electric vehicles (EVs) has the potential to reduce greenhouse gas emissions, decrease reliance on fossil fuels, and contribute to a more sustainable and environmentally friendly transportation sector. However, achieving these benefits requires strategic planning, investments, and innovations across various domains.

10.1 Key Takeaways

In summary, Electric Vehicles (EVs) offer environmental benefits, aligning with renewable energy for reduced carbon footprint. Smart charging infrastructure, including Vehicle-to-Grid (V2G), plays a vital role. Challenges include strain on grids, insufficient charging infrastructure, and lack of standardized regulations. Overcoming these requires grid modernization, strategic planning, standardized regulations, and addressing data security concerns. The transition to E-Mobility necessitates substantial investments, emphasizing the need for innovative financing models and incentives to achieve long-term sustainability goals.

10.2 Recommendations for Future Research

The shift to sustainable transportation, particularly electric mobility (e-mobility), demands focused research in electrical energy systems. Key areas include advancing EV energy storage, smart grid technologies like V2G integration, scalable charging infrastructure, wireless charging, power electronics, and grid impact assessments. Lifecycle assessments, alternative materials, and renewable energy synergies aim to enhance EV sustainability. Understanding human behavior in adopting e-mobility and addressing cybersecurity concerns are vital. Evaluating existing policies and proposing new ones will shape a resilient e-mobility ecosystem. Collaborative efforts across academia, industry, and government bodies are essential for a sustainable e-mobility future.

REFERENCES

Barman, P., Dutta, L., Bordoloi, S., Kalita, A., Buragohain, P., Bharali, S., & Azzopardi, B. (2023). Renewable energy integration with electric vehicle technology: A review of the existing smart charging approaches. *Renewable & Sustainable Energy Reviews, 183*, 113518. doi:10.1016/j.rser.2023.113518

Berkeley, N., Jarvis, D., & Jones, A. (2018). Analysing the take up of battery electric vehicles: An investigation of barriers amongst drivers in the UK. *Transportation Research Part D, Transport and Environment, 63*, 466–481. doi:10.1016/j.trd.2018.06.016

Biresselioglu, M. E., Kaplan, M. D., & Yilmaz, B. K. (2018). Electric mobility in Europe: A comprehensive review of motivators and barriers in decision making processes. *Transportation Research Part A, Policy and Practice, 109*, 1–13. doi:10.1016/j.tra.2018.01.017

Bleicher, A., & Pehlken, A. (Eds.). (2020). *The material basis of energy transitions*. Academic Press. doi:10.1016/B978-0-12-819534-5.00001-5

Brown, M. A., & Soni, A. (2019). Expert perceptions of enhancing grid resilience with electric vehicles in the United States. *Energy Research & Social Science, 57*, 101241. doi:10.1016/j.erss.2019.101241

Czerwinski, F. (2021). Current trends in automotive lightweighting strategies and materials. *Materials (Basel), 14*(21), 6631. doi:10.3390/ma14216631 PMID:34772154

Das, H. S., Rahman, M. M., Li, S., & Tan, C. W. (2020). Electric vehicles standards, charging infrastructure, and impact on grid integration: A technological review. *Renewable & Sustainable Energy Reviews*, *120*, 109618. doi:10.1016/j.rser.2019.109618

Demestichas, K., & Daskalakis, E. (2020). Information and communication technology solutions for the circular economy. *Sustainability (Basel)*, *12*(18), 7272. doi:10.3390/su12187272

Dijk, M., Orsato, R. J., & Kemp, R. (2013). The emergence of an electric mobility trajectory. *Energy Policy*, *52*, 135–145. doi:10.1016/j.enpol.2012.04.024

Feng, H., Tavakoli, R., Onar, O. C., & Pantic, Z. (2020). Advances in high-power wireless charging systems: Overview and design considerations. *IEEE Transactions on Transportation Electrification*, *6*(3), 886–919. doi:10.1109/TTE.2020.3012543

Furlan, R., & Al-Mohannadi, A. (2020). An urban regeneration planning scheme for the Souq Waqif Heritage Site of Doha. *Sustainability (Basel)*, *12*(19), 7927. doi:10.3390/su12197927

Ghosh, A. (2020). Possibilities and challenges for the inclusion of the electric vehicle (EV) to reduce the carbon footprint in the transport sector: A review. *Energies*, *13*(10), 2602. doi:10.3390/en13102602

Gilpin, R. G. (2016). *The political economy of international relations*. Princeton University Press.

Hasan, M. M., Hossain, S., Mofijur, M., Kabir, Z., Badruddin, I. A., Yunus Khan, T. M., & Jassim, E. (2023). Harnessing solar power: A review of photovoltaic innovations, solar thermal systems, and the dawn of energy storage solutions. *Energies*, *16*(18), 6456. doi:10.3390/en16186456

Hensher, D. A., Mulley, C., Ho, C., Wong, Y., Smith, G., & Nelson, J. D. (2020). *Understanding Mobility as a Service (MaaS): Past, present and future*. Elsevier.

Hoarau, Q., & Perez, Y. (2018). Interactions between electric mobility and photovoltaic generation: A review. *Renewable & Sustainable Energy Reviews*, *94*, 510–522. doi:10.1016/j.rser.2018.06.039

Hoogma, R., Kemp, R., Schot, J., & Truffer, B. (2002). *Experimenting for sustainable transport*. Taylor & Francis.

Itani, K., & De Bernardinis, A. (2023). Review on New-Generation Batteries Technologies: Trends and Future Directions. *Energies*, *16*(22), 7530. doi:10.3390/en16227530

Kapassa, E., & Themistocleous, M. (2022). Blockchain technology applied in IoV demand response management: A systematic literature review. *Future Internet*, *14*(5), 136. doi:10.3390/fi14050136

Kemp, R., Schot, J., & Hoogma, R. (1998). Regime shifts to sustainability through processes of niche formation: The approach of strategic niche management. *Technology Analysis and Strategic Management*, *10*(2), 175–196. doi:10.1080/09537329808524310

Koirala, B. P., Koliou, E., Friege, J., Hakvoort, R. A., & Herder, P. M. (2016). Energetic communities for community energy: A review of key issues and trends shaping integrated community energy systems. *Renewable & Sustainable Energy Reviews*, *56*, 722–744. doi:10.1016/j.rser.2015.11.080

Lebrouhi, B. E., Khattari, Y., Lamrani, B., Maaroufi, M., Zeraouli, Y., & Kousksou, T. (2021). Key challenges for a large-scale development of battery electric vehicles: A comprehensive review. *Journal of Energy Storage*, *44*, 103273. doi:10.1016/j.est.2021.103273

Lemian, D., & Bode, F. (2022). Battery-supercapacitor energy storage systems for electrical vehicles: A review. *Energies*, *15*(15), 5683. doi:10.3390/en15155683

Mahiban, L. N., & Emimal, M. (2023). *Longevity of Electric Vehicle Operations*. Qeios.

Mazhar, T., Asif, R. N., Malik, M. A., Nadeem, M. A., Haq, I., Iqbal, M., & Ashraf, S. (2023). Electric Vehicle Charging System in the Smart Grid Using Different Machine Learning Methods. *Sustainability (Basel)*, *15*(3), 2603. doi:10.3390/su15032603

Ming, T., Derichter, R., Shen, S., & Caillol, S. (2016). Fighting global warming by greenhouse gas removal: Destroying atmospheric nitrous oxide thanks to synergies between two breakthrough technologies. *Environmental Science and Pollution Research International*, *23*, 6119–6138. doi:10.1007/s11356-016-6103-9 PMID:26805926

Mohammadi, F., & Saif, M. (2023). A comprehensive overview of electric vehicle batteries market. *e-Prime-Advances in Electrical Engineering, Electronics and Energy*, 100127.

Mohammed, A., Saif, O., Abo-Adma, M., Fahmy, A., & Elazab, R. (2024). Strategies and sustainability in fast charging station deployment for electric vehicles. *Scientific Reports*, *14*(1), 283. doi:10.1038/s41598-023-50825-7 PMID:38168937

Montes, F., Meinen, R., Dell, C., Rotz, A., Hristov, A. N., Oh, J., & Dijkstra, J. (2013). SPECIAL TOPICS—Mitigation of methane and nitrous oxide emissions from animal operations: II. A review of manure management mitigation options. *Journal of Animal Science*, *91*(11), 5070–5094. doi:10.2527/jas.2013-6584 PMID:24045493

Muratori, M., Alexander, M., Arent, D., Bazilian, M., Cazzola, P., Dede, E. M., & Ward, J. (2021). The rise of electric vehicles—2020 status and future expectations. *Progress in Energy*, *3*(2), 022002. doi:10.1088/2516-1083/abe0ad

Muzir, N. A. Q., Mojumder, M. R. H., Hasanuzzaman, M., & Selvaraj, J. (2022). Challenges of electric vehicles and their prospects in Malaysia: A comprehensive review. *Sustainability (Basel)*, *14*(14), 8320. doi:10.3390/su14148320

Naseri, F., Farjah, E., & Ghanbari, T. (2016). An efficient regenerative braking system based on battery/supercapacitor for electric, hybrid, and plug-in hybrid electric vehicles with BLDC motor. *IEEE Transactions on Vehicular Technology*, *66*(5), 3724–3738. doi:10.1109/TVT.2016.2611655

Nikitas, A., Kougias, I., Alyavina, E., & Njoya Tchouamou, E. (2017). How can autonomous and connected vehicles, electromobility, BRT, hyperloop, shared use mobility and mobility-as-a-service shape transport futures for the context of smart cities? *Urban Science (Basel, Switzerland)*, *1*(4), 36. doi:10.3390/urbansci1040036

Nordelöf, A., Messagie, M., Tillman, A. M., Ljunggren Söderman, M., & Van Mierlo, J. (2014). Environmental impacts of hybrid, plug-in hybrid, and battery electric vehicles—What can we learn from life cycle assessment? *The International Journal of Life Cycle Assessment*, *19*(11), 1866–1890. doi:10.1007/s11367-014-0788-0

Omer, A. M. (2008). Energy, environment and sustainable development. *Renewable & Sustainable Energy Reviews*, *12*(9), 2265–2300. doi:10.1016/j.rser.2007.05.001

Peters, J. F., Baumann, M., Zimmermann, B., Braun, J., & Weil, M. (2017). The environmental impact of Li-Ion batteries and the role of key parameters–A review. *Renewable & Sustainable Energy Reviews*, *67*, 491–506. doi:10.1016/j.rser.2016.08.039

Pokhriyal, A., Rueda-García, D., & Gómez-Romero, P. (2023). To flow or not to flow. A perspective on large-scale stationary electrochemical energy storage. *Sustainable Energy & Fuels*, *7*(23), 5473–5482. doi:10.1039/D3SE00955F

Prado-Lorenzo, J. M., Rodríguez-Domínguez, L., Gallego-Álvarez, I., & García-Sánchez, I. M. (2009). Factors influencing the disclosure of greenhouse gas emissions in companies world-wide. *Management Decision*, *47*(7), 1133–1157. doi:10.1108/00251740910978340

Rahman, F. A., Aziz, M. M. A., Saidur, R., Bakar, W. A. W. A., Hainin, M. R., Putrajaya, R., & Hassan, N. A. (2017). Pollution to solution: Capture and sequestration of carbon dioxide (CO2) and its utilization as a renewable energy source for a sustainable future. *Renewable & Sustainable Energy Reviews*, *71*, 112–126. doi:10.1016/j.rser.2017.01.011

Ruggieri, R., Ruggeri, M., Vinci, G., & Poponi, S. (2021). Electric mobility in a smart city: European overview. *Energies*, *14*(2), 315. doi:10.3390/en14020315

Schiller, P. L., & Kenworthy, J. (2017). *An introduction to sustainable transportation: Policy, planning and implementation*. Routledge. doi:10.4324/9781315644486

Schot, J., & Geels, F. W. (2008). Strategic niche management and sustainable innovation journeys: Theory, findings, research agenda, and policy. *Technology Analysis and Strategic Management*, *20*(5), 537–554. doi:10.1080/09537320802292651

Soares, N., Martins, A. G., Carvalho, A. L., Caldeira, C., Du, C., Castanheira, É., & Garcia, R. (2018). The challenging paradigm of interrelated energy systems towards a more sustainable future. *Renewable & Sustainable Energy Reviews*, *95*, 171–193. doi:10.1016/j.rser.2018.07.023

Sovacool, B. K., Kester, J., Noel, L., & de Rubens, G. Z. (2019). Energy injustice and Nordic electric mobility: Inequality, elitism, and externalities in the electrification of vehicle-to-grid (V2G) transport. *Ecological Economics*, *157*, 205–217. doi:10.1016/j.ecolecon.2018.11.013

Spickermann, A., Grienitz, V., & Von der Gracht, H. A. (2014). Heading towards a multimodal city of the future?: Multi-stakeholder scenarios for urban mobility. *Technological Forecasting and Social Change*, *89*, 201–221. doi:10.1016/j.techfore.2013.08.036

Tan, K. M., Babu, T. S., Ramachandaramurthy, V. K., Kasinathan, P., Solanki, S. G., & Raveendran, S. K. (2021). Empowering smart grid: A comprehensive review of energy storage technology and application with renewable energy integration. *Journal of Energy Storage*, *39*, 102591. doi:10.1016/j.est.2021.102591

Themelis, N. J., & Ulloa, P. A. (2007). Methane generation in landfills. *Renewable Energy*, *32*(7), 1243–1257. doi:10.1016/j.renene.2006.04.020

Tian, Y., Zeng, G., Rutt, A., Shi, T., Kim, H., Wang, J., Koettgen, J., Sun, Y., Ouyang, B., Chen, T., Lun, Z., Rong, Z., Persson, K., & Ceder, G. (2020). Promises and challenges of next-generation "beyond Li-ion" batteries for electric vehicles and grid decarbonization. *Chemical Reviews*, *121*(3), 1623–1669. doi:10.1021/acs.chemrev.0c00767 PMID:33356176

World Health Organization. (2012). *Health in the green economy: health co-benefits of climate change mitigation-transport sector*. World Health Organization.

Yoro, K. O., & Daramola, M. O. (2020). CO2 emission sources, greenhouse gases, and the global warming effect. In *Advances in carbon capture* (pp. 3–28). Woodhead Publishing. doi:10.1016/B978-0-12-819657-1.00001-3

Zeng, X., Li, M., Abd El-Hady, D., Alshitari, W., Al-Bogami, A. S., Lu, J., & Amine, K. (2019). Commercialization of lithium battery technologies for electric vehicles. *Advanced Energy Materials*, *9*(27), 1900161. doi:10.1002/aenm.201900161

Chapter 3
E-Mobility as a Service and Its Economic Impact:
Regulations, Policies, Governance, and Best Practices in E-Mobility Ecosystems

Leo John
https://orcid.org/0000-0002-4071-9920
Saveeha School of Engineering, India

Rojalini Patro
KL University, India

S. Padmakala
Saveetha School of Engineering, India

R. B. Madhumala
Dayananda Sagar Academy of Technology and Management, India

ABSTRACT

This study presents a systematic review of the studies and a comprehensive assessment of the ongoing implementation of e-mobility, evaluating social, monetary, policy, and sustainability factors associated with the circular economy (CE) ecosystems framework. It also discusses the current difficulties and proposes methods to address them to improve urban quality of life and the sustainability of electro-mobility. The objective of this study is to discover the particular explanation as to why industrialized countries are able to efficiently introduce cutting-edge electric fuel automobiles (e.g., electric vehicles) while underdeveloped economies cannot find it extremely difficult to implement these vehicles, particularly in terms of infrastructure and downstream activities and how these issues can be addressed effectively.

DOI: 10.4018/979-8-3693-2611-4.ch003

INTRODUCTION

With a worldwide initiative focus on environmentally friendly progress, reducing carbon emissions, and sustainability in the coming years, the transition to more eco-friendly replacements in transportation is essential. Considering the function assumed by transportation in the economy, and the detrimental implications associated with combustion of fossil fuels, a paradigm shift is imminent (Kuzey et al., 2022). India is also gaining traction in the e-mobility field, thanks to enthusiasm from businesses and the government. In most of the developed countries, electric mobility has been the path towards success (Goel et al., 2021). The adoption and development of electric vehicles has been aided by Environmental concerns, such as weather change, advancements in alternative energy, rapid suburbanization, energy chemistry, energy assurance, etc (Haines & Frumkin, 2021). In addition to battery infrastructure, auxiliary services, and multi-modal integrated transportation, this paradigm shift may open up new business opportunities. Consequently, it may create a new job ecosystem (Koh et al., 2020). However, there are currently a number of technologies available for recycling waste batteries, such as direct recycling, e-mobility recycling, and high-temperature thermal treatments, which are capable of removing the flammability of electrolytes and decomposing organic compounds in the partitioning material. Nevertheless, high power requirements and the release of hazardous substances make waste battery recycling a challenging process. Furthermore, the E-mobility an eco-friendly technique has the following benefits, namely the reduction of energy production and the reduction of ecological impact. Additionally, CE will be monitored and evaluated every year in order to determine whether levy rates need to be adjusted.

The government or selected organizations will ensure that levy rates and rebates are in line with the objectives. A website that publishes monitoring and evaluation would facilitate transparency and investment. Conservation of precious metals in batteries is currently underway in several countries. The CE ecosystem requires a collaboration platform in order to be accelerated. This system offers to speed up the production of energy-efficient battery systems in order to meet future battery demands, improve battery industry capacity (technologies, materials, and workforce), advance battery-recycling technologies and processes, enhance battery resource reclamation, maintaining sustainability, protecting the environment from unsafe battery disposal, as well as optimize the efficiency of production and utilizing battery residue while extending its monetary importance. The European Climate Foundation, for instance, predicts that EVs will contribute to the creation of numerous jobs in research & development by 2030 with their introduction to the global market. According to another study, by 2050 more jobs will be generated in complementary business sectors such as construction, electrical appliances, and manufactured petroleum products. There is, however, a qualitative and quantitative difference between EVs and Indian job ecosystems (Brugger et al. 2021).

LITERATURE REVIEW

According to Thimm (2022), It is the purpose of this paper to develop a blueprint for the lakes coastal Continent environment in Germany as an e-tourism hotspot.i.e., as a location where travelers have ease of movement will primarily be powered by electricity in the coming years. In addition to empirical data obtained through observations and consultations with travelers and participants in 2016 and 2017, a situation evaluation carried out on factor identification is presented. A model called e-destination, which illustrates how tourism mobility will largely be based on electro-mobility (e-mobility), is included. This model encompasses optimistic, pessimistic, and realistic scenarios. An empirical study supports this particular scenario. In this study, only e-mobility is considered, so other forms of commuting such as walking and riding a bicycle are not considered. In spite of the fact that they are not strictly random, the trends they reveal are clear and are consistent with each other. In their opinion, the region is likely to Aspire to be a digital destination in the next few years, however, only through the support of the government. The study discussed the attitude-behavior gap as a possible explanation for the behavior of tourists. A limited number of studies have been conducted on tourism e-mobility. In the author's knowledge, this is the first study to analyze how e-mobility will affect tourism using a German lake as a case study. In this way, it contributes to a broader understanding of how e-mobility will affect tourism destinations in the future.

E-bike renting, e-scooter leasing, and electric carsharing are all categorized as shared e-mobility. As a result of in such offerings, cities are anticipated to minimize the unfavorable consequences of road commuting, which is predominantly supported by fossil energy production cars. Researcher performed a systematic review of published studies on the three collaborative e-mobility approaches in order to more effectively contribute to the advancement and marketing of these offerings and to identify potential areas for further investigation. Our review focused on their consumption behaviors, indications estimates, and conceivable implications. However, their differences in vehicle capabilities, we found that each of the three cooperative e-mobility products are primarily used for brief outings, and their clients are predominantly young males with moderately high incomes. In settings with enhanced performance transportation connections and many other attractions of significance, all involved e-mobility modes generate higher demand based on the same socio-demographic characteristics. A shared e-mobility service can reduce car ownership, car use, and greenhouse gas emissions, which can have a positive impact on transportation and the environment. Nevertheless, the significance of these outcomes will depend on the individual working arrangements of the organizations, such as the structure of fuel and the life of the Communal vehicles. It is also expected that each shared e-mobility mode will be impacted both by its complementarity and by its competition with other coexisting modes. In the future, multiple shared e-mobility modes should be studied in terms of competition and integration (Liao & Correia, 2022).

According to Behrendt (2018), E-velomobility is the concept proposed in this paper. Pedal power is combined Utilizing e-mobility's electric power and motor assist to accelerate a participant in electro mobility strategies, mechanisms, and developments. E-mobility, E-velomobility, and e-bikes are emphasized in this concept. In this analysis, qualitative material from an e-bike trial in the UK is analyzed to illustrate how people experience e-velocity. In this article, we demonstrate E-velomobility as an A crucial form of electronic mobility, with consequences for scientific thinking and government. By expanding the conceptualization of technologies beyond electric cars to include more active and sustainable modes of e-mobility that are less expensive, E-velomobility may Contribute to the evolution of methods and procedures.

According to Dias et al. (2021), As a sustainable means of transport, the shareable e-scooter system was first introduced in 2017 and has been gaining traction ever since. A new urban mobility strategy that focuses on nonpolluting modes, In particular, as the E.U. Ecological and Sustainable Transportation Initiative, also contributes to this mode's success. An analysis of the literature and a case study were conducted to show what can be achieved Utilizing the development of collaborative e-scooter solutions in urban mobility. A cooperative approach can assist cities in reducing pollution, reducing inequalities in access to transportation, assisting in reducing costs, and increasing mobility through collaboration. Due to the Covid-19 outbreak, public e-scooters have emerged as useful to cities across the globe in that they increase social separation and benefit local communities replace short-distance transportation with public transport. A analyze the research of Braga revealed that the city maintains that it operates on cooperative e-scooter operations to provide adaptability options during the outbreak, as well as offering special fares to encourage people to use the service.

Castillo-Calzadilla et al. (2022) studied that global's power grids are under stress from EVs (electric vehicles). Designed as prototypes of (tiny) metropolitan areas,Positive Energy Districts (PEDs) are made up of interconnected buildings, vehicles, and smart grids. With MATLAB-Simulink software, this paper simulates the impact of e-mobility on a PED. A secluded town in the north of Spain called PED is measured in five situations with varying energy efficiency, street lighting, and electric vehicle requirements. There's a connection between the emissions classification of the building (A for the least cost-effective to E for the least efficient) and the yearly energization surplus. In a PED containing multiple interacting buildings (three Including one private and three general access) and 405 EVs (as a benchmark), only high-rated buildings (A or B) are positive. Based on simulations, the PED can support nearly 9 million green kilometers, or 695 EVs in the optimum condition (highly rated properties). The district and the city as a whole could save 71% in carbon emitted from electric mobility exclusively (relative to fossil-fuel automobiles), reducing the carbon emissions of the neighbourhood and the region.

Bagheri Tookanlou et al.(2021) revealed that next-generation electric vehicles ecosystems will be a complicated process structures, with multiple Interested parties are invited to maximize their benefits and operations. Electrified automobiles, recharging facilities, and suppliers are included in this researcher analyzed driver-to-grid and grid-to-vehicle synchronization mechanisms. We define and formulate two categories of vacations, including compulsory and non-mandatory, to improve V2G functionality and to minimize congestion at CSs. To strengthen the implementation details of the time allocation framework, EV drivers' expectations are incorporated as financial requirements and an increase in usage distances. To solve the non-cooperative Stackelberg game, an elaborative methodology is suggested that determines the optimal routes, CS, retailers, and V2G prices for each EV. Using real data from San Francisco, the USA, extensive simulated experiments are performed for two alternative e-mobility frameworks, incorporating various vendors and CSs, as well as several other stakeholders electric vehicles. It is shown in the modeling suggests that the opportunistic journeys reduce the costs of EVs and PV reduced by 8.8-24.2% on average and 26.4-28.5% consequently, under Variations of situations, as well as mitigate inconvenience around particular hours while observing the preferences of EV indicators. Additionally, the simulation results revealed that the preferences of EV drivers have a significant impact on the most effective methods and cost/revenue of participants.

According to Serohi (2022), A key objective of this study is to clarify why developed countries are able to easily adopt The development of alternate fuel automobiles (e.g. battery-powered vehicles or electric vehicles) while the application in emerging countries appears to seem far-fetched in terms of Providing infrastructures and subsequent initiatives, and to propose ways to address these issues effectively. For the purpose of this study, three case studies will be examined - Tesla (USA), Mahindra and Mahindra (India) and Tata Motors - in order to highlight the problems facing Manufacturing companies engaged in research and development and developed nations. In-depth surveys have adequately represented consumers' viewpoints. Moreover, Tesla's innovation and vertical integration of upstream and downstream systems are analyzed to determine how it has acquired competitive leverage. Developing countries such as India still lack adequate EV infrastructure. Lack of research and development capabilities and the erection of power stations on residential lots and residential groups are two key areas that are largely unexplored. As with conventional vehicles, domestic EVs have lower performance metrics than their foreign counterparts, like Tesla. One of the most pressing problems among members of the population purchasers of electric vehicles in India is range anxiety. In addition to analyzing horizontal and vertical perspectives in great detail, the paper also examines the influence of manufacturing ecosystem advancements on both downstream and upward supply chain components. The general public's perceptions and viewpoints on e-mobility are determined from survey results, while Tesla's competitive leverage is analyzed via the impact analysis matrix.

Mossali et al. (2020) studied, Li-Ion batteries are high-value products with significant amounts of critical materials (e.g. Lithium and Cobalt), so an Implementing a circular economy approach is imperative. While Vehicles that use electric motors and batteries are engineered to be disassembled, tested, remanufactured, and recycled, they are not designed for easy, safe and cost-effective disassembly. This paper presents an e-mobility battery redesign study focused on circular economies in order to overcome this industry limitation. Using a structured methodology for evaluating design criteria (House of Quality), features impacting circular economy design requirements have been assessed.

Meskers et al. (2019) analysed that the society is well on its way to becoming circular and sustainable. City development is happening bottom-up, with UN Sustainable Development Goals supporting it top-down. IoT, circular economy, and e-mobility are the three trends of the day. A circular city connects people, businesses, institutions, and geographies, bringing all aspects of life together. Sustainable energy, eco-friendly, renewable products, an integrated production cycle, and smart cities are the hallmarks of circular cities. Systems meet, interact, and challenge one another in the technological, production-related, monetary, historical, and psychological aspects spheres. As society undergoes this transition, raw materials play an important role. Solutions, materials, and technologies are provided by it. Additionally, the industry must Prepare to serve as a key part of the upcoming community. In this study, current and future e-mobility (batteries) developments will be explanatory, along with the consequences and problems they pose.

Knieke et al. (2019) explained that In an upcoming Implementing a circular economy strategy, recycled commodities, waste sources, and second-hand raw commodities should be handled accordingly. Material requirements are changing dynamically as products are developed and consumer behavior changes. In order to have a highly flexible recycling system, the waste disposal procedure must also perform even more effectively and with increased consistency than it does now. The value and needs of the materials and components recovered along the process chain may change dynamically as a result. Using a novel approach, this paper develops a anticipatory and responsive attitude to continually emerging challenges in the circular economy chain, adding value to a future Advanced Circular Economy. There are three main components to the approach: an open information marketplace, an A cost-benefit comparison and optimization method, and continuous enhancement of the waste reduction and resource recovery workflow chain, which includes the arrangement of procedures and steps.

Figure 1. Conceptual diagram: E-mobility ecosystem with CE perspectives

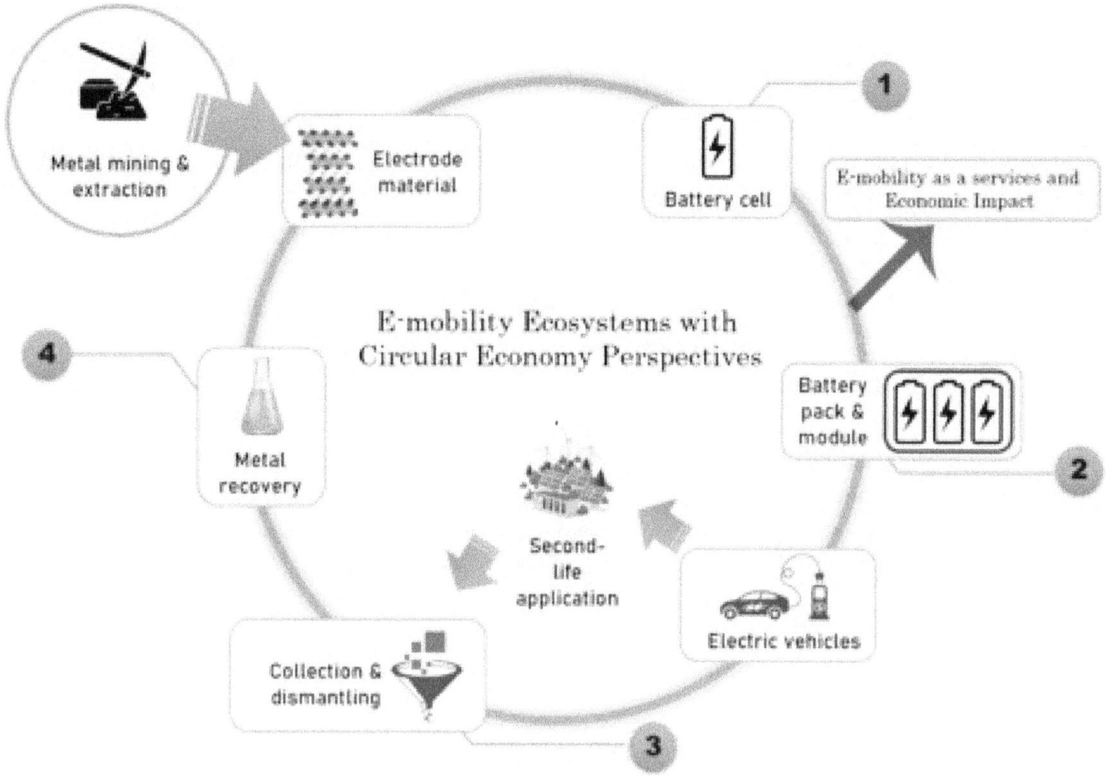

The Relationship Between Mobility and the Job Market

There are multiple sectors and stakeholder groups involved in the conversion between ICE cars and electrically powered EVs. E-mobility must therefore be integrated into Utilizing locally based mobility mechanisms, local employment, and the local labor economy in an effort to minimize the impacts on stakeholders and smooth the transition (Köhler et al. 2020). At the outset, the transformation necessitates fundamental alterations in the automotive production process, electrical supply, power technology, IT, transport networks, and transport operators and companies. Due to the possibility of the abolition of particular employment opportunities and competencies, this will disrupt the existing workforce in these sectors (Novakova, 2020). At the same time, there will be a need for new skills and jobs to promote inclusivity during an alteration. Understanding what constitutes an e-mobility environment and how it differs from ICE infrastructure is the first step in gaining a thorough understanding of the job requirements associated with the transition to e-mobility. (Sobiech-Grabka et al., 2022). ICEs or automotive ecosystems consist of the following components.

Manufacturers, dealers, and consumers make up the ICE vehicle ecosystem. As well as operations and maintenance, energy production, Besides insurance,, transport and logistics, and reuse and scrapping of automobiles, supporting members of the supply chain encompass participants in those areas. The entertainment industry and scientific breakthroughs are also transforming the ICE transportation infrastructure to include internet of things solutions (Skeete et al. 2020).

Ecosystem of E-Mobility Jobs

According to Revesz et al. (2022), On the premise of management theories, Raphael Giesecke, in his analysis, The Electric Mobility Business Ecosystem: An Initial Agenda for Future Research Needs, Based on Organization Theory, Establishes the Energy Management Organization Environmental Management Program into five primary segments: each of which is interconnected to create value for all stakeholders. As the first input provider, they provide the system with the initial hardware and software inputs, which are then Changed and gathered (e.g. EV accessories companies, electric generating equipment, etc.).

According to Mozaffari et al. (2020), Secondly, there is the integrator and aggregator, charged with for constructing or assembling materials into integrated end solutions (for example, EV Producers, EV operators, O&M providers). Lastly, there is the enabler (e.g. merchandising services, ICT systems, Research & Innovation suppliers) who deliver actual and indirect contributions to the ecosystems. Regulatory agencies, user groups, NGOs are examples of 'governors' who set and implement rules and standards for the ecosystem. In Figure 2, the connections between these elements likely depict the flow of energy, information, and regulatory oversight within the e-mobility ecosystem. It may also illustrate how different stakeholders interact and collaborate to enable the widespread adoption of EVs and associated infrastructure while meeting regulatory requirements and sustainability goals.

Figure 2. Model of e-mobility ecosystem

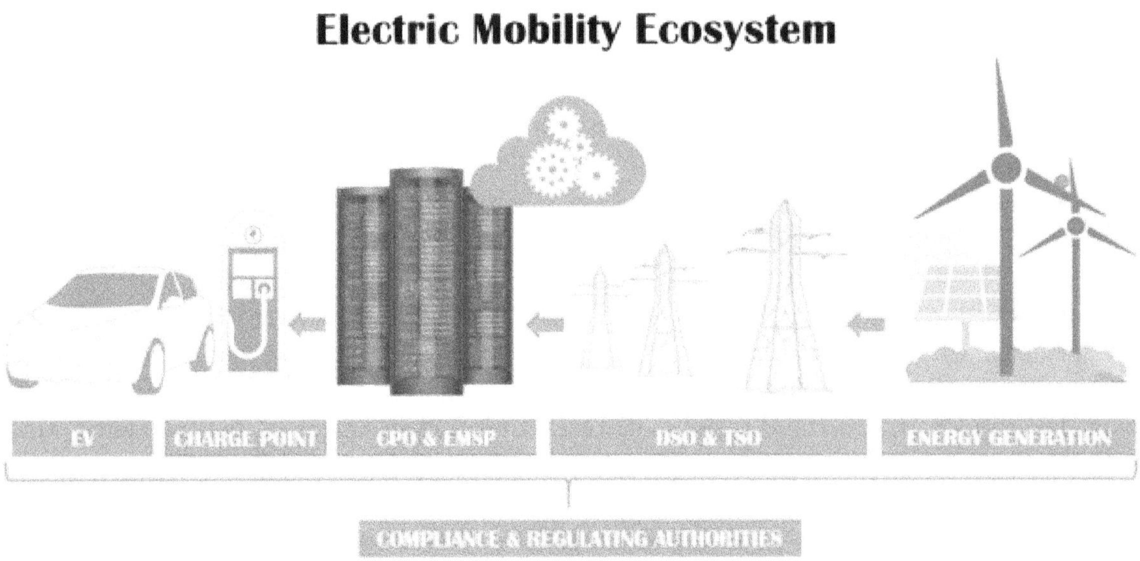

Fifth, the end customer is the primary source of sales for e-mobility, the recipient of electric mobility offerings, as well as an important link for parties involved in the EMBE (e.g., taxi service providers, mobility service vendors). Such an extensive and highly interconnected e-mobility ecosystem underscores the complexity and diversity of the e-mobility business, which encompasses well in excess of commuting (Paalosmaa & Shafie-khah, 2021)

The Role of E-Mobility in Global Level

Power, markets, skills and jobs will be redistributed in the conventional ICE vehicle sector as a result of the growth of e-mobility. In the automotive sector, jobs may be lost as manufacturing processes change, resulting in a corresponding change in skills required. New roles with specific skillsets will also be created in the manufacturing, transport, and electrical sectors as the e-mobility ecosystem gets integrated with telecommunications, transport, and electricity. Also, the reversal is probably going to result in reduced fuel utilization and a lower oil purchase contract, which will result in a Increasing economic growth and ultimately more employment in the monetary system. It is estimated that 500,000 - 850,000 jobs will be created as a result of the move, based on examples from the EU. A reduction in import dependency and a focus on indigenous manufacturing will help fuel this (Dhar et al., 2020).

According to the report, a move away from ICE or petrol-powered automobiles to plug-in electricity-powered automobiles (PEVs) has a macroeconomic impact. By replacing petroleum vehicles with electric vehicles, a significant portion of the economy will be saved. This will lead to a growth in Produces and sells products in the community's market, which in consequence will give rise to the development of new positions and growth in GDP. In addition to the growth of new jobs in the EV producing sector, studies from the USA demonstrate that shifted from to electricity-powered automobiles will add jobs along the EV production value chain and energy storage producing businesses (Showers & Raji, 2020). There will be a range of potential job profiles and skill sets. A large number of unskilled positions will be created in operations and Service, merchandising, and administrative operations. Semi-skilled occupations will be available in design and implementation, industrial production, production management, operation and preservation, allied facility development, and business management (Chenoy et al. 2019). The introduction of more advanced technology for elevating the economic effectiveness of transportation will lead to new employment opportunities in the skilled areas while their production will increase labour needs.

Figure 3 from Statista for the year 2020 illustrates the global market for plug-in electric passenger cars, showcasing China as the dominant player with 1,246,000 units sold, marking a 15% increase from the previous year. Following closely is Germany, emerging as the second-largest market with 394,943 units sold, reflecting an impressive growth rate of 263%. These figures underscore the growing significance of electric mobility on a global scale, with China's leadership driven by robust government incentives and policies, while Germany's surge is fueled by technological advancements and increasing consumer interest in eco-friendly transportation options. Together, these trends highlight the accelerating momentum towards electric vehicles and the broader transition to sustainable mobility solutions worldwide

Indian e-mobility will impact the jobs ecosystem very differently than globalized selected countries e-mobility, however. India needs approximately 2000-2500 workers to manufacture powertrains, whereas the United States needs fewer than 200. Also affected would be suppliers of auto parts and labor-intensive industries like castings, forgings, etc (Unnerstall, 2022).

According to the study by CEEW, India's existing workforce is not equipped with the required skills and capacity to prevent job losses. China and European markets account for more than two-thirds of all electronics used in OEM. In contrast, the creation of approximately 25 percent more jobs in core manufacturing could occur if 30 percent of passenger vehicle production was made up of electric vehicles. The existing workforce must be reskilled and expanded in order to meet the market's established objectives even though this area has the potential to create new jobs. As part of its mission to ensure that the electric mobility sector has an adequate workforce of skilled and trained workers, government is developing a specialized curriculum. In addition to a leading company in technology, manufacturing, evaluation, power generation products, and associated consulting. this workforce should possess other skills as well (Jacobs & Worth, 2019). Allied services technicians, supervisors, and helpers will be trained and equipped with capacity building programs by the Associated Industries Technical Coordination Organization and another association of India (Miloslavich et al., 2019).

Figure 3. Who leads the charge towards electric mobility? (Statista, 2020)

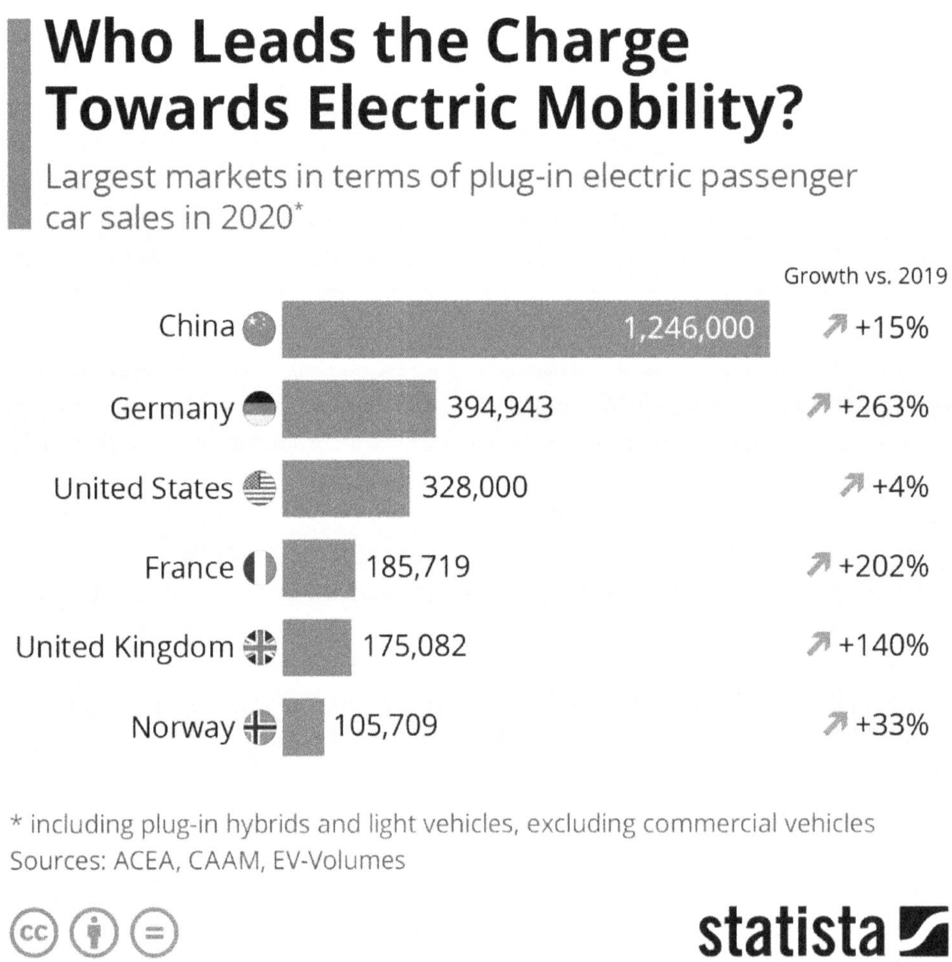

Strategies and Frameworks Related to E-Mobility and CE

Electric vehicles (EVs) are becoming more common for electromobility in developed the industrial sector and presently becoming more industrialized nations. Electro-mobility vehicles (e-mobility) are vehicles that run on electricity, either with or without an external fuel-burning generator power source. They contribute to the development of exceptionally low-emission vehicle-based transportation, primarily in urban areas. Because Although electric power is primarily a vector rather than the dominant energy supplier, the degree of electricity consumption and the quantity of electric power utilized to power automobiles and the underlying facilities along the supply chain are significant. Moreover, it is important to research the method of electrical power production (i.e., from which source does it originate).In what proportion is it derived from renewable sources and how much CO2 equivalent it emits) (Verma et al., 2021).

Due to the Due to the prevailing trend of expanding numbers of people and individual vehicles, the establishment of an e-mobility policy is necessary in cities may be able to partially address a number of sustainability issues, This includes climatic volatility, poor air standards, increased pollution, and decreased reliability. It will also help reduce fuel costs if it is connected to renewable energy (RE), developed on a large scale, has a close/closed resource loop cycle, and is developed (Householder, 2022). There are considerable quantities of non-city EV consumers, in addition to the number of non-city EV users who commute to Urbanization for job creation, domestic independence, and ecological sustainability reasons. E-mobility can contribute to a decrease in the use of fossil fuels by shifting electricity generation away from fossil fuel sources and decarbonizing transmission and distribution. According to a recently drafted comprehensive report of electric vehicles, the full potential of Electrified cars, vehicles, buses, and commercial vehicles cannot be realized until cleaner energy generation, improved recycling programs, and improved battery technology are developed. E-mobility must be a part of global sustainable mobility. By 2050, the number of electric vehicles powered by batteries is expected to surpass that of cars powered by fossil fuels (Yan & Sun, 2021). The E-Mobility Revolution and the Circular Economy can assist the United Nations in achieving its Sustainable Development Goals if they are designed and implemented in a way that is extremely sustainable. They can achieve the Sustainable Development Goals of Responsible Consumption and Production, Climate Action, Industry, Innovation, and Infrastructure, and Sustainable Cities and Communities by implementing Affordable and Clean Energy. The Sustainable Development Goals are connected to the Local Agenda of the United Nations Conference on Environment and Development, which replaced the UN's Millennium Development Goals at the international/global level in 1992. Additionally, the United Nations launched the Local 2030 Agenda. To collaborate on the development and implementation of strategies to advance the Sustainable Development Goals, this initiative brings together leaders from the private sector, civil society, philanthropic organizations, academia, local, regional, and national governments, and the United Nations (Acciarini et al., 2022).

E-Mobility Integration and Economic Impacts: Advancing Sustainability Through Smart Grids and User-Centric Approaches

E-mobility, coupled with smart grid integration, marks a pivotal advancement in sustainable transportation infrastructure. Through the seamless interaction between electric vehicles (EVs) and smart grids, bidirectional energy flow is enabled, allowing EVs to draw power but also to feed surplus energy back into the grid. This integration optimizes charging schedules based on factors such as grid load, energy demand, and renewable energy availability, ensuring efficient energy management. Moreover, Vehicle-to-Grid (V2G) technology empowers EVs to serve as distributed energy storage units, contributing to grid stability and resilience while potentially offering revenue opportunities for EV owners through grid services.

E-mobility deployment economic impact analysis encompasses a comprehensive assessment of its direct and consequential effects across various sectors. Direct economic impacts include substantial investments in EV manufacturing, charging infrastructure deployment, and revenue generation from e-mobility services. Indirect monetary effects extend to changes in consumer spending patterns, job creation in related industries like battery manufacturing and software development. In addition, indirect economic effects include potential cost savings resulting from a reduced reliance on fossil fuels. Understanding these macroeconomic dynamics is essential for policymakers, businesses, and investors to make informed decisions regarding e-mobility initiatives and resource allocation.

User experience and accessibility are paramount in ensuring widespread e-mobility adoption. Enhancing user experience involves addressing challenges such as range anxiety, seamless access to charging infrastructure, and intuitive payment systems. Moreover, ensuring accessibility requires efforts to overcome barriers to adoption for diverse user groups, including individuals with disabilities, low-income communities, and those residing in rural areas. Initiatives such as strategically deploying charging stations, offering incentives for EV purchases, and developing user-friendly mobile applications for trip planning and electric vehicle station location are instrumental in improving user experience and accessibility, thereby fostering broader acceptance of e-mobility.

Current Trends in the Circular Economy and E-Mobility

Support for the connection between CE and e-mobility has recently emerged. The German Circular Economy Initiative Deutschland has demonstrated a closed-loop method for integrating traction batteries by 2030. The CEID proposed and recommended guidelines for government, science, and industry to maximize value creation throughout the life cycle of a traction battery. One idea was to use incentives and IT systems, like battery passports supported by Avere, the Global Association of Electro mobility, to make battery data more accessible throughout and after its lifespan. The creation of a region wise battery disassembly network provides additional support for the dismantling of vehicle traction batteries. This is a significant development in light of the significant rise in the establishment of traction batteries for e-mobility in international authorities. In September 2020, the international Raw Materials Alliance's overarching objective is to diversify global supply from primary and secondary sources, increase resource efficiency, and promote circularity worldwide. The European Union has made it abundantly clear in its 2020 List of Critical Raw Materials that it will contribute to "the most effective framework for raw materials and CE worldwide."By 2030, ERMA's activities will aim to "intensify the production of raw materials and advanced materials and to address CE by increasing the recovery and recycling of "Critical Raw Materials. The activities include: (a) intensifying the development of "environmentally and socially equitable innovations and infrastructure" (b) implementing a CE of complex products such as EVs, clear (and clean) technology, and hydrogen equipment, facilitating the extraction, design, manufacturing, and recycling of raw materials for globalized and promoting innovation, strategic investment, and industrial production within specific value chains (Silvestri et al., 2021).

Researcher argue that customers should be involved in the innovation processes of companies (organizations) involved in developing a sustainable CE ecosystem. They note that so far specific preferences of users of sustainable cars like electric vehicles and biofuels are unknown in the CE context. In a CE, open innovation ecosystems represent the emerging sets of actors, activities, institutions, and relationships that facilitate innovation. A component of the automotive system for EVs that has recently been evaluated from a sustainability and sourcing perspective is tires, with some manufacturers now making

EV-specific tires, due to the high weight of electric vehicles compared to internal combustion engines (ICE). As End-of-Life tires (from ICEs in volume) are currently oversupplied, CETs and bio-based tires have been conceptually explored within the context of an acceptance model and corresponding labeling/ marketing (Wurster et al., 2020).

Figure 4. E-vehicles are more efficient than combustion engines (Model of WtT, TtW, and WtW)

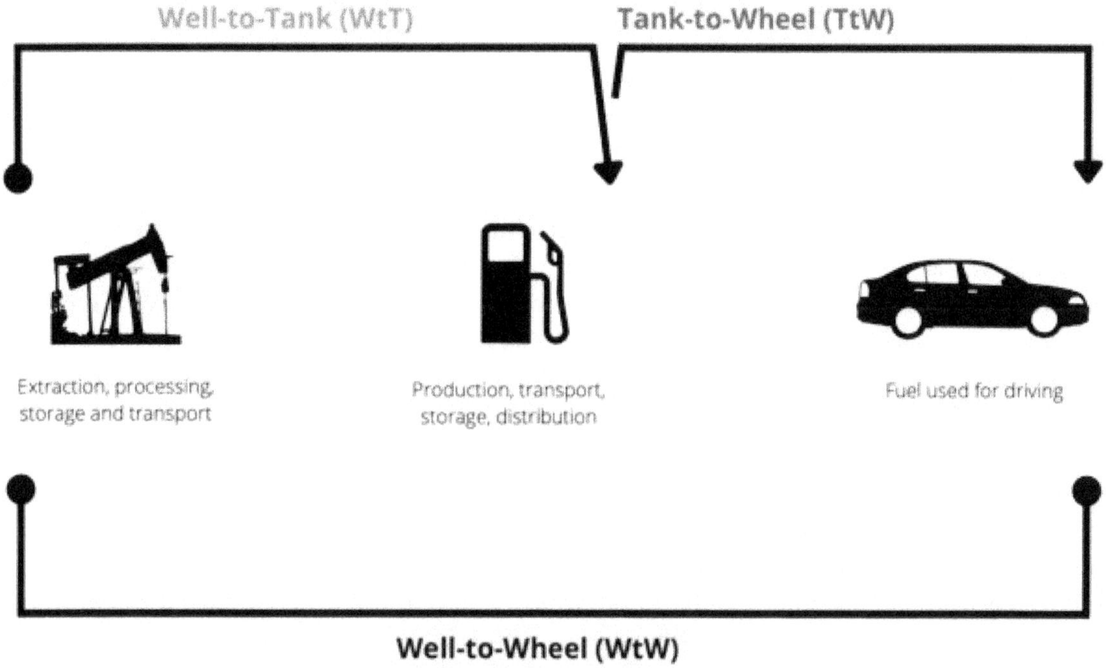

The Circular Economy, Electric Mobility, and Renewable Energy Sources

From a life cycle assessment perspective, EVs can increasingly use renewable energy sources, minimizing their carbon footprints, which is consistent with the EU's 2008 Strategic Energy Technology Plan to ensure its leadership in developing and deploying low-carbon technologies. Due to the fact that the well-to-wheel cycle consists of both the well-to-tank (WTT) and tank-to-wheel (TTW) emissions, including RE sources in the electricity production mix is vital. According to the Union of Concerned Scientists, electricity generation sources differ markedly across regions/metropolitan areas of the United States, resulting in significantly different environmental profiles for electric vehicles. It is therefore necessary to communicate with a wide range of stakeholders about the relative advantages of electric vehicles, smart charging, and bi-directional charging, as well as the net carbon savings between comparable ICE vehicles and EVs powered by plugging in, swapping out batteries, or using inductive/wireless charging. It can be a smart meter or a dashboard like the MyGridGB smart home's dashboard (displaying electricity grid data, carbon intensity, and demand/supply trends for Great Britain). According to Hoekstra, the scientific literature underestimates the potential of battery electric vehicles (BEVs) for reducing greenhouse gas emissions (GHG). Researcher identify and illustrate the following causes: overestimating battery manufacturing, underestimating battery lifetime, assuming an unchanged electricity mix for the lifetime of the vehicle, using unrealistic energy use tests, excluding fuel production emissions, and lack of system thinking (Khan et al., 2019). In the context comparing EVs with combustion engine vehicles, Figure 4 may illustrate how electric vehicles typically have a lower Well-to-Tank energy consumption compared to traditional vehicles due to the higher efficiency of electric power generation and transmission compared to the extraction and refining of fossil fuels. Additionally, it may show how electric vehicles have a more efficient Tank-to-Wheel energy conversion process, as electric motors are generally more efficient than internal combustion engines in converting energy into motion. Therefore, the overall Well-to-Wheel energy consumption of electric vehicles tends to be lower than that of combustion engine vehicles, highlighting their higher efficiency and lower environmental impact over the vehicle's lifecycle.

Protocols Supporting EVs

A variety of research domains have shown increasing interest in electric mobility (e-mobility) over the last decade. While several works have been published on e-mobility, the concept still remains fuzzy and definitions remain ambiguous. E-mobility is a new concept that requires the setup of a well-functioning ecosystem to be socially accepted. A common understanding of the term is key to achieving this. The paper at hand addresses this issue holistically by defining e-mobility as "an industry with a high degree of connectivity". As part of the sustainability concept, it uses a portable energy source and an electric drive system that can vary in degree of electrification to meet mobility needs." Moreover, this definition incorporates the development of other forms of battery-powered transportation (e.g. pedelecs) and emphasizes the importance of developing innovative business models to build new mobility partnerships and networks for the diffusion of e-mobility. A lot of attention has been paid to studies to overcome common obstacles to e-mobility over the past five years, such as achieving affordable purchase prices for electric vehicles (EVs) and reasonable total ownership costs or offering various incentives (monetary and non-monetary). Due to these studies, innovative business models and complementary mobility services are becoming more and more evident to industry and academia as means to promote e-mobility in society. It is still relatively new to research the necessity of offering complementary services (Huang, Ge, 2019). Figure 5 depicts different models of Electric Vehicle Supply Equipment (EVSE) standards, protocols, and supporting processes, specifically focusing on fault detection and protection mechanisms for both AC (alternating current) and DC (direct current) charging systems. Model 1 showcases a basic AC EVSE setup without specific fault detection or protection features, while Model 2 integrates fault detection and protection into the charging cable itself for AC charging. Model 3 addresses fault detection and protection within the AC EVSE unit, and Model 4 focuses on similar features but specifically tailored for DC charging systems. These configurations highlight the diverse approaches to ensuring safety and reliability in EVSE infrastructure, with each model targeting different aspects of fault detection and protection to enhance the overall performance and interoperability of electric vehicle charging systems.

Investments in Battery Recycling and Access to Finance

The need to close the gap between initial investments and returns presents structural financing challenges for businesses that demolish local structures. Companies that dismantle electronic waste need to make investments in salaries and infrastructure in order to properly collect and dispose of it. Investment may also be required, in addition to e-waste purchased from third parties like waste producers, waste collectors, or the informal sector. Dismantled items must be accumulated in sufficient quantities before they can be economically shipped. Returns on investment for dismantled e-waste waiting for shipment are further delayed. Dismantling companies in the area must cover the costs associated with notification and shipment in the meantime (Wu et al., 2020). Figure 6 likely illustrates the financial viability of electric vehicle battery recycling. It likely presents data or a model depicting the costs associated with recycling electric vehicle batteries compared to potential revenue or savings from recovered materials. The figure may demonstrate how factors such as material prices, recycling technologies, and government incentives impact the economic feasibility of battery recycling. It could also highlight the importance of efficient recycling processes and market demand for recycled materials in ensuring the financial viability of battery recycling operations

Figure 5. EVSE standard and protocols and supporting process

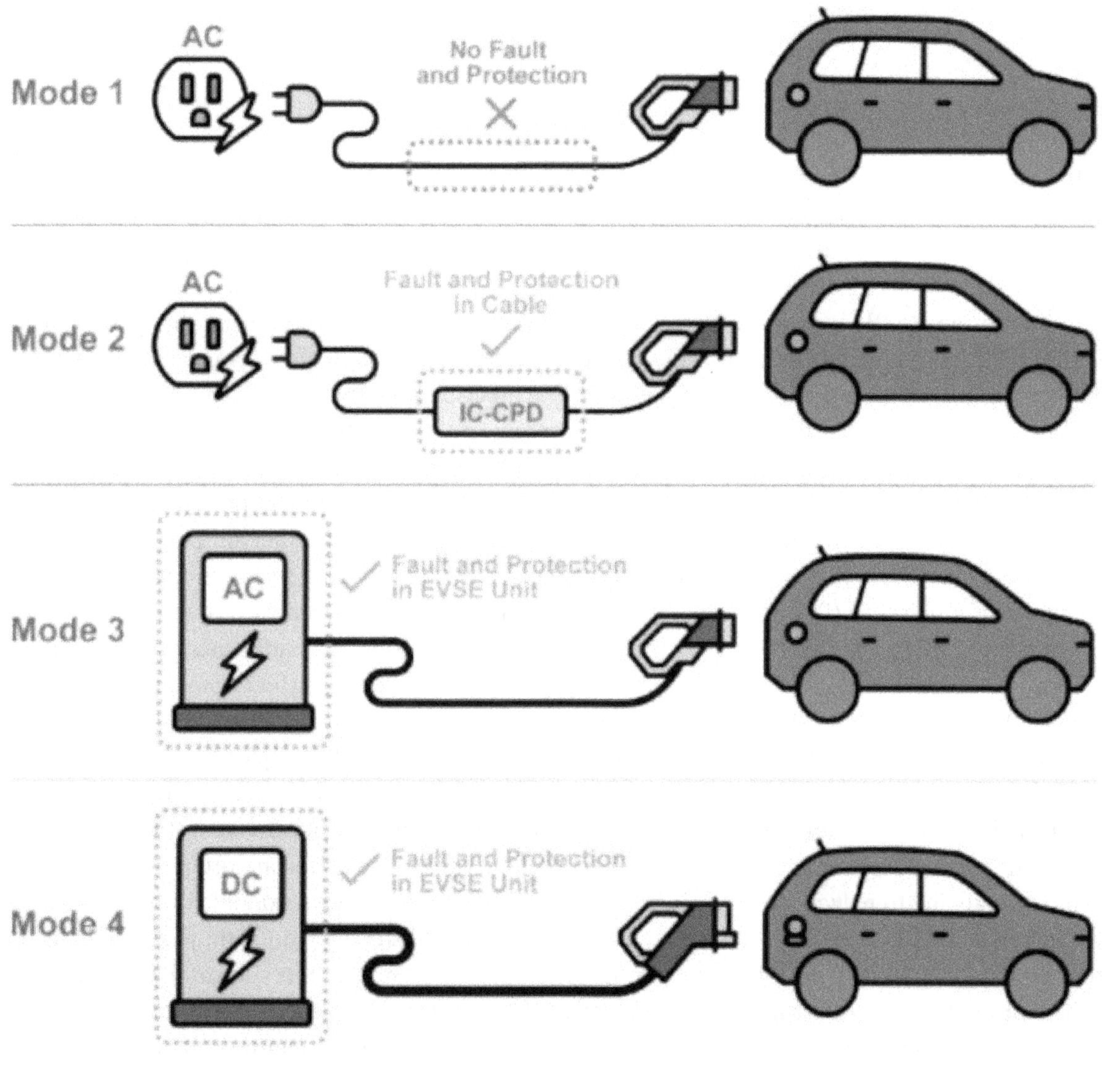

Figure 6. Financial viability of electric vehicle battery recycling

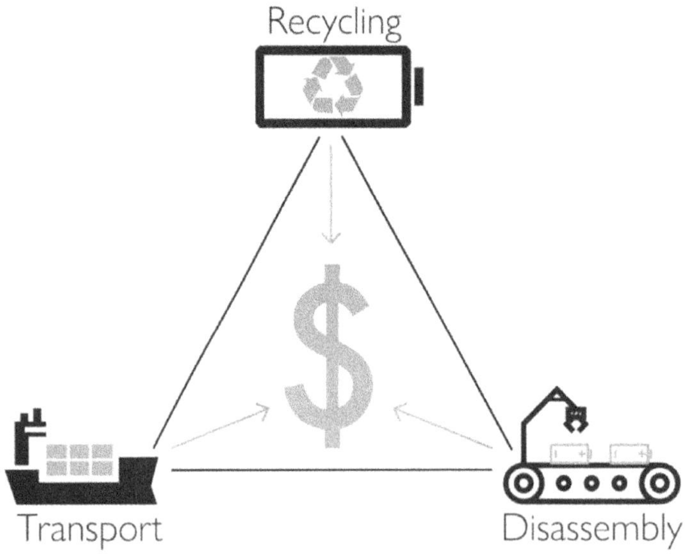

In order to determine the exact value of an e-waste shipment's waste fraction, the shipment is tested and analysed at the designated recycling plant. A shipment's value is influenced by its metal content and plastic fractions. It is only after a detailed analysis of the waste that a local dismantling company is paid.

It can take up to six months from pre-financing to transferring net proceeds from the sale of waste, according to the Bo2W experience. E-waste will continue to be processed during this period, which requires additional funding. As part of the business model, the spiraling process necessitates a line of credit rather than a single loan for a single project. The precise sum of money required for pre-financing cannot be easily calculated due to the different economic values of dismantled e-waste fractions like printed wiring boards or batteries. It is estimated that the average shipment cost between 30 and 100 thousand US dollars. In addition to developing equity for future, independent prefinancing, recycling companies should aim to reduce their dependence on bank loans (Berssaneti et al., 2019).

In the private sector, firms are usually differentiated according to their turnover, financial position, or number of employees (PWC). Corporate giants and transnational organizations and "micro, small, and medium-sized enterprises (MSME)" are also frequently classified. It is unfortunate that there are no universally accepted definitions for MSME. Ghanaian MSME definitions differ even within the country. As a result of the variable workforce size in this report, the report distinguishes between micro and SME. As defined by Samjong KPMG & KODIT, an MSME has fewer than 100 employees.

"Micro enterprises" are small, self-employed businesses with fewer than five employees or none at all. Financial services offered to small businesses, self-employed individuals, and other low-income groups are referred to as "microfinance. Banking and other related services are often provided as part of micro financing to reduce poverty (Begum et al., 2019). A very famous example of microfinance is the Grameen Bank in Bangladesh. In addition to retail shops and street vending, they also manufacture artisanal products and provide services. The majority of micro-entrepreneurs in rural areas are either farmers or food processors. (Microfinance Gateway 2015) These clients usually lack formal credit histories, collateral, and access to formal markets. A microloan that does not require collateral or a mutual guarantee can be offered by microfinance to any group. In the 1970s, non-governmental organizations led the movement to finance microenterprises in order to fight and prevent poverty. Global commercial banks have also begun to become more involved in this area in recent years (Tamanni et al., 2019).

Platforms for CE Collaboration With E-Mobility in Global Level

Automobiles consume 180 billion gallons of fuel annually, accounting for nearly 61% of global oil consumption. In 45 years, oil reserves will run out if nothing is done. Additionally, the need to reduce greenhouse gas emissions puts pressure on the automotive industry to develop environmentally friendly modes of transportation. The majority of authors are of the opinion that electric mobility, or e-mobility, is the most effective short-term solution (Andersson & Börjesson, 2021). According to the table 1, the overview of e-mobility in Europe encompasses the growing adoption of electric vehicles (EVs) and related infrastructure across the continent, driven by efforts to reduce carbon emissions and combat climate change. Various forms of e-mobility, including electric cars, buses, bicycles, and scooters, are fostering sustainability by offering cleaner transportation options and reducing dependence on fossil fuels. Within the context of the circular economy (CE), e-mobility solutions aim to minimize resource depletion and waste generation by promoting the reuse, recycling, and remanufacturing of EV components and batteries. Sustainable urban mobility planning plays a crucial role in fostering e-mobility solutions by integrating EV infrastructure into urban landscapes, promoting public transportation, prioritizing pedestrian and cyclist-friendly infrastructure, and implementing policies to incentivize the adoption of EVs. Overall, the advancement of e-mobility in Europe aligns with broader sustainability goals, with a focus on transitioning towards cleaner, more efficient transportation systems while embracing principles of the circular economy and sustainable urban planning.

Table 1. Framing electric mobility for urban sustainability in a circular economy

Themes	Data Source	Interpretation	Insights
Overview of e-mobility in Europe and forms of e-mobility	Books, journal articles, reports, organizational websites	Illustrative and descriptive	Reducing carbon and noise emission due to e-mobility. De-carbonization through RE is also required. E-bikes, e-buses, and e-scooters will play roles in achieving e-mobility in urban areas
Fostering sustainability	Books, journal articles, reports	Illustrative and descriptive	Reduction of air pollution but energy consumption might increase; the possibility of new job creation
Context of CE	Books, journal articles, reports, organizational websites, regulators	Illustrative and conceptual	Principles of a CE – preservation of materials, elongation of life cycles, reduction of distances
Perspective/principles of sustainable urban mobility planning	Books, journal articles, reports, organizational websites, news articles	Illustrative and conceptual	Electro-mobility is an important – but by no means the only – perspective to take on board, alongside other (and to be prioritized, where possible) modes of transport (active travel); inter-modal transport is important, as is spatial planning and social and organizational engagement/participation and collaboration
Fostering e-mobility solutions	Books, journal articles, reports, websites	Explanatory and analytical	Different e-mobility solutions, with trends analysis

Although viewed as a solution, moving to E-mobility also poses its own challenges. In order to decarbonize the transportation sector, electric vehicles must be mass-marketed. Battery electric vehicles (BEVs) are expected to hold a market share of two to five percent by 2020. There will be about one million to two million BEVs in circulation by then. The development of BEVs and their components as well as the construction of production lines is needed to achieve this goal. Upstream and downstream value creation processes will undergo significant changes as e-mobility moves to mass adoption. Adaptation costs are one of the major problems facing the established automotive sector. Capital intensity, cost requirements, and amortization of sunk costs are the most pressing concerns when making investments in cutting-edge automotive technology (Hsieh et al., 2020).

Despite the fact that electrochemistry and power electronics are essential to the advancement of electric vehicle technology, the majority of automobile manufacturers lack this knowledge and intellectual property. There are no expertise or economies of scale available in the automotive sector, and their employment would cost a considerable amount of money. The absence of quality and performance standards in this new supply chain further distances the automotive industry from these new technologies and only serves to increase the risk. Governments have implemented incentives to encourage e-mobility, but the automotive industry is not sufficiently motivated by the anticipated business returns. Due to this, OEMs prioritize other types of investments that have a faster return on investment in the current market. As a result, manufacturers produced less than 30 different models of electric vehicles in 2011. In addition, there is a low level of private ownership. In Germany, most of the vehicles registered are owned by businesses and are used for testing purposes. Governmental aims and societal goals are far removed from this (Das & Bhat, 2022).

Urban Sustainability Through E-Mobility in the CE Context

An e-mobility boom is currently underway, which means that future transportation systems must be redesigned to fit both the requirements and the possibilities of e-mobility. In fact, e-mobility has a good chance of reducing greenhouse gas emissions and air pollution, as well as reducing dependence on fossil fuels. There are, however, counterproductive system effects and a net expansion of private motorized transport (PMT) if lower user-specific operational expenses, certain classes of vehicles are excluded from environment-based cordons (such as low–emission zones), and 'environmental reasoning' is not applied to certain user groups. It has become a priority of various urban administration authorities to strengthen public and non-motorized transportation systems. To enable the renewal of the vehicle fleet towards e-cars, Researcher present organizational structures that do not negatively impact public transportation (PT), cyclists, or pedestrians. In this study, researcher present four different scenarios that were influenced by different transport policies (McQueen et al., 2021).

By Using Electrode Materials, Metal Can Be Mined and Extracted

A relatively limited number of studies have been conducted using electrochemical methods to reduce methane adsorption in coal. Electrode material has been identified as one of the key factors in sulfur removal, dewatering of FNE coals, and electro-osmotic strengthening of silts. Influences of electrochemistry on a variety of variables. Electrodes made of non-inert metals are typically used in electrochemical experiments. electrodes made of inert metals (like gold, silver, or platinum), electrodes made of carbon and geosynthetics with electrokinetics (EKGs). The improvement of electrode materials has been the subject of numerous comparative experimental studies. For instance, some academics believe that graphite is a superior material to non-inert metals. The electro-osmotic test revealed that, for the same amount of energy, the iron electrode performed better than the graphite electrode. In addition, it was discovered that graphite electrodes lost energy at a higher rate than metal electrodes did. Copper electrodes reduce voltage loss at the anode, are inexpensive, and can conduct electricity. This is because oxygen near copper electrodes forms copper oxide, which conducts electricity and reduces voltage loss. Te stainless steel has a high resistance to corrosion. Traditional metal electrodes, it has been argued, have numerous drawbacks, including corrosion, exhaust difficulty, and the loss of contact between electrodes and soil during electroosmosis16. According to Lefebvre and Jayasekera, mild steel and steel electrode corrosion inhibits electrochemical action. Zhou et al.'s study found that Under high potential gradients (1.58V/cm), graphite electrodes were found to be more effective than iron and copper electrodes for sludge from Hangzhou. Graphite electrodes were also found to be more effective by suits (Kumar et al., 2021).

Advancements in Electrode Materials and Circular Economy Design for Metal Recovery and Electric Vehicles

A relatively limited number of studies have been conducted using electrochemical methods to reduce methane adsorption in coal (Zhou et al., 2021).. Electrode material has been identified as one of the key factors in sulfur removal, dewatering of FNE coals, and electro-osmotic strengthening of silts (Lefebvre & Jayasekera). Influences of electrochemistry on a variety of variables have been observed, with electrodes made of non-inert metals and inert metals like gold, silver, or platinum, as well as carbon and geosynthetics with electrokinetics (EKGs), being commonly used . Electrode materials have been the subject of numerous comparative experimental studies, with graphite often considered superior to non-inert metals Graphite electrodes were found to be more effective than iron and copper electrodes under high potential gradients (1.58V/cm) for sludge from Hangzhou (Zhou et al., 2021). Additionally, graphite electrodes were found to be more effective than metal electrodes by suiting (Kumar et al., 2021). Copper electrodes, on the other hand, reduce voltage loss at the anode and are relatively inexpensive, Stainless steel, known for its high corrosion resistance, present an alternative to traditional metal electrodes (Zhou et al., 2021). However, traditional metal electrodes have drawbacks such as corrosion, exhaust difficulty, and loss of contact between electrodes and soil during electroosmosis (Lefebvre & Jayasekera).

The House of Quality (HoQ) is a well-known tool utilized in the circular economy (CE) to integrate customer expectations and engineering specifications (Hapuwatte & Jawahir, 2019). Specific guidelines have been developed so that products can be easily reusable, remanufactured, repaired, or recycled (Hapuwatte & Jawahir, 2019). During the design phase, CE suitability can be assessed and prioritized using engineering specifications (Hapuwatte & Jawahir, 2019). The columns that define the fundamental design parameters of a practical electric vehicle (EV) lithium-ion battery pack are represented by engineering specifications in this tool (Chen et al., 2021). After identifying these parameters and relating them to customer needs, design parameters will be prioritized to simplify CE tasks (Chen et al., 2021). Starting with cutting-edge design for CE criteria and defining particular requirements for EV lithium-ion battery circularity, HoQ rows correspond to guidelines for designing CE-compliant EV batteries (Bonsu, 2020). To overcome the difficulties stakeholders currently face after using e-mobility batteries, the three industrial strategies of reuse, remanufacturing, and recycling are separated into specific sub-categories (Bonsu, 2020). Through advancements in electrode materials and circular economy design, metal recovery and electric vehicle development can be achieved.

DISCUSSION

The chapter delves into the intricate ecosystem of e-mobility, highlighting its interconnected segments ranging from input providers to end customers, while noting an under-emphasis on government regulations, economic evaluation, and technological aspects. It discusses the global impact of e-mobility on job creation and economic development, recognizing regional variances such as India's larger need for powertrain manufacturing workers than the United States. Moreover, it emphasizes the necessity of reskilling and expanding the existing workforce to meet the demands of the e-mobility sector, alongside the importance of establishing e-mobility policies to address sustainability issues. The integration of e-mobility with smart grids and user-centric approaches is deemed pivotal, while the connection between the circular economy and e-mobility is explored, focusing on initiatives to promote resource efficiency and circularity in raw material production. Furthermore, the chapter proposes the inclusion of a brief section on E-mobility from an engineering perspective to address technical aspects, advancements, and limitations within the field, thereby offering a comprehensive overview of the multifaceted domain of E-mobility

FUTURE IMPLICATIONS, SUGGESTIONS AND CONCLUSIONS

In analyzing the future implications and offering suggestions for the integration of electric mobility (e-mobility) within the circular economy (CE) framework, it's crucial to recognize the multifaceted impact and interconnectedness of various sectors. As electric vehicles (EVs) continue to gain prominence globally, particularly in industrialized nations, there's a pressing need for comprehensive e-mobility policies at both city and regional levels to address sustainability challenges such as climate volatility, air quality, and fuel costs. However, achieving the full potential of e-mobility requires concerted efforts in cleaner energy generation, battery technology advancement, and improved recycling programs. Collaborative initiatives involving governments, private sector entities, academia, and civil society are essential to driving innovation and infrastructure development in line with the Sustainable Development Goals (SDGs) set forth by the United Nations. Additionally, the integration of e-mobility with smart grids offers a promising avenue for advancing sustainability goals by optimizing energy management, enhancing grid stability, and fostering revenue opportunities through Vehicle-to-Grid (V2G) technology. Moreover, enhancing user experience and accessibility through strategic deployment of charging infrastructure and user-friendly applications is pivotal in ensuring widespread adoption of e-mobility, particularly among diverse user groups such as individuals with disabilities and those in rural areas. Furthermore, the circular economy presents immense opportunities for synergies between e-mobility and resource efficiency, with initiatives like battery recycling and innovative business models playing a crucial role in maximizing value creation throughout the lifecycle of traction batteries. Moving forward, fostering collaboration between stakeholders, leveraging technological advancements, and prioritizing sustainability in policy-making will be instrumental in realizing the full potential of e-mobility within the circular economy paradigm, thereby contributing to a more sustainable and resilient future for all. In order to achieve sustainable living, mobility must be addressed because it accounts for 15% of global emissions, second only to electricity and heat. As a promising solution to climate change, the inclusion of electric vehicles is a clear EU strategy. Due to their significant increase in demand for the generation of renewable energy, the incorporation of electric vehicles into the electricity infrastructure has a significant effect on energy grids. The Joint Programming Initiative Urban regions defines a Positive Energy District as an urban area or group of connected buildings that manage an annual surplus of renewable energy at the local or regional level and do not emit any greenhouse gases. In order to achieve social, economic, and environmental sustainability, buildings, users, regional energy, mobility, and ICT systems must collaborate to ensure that everyone has access to energy and a comfortable life. An energy surplus district—or PED in general—can export its excess energy outside the district if it produces more energy than the district needs to meet its needs. In low-carbon transition pathways, some studies assert that smart charging systems for electric vehicles could significantly reduce distribution network costs. Electric vehicles can also reduce peak consumption times by storing excess energy. In urban settings, this technology can be easily implemented. Numerous studies have demonstrated that EVs increase microgrid flexibility, reduce fluctuations in the generation of renewable energy, and improve demand-side management.

REFERENCES

Acciarini, C., Borelli, F., Capo, F., Cappa, F., & Sarrocco, C. (2021). Can digitalization favour the emergence of innovative and sustainable business models? A qualitative exploration in the automotive sector. *Journal of Strategy and Management*, *15*(3), 335–352. doi:10.1108/JSMA-02-2021-0033

Andersson, Ö., & Börjesson, P. (2021). The greenhouse gas emissions of an electrified vehicle combined with renewable fuels: Life Cycle Assessment and policy implications. *Applied Energy*, *289*, 116621. doi:10.1016/j.apenergy.2021.116621

Bagheri Tookanlou, M., Pourmousavi Kani, S. A., & Marzband, M. (2021). An Optimal Day-ahead scheduling framework for E-mobility ecosystem operation with drivers' preferences. *IEEE Transactions on Power Systems*, *36*(6), 5245–5257. doi:10.1109/TPWRS.2021.3068689

Begum, H., Alam, A. S. A. F., Mia, M. A., Bhuiyan, F., & Ghani, A. B. (2019). Development of islamic microfinance: A sustainable poverty reduction approach. *Journal of Economic and Administrative Sciences*, *35*(3), 143–157. doi:10.1108/JEAS-01-2018-0007

Behrendt, F. (2017a). Why cycling matters for electric mobility: Towards diverse, active and sustainable e-mobilities. *Mobilities*, *13*(1), 64–80. doi:10.1080/17450101.2017.1335463

Berssaneti, F. T., Berger, S., Saut, A. M., Vanalle, R. M., & Santana, J. C. (2019). Value generation of remanufactured products: Multi-case study of third-party companies. *Sustainability (Basel)*, *11*(3), 584. doi:10.3390/su11030584

Bonsu, N. O. (2020). Towards a circular and low-carbon economy: Insights from the transitioning to electric vehicles and net zero economy. *Journal of Cleaner Production*, *256*, 120659. doi:10.1016/j.jclepro.2020.120659

Brugger, H., Eichhammer, W., Mikova, N., & Dönitz, E. (2021). Energy efficiency vision 2050: How will new societal trends influence future energy demand in the European countries? *Energy Policy*, *152*, 112216. doi:10.1016/j.enpol.2021.112216

Castillo-Calzadilla, T., Alonso-Vicario, A., Borges, C. E., & Martin, C. (2022). E-mobility in positive energy districts. *Buildings*, *12*(3), 264. doi:10.3390/buildings12030264

Chen, Y., Kang, Y., Zhao, Y., Wang, L., Liu, J., Li, Y., Liang, Z., He, X., Li, X., Tavajohi, N., & Li, B. (2021). A review of lithium-ion battery safety concerns: The issues, strategies, and testing standards. *Journal of Energy Chemistry*, *59*, 83–99. doi:10.1016/j.jechem.2020.10.017

Chenoy, D., Ghosh, S. M., & Shukla, S. K. (2019). Skill development for accelerating the manufacturing sector: The role of 'new-age' skills for 'make in India.' *International Journal of Training Research, 17*(sup1), 112–130.

Das, P. K., & Bhat, M. Y. (2022). Global Electric Vehicle Adoption: Implementation and Policy Implications for India. *Environmental Science and Pollution Research International*, *29*(27), 40612–40622. doi:10.1007/s11356-021-18211-w PMID:35083674

Dhar, S., Pathak, M., Shukla, P., & Gupta, A. (2020). Electric vehicles penetration in India for enhanced energy efficiency deployment in the transport sector. *Energy Efficiency in Developing Countries*, 253–270.

Dias, G., Arsenio, E., & Ribeiro, P. (2021). The role of shared e-scooter systems in urban sustainability and resilience during the covid-19 mobility restrictions. *Sustainability (Basel)*, *13*(13), 7084. doi:10.3390/su13137084

Friedel, J. N. (2008). The effect of the Community College Workforce Development Mission on Governance. *New Directions for Community Colleges*, *2008*(141), 45–55. doi:10.1002/cc.314

Haines, A., & Frumkin, H. (2021). *Planetary health: Safeguarding human health and the environment in the anthropocene*. Cambridge University Press. doi:10.1017/9781108698054

Hapuwatte, B. M., & Jawahir, I. S. (2019). A total life cycle approach for developing predictive design methodologies to optimize product performance. *Procedia Manufacturing*, *33*, 11–18. doi:10.1016/j.promfg.2019.04.003

Householder, A. (2022). *Designing Vultron: A Protocol for Multi-Party Coordinated Vulnerability Disclosure (MPCVD) (Version 1)*. Carnegie Mellon University.

Hsieh, I.-Y. L., Pan, M. S., & Green, W. H. (2020). Transition to electric vehicles in China: Implications for private motorization rate and Battery Market. *Energy Policy*, *144*, 111654. doi:10.1016/j.enpol.2020.111654

Huang, X., & Ge, J. (2019). Electric Vehicle Development in Beijing: An analysis of consumer purchase intention. *Journal of Cleaner Production*, *216*, 361–372. doi:10.1016/j.jclepro.2019.01.231

Khan, M. I., Shahrestani, M., Hayat, T., Shakoor, A., & Vahdati, M. (2019). Life cycle (well-to-wheel) energy and environmental assessment of natural gas as transportation fuel in Pakistan. *Applied Energy*, *242*, 1738–1752. doi:10.1016/j.apenergy.2019.03.196

Knieke, C., Lawrenz, S., Fröhling, M., Goldmann, D., & Rausch, A. (2019). Predictive and flexible circular economy approaches for highly integrated products and their materials as given in E-mobility and ICT. *Materials Science Forum*, *959*, 22–31. doi:10.4028/www.scientific.net/MSF.959.22

Koh, L., Dolgui, A., & Sarkis, J. (2020). Blockchain in transport and Logistics – Paradigms and transitions. *International Journal of Production Research, 58*(7), 2054–2062. doi:10.1080/00207543.2020.1736428

Köhler, J., Turnheim, B., & Hodson, M. (2020). Low carbon transitions pathways in mobility: Applying the MLP in a combined case study and simulation bridging analysis of Passenger Transport in the Netherlands. *Technological Forecasting and Social Change, 151*, 119314. doi:10.1016/j.techfore.2018.06.003

Kumar, S., Roy, D. N., & Dey, V. (2021). A comprehensive review on techniques to create the antimicrobial surface of biomaterials to intervene in biofouling. *Colloid and Interface Science Communications, 43*, 100464. doi:10.1016/j.colcom.2021.100464

Kuzey, C., Fritz, M. M. C., Uyar, A., & Karaman, A. S. (2022). Board gender diversity, CSR Strategy, and eco-friendly initiatives in the transportation and logistics sector. *International Journal of Production Economics, 247*, 108436.

Liao, F., & Correia, G. (2020). Electric carsharing and micromobility: A literature review on their usage pattern, demand, and potential impacts. *International Journal of Sustainable Transportation, 16*(3), 269–286. doi:10.1080/15568318.2020.1861394

McQueen, M., Abou-Zeid, G., MacArthur, J., & Clifton, K. (2020). Transportation transformation: Is micromobility making a macro impact on sustainability? *Journal of Planning Literature, 36*(1), 46–61. doi:10.1177/0885412220972696

Meskers, C., Caffarey, M., & Van Camp, M. (2019). Circular cities, e-mobility and the metals industry—a world in transition. *The Minerals, Metals & Materials Series*, 313–318.

Miloslavich, P., Seeyave, S., Muller-Karger, F., Bax, N., Ali, E., Delgado, C., Urban, E. (2018). Challenges for global ocean observation: The need for increased human capacity. *Journal of Operational Oceanography, 12*(sup2).

Mossali, E., Gentilini, L., Merati, G., & Colledani, M. (2020). Methodology and application of Electric Vehicles Battery packs redesign for Circular Economy. *Procedia CIRP, 91*, 747–751. doi:10.1016/j.procir.2020.01.139

Mozaffari, M., Askarian Abyaneh, H., Jooshaki, M., & Moeini-Aghtaie, M. (2020). Joint Expansion Planning Studies of EV parking lots placement and distribution network. *IEEE Transactions on Industrial Informatics, 16*(10), 6455–6465. doi:10.1109/TII.2020.2964049

Novakova, L. (2020). The impact of technology development on the future of the labour market in the Slovak Republic. *Technology in Society, 62*, 101256. doi:10.1016/j.techsoc.2020.101256

Paalosmaa, T., & Shafie-khah, M. (2021). Feasibility of innovative smart mobility solutions: A case study for Vaasa. *World Electric Vehicle Journal, 12*(4), 188. doi:10.3390/wevj12040188

Revesz, A., Dunham, C., Jones, P., Bond, C., Fenner, R., Mody, S., Nijjhar, R., Marques, C., & Maidment, G. (2022). A holistic design approach for 5th generation smart local energy systems: Project greenscies. *Energy, 242*, 122885. doi:10.1016/j.energy.2021.122885

Serohi, A. (2021). E-mobility ecosystem innovation – impact on downstream supply chain management processes. is India ready for inevitable change in auto sector? *Supply Chain Management*, *27*(2), 232–249. doi:10.1108/SCM-11-2020-0588

Showers, S. O., & Raji, A. K. (2020). Benefits of electric vehicle as Mobile Energy Storage System. *2020 IEEE PES/IAS PowerAfrica.*

Silvestri, L., Forcina, A., Silvestri, C., & Traverso, M. (2021). Circularity potential of rare earths for sustainable mobility: Recent developments, challenges and future prospects. *Journal of Cleaner Production*, *292*, 126089. doi:10.1016/j.jclepro.2021.126089

Skeete, J.-P., Wells, P., Dong, X., Heidrich, O., & Harper, G. (2020). Beyond the event horizon: Battery waste, recycling, and sustainability in the United Kingdom Electric Vehicle Transition. Energy Research &amp. *Social Science*, *69*, 101581.

Sobiech-Grabka, K., Stankowska, A., & Jerzak, K. (2022). Determinants of electric cars purchase intention in Poland: Personal attitudes v. economic arguments. *Energies*, *15*(9), 3078. doi:10.3390/en15093078

Tamanni, L., & Haji Besar, M. H. (2019). Profitability vs poverty alleviation: Has banking logic influences Islamic microfinance institutions? *Asian Journal of Accounting Research*, *4*(2), 260–279. doi:10.1108/AJAR-05-2019-0039

Thimm, T. (2020). E-destination: The future of E-mobility in the Lake Constance Region, Germany. *Journal of Tourism Futures*, *8*(1), 24–36. doi:10.1108/JTF-10-2019-0116

Unnerstall, T. (2022). *Factfulness Sustainability: What you should know about Ecological Crises and Resource Consumption.* Springer Nature.

Verma, S., Mishra, S., Gaur, A., Chowdhury, S., Mohapatra, S., Dwivedi, G., & Verma, P. (2021). A comprehensive review on energy storage in Hybrid Electric Vehicle. *Journal of Traffic and Transportation Engineering*, *8*(5), 621–637. doi:10.1016/j.jtte.2021.09.001

Wu, W., Cheung, C., Lo, S. Y., Zhong, R. Y., & Huang, G. Q. (2020). An IOT-enabled real-time logistics system for a third party company: A case study. *Procedia Manufacturing*, *49*, 16–23. doi:10.1016/j.promfg.2020.06.005

Wurster, S., Heß, P., Nauruschat, M., & Jütting, M. (2020). Sustainable circular mobility: User-integrated innovation and specifics of electric vehicle owners. *Sustainability (Basel)*, *12*(19), 7900. doi:10.3390/su12197900

Yan, X., & Sun, S. (2021). Impact of electric vehicle development on China's energy consumption and greenhouse gas emissions. *Clean Technologies and Environmental Policy*, *23*(10), 2909–2925. doi:10.1007/s10098-021-02209-6

Chapter 4
Designing Architecture for Sustainable Electric Mobility:
Ecosystems Integrating Urban Planning and Infrastructure Design

Ashish Makanadar
https://orcid.org/0000-0003-1083-5389
Indian Institute of Management, Mumbai, India

ABSTRACT

This study comprehensively evaluates the role of architectural planning in developing sustainable electric mobility infrastructure. As transport emissions rise with motorization, designing low-carbon alternatives requires nuanced, interdisciplinary solutions. Through analysis of urban form, charging access, street layouts, and public transit integration influence on electric vehicle adoption, the research assesses retrofitting challenges and opportunities afforded by forward-thinking design. Conceptualizing mobility ecosystems as sociotechnical systems, the chapter explores dimensions of accessibility, integration, and adaptability in architectural approaches. Sustainability, scalability, and flexibility are theorized as principles fundamental to resilient infrastructure planning. Empirical studies on smart grid technologies' potential to optimize renewable energy charging management are also critically appraised. Finally, analyses of how inclusive, equitable planning supports social equity and policy goals qualitatively examine best practices.

1. INTRODUCTION

Electric mobility is rapidly emerging as a viable low-emissions transportation alternative. The transport sector is the fastest-growing greenhouse gas (GHG) emitting sector, expected to reach a share of more than 30% of total GHG emissions in the future. The introduction of electric vehicles in fleets is often the first step to overcoming challenges and reducing greenhouse gas emissions and air pollution through electric mobility (de Melo et al., 2023). The global vehicle fleet is set to double by 2050, with more than

DOI: 10.4018/979-8-3693-2611-4.ch004

90% of future vehicle growth projected to take place in low and middle-income countries. The European Union has set a target for low-emission vehicles to have tailpipe emissions below 50g/km. By 2030, the sale of battery-powered cars could contribute to up to 75% of emission reduction from new vehicles, which would be crucial to meeting EU climate ambitions (Blat Belmonte et al., 2020). Developing robust electric mobility infrastructure necessitates an integrated socio-technical systems approach that accounts for the built environment. coordination between architecture, urban planning, and transportation engineering is essential for sustainable transportation ecosystem design.

This chapter comprehensively examines the role of architecture in shaping electric mobility systems. Transportation infrastructure design heavily influences the adoption rates of novel vehicle technologies. Urban form and spatial configurations impact the provision of charging access points, street layouts conducive to different modes, and public transportation integration possibilities. Retrofitting legacy auto-oriented infrastructure to accommodate electric vehicles introduces challenges. Thus, forward-thinking architectural considerations in infrastructure planning can facilitate smoother transitions to sustainable mobility.

This study explores the architectural dimensions of building electric mobility ecosystems. Accessibility analyses examine charging infrastructure placement relative to dwelling and activity densities. Network analyses evaluate street and pathway configurations' impacts on multiple transportation modes. System integration studies assess electric vehicles' potential public transportation roles. The chapter also evaluates infrastructure adaptation challenges and opportunities. Findings aim to inform architecture and urban planning practices supporting long-term electric mobility adoption.

With expanding urbanization, low-emissions transportation assumes heightened importance. Table 1 compares well-to-wheel energy usage and emissions metrics across modes, helping policymakers craft sustainability-optimized transportation systems. This research contributes to an interdisciplinary understanding of architecture and policy's roles in engineering equitable, environmentally sound urban mobility transitions.

Table 1. Comparing energy efficiency and environmental impact of transportation modes

Transportation Mode	Energy Efficiency (Passenger-Miles per Gallon of Gasoline Equivalent)	CO_2 Emissions (Grams per Passenger-Mile)
Electric Cars	118-125 MPGe	42-54 g CO_2e
Public Transit	50-100+ MPG	68-97 g CO_2e
Biking	N/A	0 g CO_2e
Walking	N/A	0 g CO_2e

2. LITERATURE REVIEW

The literature emphasizes the pivotal role of architectural design in developing a sustainable electric mobility ecosystem considering increasing demand (Figure 1). Newman and Kenworthy (1999) assert that integrating architectural planning with urban and transportation system design is essential for successfully implementing electric mobility ecosystems. They argue collaborative efforts between architects, urban planners, and transportation planners are needed to align ecosystem design with broader goals and infrastructure networks. Long-term viability depends on scalability and flexibility according. They support adaptability, scalability, and flexibility as critical design factors. Similarly, it considers future needs and technological change to ensure sustainable architectures over time.

Figure 1. Global electric passenger car stock, 2010-2020 (IEA, 2021)

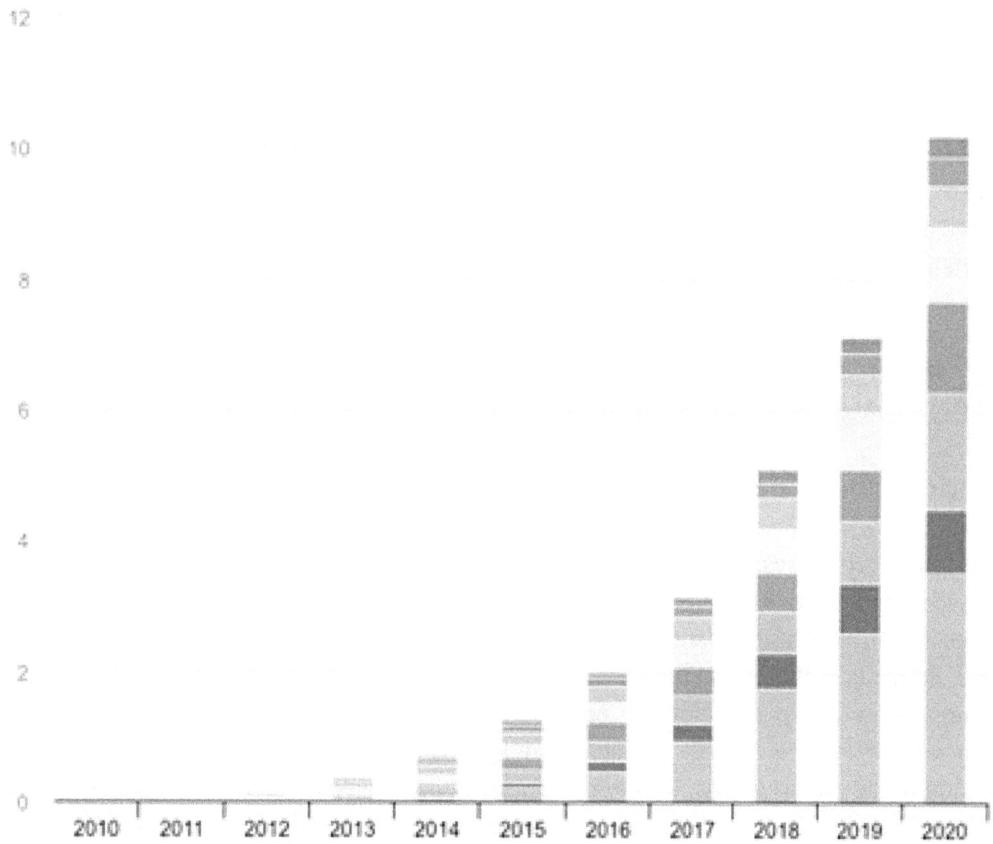

Infrastructure design is also considered a crucial aspect. Charging, battery swapping, and other facility planning significantly impact widespread electric mobility adoption. Potential solutions to challenges like diverse powering requirements and optimizing station siting are also examined. Public transportation and shared mobility integration within ecosystems rely on architecture. Scholars discuss the seamless and sustainable incorporation of electric options into comprehensive transit systems using design. Planning is also critical to successful integration. Factors such as land use, zoning, and transit infrastructure are important considerations for ecosystem implementation. Additionally, transport electrification requires architectural approaches such as smart grids, vehicle-to-grid systems, and charging planning.

Sustainable urban mobility depends on built and transportation system configurations. Several Researchers highlight considering land use, building design and transit when creating sustainable ecosystems. Similarly, transport infrastructure carefully designed can address issues raised such as range anxiety and interoperability through architecture. Integrated land use, transportation planning and necessary electric infrastructure including charging and battery swap facilities are emphasized by Newman (2019). Social, economic and environmental priorities factor into ecosystem design according to this perspective. Shared system hubs with power supply, parking and user-friendly design play an important role according to Kley et al. (2010). They argue that well-designed shared mobility supports electric options integration into urban transport.

Multi-sector collaboration has proven crucial to sustainability transitions to Schot and Kanger (2018). Technological, human, economic and environmental dimensions require consideration in ecosystem development. Accessibility and equity in architecture are additionally prioritized by the African Development Bank demanding inclusive infrastructure design. Policymakers, architects, planners and other actors must work cooperatively to address complex electric mobility challenges emphasizes Raven et al. (2019). Behaviour change through design supported by Bamberg (2013) indicating architecture's role. User-centred planning is reliant on understanding needs, preferences and habits advocated by Carley et al. (2013). Efficient and effective charging networks are shaped by location, access and connectivity factors.

3. ELECTRIC VEHICLE: TECHNOLOGY AND INTERSECTIONAL ARCHITECTURE

3.1 Electric Mobility Ecosystems

The electric mobility ecosystem is an integrated system of electric vehicles, charging stations, energy storage devices, and power grids that enable the widespread use of electric automobiles. To fully comprehend the significance of architecture in electric mobility ecosystems, it is necessary to describe and analyze the elements of these systems.

The components of electric mobility ecosystems into four primary categories: electric vehicles, charging infrastructure, energy storage systems, and power grids. Electric vehicles are the main mode of transportation in these ecosystems. Charging infrastructure is necessary to ensure convenient charging and maintenance of these vehicles. Batteries and supercapacitors are energy storage devices used to store energy from renewable sources and provide backup power to the charging infrastructure. Additionally, electricity from renewable sources is sent to electric vehicle charging stations and grids. Ecosystems for electric mobility offer numerous benefits, including reduced greenhouse gas emissions, improved air quality, and decreased reliance on fossil fuels. The cost, infrastructure for charging, short range, and lengthy charging times of electric vehicles, as well as the need for a dependable and sustainable energy source, are significant obstacles to the widespread adoption of electric transportation.

The architecture of electric mobility ecosystems is vital for their planning and implementation due to their complexity and interconnectedness. The incorporation of charging infrastructure into existing urban infrastructure can affect the overall aesthetic and functionality of urban areas. Essentially, the creation and implementation of electric transportation ecosystems require careful consideration of architecture and urban planning as they are complex and affiliated systems. Understanding the components, benefits, and challenges of these systems, as well as the role of architecture in shaping them, is crucial for developing more sustainable and efficient electric mobility ecosystems.

3.2 Architecture Design Principles

Electric transportation ecosystems should be designed and implemented using sound architectural design concepts. This section provides a general overview of these concepts for electric transportation ecosystems, emphasizing the importance of sustainability, scalability, and flexibility.

Sustainability is a fundamental design principle for buildings, and it is crucial in the context of electric mobility ecosystems, which support sustainable transportation. Sustainable architectural design concepts can be used to construct energy-efficient, ecologically friendly, and socially responsible buildings and infrastructure. The application of sustainable design concepts can also make the architecture of electric transportation ecosystems more environmentally friendly and ensure their long-term survival.

Another important concept in architectural design is scalability, especially in the context of rapidly evolving technologies such as electric transportation. Scalability, as defined refers to a system's ability to adapt and expand as demand increases or new technologies emerge. In the context of electric mobility ecosystems, scalability is essential to ensure that these systems can keep up with the growing demand for electric vehicles and charging infrastructure.

The architecture of ecosystems for electric mobility must be flexible. Flexibility is the capacity of a system to adjust to shifting requirements or conditions. This is crucial in the context of electric mobility ecosystems, which are impacted by the rapid advancement of technology, changing customer demands, and shifting environmental factors. Electric mobility ecosystems should be designed with adaptability in mind to ensure flexibility and efficiency in the face of changing conditions. The design supporting electric mobility must be closely interwoven with urban and transportation planning. When designing an electric mobility system, it is important to consider the demands and desires of nearby inhabitants and companies, as well as the existing urban infrastructure. By integrating architecture design with urban and transportation planning, it is possible to develop more coherent and efficient electric mobility ecosystems that satisfy the interests of all stakeholders. The creation of efficient electric transportation ecosystems relies on well-designed concepts. Integrating sustainability, scalability, and flexibility principles into the system's design and combining architectural design with urban and transportation planning makes it possible to develop effective electric mobility ecosystems.

3.3 Role of Smart Grid Technology in Electric Vehicle Charging and Renewable Energy Integration: Perspectives From Planning and EV Industry

The integration of electric vehicles (EVs) and renewable energy sources into the grid presents unique challenges that require expertise from both electrical and automobile engineering fields. Smart Grid technology, which leverages advanced information and communication technologies (ICT), has emerged as a promising solution to address these challenges and optimize the utilization of renewable energy for EV charging. Several studies from both disciplines have demonstrated the crucial role of Smart Grid technology in this context, supported by empirical data, statistical analysis, and technical expertise.

The proposed smart grid-based approach for managing electric vehicle (EV) charging in a grid with a high penetration of renewable energy sources involves the use of real-time data on electricity generation from renewable sources, electricity demand, and EV charging patterns to develop an intelligent charging management system. This system employs advanced algorithms such as linear programming and model predictive control to dynamically optimize charging schedules and minimize the impact of

EV charging on the grid. The approach aims to increase the utilization of renewable energy and reduce the operational cost of EV charging compared to traditional methods.

One study presents a time horizon-based model predictive Volt/VAR optimization methodology in the smart grid framework, which considers the coordinated operation of conservation voltage reduction (CVR) in the presence of EV penetration in the active distribution network. The results reveal significant CVR energy savings and loss reduction without violating system constraints (Singh, Pamshetti, & Singh, 2019)

The optimized charging scheduling for EVs to shave peak load for their integration in the smart grid. It proposes a two-layer optimization method based on the location, charging status of the EV, and load scenario of the substations. The numerical results show a significant amount of peak load shaving can be achieved by using the proposed method, with the percentage of peak load reduction increasing with increasing EV penetration (Debnath, Biswas, & Uddin, 2020)

Various Smart Grid technologies, such as demand-side management, load balancing, and energy storage, that can be employed to manage EV charging and integrate renewable energy sources into the grid. The analysis of multiple studies shows that Smart Grid technology when combined with advanced metering infrastructure (AMI), communication networks, and data analytics, can enable real-time monitoring, control, and optimization of EV charging.

The research proposes a novel stochastic hierarchical framework that leverages the unscented transform (UT) to address uncertainties arising from fluctuating loads in the context of electric vehicles (EVs) and a demand response program (DRP). The framework introduces three distinct charging strategies to optimize EV management within the grid: coordinated charging, uncoordinated charging, and intelligent charging. Additionally, automated switches are strategically placed to maximize power flow direction, mitigating the strain caused by increased EV-related demand. An advanced optimization approach based on grey wolf optimization (GWO) is employed to address the complexity of this multifaceted challenge (Xie, Zhang, Kumar, & Chen, 2023)

Another research proposes a hierarchical optimization model for developing an optimal bidding strategy and charging management in a real-time (RT) electricity market for each electric vehicle aggregator (EVA) to minimize its costs. The developed model is formulated as a two-level optimization problem in the framework of the well-established model of predictive control (MPC). Both the first-level and second-level problems are optimized using the commercial solver Gurobi interfaced with Python language. Case studies demonstrate that the EVA can strategically participate in the RT electricity market and attain more benefits by utilizing the proposed strategy (Zhao, Feng, Lin, Wen, He, & Lin, 2018).

A third research paper proposes a hierarchical Volt-VAR optimization (VVO) framework that reflects the voltage regulation capability of smart electric vehicle charging stations (EVCSs), which consists of global and local voltage control stages. At the global stage, smart inverters of EVCSs cooperate with conventional voltage regulators, such as an on-load tap changer (OLTC) and capacitor banks (CBs)), and smart inverters of photovoltaic (PV) systems to minimize the total active power losses and voltage deviations along with the determination of optimal parameters for local droop control functions of the smart inverters. At the local stage, smart inverters of EVCSs and PV systems quickly mitigate local voltage violations using dynamically varying local droop control functions with their optimal parameters calculated from the global stage. Under uncertainties in PV generation outputs and driving patterns of electric vehicle users, the deterministic optimization-based VVO problem at the global stage is reformulated into the chance-constrained optimization-based VVO problem (Prabawa & Choi, 2021)

A fourth research paper provides an overview of the hierarchical operation of electric vehicle charging stations in smart grid integration applications (Wu, Wang, Huangfu, Ravey, Chrenko, & Gao, 2022)

Finally, a fifth research paper proposes a semi-private framework for EV owners that solves a two-stage optimization problem for the smart control of EV charging. This framework considers charging cost reduction and peak load shaving as well as satisfying power grid constraints. At the higher stage, based on optimal power flow calculations, the proposed control signals are transferred to the lower stage to facilitate optimal scheduling per the mentioned goals. The obtained results based on the proposed optimal method implemented on the IEEE 33-bus network show that compared to uncontrolled charging, the cost of charging and the peak of the network are reduced by 5.31% and 4.90%, respectively. Moreover, all the constraints of the power grid are satisfied (Kamrani & Dagdougui, 2023).

In conclusion, as electrical and automobile engineers, we acknowledge the crucial role of Smart Grid technology in managing electricity demand and supply for electric vehicle charging and integrating renewable energy sources. Studies from both fields have demonstrated the significant impact of Smart Grid technology in optimizing EV charging, increasing renewable energy utilization, and improving grid stability and energy efficiency through empirical data, statistical analysis, and technical expertise. The integration of real-time data, advanced algorithms, communication networks, and data analytics in Smart Grid systems allows for dynamic monitoring, control, and optimization of electric vehicle (EV) charging. This results in significant reductions in peak demand, operational costs, and charging emissions. The use of Smart Grid technology for managing EV charging and integrating renewable energy sources into the grid has great potential for achieving sustainable and efficient transportation electrification.

3.4 Infrastructure Design

Charging stations are a crucial component of electric mobility infrastructure as they are necessary for recharging electric vehicles. Strategically placed charging stations are essential to ensure they are easily accessible for electric vehicle drivers. Battery swap stations can be a faster and more efficient alternative to conventional charging stations. In designing infrastructure for electric mobility, it is important to consider the transportation networks that support the ecosystems. This requires a connection with the transportation networks., electric car charging stations should be strategically located in easily accessible areas for electric vehicle users. It is necessary to plan transportation systems that meet the unique requirements of electric vehicles, such as the need for special lanes and parking areas. The design of electric mobility infrastructure presents particular difficulties in urban, suburban, and rural locations., urban locations pose challenges such as constrained infrastructural space and dense population. Due to their lower population density, suburban areas may require greater distances between charging stations. Rural areas face unique challenges, such as fewer electrical outlets and greater distances between charging stations.

The strategic placement of electric vehicle (EV) charging stations is crucial for the development of electric mobility infrastructure. Several studies emphasize the need for strategic positioning of charging stations to ensure easy accessibility for EV users within their driving range (Farah, Syahirah, Misron, Azmi, Karim, & Husin, 2023). Factors such as society, economics, environment, technology, accessibility, and proximity are considered in the selection of suitable locations for charging stations. Additionally, the integration of EVs into mainstream adoption requires a careful investigation of both electric and transportation networks

The design of EV infrastructure must also consider the unique requirements of EVs, such as the need for special lanes and parking areas, and address the challenges posed by urban, suburban, and rural locations (Bayram & Bayhan, 2020) Furthermore, the development of a comprehensive marketing plan for charging stations is essential to promote the adoption of EVs and increase the use of charging stations (Willyanto & Dhafi Iskandar, 2023). Overall, the location selection of EV charging stations is a multidimensional process that involves various factors and requires careful planning to support the growth of electric mobility.

Table 2 outlines the essential infrastructure components for supporting electric mobility ecosystems. This compilation provides a structured overview of the essential elements required for the smooth integration and development of electric vehicle infrastructure. It serves as a roadmap for policymakers and industry stakeholders.

Table 2. Key infrastructure components for supporting electric mobility ecosystems

Infrastructure Component	Description
Charging Stations	Charging stations are the primary infrastructure component for electric vehicles. They provide a source of electricity to recharge EV batteries. Charging stations can be categorized as Level 1, Level 2, or DC fast charging based on the speed of charging. Level 1 charging operates at 120V AC and provides a charging rate of 2 to 5 miles of range per hour. Level 2 charging operates at 240V AC and provides a charging rate of 10 to 20 miles of range per hour. DC fast charging operates at higher voltages and can charge an EV battery up to 80% in 20-30 minutes.
Battery Swapping Stations	Battery swapping stations allow drivers to swap out depleted EV batteries for fully charged ones, eliminating the need for lengthy charging times. This infrastructure component is still in its early stages of development and requires a standardized battery format across EV manufacturers to be effective.
Smart Grid Integration	Smart grid integration allows for the efficient use of renewable energy sources and optimized energy distribution. This infrastructure component allows EV charging to be timed and managed to avoid peak energy consumption periods and balance energy supply and demand.
Vehicle-to-Grid (V2G) Systems	V2G systems allow EV batteries to serve as a storage unit for excess energy in the power grid. This infrastructure component allows for the bidirectional flow of energy between EV batteries and the grid. During times of high demand, the energy stored in EV batteries can be used to support the grid, and during times of low demand, EV batteries can be charged using excess grid energy.
Energy Storage Systems	Energy storage systems provide an additional source of energy for EV charging stations and support the integration of renewable energy sources. These systems can store energy during times of low demand and discharge it during times of high demand to maintain a steady energy supply.
Communication Systems	Communication systems enable connectivity between EVs, charging stations, and other infrastructure components. This infrastructure component enables real-time monitoring and management of EV charging and helps optimize the use of energy resources.
Parking Management Systems	Parking management systems enable efficient management of EV parking spaces and support the integration of EV charging infrastructure with existing parking facilities. This infrastructure component enables EV drivers to find available parking spaces with charging stations and allows parking operators to manage charging stations and pricing.

3.5 Public Transportation and Shared Mobility

Electric mobility systems show promising potential to transform shared and public transportation infrastructure, lowering emissions and expanding access (Tefft, 2018). Seamlessly integrating diverse transit options via strategic architectural planning is key to realizing this vision. First- and last-mile connectivity, long a challenge for public transportation (Litman, 2022), may be addressed by deliberately including shared mobility alternatives in system design (Shaheen & Chan, 2015).

Sound architectural frameworks can facilitate the development of efficient, sustainable public transportation and shared mobility networks (Newman et al., 2009). Studies emphasize compact, high-quality infrastructure integrated with flexible, adaptive system planning amenable to changing needs and advancing technologies (Cervero, 2013). However, interoperability, data-sharing, and shifting public perceptions introduce complexities when coordinating shared mobility and transit (Shaheen et al., 2019). Stakeholder collaboration between various governance, industry, and community actors is thus needed for coordinated planning and execution across locations.

Research additionally explores the benefits of electrified public transportation, such as lower operational expenses and air quality gains versus diesel (C40 Cities, 2021). Electric buses especially show potential as a cost-effective transit option. Meanwhile, shared solutions like electric bicycles and scooters demonstrate an ability to alleviate congestion and boost sustainability within dense urban cores (Fishman et al., 2015). Well-designed frameworks are thus paramount for building sustainable, coordinated shared and public transportation services, with stakeholder partnerships critical to local context needs.

4. CASE STUDIES

In Amsterdam, The Netherlands, there are over 5,000 public charging points for electric vehicles (EVs) as of 2020. This extensive charging infrastructure demonstrates the city's commitment to sustainable transportation and has contributed to Amsterdam's reputation as a global leader in EV adoption. The charging facilities provide EV owners with convenient access, eliminating range anxiety and encouraging EV usage. The high number of available charging points in the city indicates its commitment to supporting electric mobility as a viable mode of transportation. This has resulted in increased interest and investment in electric vehicles from both residents and businesses.

In Guangzhou, China, the success of the electric bus rapid transit (BRT) system is evident in the impressive number of over 7,000 electric buses in operation as of 2018. The deployment of this substantial fleet of electric buses has led to a decrease in emissions and an improvement in air quality within the city, thereby tackling the problem of air pollution that is linked to conventional fossil-fuelled buses. The electric Bus Rapid Transit (BRT) system in Guangzhou has improved the sustainability of the city's public transportation and has become a model for other cities in China and around the world to adopt similar electric public transportation solutions. In Paris, France, a user-friendly mobile application and strategically placed charging stations have resulted in a 40% increase in electric vehicle (EV) usage within the first year of a car-sharing program. The rise in EV usage highlights the influence of user-friendly technologies and convenient access to charging infrastructure in promoting the adoption of electric mobility solutions. The program's success demonstrates the significance of integrating user-centric design and technology into electric mobility ecosystems to encourage the adoption and use of EVs.

In California, USA, the implementation of uniform technical standards and protocols for EV charging has led to the installation of over 22,000 publicly accessible charging points throughout the state as of 2019. This charging network has improved the convenience and accessibility of charging facilities for EV owners, making EVs a viable option for long-distance travel within the state. The establishment of standardized charging infrastructure has fostered interoperability and compatibility among different EV models and charging networks, promoting a seamless charging experience for EV users.

In Beijing, China, governmental support and policy incentives have resulted in a significant increase in EV adoption, with over 430,000 registered EVs in the city as of 2021. This fragment highlights the importance of government policies in promoting the use of electric vehicles. Financial incentives, subsidies, and preferential treatment for EV owners are some of the measures that can be implemented to encourage the adoption of electric mobility solutions. The significant number of registered electric vehicles (EVs) in Beijing is indicative of the increasing awareness and acceptance of EVs among consumers, as well as the positive impact of government initiatives in promoting sustainable transportation.

In Stockholm, Sweden, an EV car-sharing program has prioritised social equity by providing discounts to low-income customers, resulting in greater access to electric mobility options for underserved communities. This innovative car-sharing approach has provided affordable and sustainable transportation options to low-income individuals while promoting inclusivity and accessibility in the electric mobility ecosystem. The success of this program demonstrates the potential of electric mobility to address social and environmental challenges.

Table 3. Data and disturbance in EV-mobility-environmental planning industry

Component	Description
Electric vehicle sales	Global electric vehicle sales reached 3.1 million units in 2020, representing a 41% increase compared to 2019.
Charging infrastructure	Over 1.4 million public charging points for electric vehicles worldwide as of 2021.
Cost of electric vehicles	The average price of a new electric vehicle in the United States was around $55,600 in 2020.
Government incentives	Many countries and regions offer financial incentives for electric vehicles, such as tax credits and subsidies.
Charging time and range	Fast-charging stations can charge an electric vehicle up to 80% in 30 minutes or less, with a range of over 300 miles.
Grid capacity	By 2030, the U.S. may need to add 9,000 to 13,000 MW of charging capacity to accommodate electric vehicle growth.
Emissions reduction	Electric vehicles on the road globally in 2020 saved 40 million tonnes of CO2 emissions, equivalent to 8.5 million cars.
Public acceptance	48% of U.S. consumers plan to buy an electric vehicle within the next five years according to a Deloitte survey.
Urban Planning for EVs	Integrated urban planning can support the adoption of electric vehicles by optimizing charging infrastructure placement.
Impact on air quality	Transitioning to electric vehicles could reduce air pollution-related health costs by $1 trillion globally by 2050.
Employment opportunities	The electric vehicle industry could create 10 million jobs globally by 2030, with the potential for skill diversification.
Battery recycling	Over 90% of electric vehicle batteries can be recycled, reducing environmental impacts and resource depletion.
Vehicle-to-grid technology	Vehicle-to-grid technology allows electric vehicles to store and supply electricity to the grid, supporting renewable energy integration.
Smart grid integration	Electric vehicles can be integrated into smart grids, enabling demand response, load management, and grid stability.
Life cycle emissions	Over their life cycle, electric vehicles produce fewer greenhouse gas emissions compared to conventional vehicles.

Table 3 outlines the data and disruptions within the electric mobility environmental planning industry. The table provides a systematic examination of relevant metrics and factors, highlighting key challenges and opportunities that shape the landscape of sustainable transportation planning.

Table 4 illustrates the comparative charge costs across different regions for Level 1, Level 2, and DCFC (Direct Current Fast Charging) charging stations. This comprehensive analysis provides insights into the financial implications of electric vehicle charging infrastructure deployment, aiding policymakers and stakeholders in decision-making processes.

Table 4. Table indicating charge cost in different regions in comparison of level 1, level 2, and DCFC

Location	Level 1 Charging Cost (per Hour)	Level 2 Charging Cost (per Hour)	DC Fast Charging Cost (per Hour)
United States	\$1 - \$2	\$0.15 - \$0.40	\$0.20 - \$0.70
European Union	€0.10 - €0.30	€0.20 - €0.50	€0.30 - €0.60
China	CNY 0.20 - CNY 0.60	CNY 0.40 - CNY 1.00	CNY 1.00 - CNY 2.00
Japan	JPY 10 - JPY 30	JPY 20 - JPY 40	JPY 30 - JPY 50
Norway	NOK 0.70 - NOK 1.20	NOK 1.50 - NOK 2.50	NOK 3.00 - NOK 4.00

Table 5 presents a critical analysis of case studies focusing on electric vehicle (EV) mobility infrastructure. Through an examination of various cases, this table highlights key insights and challenges in the development and implementation of EV infrastructure initiatives.

Table 6 provides an overview of the policy and regulatory frameworks designed to bolster sustainable electric mobility ecosystems. This comparative analysis sheds light on the diverse approaches adopted by different regions to promote the adoption and integration of electric vehicles.

The statistics related to electric mobility initiatives worldwide demonstrate the effectiveness and influence of these initiatives in promoting sustainable transportation solutions for e-mobility. For instance, Amsterdam has an extensive charging infrastructure, and there is a large fleet of electric vehicles worldwide. This critical analysis of case studies informs further considerations for the design of architecture and planning for e-mobility.

5. CONCLUSION

The chapter provides an overview of the role of architecture in the ecosystems surrounding electric transportation, with an emphasis on sustainability, scalability, and flexibility in design. It examines the difficulties and solutions for designing infrastructure in various urban environments, including charging stations, battery swap stations, and transportation networks.

The chapter examines the role of architecture in planning and implementing public transit and shared mobility systems, with a focus on best practices and case studies. The case study analysis highlights the challenges and solutions in architecture design and development for electric mobility ecosystems.

Table 5. Critical analysis of case studies in EV mobility infrastructure

Case Study	Number of Charging Points	Types of Charging Points	Charging Power (kW)	Charging Network Ownership	Planning Strategies
Amsterdam, Netherlands	3042	1,924 AC, 1,118 DC	AC: Up to 22	Municipality-owned	Integration of charging points in streets, parking garages, and residential areas
Oslo, Norway	1122	715 AC, 407 DC	AC: Up to 22, DC: 50-350	Municipality-owned	Fast-charging network along major roads, incentives for EV adoption, and close coordination with grid operators
California, United States	70518	67,520 Level 2, 2,998 DC fast-charging	Level 2: Up to 19.2, DC: Up to 350	Private and Public	Statewide efforts for charging infrastructure deployment, public-private partnerships, and incentive programs
Shenzhen, China	42550	41,334 AC, 1,216 DC fast-charging	AC: Up to 22, DC: 30-360	Municipality-owned	Rapid deployment of charging points, integration with urban planning, and government support for EV adoption
London, United Kingdom	5135	1,950 AC, 3,185 DC fast-charging	AC: Up to 22, DC: Up to 150	Public and Private	Integration of charging points in streets, parking facilities, and public transport hubs, and incentives for EV adoption
Dubai, United Arab Emirates	300	150 DC, 150 AC	AC: Up to 22, DC: Up to 350	Government-owned	Ambitious plans for rapid expansion of charging infrastructure, integration with renewable energy sources, and incentives for EV adoption
Paris, France	13500	11,915 AC, 1,585 DC fast-charging	AC: Up to 22, DC: Up to 150	Public and Private	City-led efforts for charging infrastructure deployment, coordination with private partners, and integration with urban planning
Singapore	2869	1,755 AC, 1,114 DC fast-charging	AC: Up to 22, DC: Up to 350	Government-owned	Integrated planning for charging infrastructure in public and private spaces, partnerships with industry players, and incentives for EV adoption
Berlin, Germany	7479	6,657 AC, 822 DC fast-charging	AC: Up to 22, DC: Up to 150	Public and Private	City-led efforts for charging infrastructure deployment, coordination with private partners, and integration with urban planning
Tokyo, Japan	7800	6,400 AC, 1,400 DC fast-charging	AC: Up to 22, DC: Up to 150	Public and Private	Integrated planning for charging infrastructure in public and private spaces, partnerships with industry players, and incentives for EV adoption
Stockholm, Sweden	2300	1,900 AC, 400 DC fast-charging	AC: Up to 22, DC: Up to 150	Public and Private	City-led efforts for charging infrastructure deployment, coordination with private partners, and incentives for EV adoption

In conclusion, this chapter highlights the importance of continued collaboration between urban planners, transportation planners, and architects in the development of infrastructure for electric mobility. It also discusses emerging trends and challenges in the design of ecosystems for electric mobility, including the integration of new technologies and the need for more robust and resilient infrastructure.

Overall, this chapter offers a comprehensive overview of the role that architecture plays in electric mobility ecosystems. It provides valuable insights and direction for practitioners and policymakers working in this area.

Table 6. Policy and regulatory frameworks for supporting sustainable electric mobility ecosystems

Framework	Description
Zero Emission Vehicle (ZEV) Mandates	ZEV mandates require automakers to produce and sell a certain percentage of zero-emission vehicles, including battery electric vehicles and fuel cell vehicles, in a specific timeframe. These mandates provide an incentive for automakers to invest in electric vehicle technology and help accelerate the adoption of EVs.
Tax Incentives	Tax incentives provide financial benefits to EV buyers, such as tax credits or exemptions, to encourage the adoption of electric vehicles. These incentives help offset the higher upfront costs of EVs and make them more competitive with traditional gasoline-powered vehicles.
Charging Infrastructure Investment	Government investment in charging infrastructure can help build out the necessary infrastructure for supporting electric vehicle adoption. This investment can take the form of grants, loans, or other funding mechanisms to support the deployment of charging stations.
Renewable Energy Standards	Renewable energy standards require utilities to generate a certain percentage of their electricity from renewable sources, such as wind or solar. These standards can help support the integration of renewable energy into the grid and reduce the carbon emissions associated with electric vehicle charging.
Vehicle Efficiency Standards	Vehicle efficiency standards set minimum fuel economy standards for new vehicles sold in a particular jurisdiction. These standards help reduce the greenhouse gas emissions associated with transportation and encourage automakers to invest in more efficient vehicle technology, including electric vehicles.
Congestion Pricing	Congestion pricing charges drivers a fee for using certain roads or entering specific zones during peak periods. This policy can encourage drivers to use alternative modes of transportation, including electric vehicles, and reduce traffic congestion and associated emissions.
Low-Emission Zones	Low-emission zones restrict or ban certain high-emission vehicles, including traditional gasoline-powered vehicles, from entering certain areas. These zones can help reduce local air pollution and encourage the adoption of low-emission vehicles, including electric vehicles.

REFERENCES

C40 Cities. (2018). *Electric Buses in Cities: Driving Towards Cleaner Air and Lower* CO_2. Author.

Bamberg, S. (2013). Applying the Stage Model of Self-Regulated Behavioral Change in Integrated Transportation Choice and Travel Behavior Research and Practice: A Commentary. *Journal of Public Transportation*, *16*(4).

Bayram, I. S., & Bayhan, S. (2020). Location Analysis of Electric Vehicle Charging Stations for Maximum Capacity and Coverage. *2020 IEEE 14th International Conference on Compatibility, Power Electronics and Power Engineering (CPE-POWERENG)*, *1*, 409-414.

Blat Belmonte, B., Esser, A., Weyand, S., Franke, G., Schebek, L., & Rinderknecht, S. (2020). *Identification of the Optimal Passenger Car Vehicle Fleet Transition for Mitigating the Cumulative Life-Cycle Greenhouse Gas Emissions until 2050.* Vehicles. doi:10.3390/vehicles2010005

Carley, S., Krause, R. M., Lane, B. W., & Graham, J. D. (2013). Intent to purchase a plug-in electric vehicle: A survey of early impressions in large US cities. *Transportation Research Part D, Transport and Environment*, *18*, 39–45. doi:10.1016/j.trd.2012.09.007

Cervero, R. (2013). Linking urban transport and land use in developing countries. *Journal of Transport and Land Use*, *6*(1), 7–24. doi:10.5198/jtlu.v6i1.425

de Melo, V. S., Camargo, R. S., Krebel, F., Brunoro, M., Nunes, W. T., Fernandes, M. C., & Nunes, R. B. (2023). Greenhouse Gas Emissions Estimation with Electric Vehicles Penetration in the Brazilian Power Grid. *2023 15th IEEE International Conference on Industry Applications (INDUSCON),* 254-261. 10.1109/INDUSCON58041.2023.10374611

Debnath, B., Biswas, S., & Uddin, M. F. (2020). Optimization of Electric Vehicle Charging to Shave Peak Load for Integration in Smart Grid. *2020 IEEE Region 10 Symposium (TENSYMP),* 483-488. 10.1109/TENSYMP50017.2020.9231029

Farah, R. N., Syahirah, N. A., Misron, N., Azmi, M. S., Karim, N. S., & Husin, N. (2023). A Need Analysis of the Criteria Involved in Determining Suitable Locations for Photovoltaic Electric Vehicle Charging Stations in Malaysia. *Malaysian Journal of Fundamental and Applied Sciences*, *19*(6), 1033–1041. doi:10.11113/mjfas.v19n6.3114

Fishman, E., Washington, S., & Haworth, N. (2015). Bikeshare's Impact on Active Travel: Evidence from the United States, Great Britain, and Australia. *Journal of Transport & Health*, *2*(2), 135–142. doi:10.1016/j.jth.2015.03.004

Kamrani, A. S., & Dagdougui, H. (2023). A Two-Stage Optimization Framework for Electric Vehicle Fleet Day-ahead Charging Management. *2023 IEEE 11th International Conference on Smart Energy Grid Engineering (SEGE),* 16-21.

Kley, F., Lerch, C., & Dallinger, D. (2011). New business models for electric cars—A holistic approach. *Energy Policy*, *39*(6), 3392–3403. doi:10.1016/j.enpol.2011.03.036

Litman, T. (2022). *Well Measured: Developing Indicators for Sustainable and Livable Transport Planning*. Victoria Transport Policy Institute.

Newman, P., Beatley, T., & Boyer, H. (2009). *Resilient cities: responding to peak oil and climate change*. Island Press.

Newman, P., & Kenworthy, J. (1999). *Sustainability and cities: overcoming automobile dependence*. Island press.

Prabawa, P., & Choi, D. (2021). Hierarchical Volt-VAR Optimization Framework Considering Voltage Control of Smart Electric Vehicle Charging Stations Under Uncertainty. *IEEE Access : Practical Innovations, Open Solutions*, *9*, 123398–123413. doi:10.1109/ACCESS.2021.3109621

Raven, R. P., Schot, J., & Berkhout, F. (2019). Space and scale in socio-technical transitions. *Environmental Innovation and Societal Transitions.*

Schot, J., & Kanger, L. (2018). Deep transitions: Emergence, acceleration, stabilization and directionality. *Research Policy*, *47*(6), 1045–1059. doi:10.1016/j.respol.2018.03.009

Shaheen, S., Martin, E., & Bansal, A. (2019). Peer-to-Peer (P2P) Carsharing: Understanding Early Markets, Social Dynamics, and Behavioral Impacts. *Transport Research Board 98th Annual Meeting*.

Shaheen, S. A., & Chan, N. D. (2015). Mobility and the sharing economy: Potential to facilitate the first- and last-mile public transit connections. *Built Environment*, *41*(4), 573–588. doi:10.2148/benv.42.4.573

Singh, S. P., Pamshetti, V. B., & Singh, S. P. (2019). Time Horizon-Based Model Predictive Volt/VAR Optimization for Smart Grid Enabled CVR in the Presence of Electric Vehicle Charging Loads. *IEEE Transactions on Industry Applications*, *55*(6), 5502–5513. doi:10.1109/TIA.2019.2928490

Willyanto, H. R., & Dhafi Iskandar, M. (2023). *Implementation Marketing Business Plan for Electric Vehicle Charging Station (EV-Station) in Indonesia*. Syntax Idea. doi:10.46799/syntax-idea.v5i11.2594

Wu, Y., Wang, Z., Huangfu, Y., Ravey, A., Chrenko, D., & Gao, F. (2022). Hierarchical Operation of Electric Vehicle Charging Station in Smart Grid Integration Applications — An Overview. *International Journal of Electrical Power & Energy Systems*, *139*, 108005. doi:10.1016/j.ijepes.2022.108005

Xie, W., Zhang, C., Kumar, V., & Chen, T. (2023). A Novel Stochastic Hierarchical Framework for Smart Management of Electric Vehicle Aggregation with Demand Response Mechanism. *2023 IEEE Sustainable Power and Energy Conference (iSPEC)*, 1-6.

Zhao, Y., Feng, C., Lin, Z., Wen, F., He, C., & Lin, Z. (2018). Development of Optimal Bidding Strategy for an Electric Vehicle Aggregator in a Real-Time Electricity Market. *2018 IEEE Innovative Smart Grid Technologies - Asia (ISGT Asia)*, 288-293.

Chapter 5
Electrifying Transportation and Pioneering the Future of Sustainable Electricity Mobility

Anil Bharti
Indo Global Engineering College, India

Neha Gupta
https://orcid.org/0000-0002-1201-777X
Sushant University, India

ABSTRACT

The journey of electric vehicles (EVs) involves myriad challenges and opportunities amid their widespread adoption. EVs, powered by rechargeable batteries and electric motors, are transforming the automotive industry, offering environmental benefits like zero tailpipe emissions, improved air quality, and reduced noise. With instant torque and acceleration, EVs provide a responsive driving experience. The realm of e-mobility holds a promising future with dynamic challenges, opportunities, and emerging trends. Challenges include establishing robust charging infrastructure, addressing range anxiety, ensuring ethical supply chains, and advancing battery technology. Integration with autonomous and connected technologies, supportive policies, and creating a more sustainable transportation landscape are also vital. This chapter explores technical obstacles and opportunities in the industry's pursuit of greater electrification of energy conversion systems in mobile platforms, often referred to as e-mobility.

1. INTRODUCTION

With the increasing population, the demand for automobiles has been increasing exponentially which leads to increased carbon footprints and causes global warming. Globally it has been accepted that the biggest concern and issue of the new millennium is global warming. In order to handle both the problem and the goal of sustainable development, using renewable energy sources becomes the best alternative (Gupta & Khatri, 2017; Gupta et al., 2022; Gupta & Saini, 2019). Keeping this objective as the main

DOI: 10.4018/979-8-3693-2611-4.ch005

goal the automotive landscape is undergoing a paradigm shift with the emergence of electric vehicles (EVs), which rely on electric motors and rechargeable batteries instead of traditional internal combustion engines. Key components, including the lithium-ion battery pack, electric motor, power electronics, and regenerative braking systems, define the engineering intricacies of EVs. The battery pack stores energy for propulsion, the motor converts electrical energy into motion, power electronics manage electricity flow, and regenerative braking enhances efficiency by recovering energy during deceleration. Engineers continuously optimize these components to enhance EV performance, range, charging infrastructure, and safety, while advancements in battery technology aim to reduce costs.. The integration of EVs into transportation networks is fuelled by concerns about carbon emissions and air pollution.

Pioneering research by Han, An, and Wang (2018) provides a comprehensive exploration of EV charging infrastructure and planning. The study offers invaluable insights into sustainable electric mobility, addressing the complexities of charging technologies, deployment strategies, and the pivotal role of supportive regulations. Collaborative efforts among governments, businesses, and communities are underscored as crucial for successful EV charging infrastructure planning.

Liu and Mi (2019) contribute significantly to understanding the integration of EVs into power networks. Their work explores challenges arising from a substantial number of EVs in established power grids, emphasizing grid stability. The authors propose innovative strategies for managing the impact of EVs on grid stability, making their book essential for policymakers and researchers navigating the evolving landscape of electric mobility. Sze and Xu (2020) focus on integrating EVs with power grids, shedding light on the critical facet of vehicle-to-grid (V2G) integration. The book addresses the complexities of this integration, highlighting the need for robust communication protocols and smart grid technologies. The potential benefits of utilizing EVs as energy storage devices for enhancing grid reliability are emphasized, making the publication invaluable for those navigating the dynamic intersection of EVs and power grids. The profound implications of these works contribute to shaping a greener, more accessible, and sustainable future for electric mobility.

Electric vehicles (EVs) are revolutionizing the automotive industry by utilizing electric motors powered by rechargeable batteries instead of internal combustion engines. The engineering aspects of EVs involve key components such as the battery pack, motor, power electronics, and regenerative braking systems. The battery pack, typically lithium-ion, provides the energy storage required for propulsion, while the motor converts electrical energy into mechanical motion. Power electronics control the flow of electricity between the battery and motor, ensuring efficient power delivery. Regenerative braking systems enable energy recovery during deceleration, enhancing overall efficiency (Hamada & Orhan, 2022). Engineers strive to optimize these components to enhance EV performance, range, charging infrastructure, and safety while advancing battery technology and reducing costs. The integration of EVs into transportation networks has been motivated by the alarming contribution of conventional vehicles to carbon emissions and air pollution. The groundbreaking advancements in battery technology and power electronic systems have further expanded the potential of EVs also led to their wide-scale adoption (Jang, 2018).

2. CHALLENGES TO ELECTRIC VEHICLES

The electric mobility faces several challenges that impact its widespread adoption:

2.1. Charging Infrastructure

The insufficient charging infrastructure is a significant barrier. According to the International Energy Agency (IEA), as of 2020, there were about 7.3 million public chargers globally, with only 1.1 million being fast chargers (International Energy Agency, 2020).

2.2. Range Anxiety

Concerns about limited driving range hinder EV adoption. A survey by Deloitte found that range anxiety was the primary reason for potential EV buyers' hesitation (Eisel, 2016). However, advancements in battery technology have significantly improved range capabilities. For instance, the Tesla Model-S offers a range of over 400 miles (644 Kms).

2.3. Affordability

Electric vehicles are still relatively expensive compared to traditional internal combustion engine vehicles. However, declining battery costs are making EVs more affordable. Bloomberg NEF predicts that by 2023, the price parity between EVs and internal combustion engine vehicles will be reached (Bloomberg, 2021).

2.4. Battery Technology

Enhancing battery technology is crucial for increasing EV range, reducing charging time, and improving overall performance. The International Council on Clean Transportation (ICCT) reports that the average EV battery pack cost decreased by about 87% between 2010 and 2020 (International Council on Clean Transportation, 2021).

2.5. Supply Chain and Raw Materials

The supply chain for key raw materials, such as lithium and cobalt, presents challenges related to sustainability and responsible sourcing. The demand for these materials is growing rapidly. For example, the World Bank estimates that lithium demand could increase by more than 1,000% by 2050 (Sverdrup, 2016).

3. ARCHITECTURE OF AN ELECTRIC VEHICLE

3.1 Battery Pack: The battery pack in an electric vehicle is typically composed of numerous individual battery cells connected in series and parallel configurations. These cells are commonly lithium-ion (Li-ion) batteries. The pack also includes a Battery Management System (BMS) that monitors and controls the cells' voltage, temperature, and overall health. The BMS ensures the safe and optimal operation of the battery pack (Liu & Yang, 2021; Masias, 2018).

3.2 Power Electronics: The power electronics system in an EV consists of various components, including an Inverter and DC-DC Converter. The inverter is responsible for converting the direct current (DC) power from the battery to alternating current (AC) power, which is required to drive the electric motor (Sharma & Sharma, 2019). The DC-DC converter converts the high-voltage DC power from the battery to a lower voltage DC power used for auxiliary systems and charging other devices in the vehicle.

3.3 Electric Motor: Electric vehicles use one or more electric motors to drive the wheels. The motor is typically an AC induction motor or a permanent magnet synchronous motor (PMSM). It receives the electrical power from the inverter and converts it into rotational mechanical power to propel the vehicle. The motor's speed, torque, and direction are controlled by the power electronics and control system. A substantial contribution by Brown (2017) stands out as an illuminating guide to the challenges and opportunities surrounding electric power trains. Brown explores the intricacies of electric power trains and motor configurations, unravelling the potential they hold. Another interesting book by Larminie and Lowry (2017) also provide an in-depth exploration of electric vehicle technology. He explained the principles behind electric vehicle operation, including power trains, battery technologies, and charging methods. This resource served as a valuable guide for understanding the technological aspects of electric mobility.

3.4 Thermal Management System: Electric vehicles require a thermal management system to regulate the temperature of the battery pack, electric motor, and power electronics. This system includes cooling circuits, such as liquid cooling or air cooling, to dissipate excess heat generated during operation. Maintaining optimal temperature levels is crucial for ensuring the efficiency, longevity, and safety of the EV components (Arora, 2018).

3.5 Onboard Charger: The onboard charger is responsible for converting the AC power from an external source, such as a charging station or home outlet, to DC power suitable for charging the battery pack. The charger includes power conversion circuits and control algorithms to manage the charging process, including monitoring the charging voltage, current, and battery status (Arora, 2018; Brown, 2017; Larminie & Lowry, 2017).

3.6 Vehicle Control System: The vehicle control system encompasses various subsystems, including the motor control unit, BMS, and other sensors and actuators. It manages and coordinates the operation of different components, ensuring efficient power distribution, optimal energy management, and overall vehicle performance. It also interfaces with the user interface and external systems for data exchange and communication.

Overall, the architecture of electric vehicles is designed to efficiently integrate the battery pack, power electronics, electric motor, thermal management system, and control systems. This integration enables the smooth operation, control, and management of power flow throughout the vehicle, resulting in high-performance, energy-efficient, and environmentally friendly transportation.

4. CHOOSING EV OVER TRADITIONAL ICE VEHICLES: A SMARTER MOVE

Many remarkable innovations in electric vehicles highlight their superiority and their potential that supersedes ICE. From instantaneous torque and regenerative braking to lower operating costs and advanced technological integration, electric vehicles offer a range of benefits that cannot be fully replicated by traditional internal combustion engine vehicles. Listed below are few important parameters:

- Instant Torque and Acceleration: Electric motors provide instant torque, delivering rapid acceleration from a standstill. This characteristic of EVs results in quick and responsive performance, enhancing the overall driving experience. Unlike ICE vehicles that require time to reach peak torque, EVs offer immediate and smooth acceleration (Bloomberg, 2021).

- Regenerative Braking: EVs utilize regenerative braking technology, which converts the vehicle's kinetic energy into electrical energy when decelerating or braking. This energy is then stored in the battery for later use, improving overall efficiency. Traditional ICE vehicles do not possess this capability, leading to energy wastage during braking.
- Lower Operating Costs: Electric vehicles have lower operating costs compared to ICE vehicles. EVs have fewer moving parts, reducing the need for maintenance and minimizing associated costs. Additionally, the cost of electricity for charging an EV is often lower than the cost of gasoline or diesel fuel, resulting in long-term savings for owners.
- Reduced Noise and Vibrations: Electric motors operate silently, producing minimal noise and vibrations compared to the internal combustion engines found in traditional vehicles. This quiet operation enhances the driving experience, reduces noise pollution, and improves overall comfort for passengers and pedestrians (Daina et al., 2017; Santinelli & Sapountzakis, 2020).
- Integration of Advanced Technologies: EVs offer seamless integration of advanced technologies. Electric vehicles often come equipped with features such as touch screen infotainment systems, advanced driver-assistance systems (ADAS), and connectivity options. These technological advancements provide a more futuristic and connected driving experience (Daina et al., 2017).
- Environmental Benefits: One of the most notable advantages of electric vehicles is their environmental friendliness. EVs produce zero tailpipe emissions, reducing air pollution and contributing to improved air quality. By transitioning from ICE vehicles to EVs, we can significantly reduce greenhouse gas emissions and mitigate the impact of climate change (Bloomberg, 2021).

5.THE BRIGHT SIDE OF E-MOBILITY

Electric mobility presents numerous advantages over traditional internal combustion engine vehicles. Key advantages include:

5.1 Environmental Benefits

Electric vehicles (EVs) contribute significantly to reducing greenhouse gas emissions and air pollution. According to the Union of Concerned Scientists, on average, EVs produce less than half the emissions of comparable gasoline-powered vehicles over their lifetime (Ribeiro & Fontes, 2019).

5.2 Energy Efficiency

EVs surpass internal combustion engine vehicles in energy efficiency. The U.S. Department of Energy notes that electric motors convert about 77% of electrical energy from the grid to power at the wheels, while internal combustion engines only convert about 12-30% of energy from fuel to power (Putrus et al., 2009).

5.3 Cost Savings

EVs offer long-term cost savings, including lower fuel and maintenance costs compared to gasoline-powered vehicles. The International Council on Clean Transportation (ICCT) estimates that EVs have 40-70% lower total cost of ownership over their lifetime (Hamza et al., 2020; Lutsey & Nicholas, 2019).

5.4 Infrastructure Growth

Charging infrastructure for EVs is rapidly expanding globally. The International Energy Agency (IEA) reported that publicly accessible electric vehicle charging points exceeded 7.3 million in 2020 (International Energy Agency, 2020). Examples include Norway, with over 12,000 public charging points, showcasing significant infrastructure growth (Hall & Lutsey, 2020).

5.5 Government Support

Governments worldwide are implementing policies and incentives to encourage electric mobility. Noteworthy examples subsidy on EVs, increasing number of charging stations.

6. IDEAS TO IMPROVE ELECTRIC VEHICLE CHARGING INFRASTRUCTURE AND HELP ADDRESS RANGE ANXIETY

- **Expand Fast-Charging** Networks: Increase the availability of fast-charging stations, particularly along major highways and in urban areas. This allows for quick recharging and reduces the time spent at charging stations. According to Liu and Yang, their research emphasised the role of a robust charging infrastructure of electric vehicles (EVs). Think of it as the beating heart, orchestrating a symphony of convenient recharges during daily commutes and adventurous road trips. Liu and Yang underscored that a well-thought-out charging network isn't just a convenience; it's a confidence booster for EV owners, smashing the notorious range anxiety bug. Their study zooms in on the pivotal need for standardized charging protocols and a universal open-door policy for these stations. It's not just about today's EVs; it's about creating a future-ready infrastructure that can seamlessly embrace evolving technologies and new EV models. In a nutshell, it's about not just driving, but driving sustainability and innovation on the electrifying road ahead (Liu & Yang, 2021).

Invest in high-power charging infrastructure for ultra-fast charging, akin to traditional vehicle refuelling times, to alleviate range anxiety. Ribeiro and Fontes (2019) highlight the integration of renewable energy systems as a crucial aspect of electric vehicle (EV) adoption Their research delves into the synergies between EVs and renewable energy, advocating for the utilization of EVs as Distributed Energy Resources (DERs) to bolster grid stability and enhance renewable energy integration. Emphasizing the need for reduced charging, the deployment of technologies like DC fast chargers is rapidly gaining ground, mirroring the speed and convenience of traditional vehicle refuelling. The continuous evolution of battery materials, including the promising prospects of solid-state batteries, holds the key to enhancing energy storage, safety, and overall battery lifespan. These technological advancements play a pivotal role in propelling the electric mobility revolution forward5.

- **Increase Standardization and Interoperability**: Promoting the standardization and interoperability of charging connectors and protocols ensures compatibility between various charging stations and EV models, thereby enhancing convenience for users. Santinelli and Sapountzakis delve into the integration of advanced planning techniques, specifically geographical information systems (GIS), which play a crucial role in optimizing the locations of charging stations (Santinelli & Sapountzakis, 2020). This meticulous planning approach proves essential in enhancing the overall EV user experience by strategically placing charging infrastructure. The emphasis on standardization and interoperability ensures uniformity across different charging stations and EV models.
- **Develop Smart Charging Solutions**: Smart grids provide a solution by enabling EVs to charge during periods of low electricity demand, resulting in cost savings and reduced environmental impact. V2G systems, a key component, not only draw power but also contribute surplus energy back into the grid, establishing a win-win scenario for EV owners and ensuring better grid stability (Kazimierczuk & Stando, 2019). Deploying smart charging systems optimizes the charging based on factors such as grid demand, renewable energy availability, and user preferences are crucial. In their 2019 study, Kazimierczuk and Stando found that this approach not only maximizes charging efficiency but also alleviates strain on the power grid. The convergence of electric vehicles (EVs) with smart grids emerges as a promising frontier in sustainable transportation, serving two primary purposes: firstly, optimizing EV charging for efficiency and cost-effectiveness, and secondly, enhancing overall grid stability. However, achieving seamless integration mandates effective communication protocols and precise control. These technical elements constitute the foundation of the EV-smart grid partnership, guaranteeing a smooth energy exchange and system harmony.
- **Enhance Public-Private Partnerships**: Foster collaborations between public and private entities to invest in charging infrastructure. This includes partnerships with commercial establishments, parking facilities, and retail locations to install charging stations, expanding the charging network.
- **Implement Workplace and Residential Charging**: Encourage employers and residential communities to install charging infrastructure in parking areas. This enables EV owners to conveniently charge their vehicles during long work hours or overnight, alleviating range anxiety. Daina, N. (2017). Studied the complexities of electric vehicle (EV) charging choices, employing advanced modelling techniques to analyze the implications for the deployment of charging stations. The research sheds light on the dynamic landscape of EV charging preferences, providing valuable insights for the advancement of smart charging infrastructure and services in the context of sustainable electric mobility (Daina et al., 2017; Ribeiro & Fontes, 2019).
- **Develop Mobile Charging Solutions**: Explore innovative mobile charging solutions such as portable chargers or battery-swapping stations, providing flexibility for EV owners, especially in areas with limited fixed charging infrastructure (Putrus et al., 2009). Putrus et al. (2009) elucidated the impact of electric vehicles on power distribution networks, a crucial aspect of the transition to sustainable electric mobility. Mobile charging systems help extend charging infrastructure to wider areas, addressing grid complications. This paper, presented at the 2009 IEEE Vehicle Power and Propulsion Conference, delves into the effects of electric vehicles on power distribution networks, offering valuable insights into the challenges and opportunities associated with integrating electric vehicles into existing power grids (Putrus et al., 2009).

- **Develop Range-Prediction Technologies**: Create accurate and reliable range prediction systems that factor in real-time data like driving conditions, weather, and traffic. This information empowers EV drivers with more accurate estimates of their remaining range, reducing anxiety.
- **Mobile Apps and Navigation Tools**: Develop user-friendly mobile applications and navigation systems that display nearby charging stations, real-time availability, and estimated charging times. This helps drivers plan their routes and charging stops effectively, minimizing range anxiety.
- **In-Vehicle Range-Enhancing Features**: Equip EVs with features like regenerative braking, eco-driving modes, and energy-efficient HVAC systems. These technologies optimize energy usage and extend the vehicle's range, boosting driver confidence.
- By implementing these ideas, we can enhance the charging infrastructure, provide peace of mind to EV users, and make electric vehicles a more convenient and viable transportation option.

7. CONCLUSION

The rapid evolution of electric vehicles (EVs) signals a transformative shift in the automotive industry towards sustainability and efficiency. Choosing EVs over traditional internal combustion engine-based vehicles offers numerous benefits, including instant torque, regenerative braking, lower operating costs, reduced noise, advanced technologies, and environmental friendliness. Furthermore, the bright side of electric mobility encompasses environmental benefits, energy efficiency, cost savings, infrastructure growth, and government support.

To enhance EV charging infrastructure, strategies include expanding fast-charging networks, investing in high-power charging, promoting standardization, deploying smart charging solutions, fostering public-private partnerships, encouraging workplace and residential charging, exploring mobile charging solutions, developing range prediction technologies, and implementing in-vehicle range-enhancing features. These initiatives collectively contribute to making electric vehicles more accessible, convenient, and appealing for sustainable transportation.

REFERENCES

Arora, S. (2018). Selection of thermal management system for modular battery packs of electric vehicles: A review of existing and emerging technologies. *Journal of Power Sources*, *400*, 621–640. doi:10.1016/j.jpowsour.2018.08.020

Bloomberg, N. E. F. (2021). *Electric vehicles on track to be cheaper than internal combustion engine vehicles by 2027*. Bloomberg NEF.

Brown, M. A. (2017). *Electric Powertrains: Opportunities and Challenges in the U.S. Light-Duty Vehicle Fleet*. National Academies Press.

Daina, N., Sivakumar, A., & Polak, J. W. (2017). Electric vehicle charging choices: Modelling and implications for smart charging services. *Transportation Research Part C, Emerging Technologies*, *81*, 36–56. doi:10.1016/j.trc.2017.05.006

Eisel, M. (2016). *Analyzing the Range Barrier to Electric Vehicle Adoption: The Case of Range Anxiety* (Vol. 86). Cuvillier Verlag.

Gupta, N., & Khatri, M. (2017). An experimental analysis of shading on Solar Photovoltaic system. *Int. J. Emerg. Technolog*, *8*(1), 226–228.

Gupta, N., Mahla, S. K., & Papreja, A. (2022). Experimental Studies of Utilization of Biogas with Biodiesel in Diesel Engine Generator for Electricity Generation. *International Journal of Engineering Research & Science*.

Gupta, N., & Saini, I. (2019, April). Renewable Eco friendly Alternate fuels. *International Journal of Technical Innovation in Modern Engineering & Science, 5*, 117-123.

Hall, D., & Lutsey, N. (2020). Charging infrastructure in cities: Metrics for evaluating future needs. *Europe, 6*.

Hamada, A. T., & Orhan, M. F. (2022). An overview of regenerative braking systems. *Journal of Energy Storage*, *52*, 105033. doi:10.1016/j.est.2022.105033

Hamza, K., Laberteaux, K., & Chu, K. C. (2020). *On modeling the total cost of ownership of electric and plug-in hybrid vehicles*. Academic Press.

Han, Y., An, C., & Wang, J. (2018). A Review of Electric Vehicle Charging Infrastructure Planning. *Renewable & Sustainable Energy Reviews*, *94*, 935–946.

International Council on Clean Transportation. (2021). *Global Battery Demand and Supply Update 2021*. ICCT.

International Energy Agency. (2020). *Global EV Outlook 2020*. IEA.

Jang, Y. J. (2018). Survey of the operation and system study on wireless charging electric vehicle systems. *Transportation Research Part C, Emerging Technologies*, *95*, 844–866. doi:10.1016/j.trc.2018.04.006

Kazimierczuk, M. K., & Stando, T. (2019). Electric Vehicles Integration with Smart Grid: Opportunities, Challenges, and Solutions. *Energies*, *12*(10), 1942.

Larminie, J., & Lowry, J. (2017). *Electric Vehicle Technology Explained*. Wiley.

Liu, W., & Yang, Q. (2021). Electric Vehicle Charging Infrastructure Planning: A Review. *Energies*, *14*(2), 226.

Liu, Z., & Mi, C. (2019). *Electric Vehicle Integration into Modern Power Networks*. Springer.

Lutsey, N., & Nicholas, M. (2019). *Electric vehicle costs and consumer benefits in Colorado in the 2020–2030 time frame*. Academic Press.

Masias, A. (2018). Lithium-Ion Battery Design for Transportation. *Behaviour of Lithium-Ion Batteries in Electric Vehicles: Battery Health, Performance, Safety, and Cost*, 1-33.

Putrus, G., Suwanapingkarl, P., Johnston, D., Bentley, E., & Narayana, M. (2009). Impact of electric vehicles on power distribution networks. *Proceedings of the 2009 IEEE Vehicle Power and Propulsion Conference*, 827–831. 10.1109/VPPC.2009.5289760

Ribeiro, P., & Fontes, T. (2019). Electric Vehicles and Renewable Energy: A Review. *Renewable & Sustainable Energy Reviews*, *102*, 54–64.

Santinelli, M., & Sapountzakis, E. J. (2020). The Role of Electric Vehicles in Sustainable Urban Mobility: Challenges and Opportunities. *Sustainability*, *12*(17), 6729.

Sharma, A., & Sharma, S. (2019). Review of power electronics in vehicle-to-grid systems. *Journal of Energy Storage*, *21*, 337–361. doi:10.1016/j.est.2018.11.022

Sverdrup, H. U. (2016). Modelling global extraction, supply, price and depletion of the extractable geological resources with the LITHIUM model. *Resources, Conservation and Recycling*, *114*, 112–129. doi:10.1016/j.resconrec.2016.07.002

Sze, K. Y., & Xu, W. (2020). *Electric Vehicles and Power Grid Integration*. CRC Press.

Chapter 6
Introduction to Smart Grids and E-Mobility Ecosystem

Leo John
https://orcid.org/0000-0002-4071-9920
Saveeha School of Engineering, India

Fairosebanu Abdul
Idhaya College for Women, India

Jaffer A. Alzubi
https://orcid.org/0000-0001-6724-1421
Al-Balqa Applied University, Jordan

ABSTRACT

As transportation, communications, and energy increasingly overlap, new business models may emerge, necessitating new tools and expertise. Researchers predict these will center on three key aspects: external connectivity, internal information and communication technology (ICT) architecture, and human-machine interaction (HMI). In this chapter, the authors thoroughly examine these themes, bringing together insights from requirements analysis and prototype creation to share with the reader. This work also outlines the skills and techniques that will be essential for capitalizing on the many opportunities presented by the imminent market disruption. Data processing, data mining, and data sharing "vehicles" will play a crucial role in the future of our mobility partners. New HMI and driver-assistance apps require a stable and reliable runtime environment, which must be provided by the vehicles of the future. This study further argues that the electric car is an opportunity in and of itself for the future of product lifecycle management and the new Industry 4.0.

DOI: 10.4018/979-8-3693-2611-4.ch006

1. INTRODUCTION

Rapid urbanisation in the World could be a major contributor to the country's recent economic success. The term "electro mobility" (or "e-Mobility") refers to the concept of electric engines of cars and fleets made possible by fully electric technologies, in-vehicle technology for communication and information, and integrated infrastructures. One of the essential factors in the widespread adoption of EVs is the availability of charging stations. The charging system consists of the necessary equipment and software that monitors and controls the energy flow between the vehicle and the power grid. In a typical power electric system, the current only flows in one direction. The smart grid's two-way flow promotes inter-sectoral coordination and thereby increases the grid's adaptability, efficacy, credibility, safety, and cost-effectiveness. When these smart networks are in place, there will be better and faster two-way communication between the customer and the electric company. Everyone will have access to up-to-date data on their energy bills. New services, made possible by smart grid technology, can adjust to these transitions. The transmission of data across electrical lines, the use of electric cars, the exchange of remote sensing data, and the creation and transmission of electricity are all such services.

The introduction of EVs has the potential to affect the grid positively, and this progression towards smart grid technology is necessary to realize and regulate the dynamic interactions in smart power grids and the monitoring process. A significant change in the automobile industry's value creation cycle may result from the transition to electric mobility. The inherent energy efficiency of Electric Vehicles (EVs) makes the switch appealing. Electric vehicles convert more than 77% of the grid's electricity into battery power, compared to a normal petrol automobile's 12-30% conversion efficiency. Combining electric vehicle technology with connected cars and autonomous driving systems increases the disruptive potential of the former. The conventional automobile, infrastructure, and energy industries converge, creating a new mobility sector. The purpose of these reports (Case studies on E-bus development and operation in Chinese cities,2023) is to provide a comprehensive overview of the e-bus deployment process in a specific city, including details about the planning, implementation, and operation of the e-bus system. They may also include information about the benefits and challenges of e-bus deployment, as well as lessons learned and recommendations for future e-bus projects

The real potential of Electric Vehicles (EVs) to provide environmentally friendly and sustainable mobility is wholly dependent on the Electric Vehicle (EV) charging stations and clean power (Venugopal, 2021). Due to vehicle-to-grid (V2G) technology, electric cars can interact directly with the power grid, electric vehicles (EVs) have recently increased their prominence in electric power distribution as the preferred mode of transportation. This is mainly due to Electric Vehicles' cutting emissions, reduced fuel costs, and higher efficiency than ICE-based vehicles. More widespread adoption of EVs could pave the way for their use in power grid operations, energy storage, and ancillary service provision. This feature of electric vehicles (EVs) may be exploited further in the current smart grid (SG) context by integrating supply-side management (SSM) via EV assimilation (Mohanty et al., 2022). Figure 1 shows the components of e-mobility ecosystem.

Thus, during the golden era of the automotive industry, a stronger focus was placed on building and deploying ICE vehicles; however, the drawbacks of higher emissions and decreasing crude oil reserves have now made it infeasible to continue operating and maintaining vehicles powered by such engines. Investing in the research and development of EV technologies has been a top priority in addressing the world's pressing environmental, economic, and sustainability issues. The potential to lower greenhouse gas emissions and successful applications in the energy and transportation sectors have boosted the profile of electric mobility (Barreto et al., 2022).

Figure 1. Electric mobility ecosystem

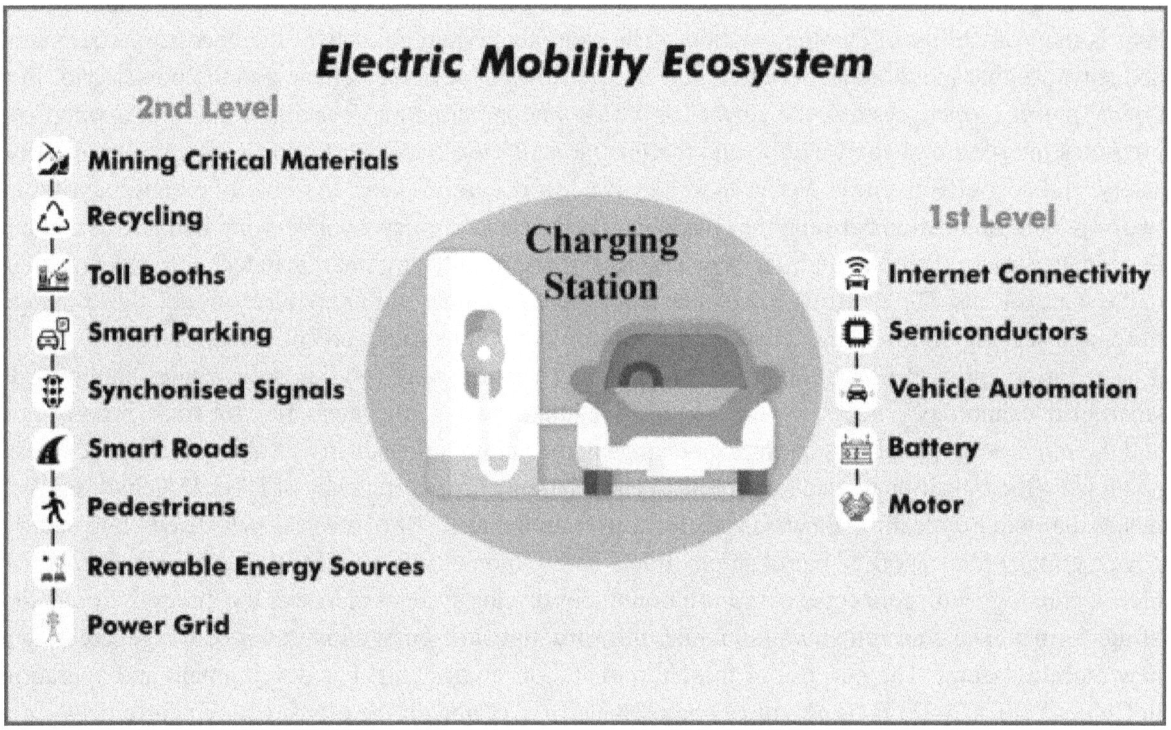

With the recent improvements to the energy infrastructure and the ultimate smart grid (SG) architecture, EV integration is now possible. Without any noticeable disruptions, both as dynamic and neutral participants in the power grid in standalone and grouped or integrated configurations. Electric vehicles' single-direction control, two way control, and power system capabilities have helped utilities realise their potential for supporting more efficient consumption patterns and providing grid services. The following are some of the many reasons that have been cited for the widespread adoption of EVs and their effective application:

1. Owners of electric vehicles might benefit financially from their involvement in a wide range of services made possible by these innovations.
2. Global warming, energy protection, and the uneven distribution of fossil fuels based on carbon are all problems that may be mitigated by electrifying the transportation sector.
3. Vehicles may stay linked to the grid's technical foundation and take part in energy flow programs utilising their battery packs as source of renewable energy courtesy of the vehicle to grid principle because most cars spend up to 90% of their time lined up at their selected locations of plug-in stations (V2G).

4. Assisting the grid in times of crisis, many EVs can provide ancillary services like peak cutting, power management, frequency control, and whirling reserve behavior to the system's entities.
5. When electric vehicles are adopted on a large scale in a technically sound and methodical manner, the smart grid's operational and economic assessment improves considerably as in the key areas of reliability, effectiveness, and improvement of power systems.
6. V2G programmes allow for a more overall energy grid load usage curve by allowing electric vehicles (EVs) to either store the surplus energy produced by renewable power plants through a variety of plug-in architectures, or to deliver electricity during deficits in the generation.

The above information provides substantial detail on many topics related to EVs. Electricity transmission and supply grid upgrades improve throughput, resilience, cost, and durability. Smart Grid technology enhances both transmission and distribution. Via the use of information and communication technology and new methods of control. Power production, distribution, transmission, and users are all part of the SG, an interconnected network of data, interaction, control systems, and technologies that connect with technology and business operations. While the SG is optimised for use with large-scale community power supplies technology, other technologies, such as microgrids and nano grids, are customised to meet the requirements of smaller communities (Abdulla and Hassan, 2022).

This chapter's main contribution is its overview of the next-generation smart grid and electro mobility system. In addition, this paper provides insights into the current SG outlook, such as Vehicle to Everything (V2E), charging and discharging, supporting services, and power management in smart grid.

2. RELATED WORKS

Plug-in EVs, including fully-electric cars and plugged-in hybrid electric vehicles (PHEVs), will present a new chance to lessen reliance on oil and be pivotal in decarbonising the transportation sector. Future grid performance and load demands may be significantly impacted. Long-term investments in the electric power system and accompanying PEV recharging infrastructure will be necessary for power providers and business investors to keep up with this rising trend. PEVs may be utilised as decentralised energy storage devices when the electrical grid and local grid technologies are combined with vehicle-to-grid (V2G) technology, allowing the grid to send power preserved within their battery packs when electricity is short supply. By providing a low-cost option for control services, rotating reserves, and peak-having capacity, PEV-powered electrical grid and local grid can contribute to cost savings in the energy system. PEV charging infrastructure and solar panel technology are two technological, environmental, and economic obstacles to this progress. Conquering these obstacles is crucial for developing V2G supply and may have far-reaching consequences for grid and load utilisation in the future (Lu and Hossain, 2015). A particular emphasis was placed on recent developments in the technology of electric vehicle batteries and charging procedures(Sanguesa et al., 2021). The researchers (Kalapsazova et al., 2023) researched the most up-to-date methods of providing electric vehicles with bidirectional onboard charging capabilities. The research (Srinivasan et al., 2021) provided, the many energy, connection, and communication system technologies that may be necessary for the incorporation of EVs in the SG of the future were brought to light. In addition, research such as (Mojumder et al.,) has provided a comprehensive breakdown of the potential effects that the incorporation of EVs into the grid will have (Sing et al., 2022)examined the present condition of other subjects, including the electric vehicle (EV) market, EV recharging infrastructures, and their respective standards. Another review instance can be found in (Ruggeri et al.,2021), which investigated the development and progression of electric mobility in six smart cities in Europe.

Techniques that rely on the regulation of distributed generation are shown in investigations such as (Clou, 2023), (Khan et al., 2022). (Meliani et al., 2021). (Torres et al., 2022) which provide approaches that enable energy management of SGs. The big problems and potential solutions associated with smart distribution grids are discussed in [Roomi et al.,2023] from the perspective of technological tools. They present the current technical level of innovative grid communications and identify gaps in this study area. The environmental, economic, and energy viability of building-integrated solar systems is provided in work (Lian et al., 2022). In this study (Wu et al., 2023), the authors propose a method for optimizing regional integrated energy systems (RIES) that considers the uncertain nature of renewable energy sources and the presence of electric vehicles. This method strikes a good balance between the system's economic viability and its low environmental impact. Uncertainty in energy generation makes it difficult to foresee future demand for decentralised power plants. Using a Long Short-Term Memory (LSTM) neural network for power generation and forecasts, the study (Chein et al.,2023) suggested a blockchain-based peer-to-peer power transaction market with automatic execution.

3. NEED FOR SMART GRIDS IN E-MOBILITY

The cost of conventional energy sources is higher than that of sustainable power plants. Due to Smart Grids(SG) technologies, sustainable power plants may be enhanced in the electrical network, which means SG can take credit for technological progress, and energy prices will go down over time.

Advantages of SG

- Manufacturing, installing, maintaining, and operating the SG and its components will provide both skilled and unskilled labourers opportunities. It will help the company expand, and it will also provide cutting-edge technological solutions.
- Improved dependability and lower costs have contributed to a rise in customer satisfaction. In addition, SG empowers consumers with a more significant say over their energy usage.
- Increased efficiency and fewer power outages are also benefits of the smart grid.
- Smart Grid has a positive effect on the environment since it facilitates the use of alternative RERs and electric vehicles, both of which contribute to the reduction of greenhouse gas emissions. It reduces oil use because of efficient electricity generation.
- Smart Grid expands upon the possibilities of the current electrical system. Features like self-repair and preventative upkeep are integrated into them as well.

Figure 2 illustrates the fundamental benefits of the smart grid, which in turn contribute to the enhancement of the smart grid's characteristics.

Figure 2. Smart grid benefits

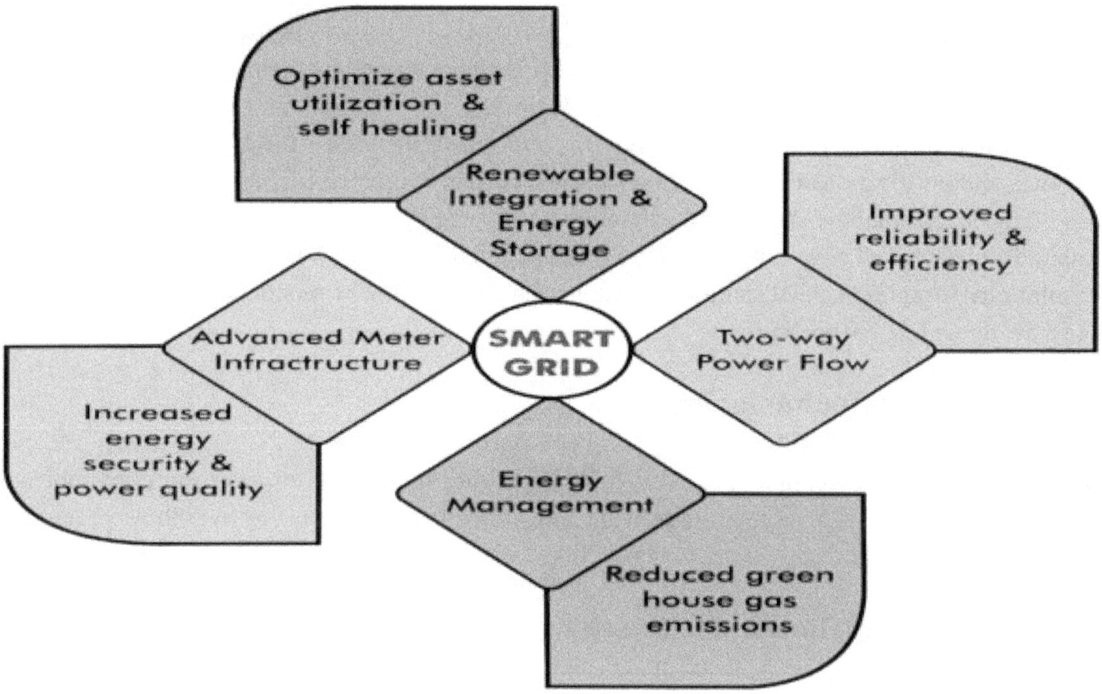

4. AUTOMATED GRID TECHNOLOGIES

The concept "smart grid" means a power transmission network that employs electronic and various emerging technology to observe and regulate the transmission of electricity from all stages of generation to meet the demands of unique customers.

4.1. Sensing and Measurement Technologies

The SG's top level focuses on sensing and measuring. Sensing and measuring in real time with high accuracy increases dependability, security, speed of analysis, and productivity. With the use of embedded sensors and automated systems, problems may be quickly identified and solved . By easing strain on the network, they also aid in preventing disruptions. This enhances the usefulness and performance of the SG these machines generate. Smart devices are used for both controlling and communicating inside the power grid system. They make sure that the home appliances they utilise are safe, reliable, and efficient. The heart of the SG system is the smart meter and accompanying software. In addition to enhancing power management, this feature also provides end users with increased agency over their own power use. Moreover, they ensure that clients are responsible for their actions by providing timely and accurate billing and information.

4.2. Intelligent Energy Monitoring and Control Systems (AMI)

Advanced Metering Infrastructure is another name for it (AMI). For a 2-way transmission of energy among industries and consumers in the intelligent grid, AMI is characterized by the pairing and connection of thousands of automated meters and transmission devices. The SG can monitor its energy use thanks to its sophisticated information management systems and sensor devices. An AMI network provides accurate data and notifications on interruptions and the corresponding rectification through smart meter sensors. Power restoration and problem-solving are both aided by this. Smart meters include software for monitoring electrical power, which can help locate and diagnose power problems more quickly and offer solutions. In addition, AMI facilitates quicker problem isolation by monitoring energy loss, making the grid available for self-healing.

4.3. Transmission Technologies

Various communication technologies exist, including those that allow for two-way communication, including complex design, and employ hardware and software. Power can flow in both ways across different demands in a power grid owing to bidirectional communication.

4.4. Smart Grid and Renewable Energy

The electric power grid may gain speed, sustainability, and dependability with renewable energy. Sustainable power plants in a local grid include photovoltaic (PV) panels, small wind generators, thermal or energy storage, integrated thermal and power, and regulated voltages. As much of the power generated by renewable sources originates from the sun and other natural resources that will remain forever, its availability is almost limitless. Research into renewable energy is a growing field, with more funding for finding and developing new renewable energy sources and improving existing ones.

4.4. Security for Intelligent Networks

Regarding SG, "protection" means securing connected devices from faults, overloads, and other accidental malfunctions. One of SG's power system protection applications, rapid fault identification, helps to minimise voltage fluctuations and blackouts. When several strategies are used to secure the country's physical infrastructures, this is known as the physical protection of SG. This sort of safety is concerned with preventing loss due to unforeseeable circumstances like mechanical breakdown, human error, natural disasters, etc. There are two main factors to think about while implementing such security. The system's reliability is one issue, and the absence of protection is another.

The SG must be well-planned, structurally sound enough to be resilient the strain, energy-efficient, environmentally friendly, reliable, innovative enough to find out where outages and problems are, cyber-attack proof, capable of real-time data monitoring and action, scalable, and interoperable with other energy sources. Many features of the automated grid need to be built in a less costly manner to foster a mutually beneficial relationship between energy producers, grid supervisors, consumers, and the environment. Many protection issues arise in SGs, including but not limited to assuring reliability, stability, and coordination. The development of efficient algorithms is the solution to all of these issues.

4.5. Features of Smart Grid

Advance Communication: The consumer's interaction with the network and the network's response to that interaction is crucial to this improved level of communication. For instance, if a consumer moves into a home, the electricity will be cut off at a specific time. Considering this, it's easy to understand the extent to which the intelligent grid's sensor and feedback devices work to collect information from the power grid.

Capability for auto-healing: in response to natural flaws or breakdowns without requiring human intervention; this is an essential feature of SG.

Ability to change and adapt: Ability to adapt to different conditions on the grid, especially in terms of the constantly shifting power and load flow direction.

Resilience: It means that intelligent grid's resistance to attacks.

Timely manner feedback and regulation: As a result of the fact that produced power can be affected by fluctuations as small as microseconds, the grid must be able to take live-time metrics and control the smart grid while keeping tabs on the central generator, the distribution generator, and any other plants connected to the electric network.

Dependable techniques: The intelligent grid should use reveal forecasts, which improve problem identification and allow the associated systems to auto-heal without intervention from a supervisor [16]. There is a consistent and dependable energy supply, and related risks are diminished.

Networking adaptability: SG technology's flexible networking makes distributed generation possible by facilitating a more efficient bidirectional energy flow. Solar photovoltaic (PV) solar power, fuel liquid cells, rechargeable batteries, wind generators, and several comparable energy suppliers are all examples.

Performance gains: Smart grids significantly improve the effectiveness of the electricity network and the power supply infrastructure. Demand-side management has profited immensely from SG. Popular algorithms include the voltage/VAR optimizer (VVO), which reduces the load on the distribution network during non-peak hours, and the smart metering infrastructure (SMI) technologies, which improve downtime management by analyzing consumption data. As a result, we are now making better use of available energy and materials than ever before.

Maximum load shedding and cost structure: At high and peak times, communications systems and monitoring technologies significantly reduce demand. Smart grids allow users to regulate power consumption. Likewise, the smart grid helps utilities reduce consumption and prevent voltage drops.

Sustainable development: Without the requirement for additional storage, the SG system is touted to be very flexible in renewable energy generation, allowing for linking several sustainable power plants such as photovoltaic solar and wind power.

The Supply and Demand System: A key feature of demand response is real-time automated interaction between consumers and power plants. This extends the useful life of electrical infrastructure and eliminates the need for costly backup generators .

Improved utilities in the coming years: The merging of powerful 2-way transmissions, combined with sophisticated sensor devices and emerging technologies, will greatly boost power production, powertrain, and distribution effectiveness, efficiency, reliability, and security.

4.6. Applications of Smart Grid

- Elevate the responsiveness and flexibility of the distribution and transmission systems.
- There will be less of an impact on the environment and the economy since distribution line and feeder faults and outages will be found and repaired more quickly.
- Makes it possible to combine and expand renewable energy systems like wind and solar power.
- Distributes work, so easing pressure on a massive scale. Figure 3 depicts the SG applications and characteristics.

Figure 3. SG features and applications

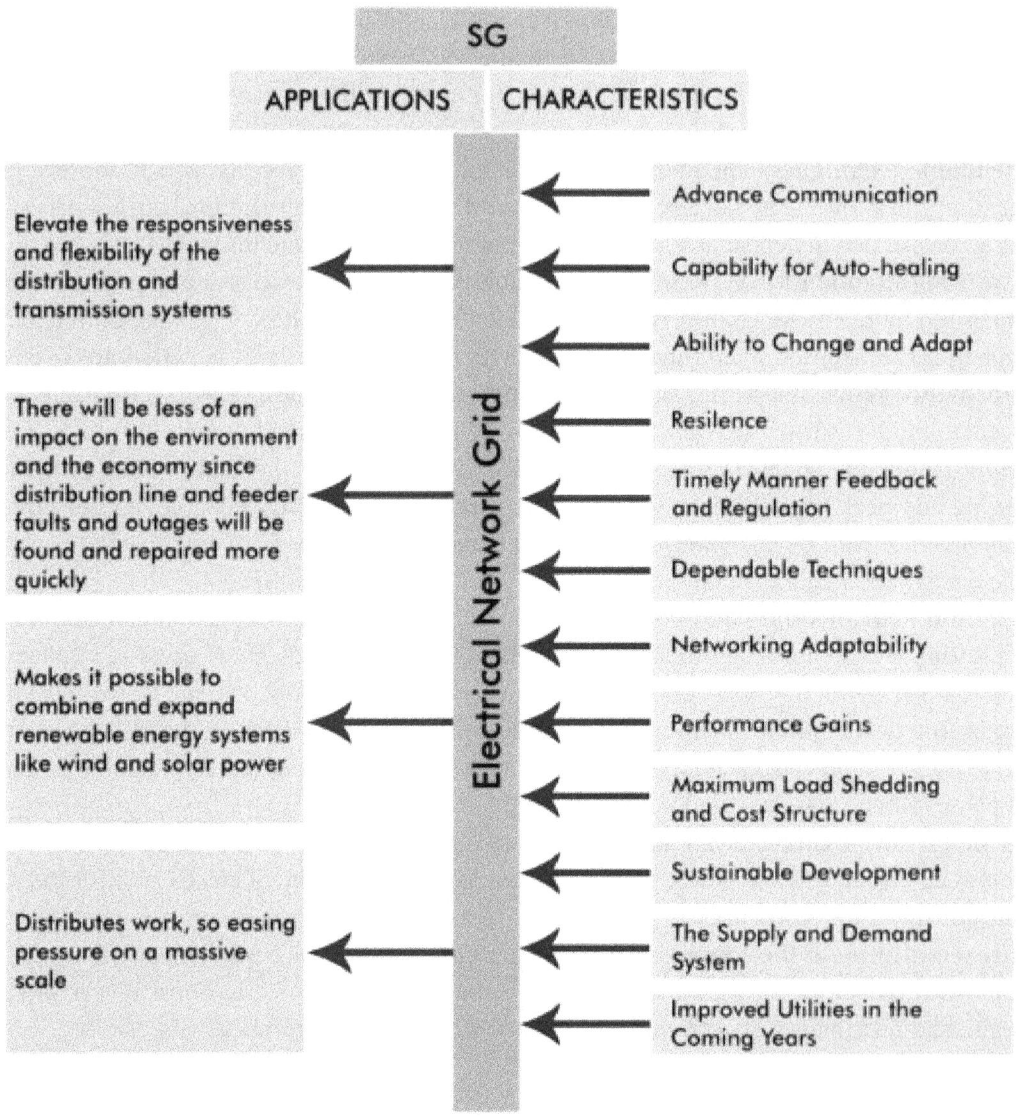

4.7. An Overview of Electric Vehicle Components

The electric engine, the smart control module (SCM), the power battery, the battery regulatory system (BRS), the adaptive battery charger, the structured wiring system, the stability control vehicle body, chassis, temperature control, braking system, and lubricating fluids, and the EV itself are the primary components. To go forward, an EV relies on an electric motor. In turn, the smart control module manages the traction motor. Basic knowledge of electromagnetism and thermodynamics is required to comprehend how electric motors work. Figure 4 presents the Figure 4 Electric Vehicle Components.

Figure 4. Electric vehicle components

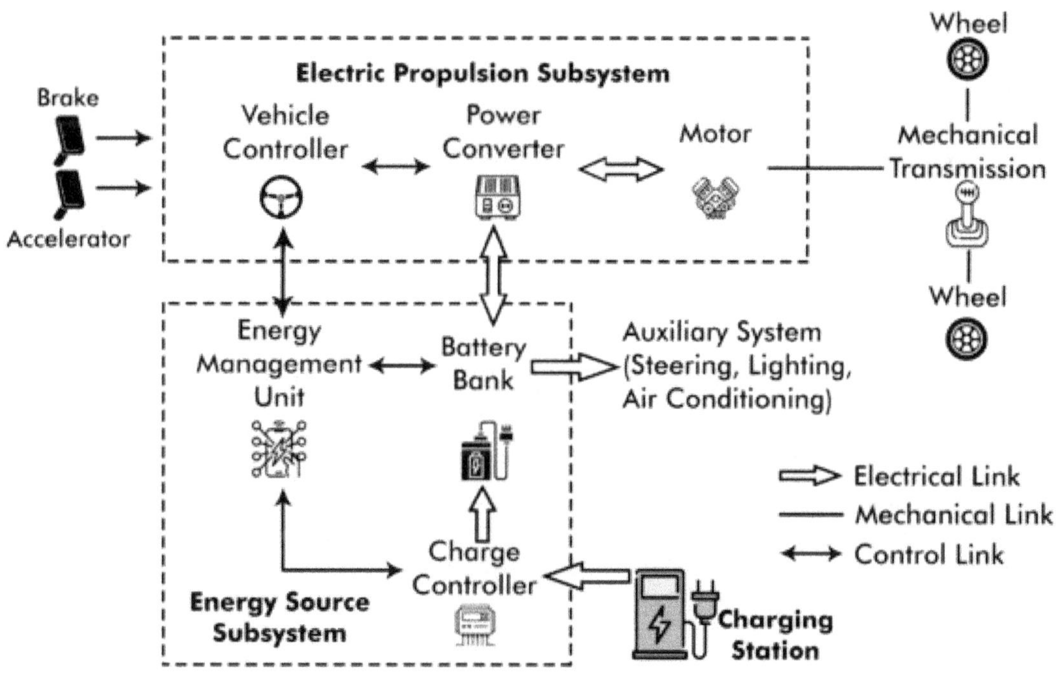

4.8. Working of an Electric Vehicle

An electric vehicle (EV) functions in much the same way as an automobile powered by an internal combustion engine. The dashboard gauges and smart control module(SCM) require the ignition key or a numeric keypad to activate.

Simply put the shifter into Drive or Reverse to start the car. The car could crawl like a gas-powered car if the brakes were freed. The SCM receives a signal transmitted by the accelerator pedal and transmits power and current from the battery pack to the electric engine in a manner proportional to the driver's foot pressure. Torque is sent to the EV wheels from the motor. Rapid acceleration is possible in an EV because electric motors have flatter power/torque curves than IC engines. When an electric vehicle driver releases their foot from the accelerator or presses the brake pedal, the vehicle's regenerative braking system kicks in.

The SCM channels the vehicle's kinetic energy to the battery pack via this function. To simulate the slowdown of an internal combustion engine, regenerative braking can be used.

One of its many advantages is the relative silence with which they travel. In most respects, commercial EVs are as easy to drive and maintain as their internal combustion engine counterparts.

5. ELECTRIC VEHICLE TECHNOLOGIES

Electric cars, often known as EVs, are driven by electromagnetism rather than internal combustion technology, which relies on combustion and pressure to move a vehicle. Internal combustion technology is known as "IC" technology. These automobiles use electric motors powered by electricity, which is usually stored in a battery. Electric motors, driven by rechargeable batteries and semiconductor-based voltage to current converters, provide propulsion for all-electric mobile energy platforms. E-mobility has several significant technological obstacles, the most important of which are the poor energy density of currently available battery technologies and the lengthy amount of time required for charging.

Figure 5. Energy storage in EV classifications

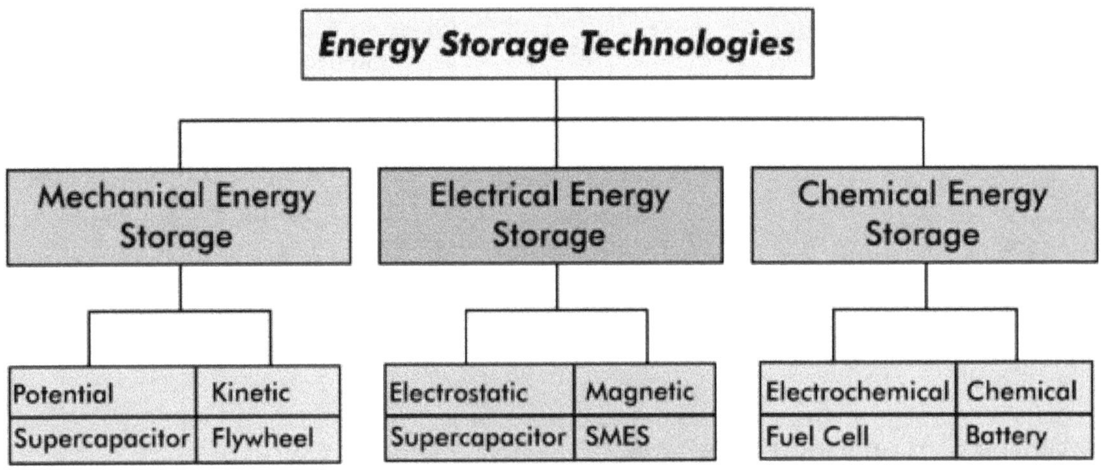

Figure 5 organizes the many energy storage methods now in use according to the type of energy being stored (electrical, mechanical, and chemical). Many technologies have diverse qualities and may be used for various tasks. When it comes to charging and discharging speed, high-power storage devices have the upper hand over their high-energy counterparts. Superconducting magnetic energy storage (SMES), supercapacitors, and high-power batteries are all examples of high-power storage technology.

5.1. PEVs Effect on the Smart Grid's Distributed Energy Resources

Information and communications technology, distributed control systems, and peak load management all promise to make it simpler and less expensive to integrate PEVs into the power grid. In peak demand, the power stored in PEVs batteries may be sent back into the grid. V2G automation's ability to provide management services, whirling backups, and peak-reduction capability cheaply can aid in lowering power costs. Charging for PEVs may be relocated to off-peak hours using intelligent grid and local grid technologies, which flattens the daily capacity curve and drastically decreases the need for investment in both generations and the distribution network. The most promising close prospect energy loss mobility is the combination of plug-in electric vehicles with renewable energy produced from sustainable power plants like solar PVs and wind generators. V2G technology enables the storage and generation of energy from car batteries to mitigate the effects of variable renewable energy sources, improve grid stability, and minimise the need for expensive peak-hour power from the grid. To solve the problems of grid demand and electric car consumer acceptance, as well as to reduce greenhouse gas emissions, the future transportation system will use V2G technology embedded in the grid and other infrastructure. The utility infrastructure can only produce so much electricity to keep up with rising demand, and most of it comes from coal and natural gas power plants. The rapid increase in the number of PEVs in use would significantly impact investment in electrical network capacity and smart grid technology.

5.2. An Overview of Electric Vehicle Batteries

A battery module is a collection of cells that function as a single electrical and mechanical unit. The electronic driving systems are powered by a rechargeable battery comprising these units. Each battery cell has five main parts: the anode, the cathode, the separator, the terminals, the electrolyte, and the casing. Figure 6 shows the components of battery.

Batteries have a positive and a negative terminal. The electrolyte might be a fluid, gel, or even a solid. The liquid electrolyte has traditionally been used in conventional batteries such as nickel-cadmium (NiCd), lead-acid (Pb-acid), and others. Depending on the battery, the electrolyte could be acidic or alkaline. The electrolyte in many modern, high-tech batteries being developed for electric vehicles, a gel, paste, or resin is utilized.

The following is a list of some of the most notable characteristics of the traction battery used in EVs:

- Long-distance travel is possible on a single charge because of the high energy density that can be achieved.
- Because of the high energy density, the EV can accelerate and increase its power output by virtue of a deep discharge that enables constant power.
- With rigorous safety standards and built-in, maintenance-free safety features, this battery can withstand more recharges and last longer.
- From an environmental perspective, this battery is often regarded as recyclable.

Figure 6. Components of battery

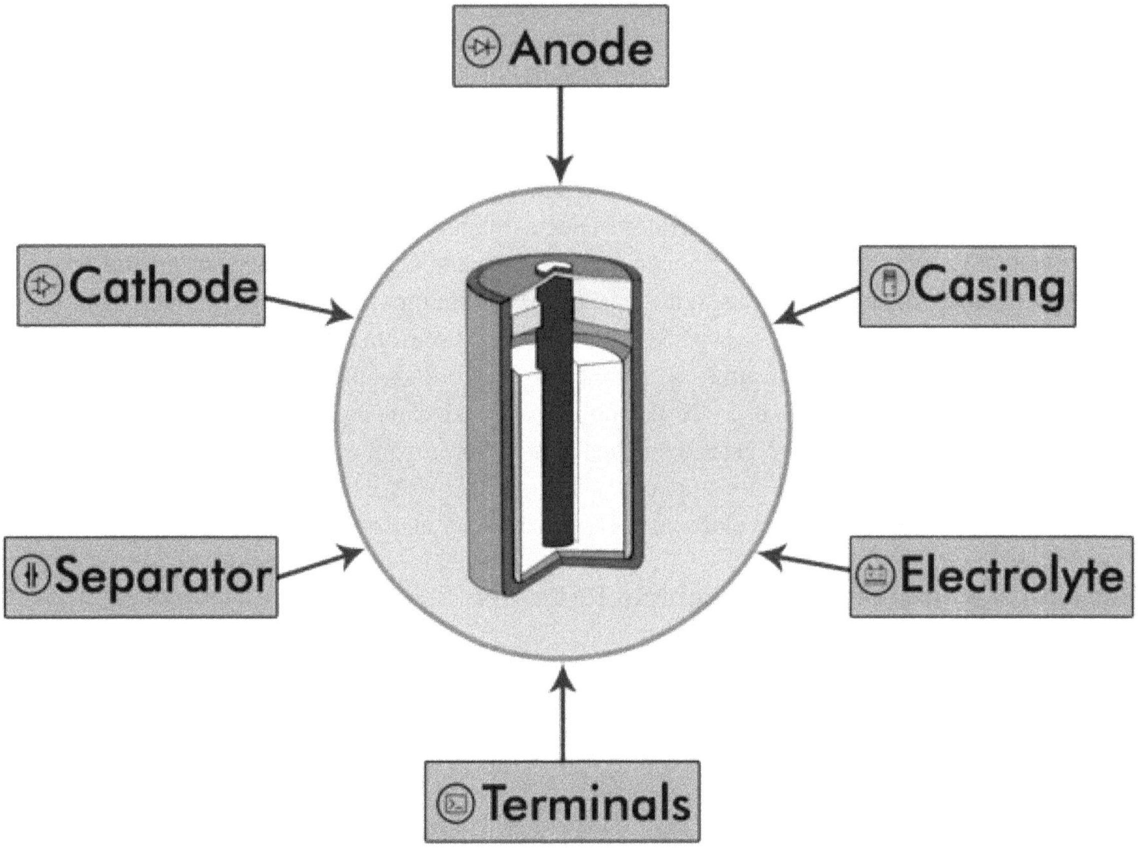

5.3. Classification of Electric Vehicle Batteries

5.3.1. Lithium-Ion Batteries

Battery packs like this are the standard for electric vehicles. The vehicles are extremely efficient because to the lithium-ion batteries' excellent power-to-weight ratio. The battery also outperforms its competitors when temperatures are really high. Very useful in batteries for electric vehicles because of its high energy density. The battery has a relatively low rate of self-discharge, making it more capable of holding a charge than competing options. Also, the majority of the materials used to construct Li-ion batteries may be reused or recycled. As electric vehicles are manufactured primarily to reduce harmful emissions, this is good news. These components are often found in plug-in hybrid vehicles and battery-power vehicles.

5.3.2. Nickel-Metal Hydride Batteries

Hybrid vehicles are a subset of electric automobiles that combine gasoline and battery power. These automobiles make up the bulk of the market for NIMH batteries, but BEVs can also utilize them. There is no need to connect this particular battery to an external power source to be recharged. The battery is charged primarily by the vehicle's speed, regenerative braking, and wheels. NIMH batteries have a longer lifespan and are less dangerous in harsh usage than Li-ion batteries.

These batteries are more costly, produce more heat in hot environments (like a vehicle), and have a greater self-discharge rate. As a result of these drawbacks, NIMH is less desirable than other battery types when it comes to cars that must be charged from an external source. This is also the primary motivation for using these batteries in hybrid vehicles.

5.3.3. Lead-Acid Batteries

Sealed Lead Acid(SLA), often known as lead-acid, is an old standby among rechargeable batteries. They are much bulkier and less long-lasting than NIMH or lithium-ion alternatives. Automakers continue to employ it primarily because of its low cost. In addition, as they have been in use for so long, improved methods for installing and utilizing the batteries have been developed, leading to a higher quality final product. Although several promising initiatives are in the works, SLA batteries are yet to be common in consumer electric vehicles. This battery type is typically used in business vehicles as a supplementary energy storage device. These batteries can be recycled, which is another perk.

5.3.4. Ultracapacitors

A polarised liquid stores energy between an electrode and an electrolyte in ultra capacitors. When more of a liquid's surface is exposed, it may store more energy. In addition to assisting with energy recovery during braking, acceleration, and hill climbing, ultracapacitors can boost automobiles' power. With their ability to aid electrochemical batteries in leveling load power, they may also be used as supplementary energy storage devices in electric-drive cars.

5.3.5. Fuel cell Technology

Similarly to batteries, fuel cells may store energy, but unlike batteries, they never die and can be used repeatedly. With a constant fuel supply, they generate electricity and thermal power. In a fuel cell, an electrolyte separates two electrodes known as the anode and cathode. Fuel, such as hydrogen, is fed to the anode, while combustion air is supplied to the cathode. It uses a catalyst on the anode to release electrons and protons from hydrogen molecules moving in opposing directions on the cathode. To create water and energy, the electrolyte allows protons to travel to the cathode, mixing with oxygen and electrons.

6. ELECTRIC VEHICLE ENGINE TYPES AND ITS FEATURES

One that is propelled by an electric motor(s) is called an electric vehicle (EV). A collector system can provide electricity from outside the vehicle, or it can run on its own batteries (Solar panels are occasionally used to charge batteries, while fuel cells or generators are used to convert fuel to power.). EVs may be anything that runs on electricity, including cars, trains, boats, planes, and even spaceships. Figure-7 shows the major types of electric vehicle. An electric vehicle gets its motion from an electric motor or motor(EV). A collector system can provide electricity from outside the vehicle or run on its batteries (Solar panels are occasionally used to charge batteries, while fuel cells or generators are used to convert fuel to power.). EVs, including cars, trains, boats, planes, and even spaceships, may run on electricity. Figure 7 shows the major types of an electric vehicles.

Figure 7. Major types of electric vehicle

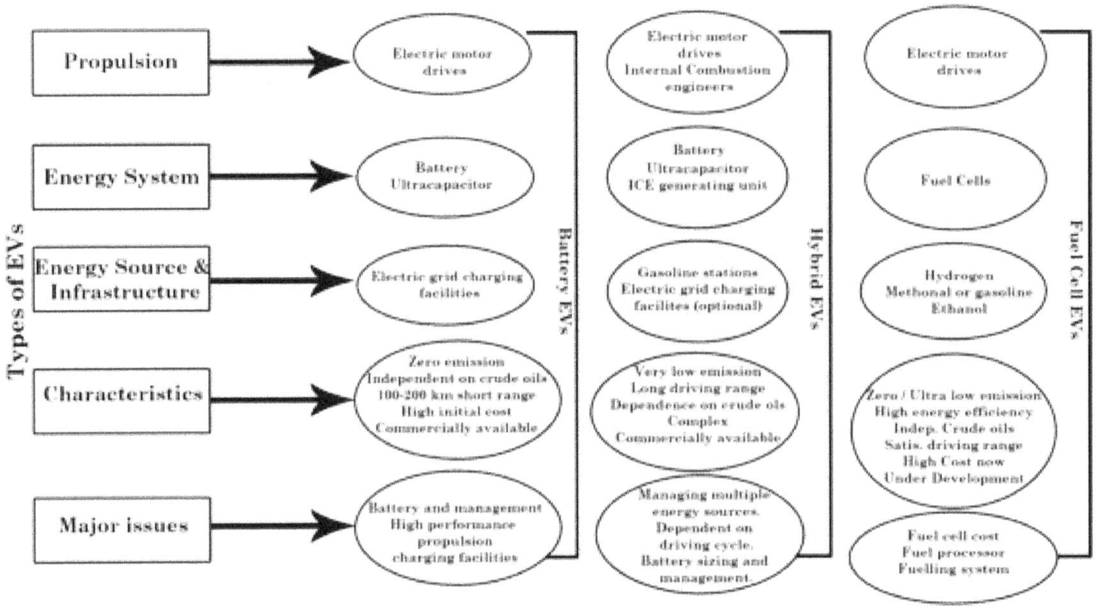

6.1. Battery EVs

The fundamental idea behind BEVs is straightforward. An electric battery stores energy for the vehicle's electric motor and controller. A battery charging equipment, which may be brought along or installed at the charging station, is often used to recharge the battery from the mains energy. In most cases, the controller will regulate the amount of energy sent to the motor and, by extension, the forward and backward velocities of the vehicle. This is what you would call a "2 quadrant controller," meaning it can regulate movement in both the forward and reverse directions. By recapturing energy and providing a smooth braking option, regenerative braking is very desirable.

6.1.1. BEV's Primary Components

Drive train, Control Module, Electric Motor, and Inverter

6.1.2. The Working of a BEV's Operation

The DC battery is changed into AC electricity to run the electric engines. The speed controller receives alert signals whenever the accelerator is depressed. The controller regulates the vehicle's speed by modulating the inverter's alternating current output to the motor at varying frequencies. The motor is connected to the turning of the wheels using a gear. As the brakes are used the electric vehicle decelerates, the engine becomes an inverter and refreshes the battery.

Several instances of BEV

Tata Nexon, MG ZS etc.,

6.2. Hybrid EVs

Many variations are available for a hybrid vehicle's combination of power sources. Most vehicles that use hybrid technology employ a combination of a traditional engine that runs on gasoline and a battery motor and generator. Hybrid vehicles use both a compression ignition engine and an electric motor. A gasoline supply powers the engine, while batteries power the motor. The engine and the electric motor work together to turn the gearbox. In turn, this powers the vehicle's wheels.

6.2.1. HEV's Primary Components

Powerplant, Generator, Motor, Power source consisting of a battery bank and a converter/controller, Gasoline storage and command center

6.2.2. The Working of a HEV's Operation

Series and parallel hybrids are the two most common configurations for hybrid cars. One or more electric motors powered by the battery, the generator unit powered by the internal combustion engine, or both propel the vehicle in a series hybrid. The wheels of a vehicle equipped with a parallel hybrid drive system can be moved by the internal combustion engine (IC) alone, by one or more electric motors, or by both the electric motors and the IC engine simultaneously. It is unnecessary to have as large of a battery in a series or parallel hybrid since it may be recharged by the vehicle's engine and generator while in motion.

Several instances of HEV

Toyota Prius(TP), Ford escape and fusion hybrid etc.

6.3. Plug-In Hybrid EVs (PHEVs)

PHEVs are frequently referred to as series hybrids: a motor and an engine power them. Traditional fuels (like petrol) and alternative fuels are also available (such as bio-diesel). The device may also be run on a rechargeable battery pack. This battery accepts external charging.

Two types of operation are available for PHEVs:

- Energy for the vehicle comes only from the electric motor and batteries in "All-electric Mode."
- Hybrid Powertrain, using a combination of electricity and gas or diesel

6.3.1. PHEV's Primary Components

Battery charger, control module, fuel tank, engine, and inverter

6.3.2. The Working of a PHEV's Operation

Plug-in hybrids run only on electric power until the battery is drained. Whenever the battery is depleted, the car reverts to normal. Plug-in hybrid vehicles (PHEVs) can get a charge from an external source of electricity, such as the vehicle's engine or brake energy regeneration. The electric motor doubles as a turbine when using the brakes, feeding that power back into the battery. The electric engine enhances the engine's power, permitting smaller engines for better fuel economy without sacrificing performance.

Several instances of HEV

Hyundai Sonata, Fiat 500e etc.[20].

6.4. Fuel Cell EVs

These vehicles, sometimes called Emissionless Vehicles, are FCEVs. Solar cell technology produces the necessary electrical energy to power the vehicle. This system immediately transforms the fuel's chemical energy into electricity.

6.4.1. Fuel Cell EV's Primary Components

An electric engine, solar battery stack, hydrogen backup tank, batteries, converter, and controller.

6.4.2. The Working of a Fuel cell EV's Operation

The fundamental concept of fuel-powered electric cars is similar to that of battery-powered electric vehicles, except fuel cells or metal air batteries are used instead of rechargeable electric batteries. In this fuel cell, liquid hydrogen is used as the power source.

Several instances of Fuel Cells EV

Honda Clarity Fuel Cell, Toyota Mirai etc.[20].

6.5. Solar EVs

The Honda Dream, which earned the international solar challenge in 1996, is a solar-powered car that is both pricey and limited to places with abundant sunlight. The efficiency of solar photovoltaic cells is continuously increasing while their cost is constantly reducing, hence although it is improbable that such a vehicle would be feasible for an everyday mode of transportation. It's not a farfetched notion to use solar battery energy that can be connected to the vehicle's body to maintain a commuter vehicle's battery power. This might become a viable proposal in the future, significantly if the cost decreases and the efficiency rises.

6.6. Schema for Vehicle-to-Everything

Vehicle-to-everything (V2E) communication refers to communication between an automated vehicle and anyone or anything the automobile can affect. It is a communication system for vehicles to integrates more specialized forms of information exchange such as V2I (vehicle-to-construction), V2V (vehicle to vehicle), V2D (vehicle to driving), V2P (vehicle to pedestrian), and V2N (vehicle to network), Figure-1.8 presents a fundamental architectural framework developed in this study to design an electric vehicle equipped with advanced communication technologies.

Figure 8. Framework model to design e-mobility ecosystem

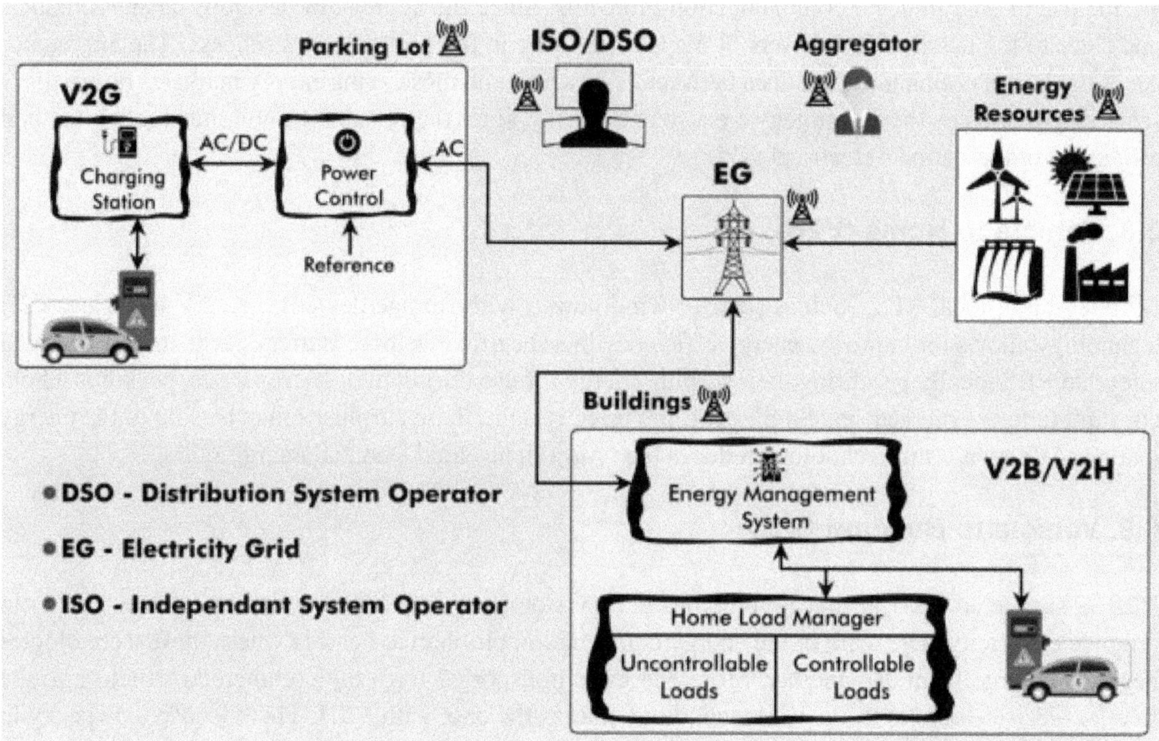

In practice, there are several benefits that society may gain from making the switch to electric transportation, including lower emissions of greenhouse gases and lower energy use. Nonetheless, there are many moving parts to consider when it comes to electric mobility, including networking, computing, architecture, protocols, battery modeling, charging, discharging, etc. Since the E-mobility system is built on numerous layers of interoperability—components, information exchange, function, and business—the transition is a complicated one.

Aggregator: The aggregator is in charge of communicating with the homes, parking lots, and power sources.

Distributed System Operator(DSO)/ Independent System Operator(ISO): The relevant Electricity grid communicates with the distribution system operator (DSO)/independent system operator (ISO).

DSOs and ISOs can coordinate with aggregators to seek optimal charging and discharging patterns for vehicles connected to their power networks. For instance, if DSO/ISO confirms the need for extra energy, it will communicate with the aggregator, who will then coordinate with the structures and parking lots to reduce the overload by making use of the available energy sources. When the grid has spare capacity, it notifies the aggregator so that electric vehicle charging may be planned around the surplus electricity.

6.7. Vehicle to Grid (V2G)

Using V2G technology, electric vehicles may send some of the energy they've stored in their batteries back to the grid, helping to keep the energy system in check. For this technology to function well, it is necessary for a number of different parts to live in harmony with one another, and this is only possible via the use of charging and communication protocols. Since the aggregator develops business models that cater to the needs of EV drivers, it plays a vital role in this technology's success. The aggregator keeps the line of communication open between EV owners and those in the energy market. The auxiliary services provided by this technology are also noteworthy, since they ensure the continued reliability and efficiency of the nation's electrical grid.

6.8. Vehicle to Home (V2H)

By integrating local RES, such as solar or wind power, with storage devices, i.e., EV batteries, V2H technology allows for improved energy efficiency in a specific dwelling. Battery packs in EVs can help integrate RES locally by storing their surplus energy for use during high energy costs, peak household consumption, or emergencies. Similarly, if the price is right, these surpluses may be sold to the energy market. As a result, this technology reduces the typical household's annual energy bill.

6.9. Vehicle to Building (V2B)

V2B is similar to V2H in that it combines energy storage with renewable energy sources (RES) to produce electricity. The scale of the energy consumption problem at hand is where these technologies begin to diverge from one another. More energy is transported from the commercial structure to the fleet of electric vehicles using V2B technology than is the case with V2H. There is also a disparity in the number of EVs used in these systems.

6.10. Vehicle to Everything (V2E)

Throughout the day, electric cars move about to take advantage of the varying rates of solar and wind power generation and energy exchange. Vehicle-to-Vehicle communication is the most popular use case for the V2E paradigm, but it isn't the only one it supports; the V2H, V2B, and V2G protocols do as well. The V2E idea is an approach to these interactions and the technology that enables them.

6.10.1. Limitations of V2E

While V2E has the potential to improve driver safety and convenience greatly, it also has several drawbacks, particularly regarding driver privacy. Hackers will inevitably find a way into V2E systems to steal data or even hold the network hostage, just as they have with every other technology that collects and stores massive amounts of data. Even if safety and efficiency are enhanced, there is still the essential privacy risk that automobile companies, governments, or those administering the infrastructure would have access to your driving data and habits.

Figure 9 provides the various features and open tools to concentrate and design a electric vehicle for the future.

Figure 9. Open-source software for designing, testing, and implementing electric vehicle features

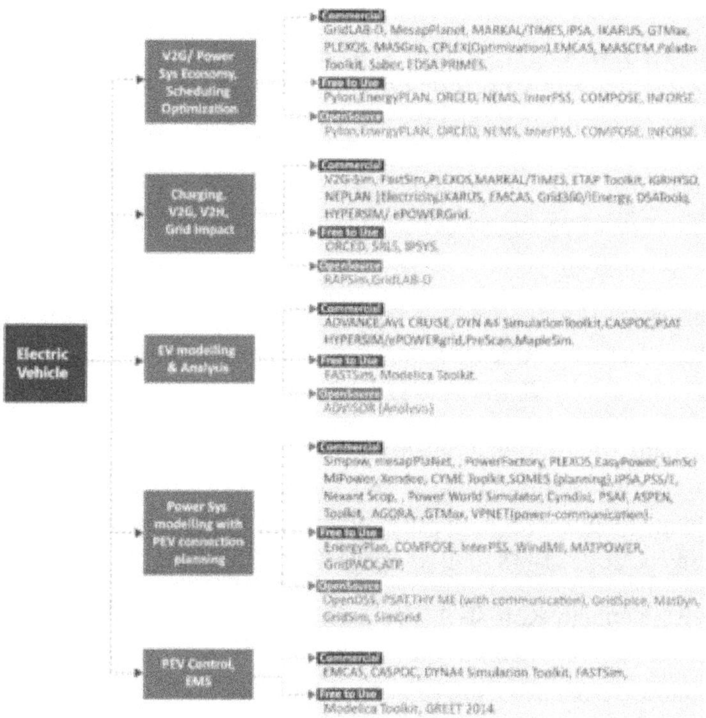

CONCLUSION

This article analyses smart grids, battery, and electric car technology. It shows how they all contribute to the overall picture of why we need a smart grid and how its many advantages may improve our daily lives. In addition, the basic architectural foundation for the design of electric vehicles is covered in detail, along with the many different types of electric vehicles and their parts. Finally, this article wraps up with various open-source tools that may serve as a tool for further assessing the qualities of electric vehicles (EVs).

REFERENCES

Abdullah, A.A., & Hassan, T.M. (2022). Smart Grid (SG) properties and challenges: An overview. *Discover Energy, 2*.

Barreto, R., Faria, P., & Vale, Z. (2022). Electric mobility: An overview of the main aspects related to the smart grid. *Electronics (Basel)*, *11*(9), 1311. doi:10.3390/electronics11091311

Case studies on E-bus development and operation in Chinese cities. (n.d.). https://repository.unescap.org/bitstream/handle/20.500.12870/6285/ESCAP-2023-RP-Case-Studies-E-Bus-Development-China.pdf?sequence=3

Chien, I., Karthikeyan, P., & Hsiung, P.-A. (2023). Peer to peer energy transaction market prediction in Smart Grids using blockchain and LSTM. *2023 IEEE International Conference on Consumer Electronics (ICCE)*. 10.1109/ICCE56470.2023.10043563

Clou, S. (2023). *Smart Grid Evolution in 2023: The Future of Energy Management*. Smart Energy International. https://www.smart-energy.com/industry-sectors/smart-grid/smart-grid-evolution-in-2023-the-future-of-energy-management

Kalapsazova, M., Kukeva, R., Harizanova, S., Markov, P., Nihtianova, D., Zhecheva, E., & Stoyanova, R. (2023). High-performance layered oxides for sodium-ion batteries achieved through combined aluminum substitution and surface treatment. *Batteries*, *9*(2), 144. doi:10.3390/batteries9020144

Khan, N., Shahid, Z., Alam, M. M., Bakar Sajak, A. A., Mazliham, M. S., Khan, T. A., & Ali Rizvi, S. S. (2022). Energy management systems using SMART GRIDS: An exhaustive parametric comprehensive analysis of existing trends, significance, opportunities, and challenges. *International Transactions on Electrical Energy Systems*, *2022*, 1–38. doi:10.1155/2022/3358795

Liang, Y., Saner, C. B., Kenneth Koh, T. J., Dou, H., Huang, Z., Wolfram, F. A., & Tan, J. (2022). *Energy-economic-environmental feasibility of building-integrated photovoltaic systems in Singapore. In 2022 IEEE PES Innovative Smart Grid Technologies - Asia*. ISGT Asia.

Lu, J., & Hossain, J. (2015). Introduction to the smart grid using pevs. *Vehicle-to-Grid: Linking electric vehicles to the smart grid*, 1–10.

Meliani, M., Barkany, A. E., Abbassi, I. E., Darcherif, A. M., & Mahmoudi, M. (2021). Energy Management in the smart grid: State-of-the-art and future trends. *International Journal of Engineering Business Management*, *13*, 184797902110329. doi:10.1177/18479790211032920

Mohanty, S., Panda, S., Parida, S. M., Rout, P. K., Sahu, B. K., Bajaj, M., Zawbaa, H. M., Kumar, N. M., & Kamel, S. (2022). Demand side management of electric vehicles in Smart Grids: A survey on strategies, challenges, modeling, and Optimization. *Energy Reports*, *8*, 12466–12490. doi:10.1016/j.egyr.2022.09.023

Mojumder, M., Ahmed Antara, F., Hasanuzzaman, M., Alamri, B., & Alsharef, M. (2022). Electric vehicle-to-grid (V2G) technologies: Impact on the Power Grid and battery. *Sustainability (Basel)*, *14*(21), 13856. doi:10.3390/su142113856

Roomi, M. M., Hussain, S. M., Mashima, D., Chang, E.-C., & Ustun, T. S. (2023). Analysis of false data injection attacks against automated control for parallel generators in IEC 61850-based Smart Grid Systems. *IEEE Systems Journal*, *17*(3), 1–12. doi:10.1109/JSYST.2023.3236951

Ruggieri, R., Ruggeri, M., Vinci, G., & Poponi, S. (2021). Electric mobility in a smart city: European overview. *Energies*, *14*(2), 315. doi:10.3390/en14020315

Sanguesa, J. A., Torres-Sanz, V., Garrido, P., Martinez, F. J., & Marquez-Barja, J. M. (2021). A review on electric vehicles: Technologies and challenges. *Smart Cities*, *4*(1), 372–404. doi:10.3390/smartcities4010022

Singh, P. P., Wen, F., Palu, I., Sachan, S., & Deb, S. (2022). Electric vehicles charging infrastructure demand and deployment: Challenges and solutions. *Energies*, *16*(1), 7. doi:10.3390/en16010007

Srinivasan, M., Stonier, A. A., Visalaxi, G., Revanth, M., Sanjeevkumar, K., Sinduja, B., & Subiksha, S. (2021). Design of bidirectional battery charger for electric vehicle. *IOP Conference Series. Materials Science and Engineering*, *1055*(1), 012141. doi:10.1088/1757-899X/1055/1/012141

Torres, B. S., Borges da Silva, L. E., Salomon, C. P., & de Moraes, C. H. (2022). Integrating smart grid devices into the traditional protection of distribution networks. *Energies*, *15*(7), 2518. doi:10.3390/en15072518

Venugopal, D. S. (2022). *Building an electric mobility ecosystem in India*. Deccan Herald.

Wu, G., Yi, C., Xiao, H., Wu, Q., Zeng, L., Yan, Q., & Zhang, M. (2023). Multi-objective optimization of Integrated Energy Systems considering renewable energy uncertainty and Electric Vehicles. *IEEE Transactions on Smart Grid*, *14*(6), 1–1. doi:10.1109/TSG.2023.3250722

Chapter 7
Study of Power Distribution System Resilience in the Presence of E-Mobility Ecosystems

Vandana Kumari
Indian Institute of Technology, Guwahati, India

Sanjib Ganguly
https://orcid.org/0000-0002-7939-7308
Indian Institute of Technology, Guwahati, India

ABSTRACT

Power distribution networks (PDNs) are most susceptible to extreme events, such as cyberattacks, hurricanes, floods, and wildfires, in a power system. Enhancing the resilience to withstand and recover after disruptive events is crucial for planners and operators of PDNs. The resilience-oriented planning and operation can enhance the resilience of PDNs. Furthermore, the significant EV integration, in recent years, provides a potential for improving resilience by restoration in PDNs. Considering the numerous studies regarding deployment of EVs for load restoration after disruptive events, this chapter provides the idea of resilience and the vulnerability of different components of PDNs due to various HILP (high impact, low probability) events. Furthermore, EV application to enhance the resilience of PDNs is reviewed with a detailed discussion of the methods, objectives, and solution strategies. In addition, a qualitative assessment of the studied literature is provided to enhance the comprehension of state-of-the-art resilience improvement methods of EV-enabled PDNs.

DOI: 10.4018/979-8-3693-2611-4.ch007

Abbreviations

CACG: Column and constraint generation
DG: Distributed generation
DSM: Demand side management
DSO: Distribution system operator
ESS: Energy storage system
EV: Electric vehicle
HILP: High impact low probability
MEG: Mobile emergency generator
MESS: Mobile energy storage system
MILP: Mixed integer linear programming
MINLP: Mixed integer non-linear programming
MIQCP: Mixed integer quadratic constraint programming
MISNLP: Mixed integer stochastic nonlinear programming
MISOCP: Mixed integer second-order cone programming
MPS: Mobile power source
PDN: Power distribution network
NR: Network Reconfiguration
TN: Transportation Network
TSROP: Two stage robust optimization problem

1. INTRODUCTION

In recent years, the rate and intensity of HILP events, such as earthquakes, hurricanes, tornadoes, floods, wildfires, and other natural or man-made attacks, have intensified worldwide. Table 1 lists the count of distinct natural calamities recorded in 1970 and 2022 (Ritchie et al., 2022). The combustion of fossil fuels, deforestation, and livestock farming are exerting growing influence on the earth's climate and temperature. This contributes substantial quantities of greenhouse gases to those that occur naturally in the atmosphere, intensifying the greenhouse effect and global warming. The primary cause of climate change is the greenhouse effect, where certain gases in the earth's atmosphere function similarly to the glass panels of a greenhouse. These glasses trap the sun's heat, preventing it from dissipating into space and thereby leading to global warming. The main factor behind the surge in the number of natural calamities is the changing climate (Panteli and Mancarella, 2015a). Natural disaster adversely affects both lives as well as critical infrastructures of society. The power system is among most vital infrastructures. The frequent natural disaster events cause grievous damage to power system infrastructure, resulting in widespread and extended power outages (Panteli et al., 2017). Therefore, to secure the power system from these unpredictable severe events and to keep light on even during and after the event, the power system's resilience comes into relevance. Power system resilience is the ability of the system to withstand disruptive events and recover quickly after the event (Gholami et al., 2018).

The PDN is one of the subsets of the power system. It is most susceptible to HILP events due to aging infrastructure, radial topology, and few reconfiguration choices (Poudel, 2020). HILP events can damage or destroy PDNs, which can cause interruption of electrical service for multiple days, weeks, or even longer. These outages can have serious consequences, including disruptions to transportation systems, communication networks, healthcare facilities, and other critical services (Lin et al., 2018). These can also lead to economic losses for businesses and individuals relying on electricity. Consequently, it is crucial to improve the resilience of PDNs. Research is mainly focused on assessing and enhancing resilience of PDNs (Kandaperumal and Srivastava, 2020). Numerous resilience metrics are available in literature to assess the resilience of PDNs. Strategies for improving the resilience of PDNs are broadly classified into planning and operational strategies (Bhusal et al., 2020). Planning strategies focus on hardening and installation of distributed generation resources to reinforce PDNs against future severe events. Operational strategies aim to develop optimization-driven methods that use the available resources of PDNs in the face of extreme events. Operational strategies give prompt and economic measures to minimize the effects of severe events on PDNs (Panteli et al., 2017). The evaluation of resilience and strategies for enhancing it are discussed in detail in section 4.

Installing energy storage systems, such as batteries, is an effective planning measure that can improve the resilience of PDNs (Nazemi et al., 2020). However, installing batteries solely to enhance resilience is not economically viable. Recent advancements in mobile power sources (MPSs), such as EVs, mobile energy storage systems (MESSs), and mobile emergency generators (MEGs), allow battery units to be transported via public transportation. These MPSs have sparked interest in exploring their potential to improve power distribution networks' resilience (Kim and Dvorkin, 2019). EVs have grown over the last decade due to their several advantages, such as reducing carbon emissions and enabling investment in renewable energy (Wu and Lin, 2021). EVs stay immobile in parking areas for multiple hours daily, allowing them to manage their charging and discharging through vehicle-to-grid technology when they are parked in parking lots (Hussain and Musilek, 2022). This can aid in the operation of the PDNs. EV storage systems can also be utilised if required during emergencies to supply power to the critical loads of PDNs. Aggregators can group multiple EVs to create a resource that can serve regional grids and offers services that would not be possible with a single EV (Razeghi et al., 2021). EVs can improve survivability just after an event and the speed of recovery of PDNs during the restoration phase of resilience.

Table 1. Reported number of natural disaster events (Ritchie et al., 2022)

Natural disaster types	Year – 1970	Year – 2022	Relative Change in the Number
All disasters	76	323	+ 325%
Earthquake	12	28	+ 133%
Flood	31	149	+ 389%
Wildfire	1	14	+ 1300%
Storm	24	96	+ 300%

2. VULNERABILITY OF DIFFERENT COMPONENTS OF PDNs TO HILP EVENTS

HILP events are those events whose probability of occurrence is low and cause a high impact on the system. HILP events, such as earthquakes, hurricanes, tornadoes, floods, wildfires, and other natural or man-made attacks, can cause extensive damage to PDNs. Generally, a PDN consists of several key components: distribution lines (overhead/underground), distribution poles, service transformers, distributed energy resources, substation transformers, measurement devices, circuit breakers and protection relay devices. Among these components, distribution lines and poles are most vulnerable to natural disaster events (Ouyang and Dueñas-Osorio, 2014). This section discusses an overview of the influence of diverse HILP events on different components of PDNs.

2.1. Hurricanes

Meteorological records indicate that hurricanes typically originate in the ocean and then move towards the continent. Once the hurricane falls on land, its intensity rapidly diminishes as it moves further inland. PDNs in coastal areas get severely affected due to hurricane events. The probability of failure lines or poles relies on factors such as the hurricane's intensity, the distance from the hurricane's centre, and the infrastructure's strength. High-speed winds can cause distribution lines and poles to break, leading to power outages (Najafi Tari et al., 2021). Trees and other vegetation uprooted or blown down during hurricanes can also damage lines and poles of PDNs (Ma et al., 2018).

2.2. Earthquakes

The severity of the impact of an earthquake on the PDN depends on the magnitude of the earthquake, the distance from the epicentre, the type of soil, and the strength of the infrastructure (Yang et al., 2020). An earthquake can damage transformers, circuit breakers, and other equipment at substations, causing them to fail and disrupt the power supply (Oboudi et al., 2021). An earthquake may also damage distribution lines and poles (Erenoglu et al., 2022).

2.3. Wildfires

During the summer season, high-temperature and forested areas are susceptible to wildfires. PDNs in those areas are vulnerable to wildfire (Trakas and Hatziargyriou, 2018). Wildfires can affect PDNs in two ways. The first one is that the fire directly burns wooden poles. The second one is that it reduces the thermal rating of distribution lines. If the distribution line's thermal limit violates, it gets tripped.

2.4. Floods

In the rainy season, cities are susceptible to severe flooding caused by continuous heavy precipitation or dam breaking. The impact of floods on the PDN depends on the extent of the flooding, the duration of the flooding, and the location of the infrastructure. Floodwaters can damage substations' transformers, switchgear, and other equipment (Movahednia et al., 2022). It can erode the soil around power poles and supports, causing them to tilt and collapse, leading to interruptions in the power supply (Souto et al., 2022). Floodwaters can damage overhead power lines by forcing them to sag, sway or break, leading to power outages. Floods can also affect the ability of utility crews to access and repair damaged equipment, which can prolong power outages.

2.5. Snowstorms

During the winter, regions with temperatures below 0°C and high moisture levels experience snowstorms. Heavy snow and ice can weigh down power lines, causing them to break and fall. This can lead to power outages and pose a safety hazard. Even if the conductor remains physically intact, it gets trips due to insulation breakdown due to ice covering (Chen et al., 2010).

2.6. Cyberattacks

The growing reliance on information and communication technologies within power systems has increased exposure to cyber-attacks (Ma et al., 2023). The power system is susceptible to cyberattacks, such as malware attacks, service denial (Lu et al., 2010), false data injection (Zhao et al., 2018), and more, leading to malfunction or widespread blackouts. Power companies and governments must take proactive measures to prevent cyberattacks and ensure the security and resilience of PDNs. Ensuring a continuous power supply is crucial for the operation of critical loads. Therefore, it is essential to diminish the influence of HILP events on PDNs and recover as quickly as possible.

3. RESILIENCE OF POWER SYSTEM

This section provided a detail explanation of the resilient power system, covering its definition, features, and resilience curves.

3.1. Definition and Features of Power System Resilience

Resilience refers to an object's ability to bounce and return to its original shape or position after experiencing stress (Panteli and Mancarella, 2015b). The idea of resilience in the power system realm is relatively recent. Several organizations have provided different definitions of power system resilience (Afzal et al., 2020). In Presidential Policy Directive 21, power system resilience is defined as the ability of the system to withstand, absorb, adapt to, and promptly recover from significant disruptions caused by natural disasters and deliberate cyber-physical attacks (Obama, 2013). Understanding the key features is crucial due to the lack of a universal consensus on how to define power system resilience. Robustness, resourcefulness, redundancy, rapidity, and adaptability are key features of resilience (Afzal et al., 2020). Figure 1 illustrates different features of resilience in various phase of resilience.

Robustness is related to the strength of system. It is ability to absorb turbulence and maintain stable power supply. Resourcefulness is the capability to mobilize resources to supply power to end users in accordance with their priorities. Redundancy involves excess capacity and a backup system. Rapidity is the capability to recover speedily. Adaptability is the capability to plan, prepare and execute an adaption plan after learning from past experiences.

Figure 1. Features of resilience
(Afzal et al., 2020)

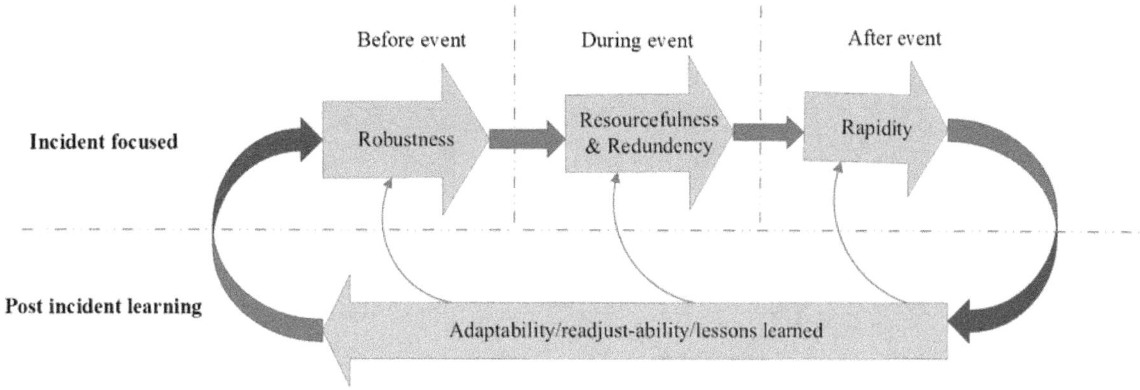

3.2. Resilience Curve

The analysis of a system's resilience is conducted through a resilience curve that shows the various stages the system goes through when the system gets exposed to an extreme occurrence. The resilience curve is a plot of power system performance over time, shown in Figure 2. In the pre-incident stage, the power system functions at a typical performance level and remains resilient. When an extreme natural disaster hits the system at time t_{es}, its performance starts to degrade and reaches to minimum performance level at time t_{de}, and this phase is the event progress phase. The system remains degraded until the time t_{rs} and in this phase damage of the system is assessed; after that, recovery starts, and the system transitions into a state of restoration. In the recovery state, critical loads are recovered by utilizing already available resources up to time t_{re}. Then repair teams are deployed to fix impaired elements in post recover state till time t_{irs}. After that, system infrastructure gets recovered up to time t_{ire} and the system's performance returned to normal performance level (Umunnakwe et al., 2021). Figure 2 also illustrates the contrast in the performance of resilient and traditional power systems. It is apparent that a resilient system exhibits consistently better system performance than a traditional system across different stages of the resilience curve.

4. ASSESSMENT AND ENHANCEMENT OF RESILIENCE OF PDNs

The PDN is one of the subsets of the power system. This particular subset is the most susceptible to being affected by HILP events. Section 2 has discussed the impacts of various HILP events on different components of PDNs. Now, this section provides a detailed discussion about assessing and enhancing the resilience of PDNs.

Figure 2. Resilience curve
(Panteli & Mancarella, 2015b)

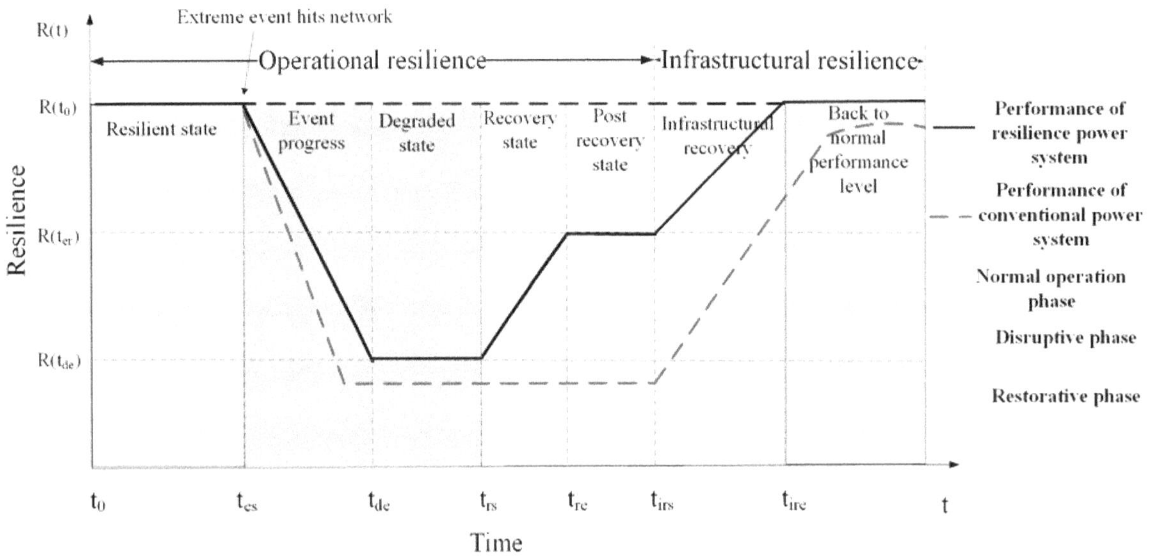

4.1. Resilience Assessment of PDNs

It is crucial to assess the influence of HILP events on the performance of PDNs to take remedial measures effectively. Resilience metrics are used to assess this impact. Resilience metrics assess the degree to which a system can sustain its performance during and after extreme events. Resilience metrics are utilized to assess the proficiency of remedial measures, justify the expenditure, and identify the weaker section of distribution networks. There are numerous planning and operational resilience metrics available in the literature but none of the metrics is universally accepted. The research related to metrics of resilience is still going on.

4.2. Resilience Enhancement Strategies for PDNs

The resilience enhancement strategies refer to actions taken to enhance the system's ability to withstand and recover from disruptive events. The strategies for improving resilience are classified into two groups, namely planning and operational strategies illustrated in Figure 3. The operational strategies are additionally classified into preventive response, emergency response, and resilience restoration.

4.2.1. Planning Strategies

The planning strategies for enhancing the resilience of PDNs involve determining the number, capacity and location of distributed generation resources (Yuan et al., 2016), such as DG (distributed generation) units, EV parking lots, automatic devices, and ESSs (energy storage systems). Hardening strategies, such as vegetation management, strengthening lines/poles (Najafi Tari et al., 2021), and substation elevation also part of the planning strategies. However, installing various local resources like batteries, diesel generators, solar panels, and wind turbines can certainly improve the resilience of PDNs. But operational strategies are more cost-effective solutions to improve the resilience of PDNs.

4.2.2. Operational Strategies

The operational strategies for enhancing the resilience of PDNs involve utilizing existing PDN resources to minimize the amount of expected energy not served during and after extreme events. These strategies include a range of cost-effective measures, such as rescheduling DG units, implementing defensive islanding, dispatching MPSs (Yang et al., 2020), deploying repair crews (Arjomandi-Nezhad et al., 2021), reconfiguring network (Arjomandi-Nezhad et al., 2021), rescheduling ESSs and implementing demand side management (Gilani et al., 2020).

Figure 3. Resilience improvement steps
(Li et al., 2021)

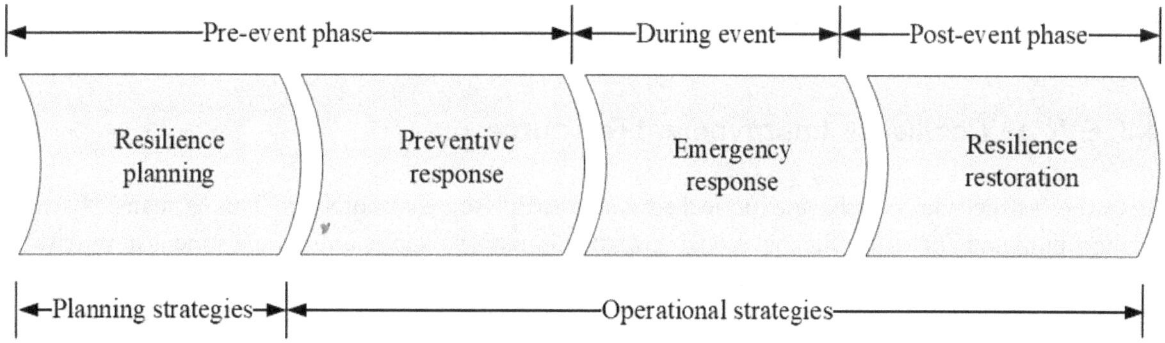

The three district categories into which the operational strategies are classified are preventive, emergency response, and resilience restoration. The preventive response involves preparation before extreme events by pre-positioning various resources. The difference between resilience planning and preventive response is listed in Table 2. Temporary measures are taken in emergency response to supply power to critical loads during extreme events. During extreme events, temporary measures are put into action as part of emergency response efforts, aiming to provide power support for crucial loads. During the post-event phase of resilience, restoration actions are undertaken with the goal of bringing the system back to its initial state.

Table 2. Comparison between resilience planning and preventive response

Resilience Planning	Preventive Response
Permanent placement of resources	Temporary placement of resources
Static resource allocation	Mobile resource allocation (MPSs, repair crews)
The scenarios are generated based on long-term historical metrological data.	The latest meteorological prediction produces the scenarios
Timing: years or months	Timing: Days or hours

One effective operational strategy for restoring PDNs is dispatching MPSs, such as EVs, MESSs, and MEGs. Compared to fixed distribution generation resources, MPSs called 'energy-on-wheel' exhibit greater flexibility. MPSs can supply power to the critical loads in areas that are hard to reach, including the islanded portions of microgrids and have spatial and temporal flexibility. Spatial flexibility is the ability of MPSs to supply power through the road network. The road network is typically resilient due to its highly interconnected structure and redundancy. Therefore, road networks can provide a path for MPSs to supply power to critical loads of islanded portions of PDNs. EVs have the following advantages compared to other MPSs, such as MEGs and MESSs (Li et al., 2021):

- The number of EVs has risen and is projected to grow further. At the same time, the electric utility owns a limited number of MPS, such as MEGs and MESSs.
- There is no need for extra investment to use EVs as emergency power supplies during power outages.
- Both EV owners, as well as electric utilities, get benefits by using EVs to supply power in emergencies.
- EVs, such as electric buses, have a large capacity of batteries. Its capacity can be reached to hundreds of kWh. Electric buses can supply power to customers for several hours.

4.3. EVs as Resilience Improvement Resource

Over the last decade, EVs have experienced a significant surge in numbers. This is mainly because of their numerous benefits, such as reducing carbon emissions and enabling investment in renewable energy. EV batteries' enhanced capacity and improved mileage efficiency can enable them to engage in various services. There exist four distinct integration modes of EVs (Alinejad et al., 2022), that is, grid-to-vehicle (G2V) mode, vehicle-to-vehicle (V2V) mode, vehicle-to-building (V2B) mode, and vehicle-to-grid (V2G) mode. Services, such as local energy resources (V2V mode), reserves (V2G and V2B mode), and backup (V2B) can improve the resilience of PDNs (Hussain and Musilek, 2022). Therefore, the integration option and the growing number of EVs present an opportunity for DSOs to enhance the resilience of the PDN by incorporating these EVs. Prior to an event, EVs have the capability to operate as a portable power source, during the event they serve as a local source for self-healing, and following the event, they can act as a black start resource. A single EV can power a single home during a disruption to serve the nearby critical loads (Shin and Baldick, 2017). Scattered household-level EVs can be clustered to assist the service restoration (Tian and Talebizadehsardari, 2021). EVs can be resilient resources for buildings, microgrids and PDNs.

The steps for the restoration of PDNs are shown in Figure 4. It is illustrated that some distribution lines get damaged due to HILP events. The unavailability of distribution lines leads to several physical islanding in the distribution network where some or all nodes get disconnected from the substation. Therefore, to restore the distribution network and enhance its resilience, damaged distribution lines are first identified, and then the EVs supply power to physical islands through charging stations.

Meanwhile, repair crews are deployed to fix the damaged distribution lines. The process is iterated until all impaired distribution lines are fixed and the substation supplies power to all nodes. This results in the complete restoration of the PDN, and its resilience reaches the maximum value.

Figure 4. EVs based PDN restoration following a HILP event
(Yang et al., 2020)

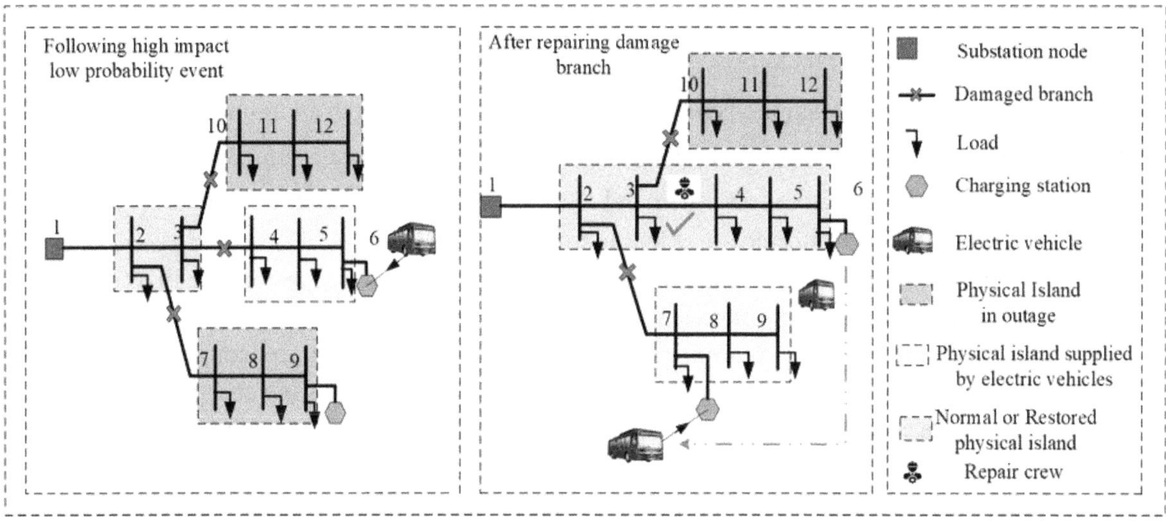

5. APPLICATION OF EV ECOSYSTEM TO IMPROVE THE RESILIENCE OF PDNs

The rise in the adoption rate of electric vehicles enables DSOs to enhance the resilience of PDNs by arranging the charging and discharging of EVs in charging stations or electrified parking lots. EVs are preferred due to their widespread availability and cost-effectiveness in restoring PDNs after natural calamities (Rahimi and Davoudi, 2018). The session offers a methodical review of available literature on the utilization of EVs as a resilience source for PDNs in various stages of power outages. Table 3 provides a glimpse of the resilience improvement measures of EV-enabled PDNs.

In (Kim and Dvorkin, 2019), an optimisation problem is provided, with two stages. The first stage aims to optimise investments in portable ESSs. In the second stage, the model re-routes the portable ESSs to form dynamic microgrids, which can prevent power shortages during disasters. The optimization problem minimizes the investment cost of portable ESSs and the expected cost of PDN operation during ordinary and emergency conditions. The mathematical expression of the objective function is given by Eq. (1).

$$min\left[\mu\left\{I_c\left(\beta_k\right)\right\} + \sum_{s\in S_{nor}} p_s\left\{O_s^{cost}\left(y_{kb}/\beta_k\right)\right\} + \sum_{s\in S_{eme}} p_s\left\{E_s^{cost}\left(\vartheta_{kbts}/\beta_k\right)\right\}\right] \quad (1)$$

$$O_s^{cost} = \sum_{t\in T, i\in N} C_i^{dg} P_{its}^{dg} + \sum_{k\in K, i\in N, t\in T}\left(\left|\frac{A}{100}\right| C_{ES}^{P}\left(P_{kits}^{char} + P_{kits}^{disc}\right)\right) \quad (2)$$

$$E_s^{cost} = \sum_{t\in T, i\in N} C_i^{dg} P_{its}^{dg} \sum_{k\in K, i\in N, t\in T}\left(\left|\frac{A}{100}\right| C_{ES}^{P}\left(P_{kits}^{char} + P_{kits}^{disc}\right)\right) + \sum_{k\in K, t\in T, i\in N} C_i^{VoLL}\left(1-\alpha_{its}^{ld}\right)P_{its}^{ld} \quad (3)$$

where μ is recovery factor; I_c is investment cost; β_k is binary variable for investment in energy storage unit; p_s is probability of s scenario; O_s^{cost} is operating cost in s scenario; E_s^{cost} is emergency cost of scenario s; y_{kb} is binary variable for stationary location of energy storage system; ϑ_{kbts} is binary variable for energy storage unit in mobile mode; C_i^{dg} is incremental cost of DG unit; P_{its}^{dg} is real power output of DG unit; A is slope of degradation of energy storage units; C_{ES}^{P} is cost of energy storage system based on power rating; P_{kits}^{char} / P_{kits}^{disc} is charged / discharge active power of energy storage unit; C_i^{VoLL} is cost of value of lost load; P_{its}^{ld} is active power demand ; α_{its}^{ld} is switch state variable for load.

Solution strategy: The optimization problem is solved by using Progressive hedging.

In (Yang et al., 2020), MPSs and dynamic reconfiguration are utilised to restore the PDN in the face of earthquakes. An optimisation problem is provided to restore distribution networks. The objective is to maximize power supply to load, minimize transportation cost of mobile power sources, minimize the cost of degradation of batteries of EVs and mobile energy storage systems during charging and discharging, and minimise the cost of power out of mobile emergency generators. The mathematical expression of the objective function is given below.

$$max\left(\sum_{t\epsilon T}\sum_{i\epsilon B}W_iP_{i,t}-\sum_{t\epsilon T}\sum_{m\epsilon M}C_m^{tr}\rho_{m,t}-\sum_{t\epsilon T}\sum_{m\epsilon\{ES,EV\}}\left|\frac{`_m}{100}\right|C_m^P\left(P_{m,t}^{ch}+P_{m,t}^{dch}\right)-\sum_{t\epsilon T}\sum_{m\epsilon EG}\beta_mP_{m,t}^G\right) \tag{4}$$

Where W_i is priority of load at node i; ES and EV is set of mobile energy storage system and EVs; EG is set of emergency mobile generators; M is set of mobile power source; B is set of node; $P_{i,t}$ is active power demand at node i; $\rho_{m,t}$ is a binary variable which is 1 if mobile power source m is traveling at time t, otherwise 0; T is set of time; C_m^{tr} is coefficient of transportation cost of mobile power source m; A_m is mobile energy storage system and EVs m degradation slop; C_m^P is cost of power (\$/kWh) of mobile power source and EVs; $P_{m,t}^{ch}$ *and* $P_{m,t}^{dch}$ are charging and discharging power of mobile power sources or EVs m at time t; β_m is coefficient of generation cost of mobile emergency generator m; $P_{m,t}^G$ is active power output mobile power source m at time t.

Solution strategy: The provided MINLP problem is first converted into the MILP and then solved with CPLEX.

In (Erenoglu et al., 2022), the restorative capability of PDNs during an emergency is improved by using electric vehicles' grid-support capacity. The outage period of critical loads is reduced by supplying power to them using EVs. An optimization problem is provided to restore the critical loads of PDN. The optimization problem aims to maximize the load served weighted by their priority, minimization of losses in branches, and minimization of EV number based on sockets available at the nodes. The mathematical expression of the objective function is given by Eq. (5).

$$max\sum_{SD}\sum_{SO}\pi_{SD}\pi_{SO}\left(\sum_{i\epsilon I}\sum_{a\epsilon A}\sum_{t\epsilon T}\left(w_av_{i,SD,SO,t}^a\right)-\varepsilon\sum_{b\epsilon B}\sum_{t\epsilon T}P_{b,SD,t,SO}^{loss}+\sum_{i\epsilon I}\sum_{e\epsilon E}\sum_{t\epsilon T}\beta_{i,e,s,v,t}\right) \tag{5}$$

Where SD is the set of location of damage; SO is the set of occurrence time extreme event; π_{SD} is the probability of damaged location scenario; π_{SO} is the probability of event occurrence time scenarios; A

is set of load types; T is set of time periods; w_a is weighting factor of load type a; I is set of buses; $v^a_{i,s,v,t}$ is binary variable it is 1 if load type a is fed by bus i in scenarios SD and SO, else 0; ε is a small value; B is set of branch; $P^{loss}_{b,SD,t,SO}$ is power loss in branch b during time t in SD and SO scenarios; E is set of electric buses; $\beta_{i,e,s,v,t}$ is 1 if electric vehicle e is not connected to candidate node at time t in scenario SD and SO.

Solution strategy: The MILP problem is solved by using CPLEX solver.

To achieve balance between the requirements of public transportation and the restoration of distribution network, an optimization problem is provided in (Li et al., 2021) that coordinated electric bus scheduling with distribution network restoration. Network reconfiguration and electric bus scheduling are utilized to restore distribution networks after extreme events. A MILP problem is formulated to coordinate distribution network reconfiguration and electric bus scheduling. The optimization problem aims to minimize the rental expense of electric buses, maximize the restored load, and minimize the departure interval of route. The mathematical expression of the objective function is given below.

$$max \sum_{i\epsilon B} W_i \beta_i P_i - C_{BC} \sum_{i\epsilon N_C} \sum_{d\epsilon D} \sum_{m\epsilon M(d)} X_i^{md} - \epsilon \sum_{r\epsilon R} t_r \tag{6}$$

Where B is set of nodes; W_i is load weighting factor; β_i is 1 if load at node $_i$ is supplied; P_i is real power load at bus i, C_{BC} is the opportunity cost of bus company, N_C set of power nodes with charging point C; D is set of electric bus depots; $M(d)$ is set of electric bus located at depot d, X_i^{md} is 1 if electric bus m parked at depot d is dispatched to charging point i; R is set of electric bus route; t_r is departure interval of route r.

Solution strategy: The MILP problem is solved using off-the-shelf optimization solvers.

In (Shin and Baldick, 2017), the aim of the paper is to optimize a vehicles-to-home system to maximize its backup duration. Backup duration refers to the amount of time that V2H can support residential loads in islanded mode without any critical load curtailment. This work is mainly focused on utilization of plug-in hybrid EVs as a backup generator. The optimization problem aims to minimize the sum of weighted unserved load over time period T.

Solution strategy: The MIQCP (mixed integer quadratic constraint programming) problem is solved by GUROBI and CPLEX solver.

In (Li et al., 2022), pre-allocation of electric buses at charging stations before hurricane occurrence is utilised to enhance the resilience of PDNs. Pre-allocation of electric buses helps to restore PDNs after the event. The model provided in this paper aims to find optimal allocation strategies of electric buses before events with minimum load loss under the worst scenario. The first term of the problem minimizes the opportunity cost for utilizing electric buses for restoration of PDN. The second term of the problem minimizes loss of load under worst scenario. The detailed formulation of the problem is given below.

$$\min_{n^{EB}_{r,cs}} \sum_r \sum_s EL^D_r n^{EB}_{r,cs} + \max_{p \in D} E_p\left[Q(x,u)\right] \tag{7}$$

$$Q(x,u)=\min_{y}\sum_{i\in N^D} w_i\left(\overline{P}_i-P_i\right)+\in\sum_{r}\sum_{s} n^{EB}_{r,cs} \tag{8}$$

Where $n^{EB}_{r,s}$ is number electric buses dispatched to charging station *cs* through route *r*; $w_i\left(\overline{P}_i-P_i\right)$ is value of lost loads;

Solution strategy: The problem is solved by using Column-and-constraint generation (CACG) algorithm.

In (Momen et al., 2020), EVs supply power to critical loads of PDN during power outages. Public and residential parking lots are used as distributed energy resources to form microgrids to restore PDNs. Two forms of objective functions are provided in this paper. One objective function is based on a technical approach that maximizes power supplied to loads. The second objective function is based on an economic approach, that is, the minimization of load interruption cost of critical loads, generation cost of DG units, and cost of EV owners. Mathematical expressions of objective functions are given below.

(1) Technical approach

$$max\left(\sum_{t\in T}\sum_{i\in B_c} P^s_{i,t}\delta_i\right) \tag{9}$$

(2) Economic Approach

$$min\sum_{t\in T}\left[\sum_{i\in B_c}\left(P^c_{i,t}-P^s_{i,t}\right)\left(IC^c_{i,t}\right)+\sum_{i\in B^{DG}_{master}}\left(P^k_{i,t}\right)\left(GC^{DG}_{t,k}\right)+\sum_{i\in B}\sum_{m\in M^{EV}_i}\left\{\left(\eta^{ch_e}_{m,t}P^{ch_e}_{m,t}C_t\right)-\left(\eta^{dch_e}_{m,t}P^{dch_e}_{m,t}R_t\right)\right\}\right] \tag{10}$$

Where $P^s_{i,t}$ is power supplied at bus *i* at time *t*; $IC^c_{i,t}$ is cost of interruption of critical loads *c* at bus *i* at time *t*; δ_i represents restoration status of node *i*; B_c is set of critical nodes; $P^c_{i,t}$ is critical load at bus *i* at time *t*; B^{DG}_{master} is set of nodes at which distributed generation units are located; $P^k_{i,t}$ is power injected at node *i*; $GC^{DG}_{t,k}$ is cost of generation of distribute generation unit *k* at time *t*; M^{EV}_i is set of EVs available at node *i*; $\eta^{ch}_{m,t}$ is charging efficiency, $\eta^{dch_e}_{m,t}$ is discharging efficiency, $P^{ch_e}_{m,t}$ / $P^{dch_e}_{m,t}$ is emergency condition charging / discharging power of EV *m* at time *t*, C_t is EV charging cost, R_t is EV discharging reward cost.

Solution strategy: The MILP problem is solved using CPLEX solver.

Resource allocation prior to event is utilized as a preventive action in (Gao et al., 2017) to improve the resilience of PDNs. Allocation of diesel oils, electric buses, and battery energy storage systems among DG units, charging stations and microgrids are used as preventive actions. An optimization problem is provided to allocate the resources prior to events. The aim of the optimization problem is to maximize the expected profit attained from critical load recovery considering the pre-positioning cost of resource. The mathematical expression of the objective function is given below.

$$max\left(\sum_{i_{cl}=1}^{N_{cl}} E\left[W_{cl}^{i_{cl}} P_{cl}^{i_{cl}} T_{cl}^{i_{cl}}\right] - C_{LR}\right) \tag{11}$$

Where N_{cl} is the overall count of critical loads; $W_{cl}^{i_{cl}}$ is the weighting factor of $i_{cl}th$ critical load; $P_{cl}^{i_{cl}}$ is the real power of $i_{cl}th$ critical load in its normal state of operation, $T_{cl}^{i_{cl}}$ denotes the duration for which service is provided to load in after extreme event, C_{LR} is the load restoration cost.

Solution strategy: A heuristic algorithm is used to solve allocation problem.

In (Lei et al., 2019), preventive and restorative actions are used to improve the resilience of PDNs. Before the occurrence of an event, mobile power sources pre-positioning in coordination with network reconfiguration are utilised to enhance the survivability of PDNs. A two-stage robust optimisation model for pre-positioning mobile power sources is provided. The optimization problem aims to maximize the weighted load supplied in the worst case, and the mathematical expression of the objective function is given by equation (12). Afterward extreme event, the MPSs are dynamically dispatched to restore the PDN. An optimisation problem is provided to dispatch mobile power sources dynamically. The objective function aims to maximize loads restored weighted by their priority, minimize transportation cost of mobile power sources, and minimize the cost of lifecycle degradation of batteries of EVs and mobile energy storage systems during charging and discharging. The mathematical expressions of the objective functions are given by Eq. (13).

First stage objective

$$\max_{X}\left[\min_{U}\left(\max_{Y} \sum_{i\epsilon N} W_i P_{i,0}^{s}\right)\right] \tag{12}$$

Second stage objective

$$max\left(\sum_{t\epsilon T}\sum_{i\epsilon N} W_i P_{i,t} - \sum_{t\epsilon T}\sum_{m\epsilon M} C_m^{tr} \rho_{m,t} - \sum_{t\epsilon T}\sum_{m\epsilon\{ES,EV\}} \left|\frac{{}^{‘}{}_{m}}{100}\right| C_m^{P}\left(P_{m,t}^{ch} + P_{m,t}^{dch}\right)\right) \tag{13}$$

Where W_i is priority of load at node i; $P_{i,0}^{s}$ is active power restored at time 0, N is set of node; T is set of time; ES and EV is set of MESSs and EVs; M is set of mobile power source; $P_{i,t}$ is active power demand at node i; $\rho_{m,t}$ is a binary variable which is 1 if mobile power source m is traveling at time t, otherwise 0; C_m^{tr} is coefficient of transportation cost of mobile power source m; A_m is mobile energy storage system and EVs m degradation slop; C_m^{P} is cost of power ($/kWh) of mobile power source and EVs; $P_{m,t}^{ch}$ *and* $P_{m,t}^{dch}$ are charging and discharging power of mobile power sources or EVs m at time t;

Solution strategy: The MPSs pre-positioning problem is solved by CACG algorithm. The second stage optimization problem is MILP that is solved by using CPLEX.

In (Erenoglu and Erdinc, 2023), multi-mobile resources, DG units, and demand side management (DSM) are used to enhance resilience in the post-event restoration phase of resilience. Microgrid formation by utilising MPSs, such as MESSs, EVs, and MEGs prevents prolonged and widespread electric outages. Optimally dispatching multiple mobile power sources can significantly decrease the duration of power interruption for critical loads. An optimization problem is provided to restore the loads of PDNs. The optimization problem aims to maximize the total load restored weighted by their priority. The mathematical expression of the objective function is given by Eq. (14).

$$max\left(\sum_{i\in I,t\in T} w_i\mu_{(i,t)}\left(P^d_{i,t}+Q^d_{i,t}\right)\Delta T\right) \tag{14}$$

Where w_i is load priority at bus i; $\mu_{(i,t)}$ represents load to nus switching status; $P^d_{i,t}$ and $Q^d_{i,t}$ is real and reactive power demand at bus i and time t.

Solution strategy: The MIQP problem is solved by using LINDO v.13 commercial solver.

Restoration model for PDN is provided in (Ding et al., 2020). This model coordinates the operation of EVs, repair crew deployment, and networked microgrid formation for load restoration. A MILP problem is provided in this article. The goal of optimization problem is to maximize the restored load weighted by their priority. The mathematical expression of the objective function is given by Eq. (15).

$$max\left(\sum_{t\epsilon T}\sum_{i\epsilon B} w_i p_{i,t}\right) \tag{15}$$

Where w_i is weighting factor of load at bus i; $p_{i,t}$ is restored load at bus i at time t.

Solution strategy: The MILP problem is solved by using an auxiliary induce function algorithm.

The quantity of EVs is growing these days. Scheduling the charging and discharging of plug-in EVs in parking lots or fast charging station can improve the resilience of PDN. A problem with two levels is provided in (Alizadeh and Jafari-Nokandi, 2023) to find the best islanding plan for PDNs. The model considers the management of charging and discharging of plug-in EVs in parking lots and variation in tour features, depending on energization status of parking lots. In upper level of problem, the goal is to maximize recovered load based on their priority. In lower level of the problem, the goal is to minimize distance between energized parking lot to the de-energized parking lot. The mathematical expressions of the objective functions are given by Eqs. (16) and (17).

$$max\sum_{t\epsilon T}\sum_{i\epsilon N} w_i p_{i,t} \tag{16}$$

$$min\sum_{b\epsilon B}\sum_{b'\epsilon B} SP_{b,b'}E_{b,b'} \tag{17}$$

where N is set of electric buses; B is set of traffic nodes; w_i is priority of load i; $p_{i,t}$ is active load restores at bus i at time t; $SP_{b,b'}$ is shortest path length between traffic nodes b and b'; $E_{b,b'}$ is a positive variable representing the change in position of parking lot b to parking lot b'.

Solution strategy: The MILP problem is solved using GAMS software.

In (Roudbari et al., 2021), a resilience-oriented operational approach is provided that aims to improve the economic benefits as well the resilience of PDNs. Network reconfiguration, EV management, demand response and distributed energy resource rescheduling are used as operational measures to enhance resilience. An optimization problem with multi-objectives is provided. The first objective is to minimize both the penalty cost of unsupplied energy and the operational expense associated with the distribution network. The second objective is the maximization of benefits to EV owners and resources.

In (Wu et al., 2022), operational plan of diesel generators, energy storage systems, and dispatch plan of EVs are utilized for optimal restoration of PDNs. EVs are deployed to assist in the restoration of PDNs, and compensation is provided based on their performance. An EV-based robust restoration model for PDNs is provided. The optimization problem aims the minimize of the operational cost of all EVs and the operational cost PDN under most unfavourable circumstances. The mathematical expression of the objective function is given by Eq. (18).

$$\min_{y} \max_{u_{1.t} u_{2.t}} \left(OC_{EV} + OC_{PDN}\right) \tag{18}$$

$$OC_{EV} = \sum_{t\in T}\sum_{m\in M} C_{ev}^{d}\left(P_{m,t}^{cha} + P_{m,t}^{disch}\right) + \sum_{t\in T} UC_{EV} N_{t}^{EV} \tag{19}$$

$$OC_{PDN} = \sum_{t\in T}\sum_{n\in N} \chi_{loss} P_{n,t}^{ls} + \sum_{t\in T}\sum_{b\in B} C_{b}^{d}\left(P_{b,t}^{cha} + P_{b,t}^{disch}\right) + \sum_{t\in T}\sum_{dg\in DG}\left(c_{dg,1}P_{dg,t} + c_{dg,2}P_{dg}^{max}\upsilon_{dg,t}\right) \tag{20}$$

where OC_{EV} is operational cost of EVs; OC_{PDN} is operational cost of PDN; C_{ev}^{d} is cost of degradation of EVs battery; $P_{m,t}^{cha/disch}$ is charging/discharging power of EV charging station *m*; UC_{EV} is EVs unit time cost; N_{t}^{EV} is total number of EVs in time period *t*; χ_{loss} is cos of unit load shedding; $P_{n,t}^{ls}$ is load shedding at node *n* at time *t*; P_{dg}^{max} is maximum power of diesel generator *dg*; C_{b}^{d} is cost of degradation of battery *b*; $P_{b,t}^{cha/disch}$ is charging/discharging power of battery *b*; $c_{dg,1/2}$ is cost factor of diesel generator *dg*; $P_{dg,t}$ is real power generated by diesel generator *dg* at time *t*; $\upsilon_{dg,t}$ state the operating condition of diesel generator d*g*.

Solution strategy: The problem is first changed into mixed-integer second-order cone model and then solved using GUROBI and optimal restoration strategy is obtained.

In (Gazijahani et al., 2022), a two-stage model is presented to enhance the resilience of PDNs. Before the hurricane arrives, the first stage involves carrying out pre-event preparedness. Once the hurricane has been identified, the second stage involves implementing certain resilience enhancement measures, such as network reconfiguration, DSM, and grid partition to minimize the vulnerability of PDNs. In first stage, objective function aims to maximize natural gas storage and SOC of EVs and minimize graph-oriented metric. In second stage, objective function aims to minimize degradation and vulnerability metrics. The mathematical expressions of objective functions of those two stages are given respectively by Eqs. (21) and (22).

$$max\left(\frac{1}{W_1}\sum_{t=1}^{T}\left\{\sum_{ev=1}^{N_{ev}}SOC_{ev,t}+\sum_{ng=1}^{N_{ng}}Q_{ng,t}\right\}-\frac{1}{W_2}\left\{\left(\frac{2}{N-1}\right)Trace\left(Z^{+}\right)\right\}\right) \tag{21}$$

$$max\left(\sum_{s=1}^{N_s}p_s\left(RM_s+SM_s\right)\right) \tag{22}$$

W_1, W_2 are base value; $SOC_{ev,t}$ is state of charge of electric vehicle *ev* at time *t*; $Q_{ng,t}$ is quantity of natural gas storage of unit *ng* at time *t*; $Trace(Z^{+})$ is summation of eigen values of given topology; *N* is total number of buses; p_s is probability of occurrence of *s* scenario; RM_s is resilience metric in *s* scenario; SM_s is restoration metric in *s* scenario.

Solution strategy: The problem is modeled as MILP and optimised using CPLEX solver.

In (Kianmehr et al., 2019), coordinated allocation of parking lots for EVs, wind farms, and energy storage systems are utilized as planning measures. DSM and dynamic reconfiguration of distribution networks are used as operational strategies to improve the resilience of PDNs. The objective aims to minimize the total expenses of power acquired from the substation and wind farms, the cost of residential load curtailment, and the compensation provided to commercial or industrial loads for their flexibility.

Solution strategy: The MINLP problem is solved using GAMS solver.

Planning measures are used in (Gan et al., 2022) to enhance the resilience of couples' PDNs and TNs (transportation networks). Planning measures include hardening and expanding distribution lines, EV charging stations, and roads. A tri-level robust problem is provided for resilient-oriented planning of coupled PDNs and TNs. The upper level aims to minimize the capital expenditure incurred for the expansion and hardening of PDNs and TNs. Middle level found worst-case scenario. The lower level aims to minimize traffic congestion and load unserved.

Solution strategy: The tri-level robust problem is solved by CACG and bender decomposition algorithm.

Mobility of EBs and network reconfiguration are complementary of each other for restoration of distribution networks. Therefore, coordinated operation of EBs and network reconfiguration is utilized in (Zhang et al., 2021) to recover critical loads of distribution networks. A bi-level post-event restoration problem is provided in this article. The upper level of the problem optimizes the topology of the PDN with the aim of maximum loads recovery using both the main grid and electric buses. The lower level of the problem optimizes routes for all electric buses with the aim of maximum load restoration using electric buses by considering the travel duration and energy consumed by electric buses during their transit.

Solution strategy: Genetic algorithm (GA) and particle swarm optimization technique are used to solve the bi-level optimization problem.

The complexity and vulnerability of PDNs and transportation networks are amplified in extreme circumstances due to the rising interdependence between the two systems and the widespread adoption of EVs. The aim of (Li et al., 2023) is to enhance the performance of transportation networks and PDNs during the emergency time after disasters. Coordinated optimization of transportation network traffic route reversing, distribution network line switching and fast charging station management is provided. The objective aims to minimize costs associated with compensation of real and reactive power shortage in PDNs and total travel cost of transportation networks.

Table 3. Glimpse of resilience enhancement measures

Ref	Problem Formulation	Resilience Enhancement Measures								Public transit service	Coupled TN and DN	Resilience Stage	Objective
		DG unit	ESS	NR	EV	DSM	Micro-grid	MESS /MEG	Repair crew				
(Kim and Dvorkin, 2019)	—						✓	✓				Planning, Response	Minimization of cost
(Shin and Baldick, 2017)	MIQCP				✓							Restoration	Minimization of unserved load
(Yang et al., 2020)	MINLP			✓	✓			✓				Restoration	Maximization of load restored, Minimization of cost
(Mehrjerdi, 2021)	MILP	✓			✓							Restoration	Minimization of energy not supplied
(Erenoglu et al., 2022)	MILP				✓							Response	Maximization of load restored, Minimization of loss
(Li et al., 2021)	MILP			✓	✓					✓	✓	Restoration	Maximization of load restored, minimization of rental cost of EV and departure interval of route
(Li et al., 2022)	—				✓					✓		Response	Minimization of unserved load, Minimization of opportunity cost
(Momen et al., 2020)	MILP				✓	✓						Response, Restoration	Maximization of load restored, Minimization of cost
(Gao et al., 2017)	MISNLP	✓	✓		✓					✓		Response,	Maximization of load restored, Minimization of cost of load recovery

Table 3. Continued

Ref	Problem Formulation	Resilience Enhancement Measures								Public transit service	Coupled TN and DN	Resilience Stage	Objective
		DG unit	ESS	NR	EV	DSM	Micro-grid	MESS /MEG	Repair crew				
(Lei et al., 2019)	—			✓	✓			✓				Response, Restoration	Maximization of load restored, minimization of cost
(Erenoglu and Erdinc, 2023)	MIQP	✓			✓	✓	✓	✓			✓	Restoration	Maximization of load restored
(Ding et al., 2020)	MILP				✓		✓		✓			Restoration	Maximization of load restored
(Alizadeh and Jafari-Nokandi, 2023)	MILP				✓							Restoration	Maximization of load restored, Minimization of distance between energized and de-energized parking lot
(Roudbari et al., 2021)	——	✓	✓	✓	✓	✓						Restoration	Minimization of cost, Maximization of benefit of resource and EV owner
(Wu et al., 2022)	MINLP	✓	✓		✓							Restoration	Minimization of cost
(Gazijahani et al., 2022)	MILP			✓	✓	✓	✓					Response, Restoration	Maximization of state of charge, Maximization of natural gas, Maximization of resilience metric
(Kianmehr et al., 2019)	MINLP	✓		✓	✓	✓						Planning, Restoration	Minimization of cost
(Zhang et al., 2021)	—			✓	✓							Restoration	Maximization of load restored

In (Mehrjerdi, 2021), solar panels and EVs are utilized continuously to supply power to crucial loads and reduce loss of energy. The provided model aims to use the existing EVs within the edifice while minimizing the incorporation of solar power. The goal of the problem is to minimize the loss of energy. The mathematical expression of the objective function is given below.

$$min\left(\sum_{h\epsilon H}\sum_{s\epsilon S}\sum_{f\epsilon F}ENS^{h,s,f}\right) \tag{23}$$

where H,S,F are set of hours, seasons, and faults respectively; ENS is energy not supplied in kWh.

Solution strategy: The MILP problem is solved by CPLEX solver.

In (Rahimi and Davoudi, 2018), hybrid EVs in vehicle-to-home integration mode are utilized to improve the resilience of PDNs. In the absence of electricity, building energy management controls the starting time and time of use of high-power consumption appliances, such as washing machine, dryer and dish washer. An optimization problem is provided to determine the optimal time for switching on flexible gadgets, with the objective of minimizing peak power consumption during the operation period of flexible appliances. The peak power is determined as the highest value obtained by summing the power consumed by all activated appliances. The mathematical expression of the objective function is given below.

$$min\left(max\sum_{a=1}^{n}P_{a,t}\right) \text{ for } t\in T_F \tag{24}$$

where n is total number of appliances; $P_{a,t}$ is power consumed by appliances a at time t; T_F is time period for which flexible appliances are controlled in a day.

Solution strategy: GA is employed to solve the optimization problem.

6. QUALITATIVE ASSESSMENT OF RESILIENCE IMPROVEMENT OF EV-ENABLED PDNs

This section presents a qualitative assessment of the literature that has been studied and analysed. The key objective of this qualitative assessment is to present a thorough overview of the existing literature, research gaps and areas that need further investigation. It also aims to highlight strategies used to enhance the resilience of EV-enabled PDNs. A comparison of the time series of the resilience curve for different extreme events is also provided in this section.

6.1. Source of Uncertainty in Assessing and Enhancing the Resilience of PDNs

It is not possible to predict event occurrence time, event clearance time, intensity of event and fault location with precision. Furthermore, EV's availability and state of charge level of EV are also uncertain. To make a practical plan, it is necessary to consider all these uncertainties while assessing and enhancing the resilience of PDNs. A fragility curve can be used to model the uncertainty associated with fault location

(Erenoglu et al., 2022). Several models have been provided to handle the uncertainty associated with line damage, including stochastic programming and robust programming (Zhang et al., 2022). The uncertainties in extreme events' occurrence time and location can be captured by a scenario-based stochastic approach (Li et al., 2022). The stochastic method (Momen et al., 2020) and learning-based (Erenoglu et al., 2022) method can be used to analyse the uncertainty of electric vehicles. Very few studies (Momen et al., 2020) have considered all these uncertainties while assessing and enhancing the resilience of PDNs.

6.2. Interdisciplinary Research for Resilience

The ultimate goal for enhancing the resilience of PDNs is to ensure the usual functioning of society in the aftermath of severe events. During HILP events, along with PDNs, other vital infrastructures, for example, communication networks, water distribution networks and transportation networks, also get affected (Panteli and Mancarella, 2015b). The electrification of transportation results in greater interdependence between PDNs and transportation networks, illustrated in Figure 5. Therefore, additional research is essential for the improvement of the resilience of coupled transportation and distribution networks by leveraging their interdependent nature. Research article (Li et al., 2023) has studied the resilience enhancement of coupled TNs and PDNs. However, most articles neglect the damage and restoration phase of transportation networks.

Figure 5. Illustration of distribution network and transportation network after a disaster
(Wu et al., 2022)

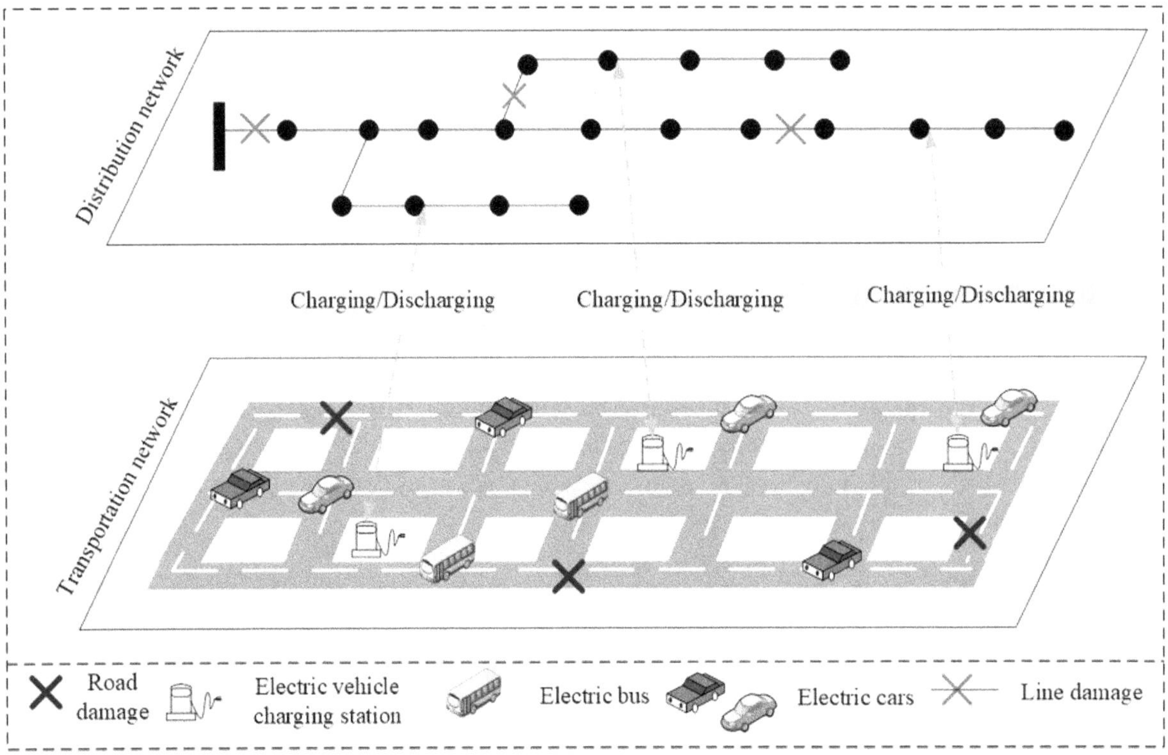

6.3. Resilience Enhancement Strategies for EV-Enabled PDNs

EVs can be employed in both pre-event operation and post-event restoration phases. The pre-event operation refers to taking preventive action when an extreme weather event is forecasted to hit a specific region within one or two days. Preventive action can aid in the efficient restoration of PDN after an extreme event. The pre-event operation problem is usually formulated as stochastic programming or robust programming due to uncertainty in fault location. Research on post-event restoration seeks to determine the best scheduling of distributed energy resources, dispatch of mobile power sources and repair crew or reconfiguration of the network under a deterministic fault scenario. Maximization of load restoration and minimization of restoration cost are the two most common objectives of restoration problems. EV-based load restoration is a cost-effective solution because there is no need for extra investment. The current research articles, such as ref. (Li et al., 2021, Yang et al., 2020), etc. have utilized EVs to restore PDNs.

The preventive response and restorative action to enhance the resilience of PDNs are illustrated in Figure 6. A two-stage optimization problem (TSROP) is provided in (Lei et al., 2019) for path selection and time allocation of MPSs. Network reconfiguration network reconfiguration together with MPSs, such as EV fleets, MESSs, and MEGs prepositioning at selected nodes are utilized as a preventive measure to enhance system survivability in first stage. EV charging stations and MESS stations considered in this work are utility-owned. The usage of utility-owned facilities prevents massive additional expenses.

Table 4. Status (stationary or travelling) and position (Bus) of MPSs in each time interval (Lei et al., 2019)

<table>
<tr><th rowspan="2">MPSs</th><th colspan="24">Time Interval</th></tr>
<tr><th>1</th><th>2</th><th>3</th><th>4</th><th>5</th><th>6</th><th>7</th><th>8</th><th>9</th><th>10</th><th>11</th><th>12</th><th>13</th><th>14</th><th>15</th><th>16</th><th>17</th><th>18</th><th>19</th><th>20</th><th>21</th><th>22</th><th>23</th><th>24</th></tr>
<tr><td>EV fleet</td><td>Bus 33</td><td colspan="2"></td><td colspan="6">Bus 8</td><td colspan="2"></td><td colspan="13">Bus 33</td></tr>
<tr><td>MESS</td><td>Bus 29</td><td></td><td colspan="4">Bus 21</td><td></td><td colspan="17">Bus 29</td></tr>
<tr><td>MEG</td><td>Bus 21</td><td colspan="2"></td><td colspan="5">Bus 15</td><td></td><td colspan="15">Bus 29</td></tr>
</table>

The preventive action plan is determined through a TSROP. The optimization problem aims to maximize the weighted sum of survived load in a worst-case scenario. The optimization model has been demonstrated on IEEE 33 bus system (total load =20672 kW), that is, shown in Figure 6 (a). And Figure 6 (b) visually presents the outcomes of the optimization problem. To evaluate the effectiveness of preventive action 10000 damage scenario is generated by Monte Carlo simulation. The load survived is calculated in each scenario, and then the average value of the survived load is evaluated, that is, 13727 kW. Without preventive action, the average value of the survived load is 5641 kW. Shortly after the event, MPSs get connected to prepositioned nodes of PDN to speedily enhance the survivability that is shown in Figure 6 (c). The damage of PDN is assessed after the event. In the second stage, restoration and infrastructure recovery in coordination with the dynamic dispatch of MPSs are used to restore the PDN after an extreme event. An optimization problem is presented for the dynamic dispatch of MPSs, aiming to maximize restored load, minimize the transportation expense of MPS and degradation expense of EV and MESS battery lifecycle. Figure 6 (c) – 6 (d) illustrate the outcomes of the TSROP for initial four-hour period.

These figures illustrated further routing of MPSs, the switch status of remotely controlled switches, the connection point of MPSs, and the repairing status of damaged lines. Table 4 provides information on the subsequent routing of MPSs during the dynamic dispatch of MPS in 24 hours. With this dynamic dispatch of MPS, 97% of the load is recovered at t = 16 hr and 100% at t = 24 hr.

Figure 6. Preventive response and restorative action using MPSs and network reconfiguration (Lei et al., 2019)

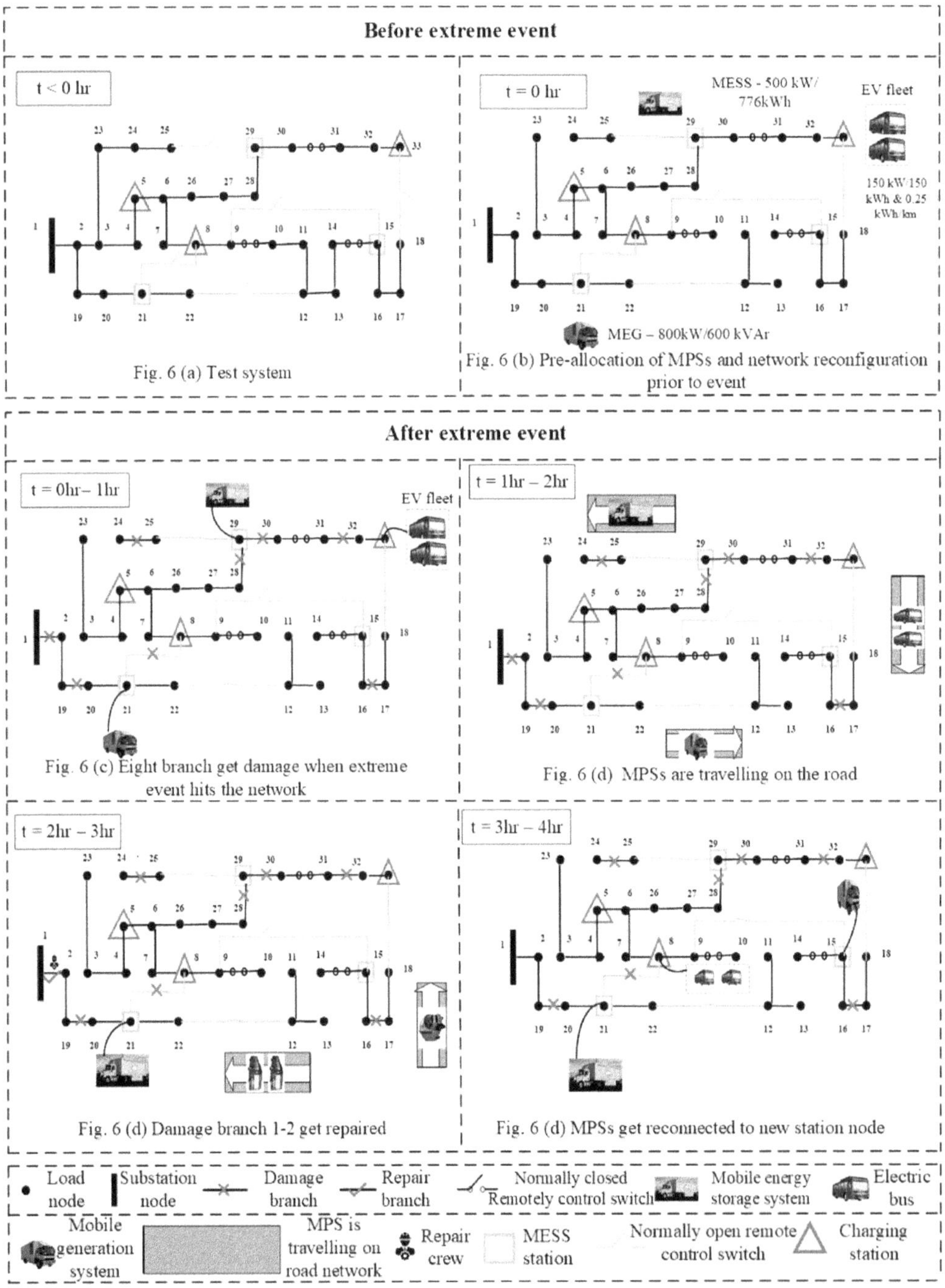

The track, time and radius of hurricanes are forecasted by the metrological department several hours or days before the occurrence of hurricanes. Therefore, the impact of the upcoming hurricanes on PDN can be estimated before hurricane occurrence. Accordingly, preparation can be done to enhance the resilience of PDN discussed above. However, in the case of an earthquake location, intensity and time of the event are known a few seconds before its occurrence. It is impossible to make any preparation within a short time. Therefore, in case of an earthquake, restorative action (Yang et al., 2020) can improve the resilience of PDN. After an earthquake, MPSs are dispatched to the different charging and MESSs stations to restore PDNs, as shown in Fig. 6 (c) – 6 (d). Most articles concentrate on dispatching MPS or repair crew to recover the PDN after an extreme event. Only a limited number of publications investigated the coordinated operation of MPS and the repair crew (Xu et al., 2019). Consequentiality, it is crucial to research coordinated dispatch of MPS and repair crew to restore PDNs.

Electric buses (EBs) are generally used for public transport. Hurricane's occurrence leads to a reduction in EBs demand for public transit. This provided an opportunity for the DSO to utilize idle EBs (Gao et al., 2017, Li et al., 2021, Li et al., 2022, Zhang et al., 2021) for the restoration of PDN. In (Li et al., 2022), a model is provided for the prior allocation of EB with a focus on restoring the PDN and public transportation after the event. A three-layer problem is provided. The first layer aims to minimize the cost of dispatch of EB and the expected load loss. In the second layer, the problem explores worst-case probability distribution.

Figure 7. Dispatch strategies for electric buses (EBs)
(Li et al., 2022)

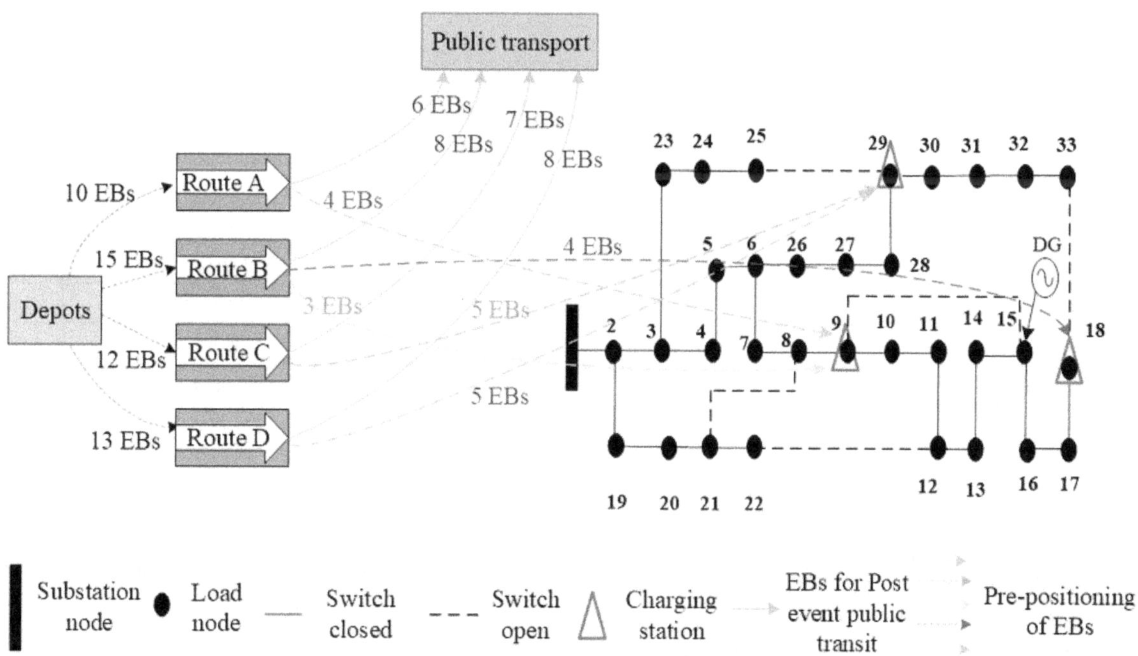

After finding out pre-allocation strategies for EB and the worst-scenario, the third layer of the problem determines the optimum operation of impaired PDN and public transport system to minimize expected load loss cost. The solution of pre-allocation and dispatch strategies of electric buses for PDN restoration and public transit through different routes is shown in Fig. 7. Similarly, in ref. (Gao et al., 2017) electric bus is used for the restoration of PDN and public transit service. Limited number of articles such as ref. (Gao et al., 2017, Li et al., 2022) have investigated both the dispatch of EB for public transportation and their potential role in restoring PDN. Socioeconomic status can significantly influence a distribution system's ability to cope with power outages. In many countries, limited access to reliable electricity infrastructure, financial resources and aged infrastructure are the challenges faced during emergencies. Some regions may lack adequate backup power sources, such as mobile power sources, or have limited access to charging station infrastructure for electric vehicles. Addressing the specific needs of these countries is crucial for equitable resilience planning, as it ensures that vulnerable communities are not disproportionately affected by power disruptions and can effectively recover from disasters.

6.4. Comparison of Time Series of Resilience Curve

The resilience curve motioned previously in section 3 is a plot of the system's performance over time, which is shown in Figure 2. The time *t*: t_0, t_{es}, t_{de}, t_{rs}, t_{re}, t_{irs}, t_{ire} are respectively the time at original state, event incidence, event end, recovery start, initial recovery state, infrastructural recovery start, full restoration. The values of the time series depend on the type of events. Table 5 presents the recorded value of the time series for resilience curve corresponding to various events, such as windstorms, hurricanes, floods, and earthquakes. The time taken to fully restore PDNs ranges from hours to days, depending on the events.

Table 5. Time series for resilience curve corresponding to various events

Events	Time Series						
	t_0	t_{es}	t_{de}	t_{rs}	t_{re}	t_{irs}	t_{ire}
Wind storm (Amirioun et al., 2019)	0	1.25 min	3.2 min	8.5 min	23.6 min	—	—
Hurricane (Hosseini and Parvania, 2020)	5 h	10 h	11 h	11 h	14 h	34 h	39 h
Flood (Liang et al., 2022)	0 h	10 h	15 h	25 h	30 h	36 h	50 h
Earthquake (Didier et al., 2017)	-10 days	0 day	0 days	1 day	20 days	90 days	—

7. CONCLUSION

This chapter presents the notion of power system resilience, incorporating its definition, features of resilience and resilience curve. Further, the vulnerability of different components of PDNs to various HILP events has been presented. Then, the resilience enhancement strategy and its classification are discussed in detail. The resilience enhancement strategy is classified into planning and operational strategies. Additionally, the application of the EV ecosystem in the restoration of PDNs has been presented, and a literature review related to the EV-based restoration of PDNs has been done. Finally, this chapter presents a qualitative assessment of different aspects of the resilience of EV-enabled PDNs. In post-event restoration phase of resilience, most of the study use repair crew, electric buses, mobile emergency generators, mobile energy storage systems-based approaches to restore the PDNs after extreme event. The coordination among these approaches and integration issue of EVs to PDNs are areas of future research.

REFERENCES

Afzal, S., Mokhlis, H., Illias, H. A., Mansor, N. N., & Shareef, H. (2020). State-of-the-art review on power system resilience and assessment techniques. *IET Generation, Transmission & Distribution*, *14*(25), 6107–6121. doi:10.1049/iet-gtd.2020.0531

Alinejad, M., Rezaei, O., Habibifar, R., & Azimian, M. (2022). A Charge/Discharge Plan for Electric Vehicles in an Intelligent Parking Lot Considering Destructive Random Decisions, and V2G and V2V Energy Transfer Modes. *Sustainability (Basel)*, *14*(19), 12816. doi:10.3390/su141912816

Alizadeh, M., & Jafari-Nokandi, M. (2023). A bi-level resilience-oriented islanding framework for an active distribution network incorporating electric vehicles parking lots. *Electric Power Systems Research*, *218*, 109233. doi:10.1016/j.epsr.2023.109233

Amirioun, M. H., Aminifar, F., Lesani, H., & Shahidehpour, M. (2019). Metrics and quantitative framework for assessing microgrid resilience against windstorms. *International Journal of Electrical Power & Energy Systems*, *104*, 716–723. doi:10.1016/j.ijepes.2018.07.025

Arjomandi-Nezhad, A., Fotuhi-Firuzabad, M., Moeini-Aghtaie, M., Safdarian, A., Dehghanian, P., & Wang, F. (2021). Modeling and Optimizing Recovery Strategies for Power Distribution System Resilience. *IEEE Systems Journal*, *15*(4), 4725–4734. doi:10.1109/JSYST.2020.3020058

Bhusal, N., Abdelmalak, M., Kamruzzaman, M., & Benidris, M. (2020). Power System Resilience: Current Practices, Challenges, and Future Directions. *IEEE Access : Practical Innovations, Open Solutions*, *8*, 18064–18086. doi:10.1109/ACCESS.2020.2968586

Chen, P., Wang, Y., Wen, X., Lan, L., Wang, C., Gan, Y., Zhang, L., Zhan, F., & Fu, P., (2010). Summary on the damaging effect of cryogenic freezing rain and snow disaster on power grid in China. *Dianwang Jishu/Power System Technology, 34*, 135–139.

Didier, M., Grauvogl, B., Steentoft, A., Ghosh, S., & Stojadinovic, B. (2017). Seismic Resilience of the Nepalese Power Supply System During the 2015 Gorkha Earthquake. *16th World Conference on Earthquake Engineering, 16WCEE 2017.*

Ding, T., Wang, Z., Jia, W., Chen, B., Chen, C., & Shahidehpour, M. (2020). Multiperiod Distribution System Restoration With Routing Repair Crews, Mobile Electric Vehicles, and Soft-Open-Point Networked Microgrids. *IEEE Transactions on Smart Grid*, *11*(6), 4795–4808. doi:10.1109/TSG.2020.3001952

Erenoglu, A. K., & Erdinc, O. (2023). Real-Time Allocation of Multi-Mobile Resources in Integrated Distribution and Transportation Systems for Resilient Electrical Grid. *IEEE Transactions on Power Delivery*, *38*(2), 1108–1119. doi:10.1109/TPWRD.2022.3206272

Erenoglu, A. K., Sancar, S., Terzi, I. S., Erdinc, O., Shafie-Khah, M., & Catalao, J. P. S. (2022). Resiliency-Driven Multi-Step Critical Load Restoration Strategy Integrating On-Call Electric Vehicle Fleet Management Services. *IEEE Transactions on Smart Grid, 13*(4), 3118–3132. doi:10.1109/TSG.2022.3155438

Gan, W., Shahidehpour, M., Guo, J., Yao, W., Pandey, S., Paaso, A., Vukojevic, A., & Wen, J. (2022). A Tri-Level Planning Approach to Resilient Expansion and Hardening of Coupled Power Distribution and Transportation Systems. *IEEE Transactions on Power Systems, 37*(2), 1495–1507. doi:10.1109/TPWRS.2021.3107402

Gao, H., Chen, Y., Mei, S., Huang, S., & Xu, Y. (2017). Resilience-Oriented Pre-Hurricane Resource Allocation in Distribution Systems Considering Electric Buses. *Proceedings of the IEEE, 105*(7), 1214–1233. doi:10.1109/JPROC.2017.2666548

Gazijahani, F. S., Salehi, J., & Shafie-khah, M. (2022). Benefiting from Energy-Hub Flexibilities to Reinforce Distribution System Resilience: A Pre- and Post-Disaster Management Model. *IEEE Systems Journal, 16*(2), 3381–3390. doi:10.1109/JSYST.2022.3147075

Gholami, A., Shekari, T., Amirioun, M. H., Aminifar, F., Amini, M. H., & Sargolzaei, A. (2018). Toward a Consensus on the Definition and Taxonomy of Power System Resilience. *IEEE Access : Practical Innovations, Open Solutions, 6*, 32035–32053. doi:10.1109/ACCESS.2018.2845378

Gilani, M. A., Kazemi, A., & Ghasemi, M. (2020). Distribution system resilience enhancement by microgrid formation considering distributed energy resources. *Energy, 191*, 116442. doi:10.1016/j.energy.2019.116442

Hosseini, M. M., & Parvania, M. (2020). Quantifying impacts of automation on resilience of distribution systems. *IET Smart Grid, 3*(2), 144–152. doi:10.1049/iet-stg.2019.0175

Hussain, A., & Musilek, P. (2022). Resilience Enhancement Strategies For and Through Electric Vehicles. *Sustainable Cities and Society, 80*, 103788. doi:10.1016/j.scs.2022.103788

Kandaperumal, G., & Srivastava, A. K. (2020). Resilience of the electric distribution systems: Concepts, classification, assessment, challenges, and research needs. *IET Smart Grid, 3*(2), 133–143. doi:10.1049/iet-stg.2019.0176

Kianmehr, E., Nikkhah, S., Vahidinasab, V., Giaouris, D., & Taylor, P. C. (2019). A Resilience-Based Architecture for Joint Distributed Energy Resources Allocation and Hourly Network Reconfiguration. *IEEE Transactions on Industrial Informatics, 15*(10), 5444–5455. doi:10.1109/TII.2019.2901538

Kim, J., & Dvorkin, Y. (2019). Enhancing Distribution System Resilience With Mobile Energy Storage and Microgrids. *IEEE Transactions on Smart Grid, 10*(5), 4996–5006. doi:10.1109/TSG.2018.2872521

Lei, S., Chen, C., Zhou, H., & Hou, Y. (2019). Routing and Scheduling of Mobile Power Sources for Distribution System Resilience Enhancement. *IEEE Transactions on Smart Grid, 10*(5), 5650–5662. doi:10.1109/TSG.2018.2889347

Li, B., Chen, Y., Wei, W., Huang, S., & Mei, S. (2021). Resilient Restoration of Distribution Systems in Coordination With Electric Bus Scheduling. *IEEE Transactions on Smart Grid, 12*(4), 3314–3325. doi:10.1109/TSG.2021.3060801

Li, B., Chen, Y., Wei, W., Mei, S., Hou, Y., & Shi, S. (2022). Preallocation of Electric Buses for Resilient Restoration of Distribution Network: A Data-Driven Robust Stochastic Optimization Method. *IEEE Systems Journal, 16*(2), 2753–2764. doi:10.1109/JSYST.2021.3123623

Li, J., Xu, X., Yan, Z., Wang, H., Shahidehpour, M., & Chen, Y. (2023). Coordinated Optimization of Emergency Response Resources in Transportation-Power Distribution Networks under Extreme Events. *IEEE Transactions on Smart Grid, 14*(6), 1–1. doi:10.1109/TSG.2023.3257040

Liang, K. T., Afzal, S., Mokhlis, H., Illias, H. A., Mansor, N. N., & Khairuddin, A. S. M. (2022). Assessing the Resilience of a Power Distribution System Considering Different Restoration Strategies. *2022 IEEE International Conference on Power Systems Technology (POWERCON),* 1–5. 10.1109/POWERCON53406.2022.9929870

Lin, Y., Bie, Z., & Qiu, A. (2018). A review of key strategies in realizing power system resilience. *Global Energy Interconnection, 1*, 70–78. https://doi.org/https://doi.org/10.14171/j.2096-5117.gei.2018.01.009

Lu, Z., Lu, X., Wang, W., & Wang, C. (2010). *Review and evaluation of security threats on the communication networks in the smart grid. In 2010 - MILCOM 2010 Military Communications Conference.* IEEE. doi:10.1109/MILCOM.2010.5679551

Ma, L., Wang, L., & Liu, Z. (2023). Soft Open Points-Assisted Resilience Enhancement of Power Distribution Networks Against Cyber Risks. *IEEE Transactions on Power Systems, 38*(1), 31–41. doi:10.1109/TPWRS.2022.3165169

Ma, S., Su, L., Wang, Z., Qiu, F., & Guo, G. (2018). Resilience Enhancement of Distribution Grids Against Extreme Weather Events. *IEEE Transactions on Power Systems, 33*(5), 4842–4853. doi:10.1109/TPWRS.2018.2822295

Mehrjerdi, H. (2021). Resilience-robustness improvement by adaptable operating pattern for electric vehicles in complementary solar-vehicle management. *Journal of Energy Storage, 37*, 102454. doi:10.1016/j.est.2021.102454

Momen, H., Abessi, A., & Jadid, S. (2020). Using EVs as distributed energy resources for critical load restoration in resilient power distribution systems. *IET Generation, Transmission & Distribution, 14*(18), 3750–3761. doi:10.1049/iet-gtd.2019.1561

Movahednia, M., Kargarian, A., Ozdemir, C. E., & Hagen, S. C. (2022). Power Grid Resilience Enhancement via Protecting Electrical Substations Against Flood Hazards: A Stochastic Framework. *IEEE Transactions on Industrial Informatics, 18*(3), 2132–2143. doi:10.1109/TII.2021.3100079

Najafi Tari, A., Sepasian, M. S., & Tourandaz Kenari, M. (2021). Resilience assessment and improvement of distribution networks against extreme weather events. *International Journal of Electrical Power & Energy Systems, 125*, 106414. doi:10.1016/j.ijepes.2020.106414

Nazemi, M., Moeini-Aghtaie, M., Fotuhi-Firuzabad, M., & Dehghanian, P. (2020). Energy Storage Planning for Enhanced Resilience of Power Distribution Networks Against Earthquakes. *IEEE Transactions on Sustainable Energy, 11*(2), 795–806. doi:10.1109/TSTE.2019.2907613

Obama, B. (2013). Presidential Policy Directive 21: Critical Infrastructure Security and Resilience. Academic Press.

Oboudi, M. H., Mohammadi, M., Trakas, D. N., & Hatziargyriou, N. D. (2021). A Systematic Method for Power System Hardening to Increase Resilience Against Earthquakes. *IEEE Systems Journal*, *15*(4), 4970–4979. doi:10.1109/JSYST.2020.3032783

Ouyang, M., & Duenas-Osorio, L. (2014). Multi-dimensional hurricane resilience assessment of electric power systems. *Structural Safety*, *48*, 15–24. doi:10.1016/j.strusafe.2014.01.001

Panteli, M., & Mancarella, P. (2015a). Influence of extreme weather and climate change on the resilience of power systems: Impacts and possible mitigation strategies. *Electric Power Systems Research*, *127*, 259–270. doi:10.1016/j.epsr.2015.06.012

Panteli, M., & Mancarella, P. (2015b). The Grid: Stronger, Bigger, Smarter?: Presenting a Conceptual Framework of Power System Resilience. *IEEE Power & Energy Magazine*, *13*(3), 58–66. doi:10.1109/MPE.2015.2397334

Panteli, M., Trakas, D. N., Mancarella, P., & Hatziargyriou, N. D. (2017). Power Systems Resilience Assessment: Hardening and Smart Operational Enhancement Strategies. *Proceedings of the IEEE*, *105*(7), 1202–1213. doi:10.1109/JPROC.2017.2691357

Poudel, S. (2020). *Quantifying and Improving Resilience in Power Distribution Systems. Doctoral dissertation.* Washington State University.

Rahimi, K., & Davoudi, M. (2018). Electric vehicles for improving resilience of distribution systems. *Sustainable Cities and Society*, *36*, 246–256. doi:10.1016/j.scs.2017.10.006

RazeghiG.LeeJ.SamuelsenS. (2021). *Resiliency Impacts of Plug-in Electric Vehicles in a Smart Grid.* doi:10.7922/G2CV4G13

Ritchie, H., Rosado, P., & Roser, M. (2022). *Natural Disasters*. https://ourworldindata.org/natural-disasters

Roudbari, A., Nateghi, A., Yousefi-khanghah, B., Asgharpour-Alamdari, H., & Zare, H. (2021). Resilience-oriented operation of smart grids by rescheduling of energy resources and electric vehicles management during extreme weather condition. Sustainable Energy. *Grids and Networks*, *28*, 100547. doi:10.1016/j.segan.2021.100547

Shin, H., & Baldick, R. (2017). Plug-In Electric Vehicle to Home (V2H) Operation Under a Grid Outage. *IEEE Transactions on Smart Grid*, *8*(4), 2032–2041. doi:10.1109/TSG.2016.2603502

Souto, L., Yip, J., Wu, W.-Y., Austgen, B., Kutanoglu, E., Hasenbein, J., Yang, Z.-L., King, C. W., & Santoso, S. (2022). Power system resilience to floods: Modeling, impact assessment, and mid-term mitigation strategies. *International Journal of Electrical Power & Energy Systems*, *135*, 107545. doi:10.1016/j.ijepes.2021.107545

Tian, M.-W., & Talebizadehsardari, P. (2021). Energy cost and efficiency analysis of building resilience against power outage by shared parking station for electric vehicles and demand response program. *Energy*, *215*, 119058. doi:10.1016/j.energy.2020.119058

Trakas, D. N., & Hatziargyriou, N. D. (2018). Optimal Distribution System Operation for Enhancing Resilience Against Wildfires. *IEEE Transactions on Power Systems*, *33*(2), 2260–2271. doi:10.1109/TPWRS.2017.2733224

Umunnakwe, A., Huang, H., Oikonomou, K., & Davis, K. R. (2021). Quantitative analysis of power systems resilience: Standardization, categorizations, and challenges. *Renewable & Sustainable Energy Reviews*, *149*, 111252. doi:10.1016/j.rser.2021.111252

Wu, C., Chen, C., Ma, Y., Li, F., Sui, Q., Lin, X., Wei, F., & Li, Z. (2022). Enhancing resilient restoration of distribution systems utilizing electric vehicles and supporting incentive mechanism. *Applied Energy*, *322*, 119452. doi:10.1016/j.apenergy.2022.119452

Wu, W., & Lin, B. (2021). Benefits of electric vehicles integrating into power grid. *Energy*, *224*, 120108. doi:10.1016/j.energy.2021.120108

Xu, Y., Wang, Y., He, J., Su, M., & Ni, P. (2019). Resilience-Oriented Distribution System Restoration Considering Mobile Emergency Resource Dispatch in Transportation System. *IEEE Access : Practical Innovations, Open Solutions*, *7*, 73899–73912. doi:10.1109/ACCESS.2019.2921017

Yang, Z., Dehghanian, P., & Nazemi, M. (2020). Seismic-Resilient Electric Power Distribution Systems: Harnessing the Mobility of Power Sources. *IEEE Transactions on Industry Applications*, *56*(3), 2304–2313. doi:10.1109/TIA.2020.2972854

Yuan, W., Wang, J., Qiu, F., Chen, C., Kang, C., & Zeng, B. (2016). Robust Optimization-Based Resilient Distribution Network Planning Against Natural Disasters. *IEEE Transactions on Smart Grid*, *7*(6), 2817–2826. doi:10.1109/TSG.2015.2513048

Zhang, B., Zhang, L., Tang, W., Wang, Z., & Wang, C. (2021). A coordinated restoration method of electric buses and network reconfiguration in distribution systems under extreme events. *CSEE Journal of Power and Energy Systems,* 1–9. doi:10.17775/CSEEJPES.2020.04320

Zhang, Z., Du, W., Lu, Y., Xu, G., & Zhao, Y. (2022). Two-Stage Robust Distribution Network Reconfiguration Against Failures of Lines and Renewable Generations. *IEEE Access : Practical Innovations, Open Solutions*, *10*, 108614–108624. doi:10.1109/ACCESS.2022.3213803

Zhao, J., Mili, L., & Wang, M. (2018). A Generalized False Data Injection Attacks Against Power System Nonlinear State Estimator and Countermeasures. *IEEE Transactions on Power Systems*, *33*(5), 4868–4877. doi:10.1109/TPWRS.2018.2794468

Chapter 8
An Overview of E-Mobility-Based Threats to the Power Grid:
Introduction to Smart Grids and Mobility Ecosystem

Neha Gupta
https://orcid.org/0000-0002-1201-777X
Sushant University, India

Nidhi Gupta
Maharaja Surajmal Institute of Technology, Delhi, India

ABSTRACT

The transportation industry is undergoing a rapid transition from fossil fuels to green energies. The emerging concept of mobility includes vehicles that not only provide environmentally friendly transportation, but also serve as storage for excess energy from renewable sources. The e-mobility infrastructure is becoming more complicated, which increases the danger of security threats. A valuable target for cyberattacks, the grid is one of the most significant vital infrastructures. Data exchange between EVs and smart grids is technically necessary. Smart grids must overcome these significant obstacles in order to guarantee security and privacy for eMobile charging. Blockchain when integrated with smart grid, as a possible solution, can solve these problems with its inherent trust building technology and makes it possible for the smart grid to manage demand response and engage in effective and reliable energy trading. The aim of this chapter is to describe the potential privacy risks of an ongoing connection between electric vehicles (EVs) and the smart grid.

DOI: 10.4018/979-8-3693-2611-4.ch008

I. INTRODUCTION

Energy being the essential factor for socio economic growth having proportional relation with the development of the country. Globally technology is depending on electricity which give rise to its demand (Pullins S, 2019). To prevent any adversity due to scarcity of energy and hampering socio economic growth, there is constant encouragement for use of renewable sources of energy. The development of the micro grid has transformed the electricity market, where renewable energy sources (RES) and distributed power generation (DEG) through green solar panels and wind turbines have gained popularity. The amount of sunlight falling on the photovoltaic panels in a given period and the speed of the wind moving the turbines have a significant impact on renewable energy, necessitating the use of batteries for continuous energy storage (Gupta N et al.,2017). As a result, the energy ecosystem has changed from a typical centralised approach to a decentralised model. These decentralized energy sources have specific properties that pose new challenges for the existing distribution system (He X et al., 2016). Renewable energy sources are regarded as the key to the success of the energy transition (Dileep, G. J. R. E., 2020), (Hossain M. S. et al., 2016).

One of the sectors is to promote transportation industry. Over the past three decades, there has been a significant global surge in the market share of hybrid and electric vehicles. The pressing concerns regarding climate change have spurred substantial efforts from both the public and private sectors (Boulanger, A. G. et al., 2017). Numerous governments and automakers have made commitments to incorporate electric or hybrid choices across their entire vehicle line-up, and some are even contemplating a complete shift to exclusively electric fleets. As the adoption of electric vehicles continues to expand, the electric grid is tasked with accommodating a higher overall demand and managing more pronounced load fluctuations. Despite the grid's overall capacity being capable of handling this demand, electric vehicles are more heavily utilized in urban settings compared to rural areas (Brown S et al.,2010). To encourage the adoption of electric mobility, micro grids have emerged as energy providers for EVs. Intelligent meters integrated with micro grids have the capability to gauge energy requirements and supply, ultimately reducing production expenses and electricity usage, thereby contributing to a more sustainable power grid (Sultan V et al., 2022).

II. EV INFRASTRUCTURES AND CONCERNS

As the ownership of electric vehicles grows, the prospect of large-scale EV charging presents significant challenges to power system planning, operation, and the electricity market. Even at relatively low levels of EV ownership, the uneven geographical distribution of EVs gives rise to issues within local power grids. The inherent, uncontrollable EV charging load leads to predicaments such as equipment overload, power loss, voltage fluctuations, decreased component lifespan, and compromised power supply reliability (Xiang Y et al., 2019).

The charging stations also do not flourish in abundance owing to the hardware expenditure, installation costs, uncertainties in utilization, lack of standards, etc. The operational costs such as maintenance purchase and local tariff also needs to be considered. There is concern of degradation of the batteries in case of fast charging. The temperature and state of the charge keeps varying rapidly in the battery due to the charging and discharging process and there exists a concern on lifetime and degradation of the batteries.

The integration of renewable energy sources (RER) or e-mobility as electric vehicles into the grid affects the performance, quality and energy efficiency of the grid. Thus, novel research of control strategies in E-mobility is essential (Gao Y et al., 2019), (Alam M et al., 2017). So far, the two most common electric vehicles (EVs) charging strategies are coordinated/controlled charging strategies and random/uncontrolled charging strategies. Since, In scenarios involving stochastic approaches, substantial adoption rates of EVs, both at medium and high levels could potentially generate significant adverse impacts on power grid quality and performance. Notably, the occurrence of considerable voltage fluctuations, particularly during peak power consumption periods, could transpire. Recent research findings indicate that the implementation of novel controlled solutions and charging strategies has the potential to mitigate the aforementioned issues, even when the proportion of EVs within the overall vehicle fleet is substantial (Fang X et al., 2011), (Dubey A and Santoso S, 2015).

The escalating intricacy of e-mobility infrastructure is giving rise to an augmented susceptibility to security threats. These threats have the potential to exert negative influences on interconnected power grid. The power grid becomes an attractive target for potential cyber-attacks. Consequently, the landscape is ripe for possible attacks on the power grid facilitated through e-mobility channels. An instance of this could involve triggering widespread blackouts through sudden surges in charging demand.

Despite the above, security and privacy concerns needs to be addressed during the realization of the charging models.

III. INTEGRATION OF SMART GRIDS AND E-VEHICLES

Viewed from the power grid's perspective, an electric vehicle (EV) emerges as a novel mobile power consumer, characterized by its relatively substantial power requirements and the intricate nature of predicting energy demands. These EVs also boast notable energy storage capacity and provide the flexibility to adjust power consumption timing according to preference. Integrating e-mobility harmoniously into the power grid necessitates a seamless alignment of power generation and consumption. This entails establishing bidirectional communication links, connecting both the EV to the e-mobility infrastructure and interlinking the e-mobility infrastructure with the power grid. These communication pathways are indispensable for overseeing the charging processes. In essence, this mandates an intelligent load management approach, enabling the dynamic adjustment of charging power to align with the available electricity supply (Kern D and Kraub C, 2021).

The examination of electric vehicle (EV) charging demands forms the foundational premise for scrutinizing the impacts on power grids and charging infrastructure planning. Typically, EVs are charged overnight using wall boxes at home, presenting a convenient integration opportunity into load management systems. Implementing load management in public spaces proves more intricate.

Charging at public facilities play a major role in catering to the changing needs of the electric vehicle users (Acharya S et al., 2020). The provision of charging during the shopping time or during their work hours enable users hassle free travels without requirement of allocating a dedicated time and effort for the purpose of charging. Shared charging is a scheme in which the private charging piles are made available to a group of electric vehicle users (Wang Y et al., 2023). This enables better utilization and profit for the pile owners whereas it provides better quality of experience and availability to the electric vehicle users.

In urban settings, suburban regions exhibit a notably higher concentration of households utilizing electric vehicles compared to their urban counterparts. These EVs adhere to a consistent charging pattern - disconnecting for daily commutes and reconnecting upon returning home. Concurrently, advancements in energy production technologies have resulted in increased clean energy generation, with solar energy gaining momentum, especially in decentralized applications such as residential and commercial installations. This decentralized energy generation transforms consumers into "prosumers," engaging in both electricity consumption and production (Sanguesa J A, et al., 2021).

As this trend continues in smart cities, households and businesses are increasingly becoming self-sufficient in electricity generation, sometimes generating surplus energy. During peak production periods, these prosumers can even generate an electricity surplus. If equipped with charging infrastructure, these consumers can sell their excess electricity (whether self-generated or not) to EV users, allowing them to charge using the prosumers' charging facilities (Grijalva S et al., 2011). However, smaller-scale prosumers often encounter challenges in pricing their electricity competitively, with their rates considerably lower than those offered by larger electricity suppliers or substantial prosumer entities (Nair A et al., 2021).

The integration of electric vehicles (EVs) into the smart grid holds considerable potential for utilities and various industry stakeholders, offering multiple advantages while facilitating the evolution of the smart grid concept (Xiao H et al., 2014). Given the intrinsic interdependency between e-mobility infrastructure and the power grid, disruptions or shifts in behaviour within one domain be able to profoundly impact the other. For example, it is widely anticipated that the continuous expansion of the Electric Vehicle (EV) market could lead to uncontrolled EV charging, which in turn could trigger detrimental repercussions for the power grid, encompassing diminished power quality, decreased transformer longevity, and heightened line losses (Verzijlbergh R A et al., 2012). This challenge can be mitigated through managed charging approaches (such as shifting the majority of EV charging to night-time hours. Furthermore, the implementation of Vehicle-to-Grid (V2G) power transfer has the potential to bolster grid stability (Tan K M et al., 2016). However, both these strategies necessitate a dependable e-mobility infrastructure capable of faithfully executing grid-aligned charging schedules and V2G strategies.

So, Blockchain in V2G plays as a virtual powerhouse for the EVs (Mwasilu F et al., 2014), (Liu C et al., 2018). This is a new technology being used in offering solutions to problems of smart grid and charging of EVs. Now the focus is on the examining scenarios involving active attackers with varying degrees of control over the e-mobility infrastructure, conventional grid stability techniques falter, giving rise to the prospect of targeted demand-side attacks.

A new type of grid users called prosumers who generate and consume electricity on the territory. Prosumers generally produces power by renewable energy sources such as solar and wind. Transmission losses are decreased by encouraging local energy production and use. Peer-to-peer electricity exchanges between prosumers and consumers should be possible. Power grid faces challenge with this discontinuous nature of renewable energy sources. To ensure the security of the power system, smart grids offer a way to combine these decentralised energy sources (Metke A R et al., 2010). Smart grids are designed to make it easier for prosumers and consumers to produce locally and consume locally (Dargan J et al., 2021), (Dewan S et al., 2021). Therefore, it would be obvious that a decentralised approach would be better.

Blockchain has advanced to a quick speed of development and application in many sectors of the modern economy, in contrast to many comparable technologies (Hua W et al., 2022), (Mengelkamp E et al., 2018) (Kumar, R et al., 2023), (Khanum, S., & Mustafa, K., 2023), (Kumar, V et al., 2023), (Solanki, A et al.,2022)

IV. VULNERABILITIES AND ATTACK ON SMART GRID DUE TO EMOBILITY

The integration of smart grid networks introduces enhancements and advancements to the traditional power grid, thereby increasing its complexity and susceptibility to a variety of potential attacks. These vulnerabilities could potentially grant unauthorized access to the network, compromise the confidentiality and integrity of transmitted data, and disrupt service availability. As outlined in (Shapsough S et al.,2010) the following vulnerabilities are particularly concerning within smart grid contexts:

1. **Customer Privacy and Security**: Smart meters gather substantial amounts of data and transmit it to utility companies, consumers, and service providers. This data often includes private consumer information that could be exploited to deduce consumer behaviours, device usage, and periods of home vacancy.
2. **Proliferation of Intelligent Devices:** Smart grids incorporate numerous intelligent devices responsible for managing both electricity supply and network demand. These devices could serve as entry points for potential attacks. Moreover, due to the considerable scale of the smart grid network (many times larger than the internet), monitoring and managing the network becomes exceptionally challenging.
3. **Physical Security**: Unlike traditional power systems confined to utility premises, the smart grid involves numerous components, many of which are situated outside the utility's controlled environment. This broader distribution of physical components increases the number of insecure locations and enhances their vulnerability to physical access.
4. **Legacy Equipment and Lifespan**: As power systems coexist with comparatively short-lived IT systems, outdated equipment remains in operation. Such equipment can become weak points in terms of security and might not be compatible with contemporary power system devices.
5. **Trust Issues in Device Communication:** Data spoofing, in which the condition of one device affects the actions of another, can occur during device communication in control systems. Manipulation of device states can lead to unintended behaviours among other devices.
6. **Diverse Backgrounds of Teams:** Inadequate communication and coordination among teams can lead to suboptimal decisions, resulting in increased vulnerabilities.
7. **Usage of Internet Protocol (IP) and Commercial Off-the-Shelf Components:** While utilizing IP standards in smart grids offers compatibility among components, it also exposes devices to various IP-based network attacks such as IP spoofing, Denial of Service and others.
8. **Increased Number of Stakeholders:** The presence of numerous stakeholders creates the potential for insider attacks, which can be particularly damaging.

In essence, the integration of smart grid systems brings about substantial benefits but also requires careful attention to these vulnerabilities to ensure the security and integrity of the network.

Drawing from these potential adversaries, the following e-mobility-based attack vectors that pose problems for grid resilience are identified:

1. **Charge Schedule Manipulation:** An adversary with access to a compromised grid operator or the capability to impersonate one can influence the operations of charging point operators (CPOs). By providing influenced grid capacity forecasts, CPOs may devise charge schedules that go beyond the real grid capacity. In the same way, if an adversary controls a jeopardized CPO or be able to

mimic its identity, they could directly instruct charging points (CPs) to alter their charge schedules, surpassing grid capacity. Furthermore, an adversary wielding a botnet of jeopardized CPs be able to swiftly modify their charge schedules, orchestrating demand-side attacks on the grid. Additionally, an adversary controlling a compromised fleet of EVs can adjust their selection of charging within permissible limits, thereby potentially destabilizing the grid (Krause T et al., 2021)

2. **Charge Price Manipulation:** Commonly, load-balancing mechanisms rely on price incentives to align the benefits of grid operators and consumers, such as offering lower prices during off-peak hours (Gan, L et al.,2012), (Crow M. L. 2016). Consequently, an adversary with the capacity to manipulate reported electricity prices could in some way effect the selection of charge schedules by EV programmers (or EVs programmed based on user preferences). This manipulation in turn affects electricity demand during specific time periods.
3. **Manipulation of Reported Data:** Analogous to false data injection threats in power grid state estimation (Liu Y et al., 2011), related vulnerability emerges in context of e-mobility integration. The grid operator requires accurate e-mobility demand data to predict and plan available and required grid capacity. This data encompasses real-time measurements from charging points, selected EV charge schedules, and reservations for charging points. Malicious tampering with these values might result in underestimating demand, effectively overestimating grid capacity and leading to scenarios of grid overload.

V. BLOCKCHAIN FOR EVS AND SMART GRID

Utilizing blockchain technology within the realm of electric vehicles (EVs) and the smart grid offers an array of advantages spanning cost reduction and the creation of novel revenue avenues for stakeholders across the ecosystem. Although coordinating vehicle-to-grid interactions might appear distinct from peer-to-peer electricity trading, the fundamental processes underlying these applications exhibit similarities. They both necessitate fundamental network services such as digital identity, permission management, data exchange, and more. At its core, blockchain establishes a foundation of trust, aligning seamlessly with the shared business interests of utilities, charging service providers, property proprietors, and EV drivers.

Moreover, blockchain empowers these entities to establish a unified data ledger, acting as the definitive source of critical information to streamline interactions. It lays the groundwork for a secure and reliable marketplace, where transactions can occur autonomously through efficient and automated smart contracts, negating the requirement for intermediaries.

As a primary smart grid application, blockchain has the potential to revolutionize trading infrastructure within this domain (Tanaka K et al.,2011). In future, consumers and prosumers within the smart grid trading electricity openly in peer-to-peer manner, without intermediaries, thereby instilling trust through blockchain. This application of blockchain-based trading infrastructure promises a host of benefits (Sabounchi M. and Wei J, 2017) and (Mannaro K et al.,2017).

Blockchain and Its Characteristics

Several industries and corporations are adopting the blockchain technology into their platforms for secured running of their operations and for faster transactions. Essential components of Block Chain are data structure, algorithm and cryptographic or security techniques. Blockchain is a distributed ledger of digital transactions where each transaction is attached to the chain as a block after proper verification by blocks previously on the chain. Each block is connected to the previous block of the chain through HASH of the previous Block. The transactions are stockpiled in Blockchain's block after validation process called Mining through consensus process. Hence there is no need for central validation agency since each block has the record of the transactions (Wang Q & Su M, 2020). Blockchain can overcome the security threat through identification, authorization and authentication (Lewis A, 2015).

Blockchain Architecture: The blockchain has distributed architecture as shown in Figure 1.

Figure 1. Centralised and Distributed Architecture

Blockchain Data Structure: In block chain data is stored in each node or block as shown in Figure 2. Each consists of three things i.e., Data, Hash function of earlier block and same block. Hash function or simply is the encrypted data of the block which is obtained by using an encrypted algorithm (SHA 256 or SHA 512). Every node is known or identified by its Hash value which is unique to every block and fingerprint of the block. Each block is connected to earlier block through HASH function of that block. Hash is calculated by adding nonce value to the data of the block i.e, (Nonce + block data). Nonce is 32bit arbitrary string which when added to block data creates unique Hash value. If at any stage, the data of any block changes or corrupts, its hash value (Alsmadi A et al.,2023) changes due to which the connection between the blocks cannot be maintained or the continuity of blockchain ceases.

Figure 2. Blockchain Data Structure

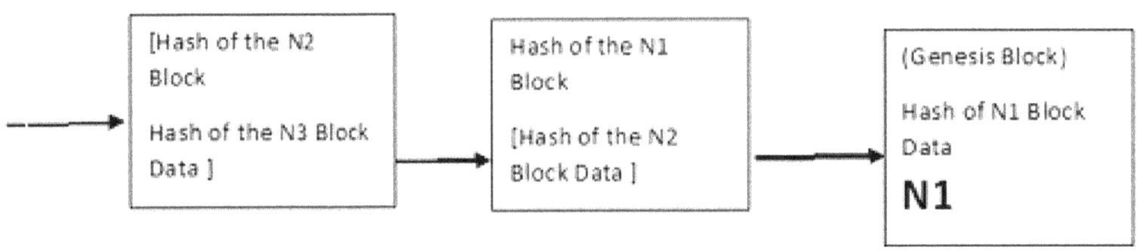

In this, a node producing a block require to evince that it has performed sufficient computation energy to resolve complex mathematical problem. The first node that solves this mathematical problem, publish the block in the network. The nodes that contribute their computing assets to determine solution to the puzzle and produce blocks are generally known as miners. Miners are awarded some fee for their efforts. In addition, the longest chain of block is often considered as a valid blockchain. Therefore, it is quite unachievable for an intruder to create a forged transaction as it will have to compute solution for all the consecutive blocks in the blockchain mathematically. This is even more impossible if the block is cryptographically associated with each other.

All the transaction occurring in the network are validated by acknowledging the consent of every node in the network by using consensus mechanism respectively and different blockchain based development platforms use different consensus mechanisms as shown in Figure 3. For instance, in the Bitcoin blockchain, all the transactions are validated in the network by using Proof-of- work consensus mechanism and nodes are rewarded for verifying the transaction in the network, this process is known as mining. The miner must solve a cryptographic puzzle to create new block. This method is known as "Proof-of-work", and it requires huge computational power to solve the puzzle. When all the transactions are verified in the network, an agreement exists amongst all nodes. Hence, new block is added to the blockchain by linking it with previous block, thus all the blocks are arranged in one consecutive chain and holds an increasing list of transaction records that are validated and confirmed by each node present in the network.

Figure 3. How Blockchain Works

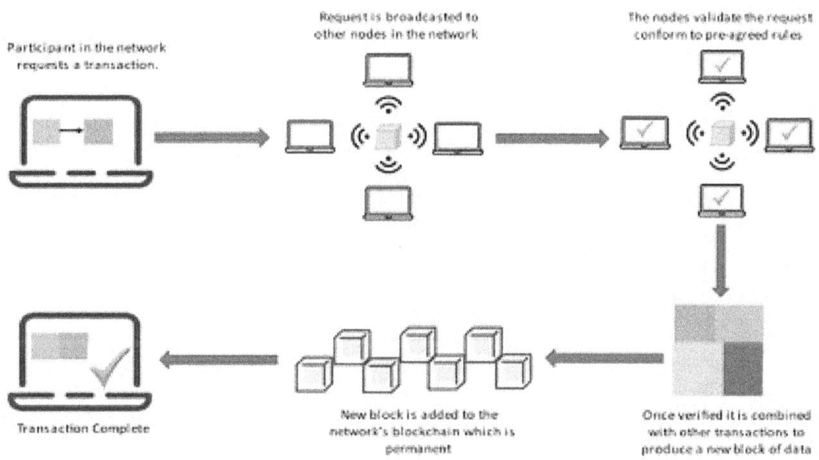

The information of every transaction ever occurred in the network is stored in the distributed ledger and shared with each node in the network. Thus, each node in the network has shared copy of data, this makes entire network transparent and increases trust among the nodes in the network. In addition, all the nodes in the blockchain are anonymous, thus it ensures privacy in the network. Another important feature of blockchain technology is Timestamping. Each block in the blockchain is timestamped and refer to the cryptographic hash of previous block. When these cryptographic hashes are linked together from the very first block (genesis block), this timestamped chain of blocks specifies an immutable record of transactions and maintains data integrity in the network. Thus, this is why blockchain has also been referred as Timechain .Furthermore, different blockchain types have different access protocols. The next section describes about different types of blockchain:

Types of Block Chain (Kang E et al.,2018): There are three types of blockchain available as shown in Figure 4. Detail description of blockchain network is shown in Table 1.

Figure 4. Types of Blockchain

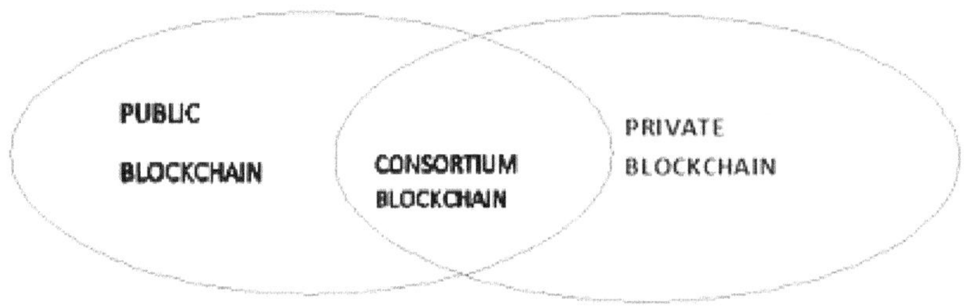

Table 1. Types of blockchain network

	Public	Private	Consortium
Network Type	Decentralized	Partially Decentralized	Partially Decentralized -Hybrid between public and private blockchains.
Functionality	Everyone can access through web. The data is checked by each node in the network.	Only authorized node can read or write in the network. The access is provided by the owner of blockchain.	The verification and permissions on the blockchain are controlled by several predetermined nodes.
Benefits	-Secure as every node verifies the transaction in the network. -Offers transparency as all the transactions can be viewed by each node in the network.	- Robust and Efficient, as verification is done by the owner only. -Retains privacy in the network as only the authorized nodes all permitted in the network.	-Efficient as few nodes verify the transactions in the network. -Retains privacy as read and write access are provided by predetermined nodes.
Challenges	Not efficient as all nodes need to verify the transaction.	Controlling power is consolidated to the owner of the organization.	

Open or Public Blockchain: This is permission less and non restrictive where anyone can join the blockchain. Examples are BITCOIN, Litcoin and Etherium etc

Private or Permissioned Blockchain: This is closed networked, restrictive and permission required type of blockchain where specific members or organizations can join the blockchain. This is more centralized than public blockchain. Examples are Hyperledger Fabric, Corda and Hyperledger Sawtooth etc.

Consortium or federated Blockchain: This is the hybrid combination of public and private type of blockchains. Here there are two or more central authorities which provide access to preselected blocks or node for specific purposes. Examples are Energy Web Foundation and IBM Food Trust etc.

Ultimately, blockchain's incorporation within the EV and smart grid landscape signifies a promising avenue for optimizing energy trading, management, and distribution, fostering a more efficient, decentralized, and secure energy ecosystem.

- Increased transparency and security in transactions.
- Facilitation of direct, efficient, and trustworthy energy exchange.
- Elimination of third-party involvement for enhanced cost-effectiveness.
- Empowerment of consumers and prosumers to take charge of their energy transactions.
- Enhanced grid resilience through distributed trading mechanisms.

Smart Contracts: Smart contracts are digital contracts that execute themselves when predetermined conditions are met. Nick Szabo introduced them in 1994. No lock intervention is required to run. Smart contract is stored as a new block of transactions on blockchain. They are stored on the blockchain in the form of software programs or logs and run once the requirements are met, without the need for intermediaries. The smart contract for energy trading verifies the accuracy and transparency of trades without government intervention (Kim S. K.and Huh J. H.,2018).

Smart Grid: The Smart Grid facilitates decentralized generation from a variety of traditional and sustainable sources and enables bi-directional energy flows (Hasankhani A et al.,2021). Smart grid technology is directing in the field of energy production and trading called "prosumer" as shown in Figure 5. Prosumers work in bidirectional exchange of energy with smart grid (Lombardi F et al.,2018)

Figure 5. Smart Grid Architecture

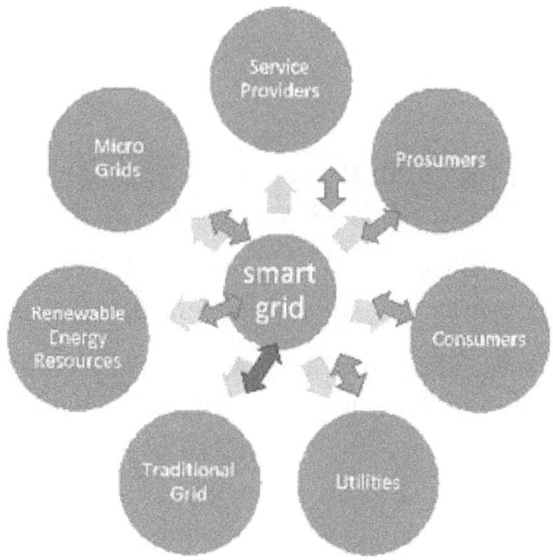

Integration of smart grid and EV might be able to solve issues of different complexity in a network with the applications built on the Blockchain.

VI. CONCLUSION

The incorporation of Information and Communication Technologies (ICT) into power grids has paved the way for efficient data exchange among various Smart Grid (SG) entities. This integration has been instrumental in enabling intelligent governance encompassing demand response management, frequency regulation, and voltage stabilization. However, this fusion also introduces a spectrum of security and privacy concerns, ranging from electricity theft and power loss to battery depletion and infrastructure mapping. These concerns gain prominence as distributed energy sources, such as electric vehicles (EVs), battery energy storage systems, and renewable energy sources, are integrated into SGs.

Given the intricate interplay between the e-mobility sector and power grids, the potential for e-mobility-based attacks targeting the grid becomes a viable threat. This concern becomes even more pronounced as the complexity of e-mobility infrastructure rises, and threat escalates in tandem with the expanding market share of EVs. In response to this challenge, the article proposes the adoption of a secure smart energy management system underpinned by blockchain technology.

In the context of e-mobility, power grids, and carbon-offset markets, blockchain technology offers a multitude of advantages. These advantages encompass cost reduction and the creation of fresh revenue opportunities for participants across the ecosystem. Whether it is optimizing energy trading, enhancing grid stability, or streamlining transaction processes, blockchain's potential is far-reaching. As e-mobility continues to gain traction, blockchain emerges as a pivotal tool to ensure security, privacy, and efficient management within this evolving landscape.

REFERENCES

Acharya, S., Dvorkin, Y., & Karri, R. (2020). Public plug-in electric vehicles+ grid data: Is a new cyberattack vector viable? *IEEE Transactions on Smart Grid*, *11*(6), 5099–5113. doi:10.1109/TSG.2020.2994177

Alam, M. M., Mekhilef, S., Seyedmahmoudian, M., & Horan, B. (2017). Dynamic charging of electric vehicle with negligible power transfer fluctuation. *Energies*, *10*(5), 701. doi:10.3390/en10050701

Alsmadi, A. A., Alrawashdeh, N., Al-Gasaymeh, A., & Alhawamdeh, L. N., & Al Hazimeh, A. M. D. (2023). Adoption of blockchain technology in supply chain. *SAGE Open*, *13*(1), 21582440231160143. doi:10.1177/21582440231160143

Boulanger, A. G., Chu, A. C., Maxx, S., & Waltz, D. L. (2011). Vehicle electrification: Status and issues. *Proceedings of the IEEE*, *99*(6), 1116–1138. doi:10.1109/JPROC.2011.2112750

Brown, S., Pyke, D., & Steenhof, P. (2010). Electric vehicles: The role and importance of standards in an emerging market. *Energy Policy*, *38*(7), 3797–3806. doi:10.1016/j.enpol.2010.02.059

Crow, M. L. (2016). Cost-constrained dynamic optimal electric vehicle charging. *IEEE Transactions on Sustainable Energy*, *8*(2), 716–724.

Dargan, J., Gupta, N., & Singh, L. (2021, November). Blockchain Based Energy Management System: A Proposed Model. In *2021 International Conference on Technological Advancements and Innovations (ICTAI)* (pp. 510-514). IEEE. 10.1109/ICTAI53825.2021.9673233

Dewan, S., Singh, L., & Gupta, N. (2021). Secure Electronic Voting System based on Mobile-app and Blockchain. *Recent Advances in Computer Science and Communications (Formerly: Recent Patents on Computer Science), 14*(9), 2992-3002.

Dileep, G. J. R. E. (2020). A survey on smart grid technologies and applications. *Renewable Energy, 146*, 2589–2625. doi:10.1016/j.renene.2019.08.092

Dubey, A., & Santoso, S. (2015). Electric vehicle charging on residential distribution systems: Impacts and mitigations. *IEEE Access : Practical Innovations, Open Solutions, 3*, 1871–1893. doi:10.1109/ACCESS.2015.2476996

Fang, X., Misra, S., Xue, G., & Yang, D. (2011). Smart grid—The new and improved power grid: A survey. *IEEE Communications Surveys and Tutorials, 14*(4), 944–980. doi:10.1109/SURV.2011.101911.00087

Gan, L., Topcu, U., & Low, S. H. (2012). Optimal decentralized protocol for electric vehicle charging. *IEEE Transactions on Power Systems, 28*(2), 940–951. doi:10.1109/TPWRS.2012.2210288

Gao, Y., Zhang, X., Cheng, Q., Guo, B., & Yang, J. (2019). Classification and review of the charging strategies for commercial lithium-ion batteries. *IEEE Access : Practical Innovations, Open Solutions, 7*, 43511–43524. doi:10.1109/ACCESS.2019.2906117

Grijalva, S., & Tariq, M. U. (2011, January). Prosumer-based smart grid architecture enables a flat, sustainable electricity industry. In *ISGT 2011* (pp. 1–6). IEEE. doi:10.1109/ISGT.2011.5759167

Gupta, N., Mahla, S. K., & Papreja, A. (2017). Experimental Studies of Utilization of Biogas with Biodiesel in Diesel Engine Generator for Electricity Generation. *International Journal of Engineering Research & Science, 3.*

Hasankhani, A., Hakimi, S. M., Bisheh-Niasar, M., Shafie-khah, M., & Asadolahi, H. (2021). Blockchain technology in the future smart grids: A comprehensive review and frameworks. *International Journal of Electrical Power & Energy Systems, 129*, 106811. doi:10.1016/j.ijepes.2021.106811

He, X., Qiu, R. C., Ai, Q., Chu, L., Xu, X., & Ling, Z. (2016). Designing for situation awareness of future power grids: An indicator system based on linear eigenvalue statistics of large random matrices. *IEEE Access : Practical Innovations, Open Solutions, 4*, 3557–3568. doi:10.1109/ACCESS.2016.2581838

Hossain, M. S., Madlool, N. A., Rahim, N. A., Selvaraj, J., Pandey, A. K., & Khan, A. F. (2016). Role of smart grid in renewable energy: An overview. *Renewable & Sustainable Energy Reviews, 60*, 1168–1184. doi:10.1016/j.rser.2015.09.098

Hua, W., Chen, Y., Qadrdan, M., Jiang, J., Sun, H., & Wu, J. (2022). Applications of blockchain and artificial intelligence technologies for enabling prosumers in smart grids: A review. *Renewable & Sustainable Energy Reviews, 161*, 112308. doi:10.1016/j.rser.2022.112308

Kang, E. S., Pee, S. J., Song, J. G., & Jang, J. W. (2018, April). A blockchain-based energy trading platform for smart homes in a microgrid. In *2018 3rd international conference on computer and communication systems (ICCCS)* (pp. 472-476). IEEE. 10.1109/CCOMS.2018.8463317

Kern, D., & Krauß, C. (2021, November). Analysis of e-mobility-based threats to power grid resilience. In *Proceedings of the 5th ACM Computer Science in Cars Symposium* (pp. 1-12). 10.1145/3488904.3493385

Khanum, S., & Mustafa, K. (2023). A systematic literature review on sensitive data protection in blockchain applications. *Concurrency and Computation: Practice and Experience, 35*(1), e7422.422.

Kim, S. K., & Huh, J. H. (2018). A study on the improvement of smart grid security performance and blockchain smart grid perspective. *Energies, 11*(8), 1973. doi:10.3390/en11081973

Krause, T., Ernst, R., Klaer, B., Hacker, I., & Henze, M. (2021). Cybersecurity in power grids: Challenges and opportunities. *Sensors (Basel), 21*(18), 6225. doi:10.3390/s21186225 PMID:34577432

Kumar, R., Jain, V., Leong, W. Y., & Teyarachakul, S. (2023). *Convergence of IoT*. Blockchain, and Computational Intelligence in Smart Cities.

Kumar, V., Jain, V., Sharma, B., Chatterjee, J. M., & Shrestha, R. (Eds.). (2022). *Smart City Infrastructure: The Blockchain Perspective*. John Wiley & Sons. doi:10.1002/9781119785569

Lewis, A. (2015). Blockchain technology explained. *Blockchain Technol*, 1-27.

Liu, C., Chai, K. K., Zhang, X., Lau, E. T., & Chen, Y. (2018). Adaptive blockchain-based electric vehicle participation scheme in smart grid platform. *IEEE Access : Practical Innovations, Open Solutions, 6*, 25657–25665. doi:10.1109/ACCESS.2018.2835309

Liu, Y., Ning, P., & Reiter, M. K. (2011). False data injection attacks against state estimation in electric power grids. *ACM Transactions on Information and System Security, 14*(1), 1–33. doi:10.1145/1952982.1952995

Lombardi, F., Aniello, L., De Angelis, S., Margheri, A., & Sassone, V. (2018). *A blockchain-based infrastructure for reliable and cost-effective IoT-aided smart grids*. Academic Press.

Mannaro, K., Pinna, A., & Marchesi, M. (2017, September). Crypto-trading: Blockchain-oriented energy market. In *2017 AEIT International Annual Conference* (pp. 1-5). IEEE.

Mengelkamp, E., Notheisen, B., Beer, C., Dauer, D., & Weinhardt, C. (2018). A blockchain-based smart grid: Towards sustainable local energy markets. *Computer Science (Berlin, Germany), 33*(1-2), 207–214. doi:10.1007/s00450-017-0360-9

Metke, A. R., & Ekl, R. L. (2010, January). Smart grid security technology. In *2010 Innovative Smart Grid Technologies (ISGT)* (pp. 1-7). IEEE.

Mwasilu, F., Justo, J. J., Kim, E. K., Do, T. D., & Jung, J. W. (2014). Electric vehicles and smart grid interaction: A review on vehicle to grid and renewable energy sources integration. *Renewable & Sustainable Energy Reviews, 34*, 501–516. doi:10.1016/j.rser.2014.03.031

Nair, A., Kapoor, C., NB, S. S., & Balamurugan, S. (2021, June). Exploring blockchain enabled smart community with electric vehicles. In *2021 10th IEEE International Conference on Communication Systems and Network Technologies (CSNT)* (pp. 745-750). IEEE.

Pullins, S. (2019). Why microgrids are becoming an important part of the energy infrastructure. *The Electricity Journal, 32*(5), 17–21. doi:10.1016/j.tej.2019.05.003

Sabounchi, M., & Wei, J. (2017, November). Towards resilient networked microgrids: Blockchain-enabled peer-to-peer electricity trading mechanism. In *2017 IEEE conference on energy internet and energy system integration (EI2)* (pp. 1-5). IEEE.

Sanguesa, J. A., Torres-Sanz, V., Garrido, P., Martinez, F. J., & Marquez-Barja, J. M. (2021). A review on electric vehicles: Technologies and challenges. *Smart Cities*, *4*(1), 372–404. doi:10.3390/smartcities4010022

Shapsough, S., Qatan, F., Aburukba, R., Aloul, F., & Al Ali, A. R. (2015, October). Smart grid cyber security: Challenges and solutions. In *2015 international conference on smart grid and clean energy technologies (ICSGCE)* (pp. 170-175). IEEE.

Solanki, A., Jain, V., & Gaur, L. (Eds.). (2022). *Applications of Blockchain and Big IoT Systems: Digital Solutions for Diverse Industries*. CRC Press. doi:10.1201/9781003231332

Sultan, V., Aryal, A., Chang, H., & Kral, J. (2022). Integration of EVs into the smart grid: A systematic literature review. *Energy Informatics*, *5*(1), 1–28. doi:10.1186/s42162-022-00251-2 PMID:35252758

Tan, K. M., Ramachandaramurthy, V. K., & Yong, J. Y. (2016). Integration of electric vehicles in smart grid: A review on vehicle to grid technologies and optimization techniques. *Renewable & Sustainable Energy Reviews*, *53*, 720–732. doi:10.1016/j.rser.2015.09.012

Tanaka, K., Nagakubo, K., & Abe, R. (2017, June). Blockchain-based electricity trading with Digitalgrid router. In *2017 IEEE international conference on consumer electronics-Taiwan (ICCE-TW)* (pp. 201-202). IEEE.

Verzijlbergh, R. A., Grond, M. O., Lukszo, Z., Slootweg, J. G., & Ilic, M. D. (2012). Network impacts and cost savings of controlled EV charging. *IEEE Transactions on Smart Grid*, *3*(3), 1203–1212. doi:10.1109/TSG.2012.2190307

Wang, Q., & Su, M. (2020). Integrating blockchain technology into the energy sector—From theory of blockchain to research and application of energy blockchain. *Computer Science Review*, *37*, 100275. doi:10.1016/j.cosrev.2020.100275

Wang, Y., Zhao, Z., & Baležentis, T. (2023). Benefit distribution in shared private charging pile projects based on modified Shapley value. *Energy*, *263*, 125720. doi:10.1016/j.energy.2022.125720

Xiang, Y., Hu, S., Liu, Y., Zhang, X., & Liu, J. (2019). Electric vehicles in smart grid: A survey on charging load modelling. *IET Smart Grid*, *2*(1), 25–33. doi:10.1049/iet-stg.2018.0053

Xiao, H., Huimei, Y., Chen, W., & Hongjun, L. (2014, June). A survey of influence of electrics vehicle charging on power grid. In *2014 9th IEEE conference on Industrial Electronics and Applications* (pp. 121-126). IEEE. 10.1109/ICIEA.2014.6931143

Chapter 9
Voltage Control of EV-Enabled Future Power Distribution Networks

Mudaser Rahman Dar
https://orcid.org/0009-0009-9418-4069
Indian Institute of Technology, Guwahati, India

Sanjib Ganguly
https://orcid.org/0000-0002-7939-7308
Indian Institute of Technology, Guwahati, India

ABSTRACT

Power distribution networks are swiftly transforming into active distribution networks (ADNs) due to increasing penetration of renewables and electric vehicles. This raises concerns among distribution system operators regarding the power quality issues. Frequent over/under voltages in ADNs are some of the major issues. The advanced devices and control techniques are to be used to prevent system voltage violations to enable efficient and reliable operation. To enhance the insight and comprehension, a precise description regarding the state-of-the-art devices used for voltage control in ADNs, their models and characteristics are provided in this chapter. Various advanced control strategies used for voltage control under different operation scenarios to highly uncertain future power networks are systematically presented, along with their merits and limitations. Finally, a voltage control approach based on dual-stage model predictive control framework is presented and tested on modified 33-bus distribution network to study the voltage control capabilities of PVs and EV charging stations.

DOI: 10.4018/979-8-3693-2611-4.ch009

Abbreviations

ADN: Active distribution network
ADMM: Alternating direction method of multipliers
ANM: Active network management
ANN: Artificial neural network
APC: Active power curtailment
AVR: Automatic voltage regulator
BESS: Battery energy storage system
CVC: Coordinated voltage control
CVR: Conservation voltage reduction
DER: Distributed energy resource
DG: Distributed generation
DNO: Distribution network operator
EV: Electric vehicle
EVCS: Electric vehicle charging station
FC: Fixed capacitor
G2V: Grid-to-vehicle
IMVBSO: Improved mixed real and binary vector based swarm optimization
LVR: Line voltage regulator
MILP: Mixed integer linear programming
MINLP: Mixed integer non-linear programming
MIQP: Mixed integer quadratic programming
MISOCP: Mixed integer second order cone programming
MPC: Model predictive control
OLTC: On-load tap changer
OPF: Optimal power flow
PCC: Point of common coupling
PV: Photo-voltaic
RTU: Remote terminal unit
SC: Switched capacitor
SOP: Soft-open-point
TSO: Transmission system operator
VSC: Voltage sources converter
V2G: Vehicle-to-grid
VVC: Volt/var control
VVO: Volt /var optimization

1. INTRODUCTION

The significant penetration of distributed generation (DG) and electric vehicles (EVs) in recent years have remarkably changed the electric distribution networks. The structural changes rendered by the DG penetration notably photovoltaic (PV) systems have turned the conventional networks into dynamic active distribution networks (ADNs). Contemplating the participation in ADN management, distributed energy resources (DERs) encourage and reassure reliable, economical, secure and environmentally safer ways to energy supply, in addition to being effective for increasing energy demand (Lopes et al., 2007). However, temporal uncertainties in renewable generation and EV loads lead to over/under voltages, stochasticity in power demand, feeder power variations and reverse power flow, thus raising technical and operational difficulties to distribution network operators (DNOs). The stochastic nature of ADNs drives the system to insecure situations and unforeseen incidents, which may include incompetent system protection, thermal limit violation of equipments, voltage regulation issues and change in fault levels (Bidgoli and Cutsem, 2018). This necessitates the incorporation of advanced control and communication setups to manage the evolved power networks for efficient and reliable operation (Hidalgo et al., 2010; Walling et al., 2008). Voltage rise issues form one of the significant concerns in ADNs, restricting the wide integration of distributed generation.

The node voltage violations due to uncertain generation interfere with operation of conventional voltage control equipment (on load tap changer (OLTC)) leading to increased wear and tear and improper operation. The problem of voltage regulation is deteriorated in situations of simultaneous under/over voltages due to uncertainty in generation (PVs) and demand (EVs), thus restraining the use of only traditional device and control strategies. The existing networks lack the infrastructure for topological changes brought about by expansive integration of PVs and electric vehicle charging stations (EVCS), forcing steps towards active network management (ANM) (Ranamuka et al., 2016). ANM involves deployment of advanced control and communication to control generation and demand, which are potentially effective for voltage control and optimal network operation, utilizing the fast acting capabilities of advanced control devices. The power electronic based devices and advancements in information and communication technology promise to mitigate issues raised by DERs, by employing their reactive power capabilities for effective volt/var control (VVC). This significantly enhances connected capability and thus helps to avoid the DG or load disconnection.

Future power networks will see immense rise in plug-in electric vehicles (PEVs) motivated by the efforts on reducing carbon footprint. The uncertainty in their charging /discharging pattern, state-of-charge, and randomness in their spread over the power network pose issues of voltage deviations, power loss, and localized system overloading, if uncoordinated (Ma et al., 2018). PEVs are integrated either in grid-to-vehicle (G2V) mode (with EVs as only load) or vehicle-to-grid (V2G) mode, wherein EV batteries can be employed for grid services such as peak demand matching. Participation of V2G technology as auxiliary services and smart charging schemes can be utilized for smoothing voltage fluctuations and keeping the system voltage stable. With the concept of active distribution network things like smart meters, control equipments, bidirectional communication links forming a complex communication network, and adjustable loads increase the computation burden as well as communication burden to heavy extent. Furthermore, data acquisition and metering become essential components in ADN scenario, with increased demand for data collection, which may raise concerns among the users regarding data privacy protection (Tang et al., 2021).

2. VOLTAGE FLUCTUATION IN ADNS

Voltage fluctuation in distribution networks is largely attributed to the integration of PVs and PEVs. PVs raise voltage particularly at feeder ends under lower load and higher generation, affecting the voltage profile of nodes which are relatively more sensitive to the power variations at PV node. Similarly, EV charging adds to the power demand leading to localized voltage dips. In certain scenarios, simultaneous over voltage and under voltage can aggravate the voltage profile, wherein, large dips may lead to instability in weak networks. The unaddressed issues related to voltage regulation may result in voltage collapse. A typical PV source induced voltage rise problem in an ADN, considering PV generation greater than local demand is shown in Figure 1, wherein, reverse flow of power from PV source back to utility grid elevates the voltage at PV node. From Figure 1, the voltage rise $\Delta\overline{\upsilon} = \Delta\upsilon_r + j\Delta\upsilon_x = \overline{\upsilon}_{DG} - \overline{\upsilon}_g$ is voltage difference between PV node and node connected to utility grid, linked through line impedance $z_l = r_l + jx_l$ and expressed as:

$$\Delta\overline{\upsilon} = \frac{(P - jQ)}{\overline{\upsilon}_g^*}(r_l + jx_l) = \frac{(r_l P + x_l Q)}{\overline{\upsilon}_g^*} + j\frac{(x_l P - r_l Q)}{\overline{\upsilon}_g^*} \tag{1}$$

Where, $P = P_{DG} - P_L - P_{loss}$ is active power and $Q = Q_{DG} - Q_L - Q_{loss}$ is reactive power flowing to the grid. In LV distribution systems R/X is high, thus assuming $\overline{\upsilon}_g = 1\angle 0^0$, $\Delta\overline{\upsilon}$ can be approximated to real component $\Delta\upsilon_r$. Neglecting reactive component of voltage deviation ($\Delta\upsilon_x$), voltage deviation reduces to the equation:

$$\Delta\overline{\upsilon} \simeq r_l P + x_l Q \tag{2}$$

Figure 1. Grid connected PV system

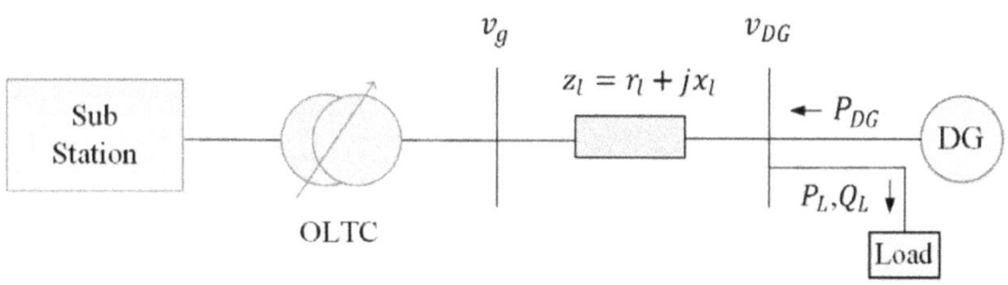

For the voltage regulation of distribution system approaching nominal value, $\Delta\overline{\upsilon} = 0$, thus (Shuvra and Chowdhury, 2018):

$$\frac{r_l}{x_l} \simeq -\frac{P}{Q} \tag{3}$$

It is noteworthy, that the voltage has more sensitivity towards active power change in low voltage (LV) networks, on account of high line resistance (r_l) to reactance (x_l) ratio ($r_l > x_l$), thus, active power control forms a vital component of voltage regulation strategy. Under the reactive power insufficiency, active power control utilizing battery storage, heat pumps and controllable loads such as V2G services of EVs provide alternatives. To mitigate the over voltages, active power curtailment (APC) of PVs provides another way of voltage control. However it is mainly abstained and kept as a last resort to avoid loss of renewable energy.

The permissible voltage deviation in a distribution network is governed by the national standard. For instance, in accordance with definition of range A in standard ANSI C84.1-2016, the medium voltage deviation is restricted to upper bound and lower bound of +5% and –2.5%, respectively, while as LV network voltage bounds are limited to ±5% (ANSI, 2020). Similarly, in Germany the grid voltage is required to stay within ±10% of the rms voltage averaged over 10 minutes as per European standard EN50160 (Troester, 2009). The voltages should remain in this range for greater than 95% of all 10 minute time steps of a week, and for less than 5% of all 10 minute intervals in a week the voltage can go below 90% of the nominal value but should stay above 85% of nominal voltage. In power networks with high PV penetration, more strict requirements are demanded during peak power generations. For example, in Hawaii (USA) it is demanded by power grid that the voltage during peak generation hours in day must stay inside 0.975 p.u. and 1.025 p.u., these requirements are relaxed to maximum deviation of ±5% in off peak hours. The service voltage variation range of some countries are provided in Table 1.

Table 1. Standard service voltage variation ranges in different countries (Lee & Kim, 2017)

Country	Nominal Voltage	Mandatory Regulation Limits (min/max)
Japan	100 V	-6%/+6%
U.S.	120 V	-5%/+5%
Germany	230V	-10% /+10%
Australia	230V	-6.1%/+10%
Korea	220V	-5.9%/+5.9%
U.K.	230V	-6%/+10%
Canada	230V	-8.3%/+4.2%

Traditionally devices like OLTC, line voltage regulators (LVRs), switched capacitors (SCs) and feeder capacitors (FCs) are used for voltage control. OLTC and LVRs worked on the notion of unidirectional flow of power. The DER integration and evolution to active distribution networks affect the operation of these conventional devices due to bidirectional power flow. The traditional line voltage regulators (LVRs) are operated based on unidirectional power flow, by sensing voltage on load side. The voltage may rise sufficiently at DER connected node. Thus, it may render LVR insufficient to regulate voltage due to its limited tap operations. Under line drop compensation operation mode of LVR, DER can further aggravate the voltage elevation problem under the reverse power-flow. In this way DER integration interferes differently with the LVR in its all modes of operation. Traditional distribution networks engaged manual tap settings or automatic tap changing by employing OLTC to limit the voltage within pre-specified limits. The OLTC changes its tap setting by measuring the feeder current and thus voltage drop of the feeder, assuming unidirectional flow. Bi-directional power flow due to active generation makes the operation of OLTC unforeseeable and uncontrollable (Sun et al., 2019). Moreover, OLTC is not able to regulate voltage for fast fluctuations of renewable sources due it its comparatively slow operation. Furthermore, the capacitor bank reduces the reactive power flow towards load and further improves voltage at common coupling point (PCC). Uncoordinated DER off-set currents in the feeder, steering bidirectional currents, thus interacting with capacitor switching operations. The switching of capacitors in traditional networks is controlled assuming downstream currents in the feeders. It is noteworthy that in conventional control, voltage is controlled locally by OLTC and shunt capacitors, thus, there is no communication or use of remote-control facility.

The conventional voltage control devices are not competent with the rapidly fluctuating voltages due their discrete and slow response, thus necessitating the deployment of additional voltage control device in the vicinity of voltage violated nodes (Yan and Saha, 2012). Incorporation of DERs for fast voltage control seems a promising solution (Shuvra and Chowdhury, 2018). New control equipment should operate in coordination with the conventional devices to carry through proper voltage regulation and mitigate network losses. The complexity of coordinated regulation enforces the utilities to adopt stringent requirements for grid connection of DERs. Previously, DG units were not permitted to actively regulate the voltage in distribution network, in order to prevent interference with existing voltage balancing schemes and overloading of equipments, due to additional reactive current (Arshad et al., 2018). The revised standards permit the DER owners to control voltage by active/reactive power management (IEEE, 2018). With proper cooperation between distribution system operator (DSO) and DG operator, DG can provide regulated active and reactive power as per IEEE 1547-2018 (IEEE, 2018), with active power adjustments mandated for supporting the proper voltage regulation. To address the voltage control issue various control devices and strategies are suggested by the researchers which comprise of (a) reactive power control (VAR compensation) (b) curtailment of active power (c) reactive power control by DG inverters (d) area based voltage control operation of OLTC (e) distributed energy storage devices (f) load control (g) reactive power control by inverters at EVCSs and (h) V2G services of PEVs.

Recent years have seen enormous work carried out for VVC in ADNs utilizing control capabilities of PVs, PEVs, and custom power devices. Model based control strategies utilize the sensitivities of node voltage deviations. These sensitivities are calculated with respect to the change in control variables of various control devices operated in coordination with each other. The sensitivity terms with respect to active as well as reactive power maybe numerically extracted by inverting jacobian matrix or can be calculated using intuitive analytical methods from power flow equations. For a network having radial topology, the VVC control can be simply implemented using linearized load flow (Distflow) or branch flow model, using simple node equations. Considering a radial network shown in Figure 2, having N nodes and E branches, the power flow in the network can be written by following equations:

$$P_i = \sum_{E^{sub}} P_{ij} + \frac{P_{hi}^2 + Q_{hi}^2}{v_h^2}.r_{hi} - P_{hi} \quad (4)$$

$$Q_i = \sum_{E^{sub}} Q_{ij} + \frac{P_{hi}^2 + Q_{hi}^2}{v_h^2}.x_{hi} - Q_{hi} \quad (5)$$

$$v_h^2 - v_i^2 = 2(r_{hi}P_{hi} + x_{hi}Q_{hi}) - (r_{hi}^2 + x_{hi}^2)\frac{P_{hi}^2 + Q_{hi}^2}{v_h^2} \quad (6)$$

Where, $(i,j)=(i,m)$, (i,p), (i,l), $\in E^{sub} \subset E$ is the branch set connected to node i in downstream of power flow; (h,i), $(i,j) \in E$ signify branches connected between nodes h and i, and nodes i and j, respectively; v_h, v_i, P_i, Q_i are voltage at network node h, voltage at node network i, injected active power at network node i and injected reactive power at network node i, respectively. Powers P_{hi}, Q_{hi} are the power flows in branch from connected node h to connected node i; $r_{hi}+jx_{hi}$ is the impedance of branch joining nodes h and node$i$.

Figure 2. A radial distribution network (illustration)

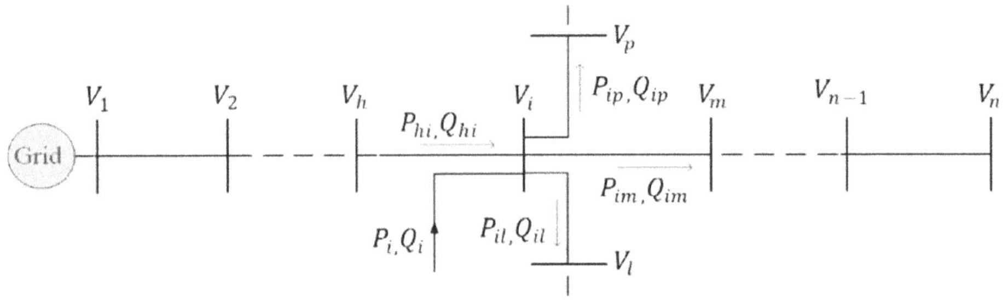

3. VOLTAGE CONTROL DEVICES EMPLOYED IN ADNS

In this section, the voltage control devices used for control in ADNs, their models and operation constraints are presented, which include conventional slow acting devices and power electronic based fast acting devices.

3.1. Distributed Energy Resources (DERs)

The increasing demand stress on the distribution system demands more generation from the supply authority. The integration of DERs provides a promising way to avoid or put off the expenditure for the installation of large power plants (Tanaka, 2009). Apart from supporting the increasing active power demand, DERs have numerous capabilities in dealing with power quality issues like voltage regulation.

3.1.1. Distributed Generation (DG)

With reference to active distribution networks, DG units can be used for frequent voltage deviations and to avoid network reinforcements. The reactive as well as active power exchange capability of smart inverter equipped DG units like solar PVs can be utilized for voltage control, subjected to the capacity of interfacing inverters (Martinez et al., 2017). DER inverters mostly operate at lower powers than their ratings which establish provision for operating them to inject/absorb reactive power for maintaining voltages, by utilizing the reserve capacity (Du and Lu, 2014). DG units when operated for voltage control can be operated in a mode in which power factor is constant. Therefore, the reactive power injection from the inverter changes dynamically in accordance with active generation. The other way around called voltage adaptive control (VAC) employs the droop in voltage (P/V and Q/V droop) (Olival et al., 2017).

While dispensing DG units for voltage control, utilities may need inflexible operational preconditions from DG owners. For example, as per standard VDE-AR-N 4105, power grid in Germany requires DG units of capacity between 3.68 kVA and 13.8 kVA to maintain power factor 0.95 (lagging or leading) while operation at MV level. For LV level with DG units of capacity above 13.8 kVA, the power factor is demanded to remain within 0.90 leading or lagging (VDE, 2011). Under changes in active power, the reactive power support may be through active power /power factor characteristic or voltage/reactive power characteristic or through power factor/reactive power set point (Schwartfeger et al., 2014). The latest PV smart inverters are able to participate in VVC in centralized manner in which the operating power factor is determined by communication based central control algorithm, utilizing reactive power/ voltage characteristic. As per revised IEEE standard 1547-2018, DERs are required to provide following four reactive power support functions (i) constant reactive power mode (ii) active power-reactive power mode (iii) constant power factor mode (iv) voltage-reactive power mode. DERs shall be able to supply reactive power under the situations where active power generation is 5% or above of the rated output. Under maximum power generation DERs are required to exchange reactive power of at least 44% of the apparent power rating. When the active power generation is within 5% - 20%, the DERs should be able to supply reactive power equal to 44% of rated value multiplied by the fraction of active generation output to the 20% of rated value, as per IEEE standard 1547-2018 (IEEE, 2018). Renewable DG units (PVs) operate under maximum power point tracking, thus, DG inverters need to be slightly over rated by around 10%, to be able to keep up reactive power supply of 44% under maximum power generation. The inverter interfaced DG units are generally used to provide reactive power injection/absorption instantly according to the piece-wise linear Q (V) curves shown in Figure 3. The reactive power injection of DG inverter from QV curve in Figure 3, can be written as:

$$Q_{inj} = \begin{cases} 0 & V_{low1} \le V \le V_{up1} \\ Q_{inj}^{\max} & V \le V_{low2} \\ -Q_{inj}^{\max} & V \ge V_{up2} \\ \dfrac{V - V_{low1}}{V_{low2} - V_{low1}} Q_{inj}^{\max} & V_{low2} \le V \le V_{low1} \\ -\dfrac{V - V_{up1}}{V_{low2} - V_{low1}} Q_{inj}^{\max} & V_{up1} \le V \le V_{up2} \end{cases} \tag{7}$$

The operational constraints of DG inverters can be expressed as:

$$Q_{DGU}^{\min}(t) \le Q_{DGU}(t) \le Q_{DGU}^{\max}(t) \tag{8}$$

$$Q_{DGU}^{\max}(t) = \sqrt{S_{DGU}^2 - P_{DGU}^2(t)} \tag{9}$$

$$Q_{DGU}^{\min}(t) = -\sqrt{S_{DGU}^2 - P_{DGU}^2(t)} \tag{10}$$

Where, $Q_{DGU}^{\min}(t)$ and $Q_{DGU}^{\max}(t)$ are the reactive power limits available at time *t*, corresponding to the active power generation ($P_{DGU}(t)$), and S_{DGU} is the *VA* rating of the inverter of DG unit.

Figure 3. Q (V) curve of DG inverters for local control

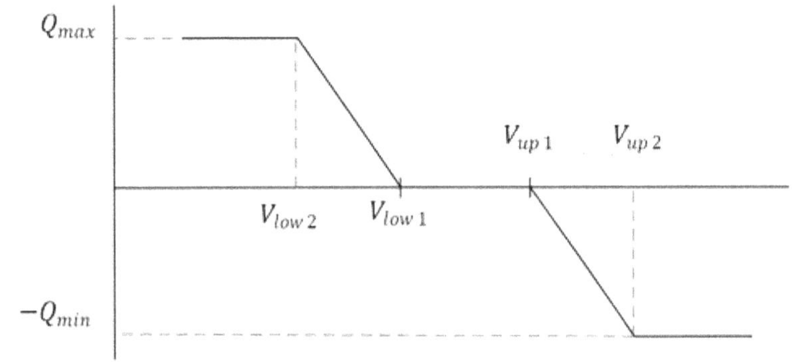

Recent years have seen a considerable attention towards the VVC capabilities of PVs among all DERs. A substantial amount of research is devoted to the studies of coordination of distributed PV sources such as in (Amanipoor et al., 2023; Anurag et al., 2015; Chai et al., 2021; Ding et al., 2017; Safayet et al., 2017; Wang et al., 2020; Zhang and Xu, 2020) to provide reactive power support both from grid operator end as well as from customer end. Voltage regulation of radial LV networks is studied in (Safayet et al., 2017) to exploit the available reactive power of PV inverters, wherein, voltage sensitivity to critical buses in the network are utilized. The randomness of the rooftop PV integration significantly increases the imbalance in power in LV distribution network thus limiting efficacy of most VVC techniques, devised assuming balanced phases. To address the issue, a consensus based distributed control mechanism is provided in (Wang et al., 2020) considering unbalanced phase powers. The PVs are coordinated centrally in (Zhang and Xu, 2020) to dispatch reactive power and provide droop functions for real time local control, to have effective coordination at both hierarchical stages. Authors in (Ding et al., 2017) propose a two-stage model for optimal dispatching of PV inverters to maximize the PV power surplus and to decide the optimum PV sub-set for reactive power support under uncertainties. It is noteworthy to mention that conventional volt/var control using PV inverters dispatch required reactive power subjected to the VA rating of inverter. The full utilization in real practice depends on rated current and thus limits the full capability utilization, available as per apparent power rating. Furthermore, most of the techniques do not provide the efficacy during transient voltage overshoots due to sudden generation change. To address these issues a decentralized volt/var control scheme, using a voltage-oriented control technique is provided in (Amanipoor et al., 2023), to effectively eliminate transient overshoots in voltage. The authors provide control strategy to fully utilize the inverter capabilities based on rms value of rated current. Furthermore, The VVC control methods utilizing reactive power capability of inverters widely ignore the reliability assessment of inverters. Utilizing inverters for reactive power support significantly reduces their reliability and thus life span, which is usually less than other components of distribution network (Anurag et al., 2015). Since, existing reliability assessment measures are difficult to implement in conjunction with VVC, authors in (Chai et al., 2021) have attempted to bring forth detailed reliability modelling and have proposed VVC optimization considering the reliability constraints, to enhance the life of inverters.

3.1.1.1. Generation Curtailment

Voltage rise in ADN can also be alleviated by bringing down or curtailing the DG output active power. The reactive power compensation and active power curtailment (APC) of the smart inverter interfaced PVs are the major techniques used for ADN voltage control (Luo et al., 2021). The active power can be curtailed much easily by disconnecting the specific DG or group of DG units. If the control for mitigating the DG output is available like the pitch control of wind turbines, then the complete detachment of DG is not necessary. Renewable generation curtailment lead to the loss of green energy. Further, being uneconomical DG owners discourage the active power curtailment. Therefore, this control strategy is reserved as last availing resource or for emergency conditions, when all other control resources are exhausted.

3.1.2. Distributed Storage

As far as the distributed energy storage is considered the battery energy storage systems (BESS) and PEVs are mainly employed for voltage regulation, owing to their operation at lagging or leading power factors. Due to recent advancements in lithium-ion battery technology such as low self-discharging tendencies and their high energy density, they are affordably employed in the PEV applications (Bahramipanah et al., 2016; Enang and Bannister, 2017). The effective response of battery storage to transients, ability to meet peak demand, application in energy management and ability for voltage regulation (using its DC/AC bidirectional converter) have set about the wide integration of BESS in the LV distributions networks (Amini et al., 2016; Li et al., 2019). A simple battery model considering the power flowing in and out, is given by equations:

$$\frac{d}{dt}SOC = -\frac{\eta . P_b}{J_{rated}} \tag{11}$$

$$-P_b^{rated} \le P_b \le P_b^{rated} \tag{12}$$

$$0 \le SOC \le 1 \tag{13}$$

Where, SOC means the state-of-charge; J_{rated} is the energy capacity; P_b is instantaneous power flowing in or out of the battery; η is the efficiency, where $\eta = \eta_{out}$ while discharging ($P_b > 0$) and $\eta = \eta_{in}$ while charging ($P_b < 0$). Furthermore, BESS are suitable storage devices for high energy and long duration applications as compared to the non-battery type storage systems, which are applicable for short durations (Bahramipanah et al., 2016). In similar fashion, the PEVs can be employed for voltage control during high demand using V2G technology. The batteries used in PEV can be used to provide V2G services by proper management of charging period and discharging services needed under heavy demands. Even though the PEV integration gives rise to additional demand in the power distribution network, using advanced communication technologies and metering devices, they can be effectively utilized to provide services for grid stability (Elkhatib et al., 2011; Kulmala et al, 2010). The charging strategy can be centralized in which the utility controls the charging scheme and in decentralized way in which users have control based on the incentives in price.

3.2. OLTC and Capacitor Banks (CBs)

The OLTC uses the local current and voltage measurements to regulate its tapping operation based on unidirectional power flow, to maintain voltage locally and does not require any communication network (Kulmala et al., 2014). The OLTC operation can be integrated in the centralized control scheme where the set points of all the control equipments are coordinated in a centralized manner (Enang & Bannister, 2017). A simple model of an OLTC equipped tap changing transformer connected at the substation with OLTC operation on secondary coil is shown in Figure 4. If the primary side and secondary side voltages are VP a_nd VS $r_e$spectively, the relation of the secondary voltage corresponding to the different tap positions and operational constraints can be expressed as (Guo et al., 2020):

$$\frac{V_S}{V_P} = (1 + n_{tap}.\Delta V_{tap})\frac{V_{oP}}{V_{oS}} \tag{14}$$

$$n_{tap}^{\min} \le n_{tap}(k) \le n_{tap}^{\max} \tag{15}$$

$$\Delta n_{tap}^{\min} \le \Delta n_{tap}(k) \le \Delta n_{tap}^{\max} \tag{16}$$

Where, n_{tap} and ΔV_{tap} are the tap position and discrete voltage change per step, respectively. Voltages V_{oP} and V_{oS} are the nominal values of transformer voltages on primary and secondary side.

Figure 4. Eq. circuit of an OLTC transformer

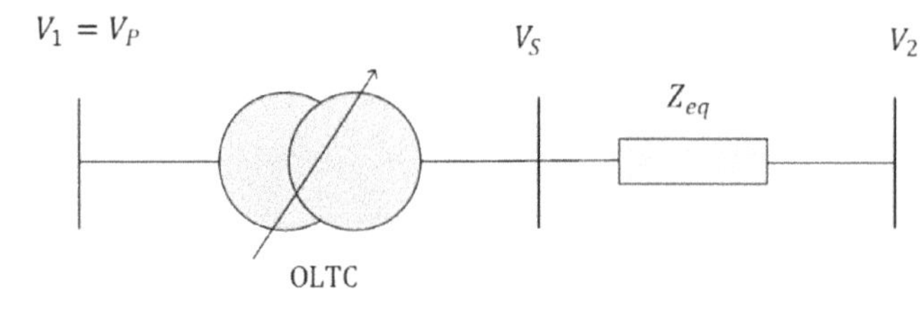

The reactive power injection of capacitor banks change in discrete steps, which can be expressed as:

$$Q_{cap}(t) = n_{cap}(k).q_{cap} \tag{17}$$

Where, n_{cap} and q_{cap} are number of CB units in use and maximum reactive power limit of each unit, respectively. The reactive power injection of CBs is restricted by the number of units and maximum change in reactive power in certain time interval, which can be expressed using the constraints:

$$0 \leq n_{cap}(k) \leq n_{cap}^{max} \tag{18}$$

$$\Delta n_{cap}^{min} \leq \Delta n_{cap}(k) \leq \Delta n_{cap}^{max} \tag{19}$$

3.3. Electric Vehicles (EVs)

The widespread incorporation of Electric Vehicles (EVs) in the last decade has added significantly to the emerging distribution system as controllable loads, thus necessitating the study of advanced techniques and mechanisms for EV integration and its impacts. The large scale deployment of EVs (variable load) pose different challenges like randomness in charging/discharging patterns due to the stochasticity associated with their frequency in visiting parking lots, charging capacities, and charging demand. However, EV integrations potentially have numerous advantages to distribution network. Different from conventional loads, EVs can at times be deployed in distribution grid for backing reliable, efficient, stable and resilient operation (Z. Li et al., 2019). Using V2G operation, active and reactive power from EVCS is used for grid services. EVs can act as energy storage systems, which possess capability to provide energy to the grid or customers under needed circumstances, employing V2G technology. EV integration to the ADN and their contributions in inertia and frequency regulation is studied in (Amini et al., 2016). Moreover, EV charging is made possible by power electronic interface, by which reactive power capability of EV chargers (inverters) are used for energy control and voltage regulation. Different operation modes depend upon the active/reactive power exchange from grid to EV charger or EV charger to the utility grid. The possible operational modes of smart inverter interfaced EVs provided in (Singh et al., 2019) are depicted in Figure 5. Mode I: $P>0$ and $Q=0$, Mode II: $P>0$ and $Q>0$, Mode III: $P=0$ and $Q>0$, Mode IV: $P<0$ and $Q>0$, Mode V: $P<0$ and $Q=0$, Mode VI: $P<0$ and $Q<0$, Mode VII: $P=0$ and $Q<0$, Mode VIII: $P>0$ and $Q<0$.

Generally, EV operates in Mode I (power factor=1), for only battery charging. Mode II and mode VIII empower the reactive power injection/absorption capability in addition to battery charging. Mode IV and mode VI operations enable the V2G operation for active power support as well, in addition to reactive power. An operation scenario is created by mode III and mode VII in which the EVCS can exchange reactive power when no battery is connected. If the EV interfacing inverter rating is S_{EV}, the operational constraints of EV inverters for VVC are:

$$Q_{EV}^{min}(t) \leq Q_{EV}(t) \leq Q_{EV}^{max}(t) \tag{20}$$

$$Q_{EV}^{max}(t) = \sqrt{S_{EV}^2 - P_{EV}^2(t)} \tag{21}$$

$$Q_{EV}^{min}(t) = -\sqrt{S_{EV}^2 - P_{EV}^2(t)} \tag{22}$$

Where, $P_{EV}(t)$ is the power absorbed for charging at instant t or discharging power when V2G provision is enabled. $Q_{EV}(t)$ is reactive power exchange at instant t, and $Q_{EV}^{min}(t), Q_{EV}^{max}(t)$ are minimum and maximum available power at time instant t, respectively.

Figure 5. EV inverter capability for active/reactive power

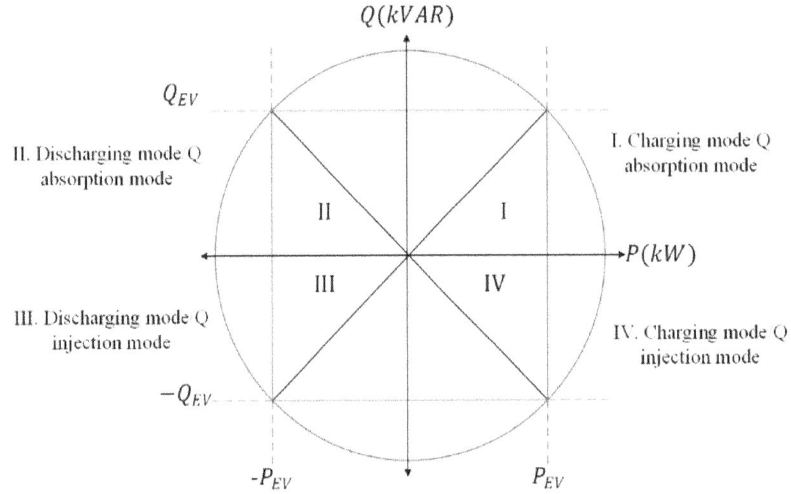

A hybrid distributed and local control is provided in (Wang et al., 2019) for LV distribution networks using sparse communication. Wherein, distributed control at higher level determines the voltage regulation inputs while as at lower level the local control decides the active/reactive power exchange of EV chargers, based on the operational requirements. Uncertainty in EVs is considered in (Hemmatpour et al., 2022) in studying the effects of EVs on energy as well as voltage regulation in LV networks. An effective EV integration demands customer satisfaction in terms charging and discharging, when V2G services are needed. A transactive energy based control frame is provided in (Hoque et al., 2022) to simultaneously consider the economics and voltage control based on coordinated charging/discharging of EVs while preferring customer needs. Ref. (Wang et al., 2023) provide a model-free and deep reinforcement learning framework using multi-agents for active/reactive power control of DG units together with coordination of EVs for voltage control. Furthermore, a distributed coordination for EVs is based on partially observable Markov process and the proposed strategy well encompasses the uncertainties in DG units, EVs and market prices.

3.4. Soft Open Point (SOP)

Soft open point (SOP) is deployed to connect two nodes in distribution network, where normally open point (NOP) exist. It is a power electronic based device generally using two back-to-back connected VSCs, to control active and reactive power exchange between two branches or feeders. A schematic of using SOP is depicted in Figure 6. SOP model generally includes active power transmission constraints and reactive compensation capability constraints (Li et al., 2017). The constraints for SOP shown in Figure 6, can be expressed as:

Active power transmission constraints:

$$P_{t,m}^{sop} + P_{t,n}^{sop} + P_{t,m}^{sop,loss} + P_{t,n}^{sop,loss} = 0 \tag{23}$$

$$P_{t,m}^{sop,loss} = C_m \sqrt{(P_{t,m}^{sop})^2 + (Q_{t,m}^{sop})^2} \tag{24}$$

$$P_{t,n}^{sop,loss} = C_n \sqrt{(P_{t,n}^{sop})^2 + (Q_{t,n}^{sop})^2} \tag{25}$$

Reactive power exchange constraints:

$$Q_{m,\min}^{sop} \leq Q_{t,m}^{sop} \leq Q_{m,\max}^{sop} \tag{26}$$

$$Q_{n,\min}^{sop} \leq Q_{t,n}^{sop} \leq Q_{n,\max}^{sop} \tag{27}$$

SOP capacity constraints:

$$\sqrt{(P_{t,m}^{sop})^2 + (Q_{t,m}^{sop})^2} \leq S_m^{sop} \tag{28}$$

$$\sqrt{(P_{t,n}^{sop})^2 + (Q_{t,n}^{sop})^2} \leq S_n^{sop} \tag{29}$$

Where, $P_{t,m}^{sop}, Q_{t,m}^{sop}$ are injected active/reactive power by SOP at node m at time t. $P_{t,n}^{sop}, Q_{t,n}^{sop}$ are injected active/reactive power at network node n at instant t, $P_{t,m}^{sop,loss}$ and $P_{t,n}^{sop,loss}$ are active power losses at node m and node n, respectively and C_m and C_n are SOP power loss coefficients. The capacity limit of SOP at node m and n are S_m^{sop} and S_n^{sop}, respectively.

Figure 6. Schematic of SOP placement

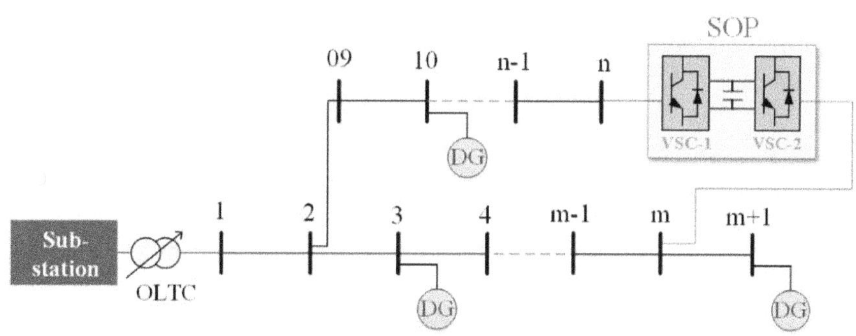

SOP operation in active distribution networks has been carried for real time and cost-effective voltage control in (Li et al., 2017). SOPs are generally coordinated in centralized manner. In articles (Ji et al., 2019; Zhao et al., 2020) SOP is used for voltage control in decentralized manner. To address voltage issues due to frequent DG power variation, a data-driven adaptive voltage control is carried out in ref. (Li et al., 2022) utilizing the capabilities of multi-terminal soft open points.

3.5. Distribution Static Compensator (DSTATCOM)

DSTATCOM, a shunt connected power electronic based custom power device of various topologies have been used for controlling voltages in active distribution networks. DSTATCOM has the capability for injection and absorption of reactive power by current injection in quadrature with the bus voltage at connected point, thus capable of providing dynamic voltage support (Yan et al., 2014). In addition to reactive power compensation for voltage support DSTATCOM has harmonic elimination and load balancing capabilities as well. During reactive power exchange with the grid, current stays in quadrature with grid voltage to avoid active power exchange. A current injection model for DSTATCOM operation for voltage control is used in (Dutta et al., 2022b). Differential equation for dynamic model shown in Fig. 7, is represented as:

$$i'_{stat} = [K_{stat}(v_{stat} - v_g) - i_{stat}] / T_{stat} \tag{30}$$

Where, T_{stat} and K_{stat} are time constant and controller gain of DSTATCOM. v_{stat} and v_g are DSTATCOM voltage and voltage of the bus at which DSTATCOM is connected, respectively. The injection of reactive power to the grid is represented by the equation:

$$Q_{stat} = v_g \cdot i_{stat} \tag{31}$$

Direction of current flow (i_{stat}) determines whether reactive power is injected or absorbed.

Figure 7. Current injection model of STATCOM

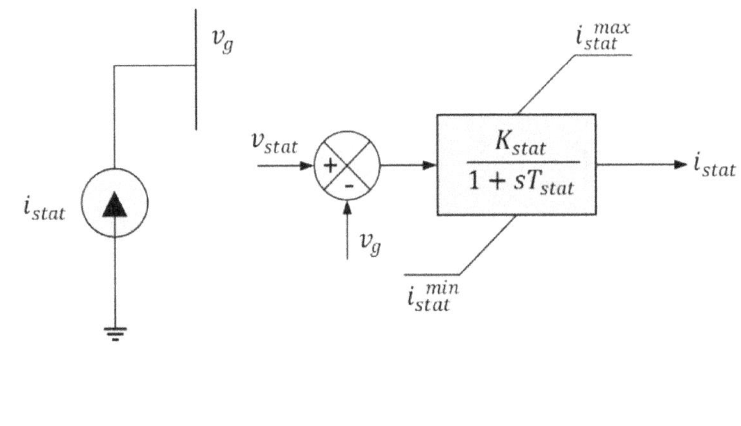

Authors in ref. (Yan et al., 2014) have carried a study on the influence of PV penetration on on-load tap changer in LV networks, and a mitigation approach to overcome the consequences using the capabilities of DSTATCOM is developed. The interaction of DSTATCOM with OLTC is studied to enable determination of technical provisions and optimal size of DSTATCOM for voltage regulation. A centrally coordinated strategy for voltage balancing in ADNs is carried out in (Dutta et al., 2022b) using model predictive control (MPC) approach. The MPC approach works for dual stage coordination, to control resources of varying temporal characteristics, wherein, the control capabilities of DSTATCOM for dynamic fast voltage control is studied in coordination with PVs and EVs.

4. VOLTAGE CONTROL METHODS IN ACTIVE DISTRIBUTION NETWORKS

Over the years, voltage control technologies have significantly advanced for reliable operation of power networks. The advent of improved technologies have been utilized in voltage control with network side management as well as demand side management (Gholami et al., 2022). From network end the DSOs use devices like capacitors, OLTCs, PV inverters etc., to control grid voltage and from the demand side the consumers mainly in industrial and commercial sector are actively participating in network management by utilizing roof top PVs, smart inverters, conservation voltage reduction for energy conservation and other measures for demand response. In current distribution networks, voltage control for active network management has been studied broadly in three categories which are: (i) local control (ii) centralized control and (iii) distributed control. The different categories for voltage control are based on control strategies used in the network. Local control is simplest communication free control mechanism while centralized and distributed control strategies employ communication setup in the network for coordination.

Figure 8. Voltage control methods (classification)

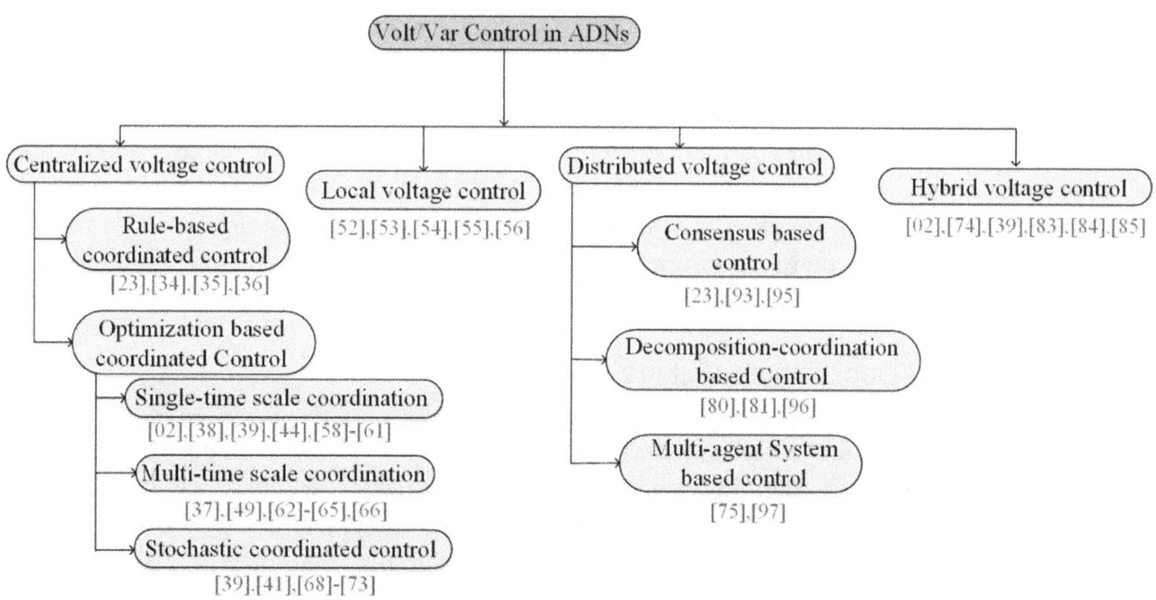

4.1. Local Voltage Control

The simplest voltage regulation method in active distribution networks can be thought to be the local control. The fast control actions are achieved, wherein, the control is provided based on the locally measured system state. Control resources in local control approach are mainly steered through, based on pre-specified rules. Since communication facility is not availed, the control resources are dispatched using local measurements, without any coordination among the control resources (Dar and Ganguly, 2023). A local voltage control strategy is studied in (Martinenas et al., 2017) to demonstrate the voltage regulation capabilities of EVs using droop control. Applied droop controller is designed based on voltage limit thresholds. The intelligent charging strategy can effectively regulate voltage even without the use of V2G services. A coordinated, decentralized and MPC based voltage control approach is proposed in (Li et al., 2020) based on local tuning of DG-QV curves. The ADN is partitioned using dynamic network partition to relieve burden of computation and communication. DG units react fast locally to correct any voltage violation while, a decentralized coordination is performed based on ADMM to perform optimal operation by tuning the local DG curves, thus, forming a two-level control strategy. Local voltage control techniques with capacitor banks in addition to other devices such as OLTCs, BESSs and DG units and network reconfiguration have been provided in various research articles. For example, ref. (Islam et al., 2021) simultaneously allocated storage devices and CBs for voltage drop mitigation, authors in ref. (Singh et al., 2020) use DG units, OLTC and CBs to minimize voltage violations and energy loss. Network reconfiguration is incorporated in ref. (Navesi et al., 2022) for voltage control and network power loss mitigation to enhance network security. Local control is resilient and economical but the control resources may be unnecessarily over-utilized, which leads to sub-optimal operation. The over-utilization of resources lead to insufficient reactive power, due to which network may demand active power curtailment from DG units frequently. BESSs and EV charging can be employed in integration with DG units for local voltage control, to mitigate need for APC, where reactive power control is rendered insufficient.

4.2. Centralized Voltage Control

Centralized voltage control strategies are widely studied in literature owing to their coordination capabilities in small or medium sized networks and relatively ease in implementation as compared to distributed control methods. The control resources are coordinated using a wide range communication network with central controller generally at substation. In literature the centralized voltage control in ADNs with multiple control variables is achieved by coordinating the control inputs in two ways; (i) rule-based coordination of control resources (ii) optimization-based coordination of control resources. A brief taxonomy of voltage control in ADNs is provided in Table 2.

4.2.1. Rule-Based Coordinated Voltage Control

Rule based strategies are suited for networks which possess only handful of control possibilities. Defining rules for large number of control possibilities becomes significantly complex task in networks in which large number of control devices are present. A coordinated voltage control is provided in (Elkhatib et al., 2011) to regulate voltages within prescribed limits by centrally controlling the tap positions of OLTC. It is established that maximum and minimum voltage estimation in the network are sufficient to keep the voltages inside the specified bounds. However, this can be true only for controlling voltages with incapability to carry out optimal operation. The remote terminal units (RTUs) are employed at DG locations for estimation so that the voltage extremes are eliminated with proper control. A rule-based voltage control approach utilizing OLTC and a DER is carried out in (Kulmala et al., 2010). The control mechanism is further improved in (Kulmala et al., 2014) to facilitate the controlling of several DERs coupled with voltage control at substation. A rule-based distributed voltage regulation using consensus algorithm is provided in (Wang et al., 2020), to effectively coordinate the unbalanced phases in LV distribution networks.

4.2.2. Optimization-Based Coordinated Voltage Control

Unlike traditional power distribution networks ADNs have numerous control possibilities. It becomes suitable to employ optimization based coordinated control to coordinate all resources with one or multiple objectives. It is evident that the control resources operate on different time scales. OLTC operation is slow and discrete while as PV inverters act fast and provide continuous range reactive power. Coordination of control resources with varying temporal characteristics becomes a complex task, thus necessitating the coordination of device group with same temporal characteristics. To deal with this, centralized control is studied on single time-scale coordination or multi-time scale coordination in literature.

4.2.2.1. Single-Time Scale Coordinated Voltage Control

Based on single-time scale coordination, real-time centralized voltage is carried out in articles (Alzaareer et al., 2020; Bidgoli and Cutsem, 2018; Dutta et al., 2020; Kryonidis et al., 2018; Li et al., 2017; Nguyen et al., 2021; Singh et al., 2019; Valverde and Cutsem, 2013). A centralized voltage control based on SOP is proposed in (Li et al., 2017) for real time voltage control of ADN. The mixed integer non-linear programming (MINLP) developed for loss minimization is converted into cone programming problem by linearization of the model and cone relaxation, so as to remove convexity of objective formulation, for proper real time implementation. Authors in (Kryonidis et al., 2018) propose a control strategy for LV networks by carrying out uniform active power curtailment in a centralized and optimal way. The droop curve set points are determined by optimization using interior point method. In (Alzaareer et al., 2020) a coordinated control strategy is developed by identifying the global set of effective and low-cost control variables based on Correction Index (CI), which in turn is developed using the notion of electrical distances. The method is resilient and works under the network structural changes. The proposed technique uses neither rule-based voltage control nor optimization-based voltage control. However, in (Alzaareer et al., 2020), the algorithm based on the correction indices with respect to control variables identifies the effective and economical control devices to be put in use during voltage violations. Different from the existing MPC based centralized control techniques, authors in (Nguyen et al., 2021) have proposed an MPC based centralized control framework, in integration with PI controller, to address the need of reactive power demand by transmission system operator (TSO) at PCC. The reactive power demand by TSO is uniformly distributed among all the control devices of ADN. Furthermore, the sensitivities of losses are calculated by artificial neural network (ANN). Since, the centralized control schemes solely rely on efficient communication with central controller, researchers have made attempts for combined application of centralized and local control to provide resilient and reliable control under the circumstances of any failures or mishaps in communication setup. Furthermore, the incorporation of local control mechanism results in fast control signals as compared to the centrally allocated control signals, thus enhancing the control efficiency (Singh et al., 2019). An MPC based coordinated two-level voltage control is proposed in (Bidgoli and Cutsem, 2018) to control voltage by optimal operation of DG units. The control scheme developed comprises of a central controller and local controllers. The local controllers act continuously for any voltage violations, based on piece-wise linear reactive power curves of DG units. The local control scheme acts as backup in case of communication failures of the central controller.

4.2.2.2. Multi-Time Scale Coordinated Voltage Control

The temporal features of control equipment can be addressed by disintegrating control problem on multi-time scale basis, so as to avoid inefficient operation and over-utilization. However, the computational complexity is not reduced significantly. Various research articles, such as (Dutta et al., 2022b, 2022a, 2022c; Guo et al., 2020, 2019; Zhang et al., 2021) provide multi-time scale cooperation of volt/var optimization devices, both in centralized as well as distributed control strategies. Authors in (Dutta et al., 2022b; Guo et al., 2020; Zhang et al., 2021) and (Dutta et al., 2022c) have carried out the coordination by decomposing the coordinated VVC optimization into slow coordination problem and fast problem, with optimization at both the stages. A third stage is added in (Dutta et al., 2022a). A coordinated dual-stage deterministic MPC based voltage control framework is developed in (Dutta et al., 2022b) and (Dutta et al., 2022c), considering different time based characteristics of control equipment used in ADNs. Considering varying time based characteristics of control devices, double-time scale MPC based voltage control is proposed in (Guo et al., 2020). ADMM based fast QP solver is used in fast time scale (FTS) and branch and bound based tailored exact solution with ADMM solves in slow time scale (STS). A detailed MPC based coordinated voltage control strategy utilizing PVs and BESSs for voltage regulation of ADNs is proposed in (Guo et al., 2019) employing two operational modes (preventive and corrective). Furthermore, hierarchical clustering method (based on electrical distances) is used for zoning of the network, so that critical limited buses are selected for measurements. The authors in (Zhang et al., 2021) have proposed a centrally coordinated control framework combining the merits of stochastic programming with MPC, wherein, the control strategy is carried out in two layers by time scale decomposition, based on device response speeds. To study the coordinated control of EVs together with the DER based voltage control devices, a three-stage deterministic MPC based centralized voltage control framework is proposed in (Dutta et al., 2022a). The discrete and slow operating devices like OLTC are coordinated after every 1 hour, while as fast response devices (DGs, EVs) are regulated after every minute, considering the temporal characteristics volt/var optimization (VVO) devices. Furthermore, EVs are dispatched after every 30 minutes in third stage to account for the customer satisfaction and operating cost based on the real-time electricity price.

4.2.2.3. Stochastic CVC

Many of the research investigations conducted rely on the anticipation of the expected output from PVs, and do not take into account the unavoidable inaccuracies in such predictions. Stochastic programming is a commonly employed technique for examining the randomness or unpredictability in power generation and load consumption (Kekatos et al., 2014). A detailed analysis of the impacts of EVs on energy as well as voltage control is studied in (Hemmatpour et al., 2022), considering fast and slow charging of EVs. The stochasticity in charging /discharging pattern of EVs are assumed to be randomly distributed. The optimization for optimal cost of operation of control devices is carried out using improved swarm optimization technique (IMVBSO). In (Singh et al., 2019) the authors have proposed an MPC based volt/var optimization (MP-VVO) for voltage in ADNs by employing the OLTC, CBs, EVs and PV-DGs in presence of conservation voltage reduction (CVR). Furthermore, the authors have considered the stochasticity using probability distribution functions (PDFs) for generation using beta distribution and for load using normal distribution. The Monte-Carlo simulation is used for scenario generation and the scenarios are reduced using K-means clustering. The stochastic optimization is carried out using discrete particle swarm optimization. For restraining the voltages within bounds in LV distribution network, a coordinated voltage control using PEV inverters and BESS is studied in (Kabir et al., 2014), for urban and rural scenarios (having different *R*/*X* ratios). The work carried out establishes a voltage regulation strategy in which the voltage violation is checked after every hour. For urban scenario (i) $R/X=1$, the PV inverter reactive power capabilities prove to be sufficient for regulating the voltage, however for the rural scenario (ii) $R/X=2.8$, the coordination of droop-based BES systems and PV inverters is required (Kabir et al., 2014) . The uncertainties in the load /generation are also considered by employing probabilistic estimation models, and simulations are carried out to find out the optimal battery size and critical *R*/*X* ratio for efficient performance. Under high degree of penetration and uncertainty, the solution tends to deviate from optimum, thus, the coordinated control scheme is better suited for lower control time steps. To account for the uncertainties, robust constrained model predictive control (RC-MPC) for voltage regulation in ADNs is proposed in (Maharjan et al., 2021). Knowing the boundary values of uncertainty, a method called direct method is proposed to quantify the worst-case transitions of stochastic resources, which avoids the requirement for computationally heavy Monte-Carlo simulations. The use of robust MPC, unlike stochastic MPC, relaxes the need for probability distribution functions (PDFs) for incorporating the uncertainties in generation or load, and instead modifies the constraints of the optimization. Unlike uncertainties considered in literature, the study conducted in (Christakou et al., 2018) focuses on voltage control in ADNs while taking into account the uncertain line parameters, specifically the admittance matrix. Voltage sensitivity bounds for robust optimization are determined using Monte-Carlo simulation. Furthermore, considering the impact on the power quality of sensitive loads, due to voltage fluctuations, authors in (Ma and Zhang, 2019; Zhao et al., 2015) have studied the methods for voltage regulation and optimal operation of ADNs considering the voltage dependence of loads. The work is extended in (Yong et al., 2020) to incorporate the time-varying features of voltage sensitive loads based on load models with constant power, impedance and current (ZIP model).

Table 2. Brief taxonomy of voltage control in ADNs

Ref.	Control Strategy	Control Methodology	Optimization Formulation	Control Devices	Objective
(Arshad et al., 2018)	Centralized	MPC	Quadratic programming	OLTC, PV inverters	Minimization of voltage control actions
(Bidgoli and Cutsem, 2018)	Hybrid (centralized + Local)	MPC	Quadratic programming	DG units	Minimization of voltage control actions
(Wang et al., 2020)	Distributed	Branch flow model/ Consensus algorithm	Rule-based coordination	PV inverters	Inter-phase coordination/ Over-voltage mitigation
(Valverde and Cutsem, 2013)	Centralized	MPC	Quadratic programming	OLTC, DG units	Minimization of losses, OLTC operation, active power curtailment
(Li et al., 2017)	Centralized	Dist. load-flow (BFS)	MINLP reduced to MISOCP	SOP, OLTC, CBs	Minimization of losses (network losses, switching losses, SOP losses)
(Islam et al., 2021)	Local	Optimal placement of CBs & BESSs (Load flow)	---	CBs and BESSs	Mitigation of losses and voltage drop
(Singh et al., 2019)	Hybrid (centralized + Local)	MPC	NL stochastic optimization	OLTC, CBs, EVs, PV inverters	Minimization of cost of (CVR, losses, control actions)
(Alzaareer et al., 2020)	Centralized	CI (correction index) based VVC	----	CBs, VRs, DG units	Minimization of voltage deviations
(Nguyen et al., 2021)	Centralized	MPC	Linear Programming	OLTC, DG units	Minimization of losses and control actions
(Navesi et al., 2022)	Local	Load flow approach	Non-linear programming	CBs and reconfiguration switches	Voltage violation reduction and minimization of power loss
(Maharjan et al., 2021)	Centralized	Robust constrained MPC (RC-MPC)	Mixed integer QP	OLTC, controllable PVs, Uncontrollable PVs	Minimization of control actions, APC and voltage deviations
(Luo et al., 2021)	Distributed	Sensitivity analysis	MILP	PV-DG units (active/ reactive power)	Minimization of control actions and APC
(Li et al., 2020)	Hybrid (decentralized + Local)	MPC	MINLP reduced to MISOCP	DGs (PVs and WTs)	Minimization of cost and voltage deviation
(Kryonidis et al., 2018)	Centralized	OPF/Sensitivity analysis	Non-linear programming	DG units (active power only)	Minimization of active power curtailment
(Hu et al., 2022)	Distributed	MPC	Quadratic programming	EVs (reactive power)	Minimization of voltage deviations
(Yong et al., 2020)	Centralized	Power flow based	NLP	OLTC, SVCs, DG units	Minimization of losses, imported power, loss of sensitive loads, cost of DG operation
(Wang et al., 2019)	Hybrid (distributed + Local)	Consensus algorithm	----	EVCs in LV networks	Minimization of voltage deviations
(Guo et al., 2020)	Centralized	MPC	MIQP	OLTC, SVR, CBs (STS) ; DG units (FTS)	Minimization of voltage deviations and control actions
(Dou et al., 2018)	Distributed	Distributed MPC	MIQP	CBs, DG units (active/reactive power)	Minimization of operation cost, voltage deviation and APC
(Zhang et al., 2021)	Centralized	Stochastic MPC /linear Distflow	MIQP	OLTC, CBs, DG units (active/ reactive power)	Minimization of control actions, APC, network losses

Table 2. Continued

Ref.	Control Strategy	Control Methodology	Optimization Formulation	Control Devices	Objective
(Wang et al., 2023)	Decentralized	Multi-agent deep reinforcement learning	MILP	OLTC, DG units (PVs, WTs) and EVs	Minimization of cost of active/reactive power of DG units
(Christakou et al., 2018)	Centralized	Sensitivity analysis	MILP	OLTC, DG units (active/reactive power)	Maximization of utility cost function
(Hemmatpour et al., 2022)	Coordinated	Load flow approach	Nonlinear stochastic programming	OLTC, CBs, controllable loads and EVs	Minimization of operation cost of control devices, energy and voltage violations
(Hu et al., 2022)	Hybrid (centralized + distributed)	MPC	MINLP	OLTC, CBs, BESS, DG units (active/reactive power)	Minimization of voltage deviation, losses, APC and control actions
(Wang et al., 2019)	Distributed	Distflow approach	----	EV chargers (EVCs)	Minimum power exchange from EVCs
(Guo et al., 2019)	Centralized	MPC	Quadratic programming	OLTC, DG units, BESSs	Minimization of voltage deviations and control actions
(Tang et al., 2021)	Distributed	Distflow approach	MINLP	OLTCs, DERs (reactive power)	Minimization of cost of control actions
(Dutta et al., 2022c)	Centralized	MPC	Quadratic programming	OLTC, DSTATCOM, DG units	Minimization of control actions, voltage deviations and energy demand
(Romvary et al., 2021)	Hybrid (centralized + distributed)	OPF, (PAC) proximal atomic coordination algorithm	---	PVs, CBs	Minimization of voltage deviations and power loss
(Dutta et al., 2022a)	Centralized	MPC	Quadratic programming	OLTC, EVs, DG units (active/reactive)	Minimization of control actions, voltage deviations and APC
(Dutta et al., 2022b)	Centralized	MPC	Quadratic programming	OLTC, DSTATCOMs, EVs, DG units	Minimization of control actions, voltage deviations and APC

4.3. Distributed Voltage Control

Under heavy DG integration and large scale control, centralized control schemes become significantly difficult and complex. For large networks, centralized control appears to be unfeasible due to the need for heavy reliable communication, transparency in information, computational complexity and increased cost for implementation (Liu et al., 2015). Furthermore, centralized control lacks scalability or plug-and-play functionality for further advancements in ADNs, and seems unsuccessful in dealing with the weak communication in present distribution networks (Tang et al., 2021). Distributed control schemes seem to emerge as promising techniques for achieving real-time coordinated control of various control equipment. These strategies involve sharing the communication and computation burden among multiple distributed controllers, with only neighbour to neighbour communication, thus, resilient to communication failures. However, the solutions provided may be sub-optimal and may require more execution time than centralized control. Distributed control based automatic voltage control has gain wide attention in recent years. For distributed control the partition of network is preliminary, to allocate controllers in different zones. The partition needs to be done sensibly so that the nodes in a zone are strongly correlated and the zones are weakly coupled with each other. Numerous kinds of network zoning/partitioning methods and zoning evaluation indices are proposed in literature, based on spatial arrangement of control devices, based on evaluating electrical distances among network nodes or based on some other proximity measure for clustering (Zhao et al., 2018). Electrical distance quantifies the coupling between the nodes of network based on voltage sensitivities. The electrical distance (D_{ij}) between network nodes i and network node j is given by:

$$D_{ij} = -\log(\alpha_{ij}.\alpha_{ji}) \tag{32}$$

Where, $\alpha_{ij}=\Delta V_i/\Delta V_j$ and $\alpha_{ji}=\Delta V_j/\Delta V_i$ are attenuation of voltage deviations caused due to control variable at node j and node i, respectively. Numerous clustering algorithms are used for zoning/clustering of network nodes, such as hierarchical clustering algorithm, spectral K-way clustering algorithm, fuzzy k-means clustering algorithm and affinity propagation clustering algorithm (Alimisis and Taylor, 2015). Hierarchical clustering algorithm forms one of the popular clustering algorithms as explained:

Hierarchical clustering method: The clustering is achieved using bottom up approach. Initially, all nodes are assumed as individual zones/clusters, the criterion for subsequent merging of clusters C_m and C_n is given by:

$$C_m \cup C_n = \min(cd_{m,n}) \tag{33}$$

Where, $cd_{m,n}$ is the inter-cluster distance between zones m and n, which is the maximum of electrical distance between the nodes of zone m and zone n.

$$cd_{m,n} = \max[D_{ij}\text{: } i\in C_m \text{ and } j\in C_n] \tag{34}$$

The clusters merge to form one cluster finally, unless a criteria is set to identify appropriate number of clusters. The appropriate clusters are identified based on the criterion of relative diameter of the network at a clustering stage (Alimisis and Taylor, 2015). Diameter (dia_C) of a cluster and relative diameter (Rd) of network, when k number of clusters are identified, is given as:

$$dia_C = \max\{D_{ij}\text{: } i,j\in C\} \tag{35}$$

$$Rd = \frac{\sum_{i=1}^{k}(dia_C^i)}{k} \tag{36}$$

The appropriate clustering stage is chosen (with k clusters), where, the relative diameter change between successive stages is maximum, i.e.:

$$k := \max\{|Rd_2 - Rd_1|, |Rd_3 - Rd_2|, |Rd_4 - Rd_3|, \ldots, |Rd_{end} - Rd_{end-1}|\} \tag{37}$$

In VAR control space the clusters are linked based on average linking criterion, wherein, the inter-cluster distance is average of electrical distances between the nodes of two clusters, given as:

$$cd_{mn} = \frac{1}{\sum_{k=1}^{m_{max}} (n_m^k) \sum_{k=1}^{n_{max}} (n_n^k)} \sum_{i \in C_m} \sum_{j \in C_n} D_{ij} \tag{38}$$

Where, $n_m^k \in C_m$ and $n_n^k \in C_n$ are k^{th} nodes in cluster C_m and C_n, respectively. m_{max} and n_{max} are total number of nodes in cluster C_m and C_n, respectively. The appropriate clustering (with k clusters) is decided based on maximum value of the average zone distance (*Ad*) of network at each stage, given by:

$$Ad = \sqrt{\frac{\sum cd_{mn}^2}{k}} \qquad m, n = 1, 2, 3, \ldots, k-1 \tag{39}$$

$$k := \max\{Ad_1, Ad_2, Ad_3, \ldots, Ad_{end}\} \tag{40}$$

For optimal partition, various indices, such as modularity index, improved modularity index, comprehensive index and correction index are proposed in literature. Distributed coordination is studied in literature using consensus based algorithms in (Haque et al., 2019; Mai et al., 2019; Wang et al., 2020; Zeraati et al., 2019) or multi-agent system based coordination algorithms in (D. Hu et al., 2022; Liu et al., 2015) or decomposition coordination based algorithms in (Dou et al., 2018; Yu et al., 2012; J. Hu et al., 2022). In consensus based algorithms, a common consensus is reached among the various control resources for a desired voltage level. Multi-agent based coordination is achieved based on local decisions by the agents. The agents collaborate and communicate with each other to exchange information regarding their local decisions. Furthermore, decomposition coordination is achieved by dividing the control problem into manageable sub-problems/subsystems, which are solved independently.

Distributed voltage control based on network partition has been studied on single-time scale coordination in (Dou et al., 2018; J. Hu et al., 2022; Ruan et al., 2020; Y. Wang et al., 2019; Zhao et al., 2018) and based on double-time scale coordination in (Tang et al., 2021). Authors in (Dou et al., 2018) have studied a cooperative distributed MPC based voltage control framework of a multi-microgrid distribution system which works under active supervisory control of automatic voltage regulator, for acquiring voltage sensitivity matrices, while each microgrid communicate with neighbouring parts for distributed coordination of control resources. A distributed MPC (DMPC) based regulation strategy is studied in (J. Hu et al., 2022), considering the V2G capabilities of EV chargers. Each charger solves its own optimization, utilizing data from neighbouring chargers, using peer-to-peer communication network. A modularity function based index called comprehensive index (using electrical distances) is used for network partitioning in (Ruan et al., 2020), to reduce computational complexity. The proposed index has greater efficacy compared to other network partition methods. Taking into consideration of EV application and V2G services, distributed control mechanism still lake sufficient studies. To study high penetration of EVs, a multi-level distributed control strategy, taking advantage of the active/reactive power injection/absorption capacities of EVCs is proposed in (Y. Wang et al., 2019). To establish a compromise between decentralized control and centralized control in ADNs, authors in (Tang et al., 2021) propose distributed voltage control strategy with two control stages working in complementary manner. The network partitioning for implementation of distributed MPC based control strategy is carried out based on position of OLTCs. This work considers multiple OLTCs for discrete slow control mechanism at the first stage and DERs as continuous fast controllers at the second stage, using ADMM. On contrary to the various clustering methods, in which the optimally chosen network zones are not contemplated, a distributed strategy is studied in (Zhao et al., 2018). The network is dynamically subdivided into partitions. The network partition provides optimal number of zones, unlike various clustering techniques studied in the literature, using improved modularity index and enhanced community detection algorithm. The community with the highest violated voltage node serves as the first zone for voltage control. The optimization problem formulated for each zone/community is solved using particle swarm optimization (PSO). Among the various distributed algorithms used in ADNs, ADMM is commonly used for voltage control. Some other notable algorithms are Lagrange dual relaxation and auxiliary problem principle.

4.4. Hybrid Voltage Control

Control methods mentioned above carry their own supremacy and downsides as far as diverge scenarios of ADNs are considered. To subjugate the shortcomings, some researchers in recent years have been inclined towards the hybrid control models. Hybrid control seems a potentially attractive strategy in which combination of centralized, local and distributed methods can be simultaneously used in same network, using multi-layer control scheme. For example, centralized control can be used to provide optimal control inputs hourly while as a local control can be integrated to dispatch DG inverters, locally, to respond at shorter time intervals of few seconds. A hybrid centralized and local control is proposed in (Bidgoli and Cutsem, 2018), wherein, the local control mechanism runs continuously while as the central controller dispatches optimal set-points to local controllers after longer time durations. Coordination of diverse voltage regulation devices (OLTC, DG units, CBs and BESSs) for large scale networks is proposed in (R. Hu et al., 2022), in which the control devices (OLTC, DG units and CBs) are centrally controlled based on MPC framework. A real-time multi-agent distributed control is used at second level for active/reactive power control and coordination of PVs and BESSs. Hybrid control using centralized control and local control is used in (Singh et al., 2019) based on model predictive control, to optimally control resources in PV and EV integrated distribution networks. It is quite apparent that centralized control schemes find it hard and intractable to coordinate resources for real time control in large networks, particularly with heavy distributed penetration of PVs. A hybrid of centralized and distributed voltage control architecture is presented in (Haider and Annaswamy, 2022), wherein, an OPF problem is solved at centralized control level, after each hour. The convex relaxation of OPF problem of the network relies on current injection (CI) and McCormick Envelopes for unbalanced distribution networks. At second level, a distributed volt/var optimal control is performed under various DER penetration levels, after every 5 minute time interval. An attempt has been made in (Chakraborty et al., 2023) for simultaneous frequency and voltage control using hybrid control architecture, wherein, preference is given to the voltage control. To address the issue of communication failures authors in (Liu et al., 2017) present a hybrid voltage control model which employs distributed as well as local control to coordinate DERs throughout the network. The control scheme is communication cognizant due to utilization of attractive features of both the control schemes, to effectively manage the resources.

In addition to the above mentioned voltage control techniques, there are other techniques which have acquired a lot of interest among researchers. The conventional and most of the advanced techniques require distribution network model and line parameters, which may not be available. To address these issues deep reinforcement learning based techniques have gained interest in distribution voltage control, utilizing the coordination capabilities of above mentioned methods as in research articles (Jiajie et al., 2022; Kabir et al., 2023; Li et al., 2022; Yang et al., 2020; Zhang and Xu, 2020). Moreover, there are control strategies which employ VVC coordination jointly with network reconfiguration such as in (Chen et al., 2016; Gao et al., 2020; Navesi et al., 2020). Network reconfiguration changes the topology of flexible distribution networks to alter power flow paths in order to minimize losses, reduce voltage deviations, promote DG integration and enhance economical operation.

5. MODEL PREDICTIVE CONTROL-BASED REAL-TIME VOLTAGE CONTROL

Since, most of the coordinated control techniques use predicted data of renewable DG units, the uncertainty significantly rises under large DG penetration. If the time step for optimization is large, the uncertainties may lead to sub-optimal solutions. Under this scenario, day-ahead coordination becomes infeasible, raising the need for real-time control and coordination. For real time control in ADNs model predictive control technique has received significant attention in recent time, as can be seen in Table 2. At the heart of MPC lies multi-step optimization which provides optimal control sequence over multiple steps in future. MPC can effectively tackle model inaccuracies due to its measurement based closed loop operation, and has ability to coordinate multiple control devices having diverse response times (Dou et al., 2018). Model predictive control approach is being studied in various forms. Notably, deterministic MPC is used in most of the centralized control strategies. However, other variants such as stochastic MPC and robust MPC have also been studied for voltage control in ADNs. The uncertainties in generation and load are neglected in deterministic MPC formulation, which are somehow mitigated assuming small time steps and closed loop performance. The uncertainties for stochastic MPC are generally modelled for generation employing beta distribution and for modelling load uncertainty normal distribution is generally used. Single-time scale coordinated control based on MPC has been studied in (Bidgoli and Cutsem, 2018; Guo et al., 2019) and (Valverde and Cutsem, 2013). The coordination of OLTC and DG units is carried out using MPC based central controller in (Valverde and Cutsem, 2013), a local control scheme is integrated in (Bidgoli and Cutsem, 2018) to have combined centralized and local control, while as (Guo et al., 2019) coordinates DG units and BESSs. In addition, EVs have also been studied for voltage control in (Dutta et al., 2022a, 2022c) in coordination PVs, OLTC and custom power devices. Furthermore, conservation voltage reduction (CVR) and demand response has been incorporated to achieve effective control. The slow control devices are dispatched after long time (after every hour) and fast control devices are dispatch after short time period of 1 minute or less in multi-time time scale control in (Dutta et al., 2022a, 2022b, 2022c).

5.1. Double-Time Scale Model Predictive Control of ADNs

A double-time scale centralized voltage control of active distribution networks using MPC framework, is presented in this section, to study the control capabilities PVs and EVs.

Figure 9. Power profile: (a) PV generation and EVCS demand, (b) load demand

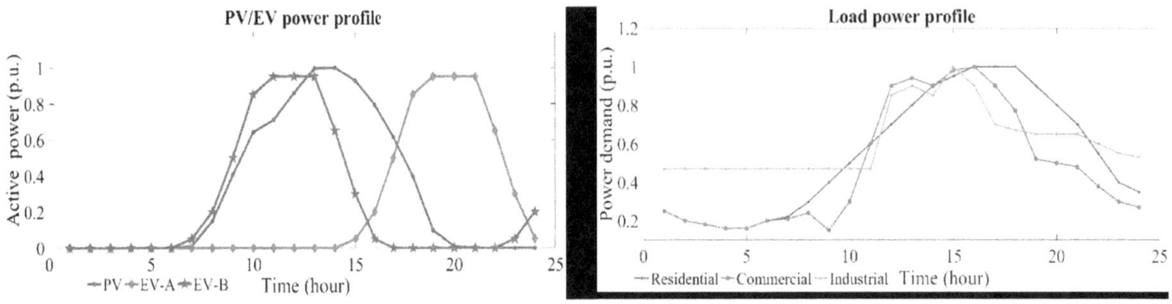

5.1.1. System Description of Active Distribution Network

A balanced three-phase radial active distribution network with PVs as active sources is considered with N nodes and E branches. The network branches e_{ij} represent distribution lines, modelled by series admittance $y_{ij}=1/z_{ij}$, where $z_{ij}=r_{ij}+jx_{ij}$ is the line impedance connecting network node i and network node j, r_{ij} and x_{ij} are resistance and reactance of line e_{ij}, respectively. For the system with N as the set of nodes, $e_{ij}\in E\subset N\times N$ is the set of branches of the network. The active sources are distributed throughout the network, an OLTC connects the substation to the network at node 1. The control resources considered are OLTC, PV and EV inverters. The OLTC can step up or step down voltage by 10%, while as PV and EV inverters have reactive power injection/absorption capabilities to regulate voltage. The PV generation depends on the irradiance at the locations where PVs are located. A uniform generation profile of PVs are considered throughout the network as shown in Figure 9 (a). EVs are integrated for voltage support using reactive power exchange from or to the grid, however the active power is only absorbed for charging with no service enabling active power flow back to the grid. Enabling V2G services for active power support degrades battery life, thus needs willingness of EV users. Incentive based demand response can serve the purpose but needs detailed modelling of uncertainties due to EV arrival/departure and battery state of charge. Furthermore, EV charging can be controlled by DSO for grid services, however, we assume the EVs start absorbing charging power as soon as they reach the parking lot, and the charging power profile is shown in Figure 9 (a). Apart from EV demand, three kinds of loads such as residential, commercial and industrial loads are considered in the test network (Dutta et al., 2022b). The load profile is shown in Figure 9 (b).

5.1.2. Voltage Control Based on MPC Framework

The system is represented by a linearized model with following equations:

$$x(k+1) = Ax(k) + B\Delta u(k) \tag{41}$$

$$y(k) = Cx(k) \tag{42}$$

Here, matrix A is system matrix based on the system modelling and C is the model output matrix. Matrix B provide the sensitivities of monitored quantities with respect to the control inputs of the system considered. The elements of B are obtained by inverting jacobian matrix which in turn is calculated using an offline load flow. The sensitivity elements with respect to the tap positions of OLTC are determined by observing voltage deviations due to two consecutive tap operations. It is to be noted that for our system the control inputs (variables) are the reactive power inputs from PV and EV inverters and the OLTC tap position. The node voltages form the state variables and outputs of the MPC controller. Since, MPC incorporates multi-step optimization, the control input sequence is provided up to N_C time steps in future called control horizon and the corresponding output is predicted over N_P steps called prediction horizon. The control sequence and predicted output sequence over time horizon N_C and N_P, respectively, at time instant k, is represented by following equations:

$$\Delta U = [\Delta u(k)^T \ \Delta u(k+1)^T \ldots \Delta u(k+N_C-1)^T]^T \tag{43}$$

$$Y = [y(k+1\,|\,k)^T \ y(k+2\,|\,k)^T \ldots y(k+N_P\,|\,k)^T]^T \tag{44}$$

Based on the above linearized discrete model of the system the state variables are obtained sequentially as:

$$\begin{aligned} x(k+1\,|\,k) &= Ax(k) + B\Delta u(k) \\ x(k+2\,|\,k) &= A^2x(k) + AB\Delta u(k) + B\Delta u(k+1) \\ &\vdots \\ x(k+N_P\,|\,k) &= A^{N_P}x(k) + A^{N_P-1}B\Delta u(k) + A^{N_P-2}B\Delta u(k+1) \\ &+ \cdots + A^{N_P-N_C}B\Delta u(k+N_C-1) \end{aligned} \tag{45}$$

The predicted output from these states can be obtained as:

$$\begin{aligned} y(k+1\,|\,k) &= CAx(k) + CB\Delta u(k) \\ y(k+2\,|\,k) &= CA^2x(k) + CAB\Delta u(k) + CB\Delta u(k+1) \\ &\vdots \\ y(k+N_P\,|\,k) &= CA^{N_P}x(k) + CA^{N_P-1}B\Delta u(k) + CA^{N_P-2}B\Delta u(k+1) \\ &+ \cdots + CA^{N_P-N_C}B\Delta u(k+N_C-1) \end{aligned} \tag{46}$$

Using above equations, the MPC framework reduces to the following equation, providing the output as:

$$Y = Fx(k) + \Phi\Delta U \tag{47}$$

Where,

$$F = \begin{bmatrix} CA \\ CA^2 \\ CA^3 \\ \vdots \\ CA^{N_P} \end{bmatrix} \; ; \quad \Phi = \begin{bmatrix} CB & 0 & 0 & \cdots & 0 \\ CAB & CB & 0 & \cdots & 0 \\ CA^2B & CAB & CB & \cdots & 0 \\ \vdots & \vdots & \vdots & \ddots & \vdots \\ CA^{N_P-1}B & CA^{N_P-2}B & CA^{N_P-3}B & \cdots & CA^{N_P-N_C}B \end{bmatrix} \tag{48}$$

The output is predicted based on the change in control inputs. To obtain the control inputs at instant k we can write;

$$\begin{aligned} u(k) &= u(k-1) + \Delta u(k) \\ u(k+1) &= u(k-1) + \Delta u(k) + \Delta u(k+1) \\ &\vdots \\ u(k+N_C-1) &= u(k-1) + \cdots + \Delta u(k) + \Delta u(k+N_C-1) \end{aligned} \tag{49}$$

Which reduces to the compact form:

$$u(k) = Gu(k-1) + \lambda \Delta \mathrm{u} \tag{50}$$

Where,

$$G = \begin{bmatrix} I \\ I \\ I \\ \vdots \\ I \end{bmatrix} \; ; \lambda = \begin{bmatrix} I & 0 & 0 & \cdots & 0 \\ I & I & 0 & \cdots & 0 \\ I & I & I & \cdots & 0 \\ \vdots & \vdots & \vdots & \ddots & \vdots \\ I & I & I & \cdots & I \end{bmatrix} \tag{51}$$

Where, I and 0 are identity matrix and null matrix, respectively. The MPC framework is applied centrally at two stages, based on the temporal characteristics of the control devices used. OLTC control is dispatched at first stage after every hour while at the PVs and EVCS are dispatched after every minute at second stage. Therefore, the input changes at an arbitrary time instant k, at first stage and second stage are represented as:

$$\Delta u_{stage1}(k) = [\Delta v_{tap}(k)]^T \tag{52}$$

$$\Delta u_{stage2}(k) = [\Delta Q_{pv}(k)^T \; \Delta Q_{ev}(k)^T]^T \tag{53}$$

5.1.3. Optimization Problem formulation

MPC works with receding horizon control with optimization at each time step. Since, the control sequence is established over control horizon, only the control inputs at first step are applied for action and rest of the control inputs of obtained optimal sequence are discarded. Due to closed loop performance and real time measurements, receding horizon control can effectively tackle the model uncertainties. The objective function is a quadratic programming problem to optimally reduce voltage deviations and minimize control actions, and can be represented by:

$$\min\left[\sum_{i=1}^{N_P}(\| y_{ref} - y(k+i) \|_Q^2 + \| \sigma \|_S^2) + \sum_{i=1}^{N_C}\| \Delta u(k+i) \|_R^2\right] \tag{54}$$

Here, y_{ref} and y represent the output reference and output variables, respectively. The output variables are the voltages at the output nodes. Δu is the change in control actions of control devices and σ is slack variable used to relieve constraints in case of solution in-feasibility. Matrices Q, R and S are the objective weights of voltage deviations, control actions and slack variables, respectively. The weights can be dynamically changed to achieve desired objectives.

Based on temporal characteristics of OLTC and inverter based devices, the control actions are carried out with optimization at two stages. At slow time scale only the actions of OLTC are determined, while as at shorter time scale the PV and EV inverters are dispatched to achieve fast control. The objective function at first stage is represented by quadratic formulation as:

$$\min\left[\sum_{i=1}^{N_P}(\| v_{ref} - v(k+i) \|_Q^2 + \| \sigma \|_S^2) + \sum_{i=1}^{N_C}\| \Delta v_{tap}(k+i) \|_{R1}^2\right] \tag{55}$$

Similarly, the objective function at second stage of MPC controller is:

$$\min\left[\sum_{i=1}^{N_P}(\| v_{ref} - v(k+i) \|_Q^2 + \| \sigma \|_S^2) + \sum_{i=1}^{N_C}\left(\| \Delta Q_{pv}(k+i) \|_{R2}^2 + \| \Delta Q_{ev}(k+i) \|_{R3}^2\right)\right] \tag{56}$$

The optimization is carried under the following system and input constraints:

$$\Delta u_{\min} \le \Delta u(k+i) \le \Delta u_{\max} \tag{57}$$

$$u_{\min} \le u(k+i) \le u_{\max} \tag{58}$$

$$-\sigma 1 + v_{\min} \le v(k+i) \le v_{\max} + \sigma 2 \tag{59}$$

$$v(k+i) = v(k+i-1) + \frac{\delta v}{\delta u}\Delta u(k+i-1) \tag{60}$$

Figure 10. Modified 33 node distribution network with PVs and EVCSs

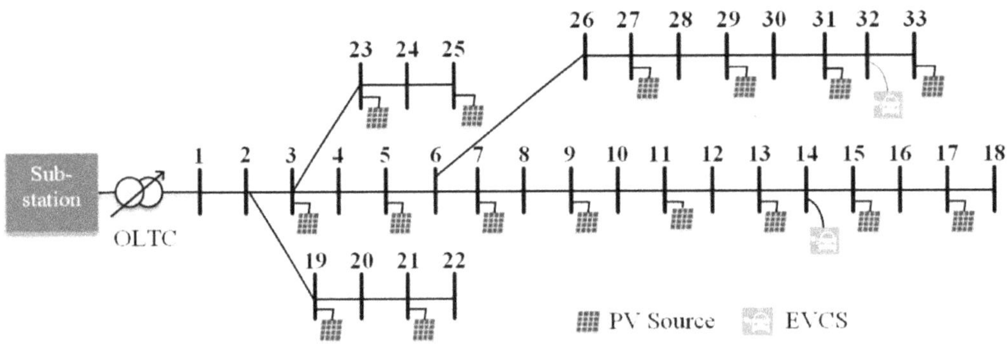

5.1.4. Results and Discussion

The above discussed dual-stage MPC based framework for voltage control is tested and verified on a modified 33-node balanced distribution network as shown in Figure 10. An OLTC connects the substation to node 1. OLTC operation is assumed to step up/step down voltage by 10%, with total number of 32 tap operations, and each tap change operation steps the voltage at node 1 by 0.00625 p.u. At a given time instant, the maximum tap change is limited to 3 steps. Three sets of hourly varying loads such as industrial load from node 26 to node 33, residential load from node 2 to node 18 and commercial load from node 19 to node 25 are considered. A high PV penetration is considered where 16 PV sources are placed at all odd numbered buses from bus number 3 to bus number 33. In addition to PV sources, two electric vehicle charging stations are assumed to be located at bus number 14 and bus number 32. PVs are considered as VAR controllable sources with peak power generation of 1 MW. The PV inverters are assumed to be slightly over-rated by 10%, to follow the requirements as per IEEE-1547-2018 standard, so as to enable them to supply at least 440 kVAr under peak active power generation. The simulations are carried over 24 hours with base MVA 10 and base voltage as 12.66 kV. PV generation varies after 15 minute interval while as EV load demand is assumed to vary after each hour as shown in Figure 9 (a). All simulations are carried out in MATLAB platform. The distribution system modelling is done using PSAT toolbox and all optimizations are carried out using CPLEX solver.

The controller optimally decides the control actions to bring voltages to the specified nominal value of 1 p.u., however, the control actions are relaxed when the voltages are within permissible operations limits i.e., 0.95 p.u.-1.05 p.u. The control actions are based on pre-defined rules depending on the voltage deviations from set point. This is achieved by dynamically changing the weight matrices Q, R and S associated with the control objectives. Two case scenarios are studied (i) Case I: EV charging station at node only 14 (ii) Case II: EV charging station at node 14 and node 32 (both).

Case 1: The voltage profile of test system for 24 hours is shown in Figure 11 (a), when no control actions are applied. The severe over and under voltages can be seen at most of the nodes due to PV generation and varying load and varying EV demand. The voltage profile at nodes 6, 7, 8, 9, 10, 11, 12, 13, 14, 15, 16, 18, 26, 27, 28, 29, 30, 31, 32 and 33 go beyond 1.05 p.u, with maximum violation at node 17, where the voltage rises up to 1.1179 p.u. Furthermore, it can be seen that during the late hours of day the PV generation is very low. Due to high power demand and low PV generation, the under-voltages are seen on nodes 6, 7, 8, 9,10, 11, 12, 13, 14, 15, 16, 17, 18, 26, 27, 28, 29, 30, 31, 32 and

33 with maximum violation at node 18, where the voltage falls up to 0.8788 p.u. The voltage profile at node 18 and node 14 are chosen for observations because of maximum voltage violation and EV station hosting node, respectively. The voltage profile of node 18 and node 14 is shown in Figure 11 (b). To correct the voltage violations, OLTC is used hourly using the centralized controller, to bring the voltages back to normal operation limits. The voltage profile after OLTC control is shown in Figure 12. It is observed that OLTC is not able to bring voltages back to normal operation limits, due to limited tap operations of only 3 steps at a given time instant and only 24 tap operations allowed over a day. However the maximum voltage deviation is brought down to 1.1078 p.u. Tap operations of OLTC are depicted in Figure 13 (a) with 10 tap operations over 24 hours. The voltage violations are eliminated when PV and EV inverters are used for voltage control in addition to OLTC. From Figure 13 (b), it can been that there are no voltage violations, in the meantime tap operations are considerably reduced to just 1 operation, in the beginning hour as shown in Figure 14 (b). Since, at both the optimization stages the voltage deviations from target value of 1 p.u. are minimized, it is seen that for most of the time the voltages at all nodes stay within 0.98 p.u.-1.02 p.u., except at few instants. It is worth to mention that it may not be necessary to bring voltages close to nominal value when some other objectives are also incorporated in optimization such as minimization of power loss.

Figure 11. Case I: Voltage profile under no control action: (a) all nodes, (b) node 18 and node 14

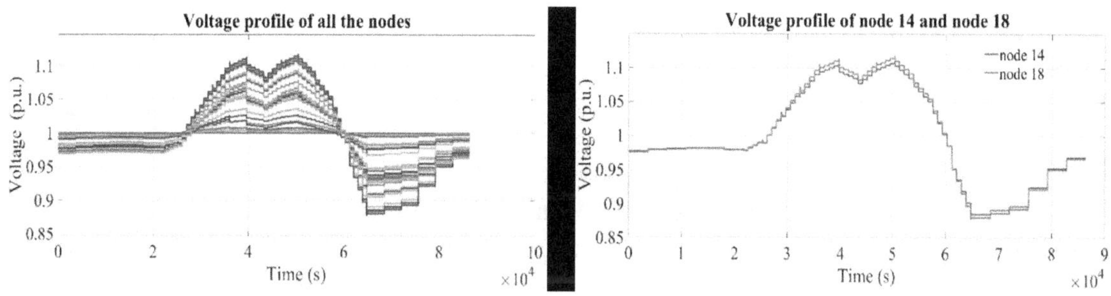

Figure 12. Case I: Voltage profile after OLTC control: (a) all nodes, (b) node 18 and node 14

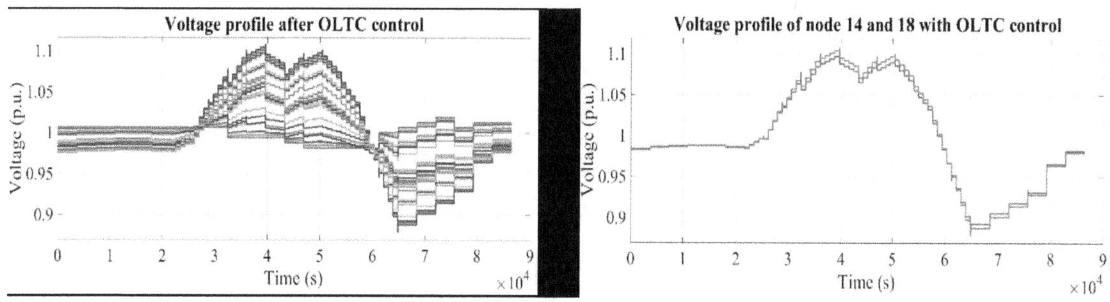

Figure 13. Case I: (a) tap operation under only OLTC control, (b) voltage profile of all nodes when all control actions are applied

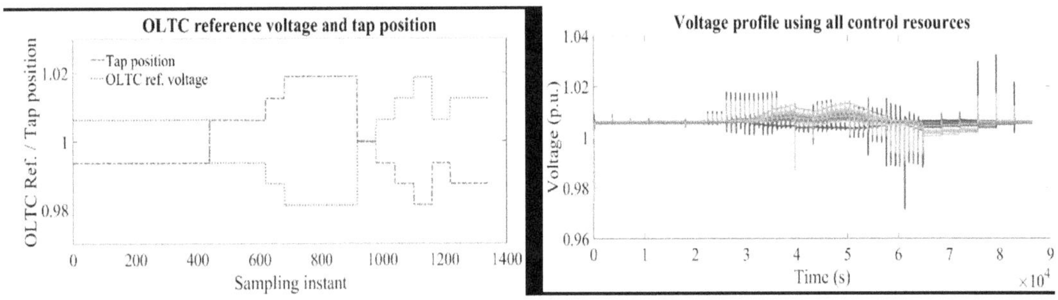

Figure 14. Case I: All control actions are applied: (a) voltage profile of node 18 and node 14, (b) OLTC operation

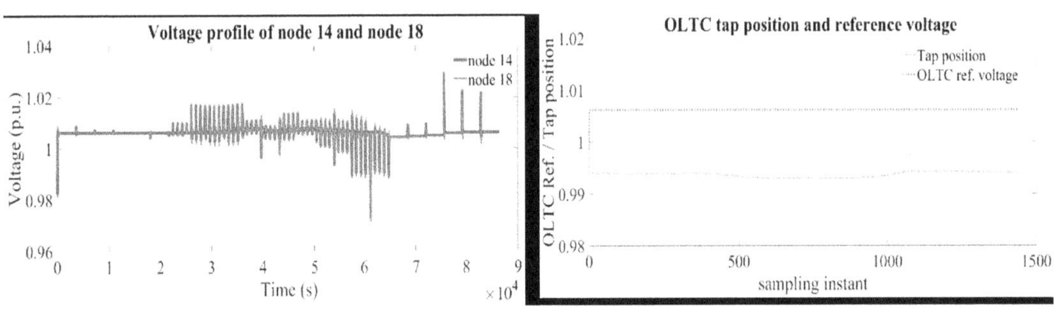

Case 2: In this case, the EV stations are situated in residential area at node 14 as well as in industrial area at node 32. The voltage profile without any control enabled, is shown in Figure 15, with maximum and minimum voltages of 1.1110 p.u and 0.8788 p.u. at node 17 and node 18, respectively. The maximum voltage rise is slightly reduced as compared to case I, due to additional EV demand at node 32. With implementation of OLTC control only, the voltage profile is shown in Figure 16. A total of 13 tap operations of centrally controlled OLTC are observed. The OLTC tap operations are shown in Figure 17 (a). With the implementation of control actions from PVs and EVs, the network node voltages are kept within normal limits and close to nominal value of 1 p.u. for major portion of the time. The voltage profile with all the control devices is shown in Figure 17 (b).

Figure 15. Case II: Voltage profile under no control action: (a) all nodes, (b) node 18 and node 14

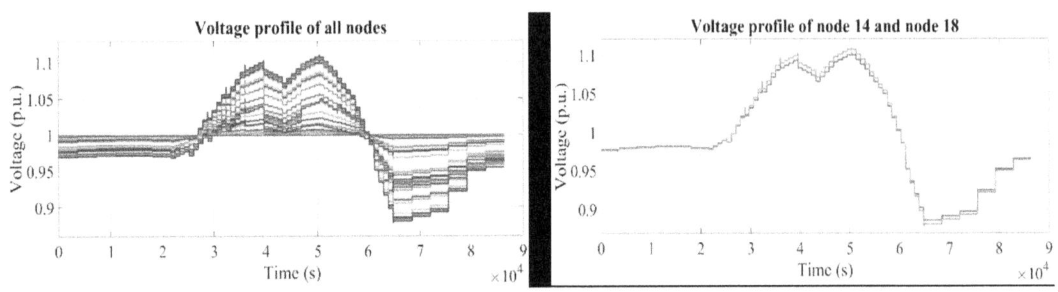

Figure 16. Case II: (a) Voltage profile using OLTC control: (a) all nodes, (b) node 18 and node 14

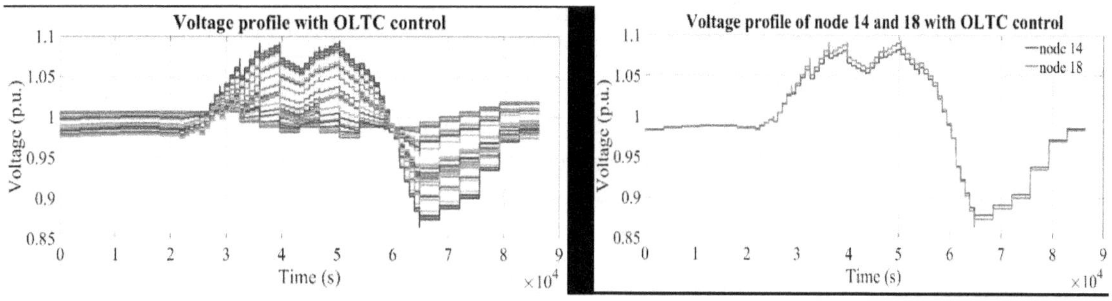

Figure 17. Case II: (a) Tap operation under only OLTC control, (b) voltage profile of all nodes when all control actions are applied

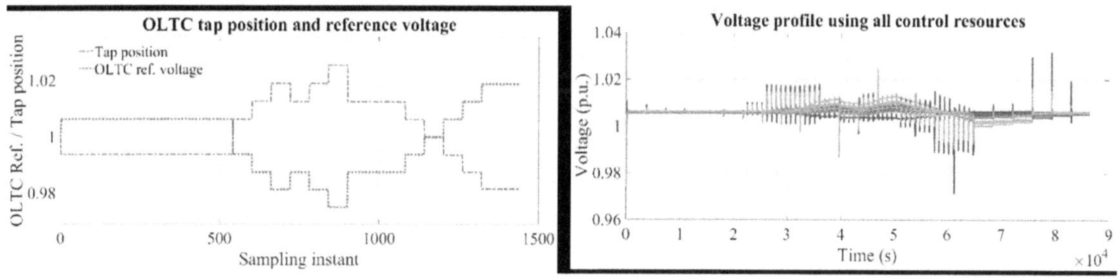

Figure 18. Case II: All control actions applied: (a) voltage profile of node 18 and node 14, (b) OLTC operation

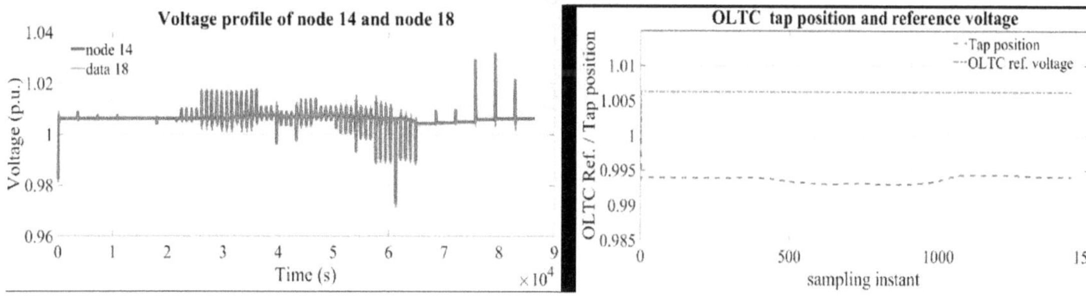

To show the validation of dual-stage MPC framework used in this chapter, a modified 33-node system used in ref. (Dutta et al., 2020) is simulated. The system has PVs located at all nodes except the substation. The apparent power rating of each PV interfacing inverter is 347 kVA.

The PV power generation profile shown in Figure 19 (a) is assumed on all the PVs and loads are assumed to be residential load from node 2-18, commercial load from node 19-25 and industrial load from node 26-33, as shown in Figure 9 (b). To observe the impact of varying generation and load a simulation is carried out over the period of 24 hours. The control horizon and prediction horizon are set to 3 and the sampling time step for each optimization is 10 seconds as in (Dutta et al., 2020). The voltage profile under no voltage control mechanism is shown in Figure 19 (b). It is remarkable that the voltages of most of the nodes such as node 6, 7, 8, 9, 10, 11, 12, 13, 14, 15, 16, 17, 18, 26, 27, 28, 29, 30, 31, 32 and 33 rise beyond permissible value of 1.05 p.u., due to high PV generation during day time. Towards the end of the day PV generation reduces considerably, the high load demand causes some under-voltages in later part of the day. Under no control actions, the maximum over-voltage and under-voltage occurs at node 18 (feeder end), where the voltage reaches 1.1391 during high PV generation and 0.9436 during low PV generation. The under-voltages can been seen on nodes 13, 14, 15, 16, 117, 18, 31, 32 and 33. To compare the effectiveness of used control strategy node 15 is chosen for reference, it is noticed that there are many voltage violations at the chosen observation node in (Dutta et al., 2020), though the violations are for very short durations of few seconds, however, there are only 2 violations at beginning of 10th and 20th hour for just around 10 seconds in the control strategy used in this chapter, as seen from Figure 20 (a) with 3 tap operations in 24 hours (Figure 20 (b)). The performance comparison of control strategy is shown in Table 3. The voltage profile and the reactive power injection/absorption of the PV inverter at network node 15 are shown in Figure 21 (a) and Figure 21 (b), respectively. It is seen that voltages remain within permissible limits for almost all the time over 24 hour duration. The voltage correction of node 15 during over-voltage at 10th hour and under-voltage at 20th hour are shown in Figure 22 (a) and Figure 22 (b), respectively. It is marked that the voltage violations are rectified within 10.5 seconds for both the cases of voltage violation.

Figure 19. (a) PV power generation profile, (b) voltage profile when no control action is applied

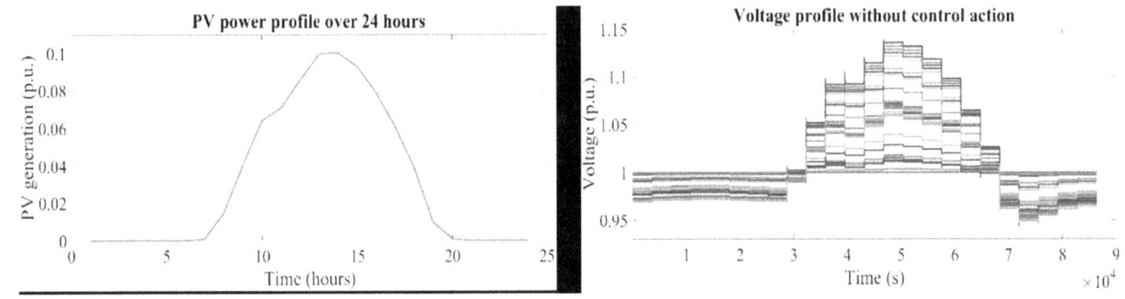

Table 3. Performance comparison of the simulations

Considered Work	Total Number of Voltage Violations	Max. Voltage at Node 15	Time Duration of Voltage Violation
(Dutta et al., 2020)	24	1.1890	76.6 s
Present work	02	1.0578	10 s

6. CONCLUSION

Considering high DG penetration and EV integration, a systematic classification and discussion of various control devices and control strategies has been presented in detail in the chapter, to gain insight about current voltage quality issues in ADNs. A detailed classification of voltage control techniques used in the recent literature for voltage regulation in ADNs is provided. It is seen that the majority of the control strategies used in ADNs are based on centralized voltage control techniques. Local voltage control strategies have limited applications due to limited control resources. For large distribution networks, distributed control has seen significant attention and developments in recent time, however distributed control techniques are hard and complex in implementation in real distribution networks. The high uncertainty in distribution networks due to PVs and EVs demand for real time control by using real time measuring infrastructure and data acquisition systems. For real time control, the capabilities of predictive control based techniques in future ADNs, have gained wide attention in recent years. An MPC based control strategy carried out in this chapter shows the effectiveness of PVs and EVs for voltage control in high penetrated power networks. Furthermore, the time scale decomposition of the studied optimization problem effectively helps to reduce tap operation of OLTC, while restraining the voltage within the defined limits, for almost all the time intervals in a day.

REFERENCES

Alimisis, V., & Taylor, P. C. (2015). Zoning Evaluation for Improved Coordinated Automatic Voltage Control. *IEEE Transactions on Power Systems*, *30*(5), 2736–2746. doi:10.1109/TPWRS.2014.2369428

Alzaareer, K., Saad, M., Mehrjerdi, H., Asber, D., & Lefebvre, S. (2020). Development of New Identification Method for Global Group of Controls for Online Coordinated Voltage Control in Active Distribution Networks. *IEEE Transactions on Smart Grid*, *11*(5), 3921–3931. doi:10.1109/TSG.2020.2981145

Amanipoor, A., Golsorkhi, M. S., Bayati, N., & Savaghebi, M. (2023). V-Iq Based Control Scheme for Mitigation of Transient Overvoltage in Distribution Feeders with High PV Penetration. *IEEE Transactions on Sustainable Energy*, *14*(1), 283–296. doi:10.1109/TSTE.2022.3211179

American National Standard for Electric Power Systems and Equipment-Voltage Ratings. (2020). *60 Hertz*. National Electrical Manufacturers Association.

Amini, M. H., Kargarian, A., & Karabasoglu, O. (2016). ARIMA-based decoupled time series forecasting of electric vehicle charging demand for stochastic power system operation. *Electric Power Systems Research*, *140*, 378–390. doi:10.1016/j.epsr.2016.06.003

Anurag, A., Yang, Y., & Blaabjerg, F. (2015). Thermal Performance and Reliability Analysis of Single-Phase PV Inverters with Reactive Power Injection outside Feed-In Operating Hours. *IEEE Journal of Emerging and Selected Topics in Power Electronics*, *3*(4), 870–880. doi:10.1109/JESTPE.2015.2428432

Arshad, A., Ekström, J., & Lehtonen, M. (2018). Multi-agent based distributed voltage regulation scheme with grid-tied inverters in active distribution networks. *Electric Power Systems Research*, *160*, 180–190. doi:10.1016/j.epsr.2018.02.015

Bahramipanah, M., Cherkaoui, R., & Paolone, M. (2016). Decentralized voltage control of clustered active distribution network by means of energy storage systems. *Electric Power Systems Research*, *136*, 370–382. doi:10.1016/j.epsr.2016.03.021

Bidgoli, H. S., & Van Cutsem, T. (2018). Combined Local and Centralized Voltage Control in Active Distribution Networks. *IEEE Transactions on Power Systems*, *33*(2), 1374–1384. doi:10.1109/TPWRS.2017.2716407

Chai, Q., Zhang, C., Xu, Y., & Dong, Z. Y. (2021). PV inverter reliability-constrained volt/var control of distribution networks. *IEEE Transactions on Sustainable Energy*, *12*(3), 1788–1800. doi:10.1109/TSTE.2021.3065451

Chakraborty, R., Chakrabortty, A., Farantatos, E., Patel, M., Hooshyar, H., & Darvishi, A. (2023). Hierarchical frequency and voltage control using prioritized utilization of inverter based resources. *International Journal of Electrical Power & Energy Systems*, *144*, 108527. doi:10.1016/j.ijepes.2022.108527

Chen, S., Hu, W., & Chen, Z. (2016). Comprehensive Cost Minimization in Distribution Networks Using Segmented-Time Feeder Reconfiguration and Reactive Power Control of Distributed Generators. *IEEE Transactions on Power Systems*, *31*(2), 983–993. doi:10.1109/TPWRS.2015.2419716

Christakou, K., Paolone, M., & Abur, A. (2018). Voltage control in active distribution networks under uncertainty in the system model: A robust optimization approach. *IEEE Transactions on Smart Grid*, *9*(6), 5631–5642. doi:10.1109/TSG.2017.2693212

Dar, M. R., & Ganguly, S. (2022.) A review of voltage control strategies in active distribution networks. *2022 IEEE Calcutta Conference (CALCON),* 145-149. 10.1109/CALCON56258.2022.10060319

Ding, T., Li, C., Yang, Y., Jiang, J., Bie, Z., & Blaabjerg, F. (2017). A Two-Stage Robust Optimization for Centralized-Optimal Dispatch of Photovoltaic Inverters in Active Distribution Networks. *IEEE Transactions on Sustainable Energy*, *8*(2), 744–754. doi:10.1109/TSTE.2016.2605926

Dou, X., Xu, P., Hu, Q., Sheng, W., Quan, X., Wu, Z., & Xu, B. (2018). A distributed voltage control strategy for multi-microgrid active distribution networks considering economy and response speed. *IEEE Access : Practical Innovations, Open Solutions*, *6*, 31259–31268. doi:10.1109/ACCESS.2018.2837082

Du, P., & Lu, N. (2014). *Energy Storage for Smart Grids: Planning and Operation for Renewable and Variable Energy Resources (VERs).* Academic Press.

Dutta, A., Ganguly, S., & Kumar, C. (2020). Model predictive control-based optimal voltage regulation of active distribution networks with OLTC and reactive power capability of PV inverters. *IET Generation, Transmission & Distribution*, *14*(22), 5183–5192. doi:10.1049/iet-gtd.2020.0378

Dutta, A., Ganguly, S., & Kumar, C. (2022[a]). Coordinated control scheme for EV charging and volt/var devices scheduling to regulate voltages of active distribution networks. *Sustainable Energy, Grids and Networks*, 31.

Dutta, A., Ganguly, S., & Kumar, C. (2022b). Coordinated Volt/Var Control of PV and EV Interfaced Active Distribution Networks Based on Dual-Stage Model Predictive Control. *IEEE Systems Journal*, *16*(3), 4291–4300. doi:10.1109/JSYST.2021.3110509

Dutta, A., Ganguly, S., & Kumar, C. (2022c). MPC-Based Coordinated Voltage Control in Active Distribution Networks Incorporating CVR and DR. *IEEE Transactions on Industry Applications*, *58*(4), 4309–4318. doi:10.1109/TIA.2022.3163108

Elkhatib, M. E., El-Shatshat, R., & Salama, M. M. A. (2011). Novel coordinated voltage control for smart distribution networks with DG. *IEEE Transactions on Smart Grid*, *2*(4), 598–605. doi:10.1109/TSG.2011.2162083

Enang, W., & Bannister, C. 2017. Modelling and control of hybrid electric vehicles (A comprehensive review). In Renewable and Sustainable Energy Reviews (Vol. 74). Academic Press.

Gao, H., Zhang, Y., & Ji, X. (2020). Scenario clustering based distributionally robust comprehensive optimization of active distribution network. *Dianli Xitong Zidonghua*, *44*(21), 32–41.

Gholami, K., Islam, M. R., Rahman, M. M., Azizivahed, A., & Fekih, A. (2022). State-of-the-art technologies for volt-var control to support the penetration of renewable energy into the smart distribution grids. *Energy Reports*, *8*, 8630–8651. doi:10.1016/j.egyr.2022.06.080

Guo, Y., Wu, Q., Gao, H., Chen, X., Ostergaard, J., & Xin, H. (2019). MPC-Based Coordinated Voltage Regulation for Distribution Networks with Distributed Generation and Energy Storage System. *IEEE Transactions on Sustainable Energy*, *10*(4), 1731–1739. doi:10.1109/TSTE.2018.2869932

Guo, Y., Wu, Q., Gao, H., Huang, S., Zhou, B., & Li, C. (2020a). Double-Time-Scale Coordinated Voltage Control in Active Distribution Networks Based on MPC. *IEEE Transactions on Sustainable Energy*, *11*(1), 294–303. doi:10.1109/TSTE.2018.2890621

Haider, R., & Annaswamy, A. M. (2022). A hybrid architecture for volt-var control in active distribution grids. *Applied Energy*, *312*, 118735. doi:10.1016/j.apenergy.2022.118735

Haque, A. N. M. M., Xiong, M., & Nguyen, P. H. (2019). March. Consensus algorithm for fair power curtailment of PV systems in LV networks. *2019 IEEE PES GTD Grand International Conference and Exposition Asia (GTD Asia)*, 813-818.

Hemmatpour, M. H., Rezaeian Koochi, M. H., Dehghanian, P., & Dehghanian, P. (2022). Voltage and energy control in distribution systems in the presence of flexible loads considering coordinated charging of electric vehicles. *Energy*, *239*, 121880. doi:10.1016/j.energy.2021.121880

Hidalgo, R., Abbey, C., & Joós, G. (2010). A review of active distribution networks enabling technologies. *IEEE PES General Meeting, PES 2010*. 10.1109/PES.2010.5590051

Hoque, M. M., Khorasany, M., Razzaghi, R., Wang, H., & Jalili, M. (2022). Transactive Coordination of Electric Vehicles with Voltage Control in Distribution Networks. *IEEE Transactions on Sustainable Energy*, *13*(1), 391–402. doi:10.1109/TSTE.2021.3113614

Hu, D., Ye, Z., Gao, Y., Ye, Z., Peng, Y., & Yu, N. (2022). Multi-Agent Deep Reinforcement Learning for Voltage Control with Coordinated Active and Reactive Power Optimization. *IEEE Transactions on Smart Grid*, *13*(6), 4873–4886. doi:10.1109/TSG.2022.3185975

Hu, J., Ye, C., Ding, Y., Tang, J., & Liu, S. (2022). A Distributed MPC to Exploit Reactive Power V2G for Real-Time Voltage Regulation in Distribution Networks. *IEEE Transactions on Smart Grid*, *13*(1), 576–588. doi:10.1109/TSG.2021.3109453

Hu, R., Wang, W., Wu, X., Chen, Z., Jing, L., Ma, W., & Zeng, G. (2022). Coordinated active and reactive power control for distribution networks with high penetrations of photovoltaic systems. *Solar Energy*, *231*, 809–827. doi:10.1016/j.solener.2021.12.025

IEEE Power & Energy Society. (2019). *Big shift in power & energy*. IEEE PES GTD 2019 Grand International Conference & Exposition Asia. Bangkok International Trade & Exhibition Centre (BITEC), Bangkok, Thailand.

Islam, F. R., Lallu, A., Mamun, K. A., Prakash, K., & Roy, N. K. (2021). Power Quality Improvement of Distribution Network Using BESS and Capacitor Bank. *Journal of Modern Power Systems and Clean Energy*, *9*(3), 625–632. doi:10.35833/MPCE.2019.000304

Ji, H., Yu, H., Song, G., Li, P., Wang, C., & Wu, J. (2019). A Decentralized Voltage Control Strategy of Soft Open Points in Active Distribution Networks. *Energy Procedia*, *159*, 412–417. doi:10.1016/j.egypro.2018.12.067

Jiajie, Z., Xinru, J., Jia, F., Wenbo, Z., Penghui, L., Chen, L., & Zheng, L. (2022). Online Reactive Power Optimization of Distribution Network with Soft Open Point Based on Hierarchical Reinforcement Learning. *Proceedings - 2022 4th International Conference on Electrical Engineering and Control Technologies, CEECT*, 66–72. 10.1109/CEECT55960.2022.10030151

Kabir, F., Yu, N., Gao, Y., & Wang, W. (2023). Deep reinforcement learning-based two-timescale Volt-VAR control with degradation-aware smart inverters in power distribution systems. *Applied Energy*, *335*, 120629. doi:10.1016/j.apenergy.2022.120629

Kabir, M. N., Mishra, Y., Ledwich, G., Dong, Z. Y., & Wong, K. P. (2014). Coordinated control of grid-connected photovoltaic reactive power and battery energy storage systems to improve the voltage profile of a residential distribution feeder. *IEEE Transactions on Industrial Informatics*, *10*(2), 967–977. doi:10.1109/TII.2014.2299336

Kekatos, V., Wang, G., Conejo, A. J., & Giannakis, G. B. (2014). Stochastic reactive power management in microgrids with renewables. *IEEE Transactions on Power Systems*, *30*(6), 3386–3395. doi:10.1109/TPWRS.2014.2369452

Kryonidis, G. C., Kontis, E. O., Chrysochos, A. I., Demoulias, C. S., & Papagiannis, G. K. (2018). A Coordinated Droop Control Strategy for Overvoltage Mitigation in Active Distribution Networks. *IEEE Transactions on Smart Grid*, *9*(5), 5260–5270. doi:10.1109/TSG.2017.2685686

Kulmala, A., Mutanen, A., Koto, A., Repo, S., & Järventausta, P. (2010.) RTDS verification of a coordinated voltage control implementation for distribution networks with distributed generation. *2010 IEEE PES Innovative Smart Grid Technologies Conference Europe (ISGT Europe),* 1-8.

Kulmala, A., Repo, S., & Järventausta, P. (2014). Coordinated voltage control in distribution networks including several distributed energy resources. *IEEE Transactions on Smart Grid*, *5*(4), 2010–2020. doi:10.1109/TSG.2014.2297971

Lee, J., & Kim, G. H. (2017). Comparison analysis of the voltage variation ranges for distribution networks. *2017 IEEE International Conference on Environment and Electrical Engineering and 2017 IEEE Industrial and Commercial Power Systems Europe (EEEIC/I&CPS Europe),* 1-3. 10.1109/EEEIC.2017.7977500

Li, P., Ji, H., Wang, C., Zhao, J., Song, G., Ding, F., & Wu, J. (2017). Coordinated Control Method of Voltage and Reactive Power for Active Distribution Networks Based on Soft Open Point. *IEEE Transactions on Sustainable Energy*, *8*(4), 1430–1442. doi:10.1109/TSTE.2017.2686009

Li, P., Ji, J., Ji, H., Jian, J., Ding, F., Wu, J., & Wang, C. (2020). MPC-Based Local Voltage Control Strategy of DGs in Active Distribution Networks. *IEEE Transactions on Sustainable Energy*, *11*(4), 2911–2921. doi:10.1109/TSTE.2020.2981486

Li, P., Wei, M., Ji, H., Xi, W., Yu, H., Wu, J., Yao, H., & Chen, J. (2022). Deep Reinforcement Learning-Based Adaptive Voltage Control of Active Distribution Networks with Multi-terminal Soft Open Point. *International Journal of Electrical Power & Energy Systems*, *141*, 108138. doi:10.1016/j.ijepes.2022.108138

Li, Z., Khajepour, A., & Song, J. (2019). A comprehensive review of the key technologies for pure electric vehicles. *Energy*, *182*, 824–839. doi:10.1016/j.energy.2019.06.077

Liu, H. J., Shi, W., & Zhu, H. (2018). Hybrid voltage control in distribution networks under limited communication rates. *IEEE Transactions on Smart Grid*, *10*(3), 2416–2427. doi:10.1109/TSG.2018.2797692

Liu, Y., Gao, H., Liu, J., Ma, Z., Chen, J., & Yang, Y. (2015). Multi-agent based hierarchical power scheduling strategy for active distribution network. *Proceedings - 2015 International Symposium on Smart Electric Distribution Systems and Technologies, EDST 2015, 1*, 151–158.

Lopes, J. A. P., Hatziargyriou, N., Mutale, J., Djapic, P., & Jenkins, N. (2007). Integrating distributed generation into electric power systems: A review of drivers, challenges and opportunities. *Electric Power Systems Research*, *77*(9), 1189–1203. doi:10.1016/j.epsr.2006.08.016

Luo, C., Wu, H., Zhou, Y., Qiao, Y., & Cai, M. (2021). Network partition-based hierarchical decentralised voltage control for distribution networks with distributed PV systems. *International Journal of Electrical Power & Energy Systems*, *130*(5), 4087–4098. doi:10.1016/j.ijepes.2021.106929

Ma, T., & Zhang, T. (2019). Research on Reactive Power and Voltage Characteristics Analysis and Reactive Power Configuration Strategy of Active Distribution Network. *2019 12th International Conference on Intelligent Computation Technology and Automation (ICICTA)*, 108-111.

Ma, Z., Yang, N., Zou, S., & Shao, Y. (2018). Charging coordination of Plug-In electric vehicles in distribution networks with capacity constrained feeder lines. *IEEE Transactions on Control Systems Technology*, *26*(5), 1917–1924. doi:10.1109/TCST.2017.2730825

Maharjan, S., Khambadkone, A. M., & Peng, J. C. H. (2021). Robust constrained model predictive voltage control in active distribution networks. *IEEE Transactions on Sustainable Energy*, *12*(1), 400–411. doi:10.1109/TSTE.2020.3001115

Mai, T.T., Haque, A.N.M.M., & Nguyen, P.H. (2019, June). Consensus-based distributed control for overvoltage mitigation in LV microgrids. *2019 IEEE Milan Power Tech,* 1-6.

Martinenas, S., Knezovic, K., & Marinelli, M. (2017). Management of Power Quality Issues in Low Voltage Networks Using Electric Vehicles: Experimental Validation. *IEEE Transactions on Power Delivery*, *32*(2), 971–979. doi:10.1109/TPWRD.2016.2614582

Martinez, C. M., Hu, X., Cao, D., Velenis, E., Gao, B., & Wellers, M. (2017). Energy Management in Plug-in Hybrid Electric Vehicles: Recent Progress and a Connected Vehicles Perspective. *IEEE Transactions on Vehicular Technology*, *66*(6), 4534–4549. doi:10.1109/TVT.2016.2582721

Navesi, R. B., Nazarpour, D., Ghanizadeh, R., & Alemi, P. (2022). Switchable Capacitor Bank Coordination and Dynamic Network Reconfiguration for Improving Operation of Distribution Network Integrated with Renewable Energy Resources. *Journal of Modern Power Systems and Clean Energy*, *10*(3), 637–646. doi:10.35833/MPCE.2020.000067

Nguyen, H. M., Torres, J. L. R., Leki, A., & Pham, H. V. (2021). MPC Based Centralized Voltage and Reactive Power Control for Active Distribution Networks. *IEEE Transactions on Energy Conversion*, *36*(2), 1537–1547. doi:10.1109/TEC.2021.3054844

Olival, P. C., Madureira, A. G., & Matos, M. (2017). Advanced voltage control for smart microgrids using distributed energy resources. *Electric Power Systems Research*, *146*, 132–140. doi:10.1016/j.epsr.2017.01.027

Photovoltaics, D. G., & Storage, E. (2018). IEEE standard for interconnection and interoperability of distributed energy resources with associated electric power systems interfaces. *IEEE Std.*, *1547*, 1547–2018.

Ranamuka, D., Agalgaonkar, A. P., & Muttaqi, K. M. (2016). Online Coordinated Voltage Control in Distribution Systems Subjected to Structural Changes and DG Availability. *IEEE Transactions on Smart Grid*, *7*(2), 580–591.

Romvary, J. J., Ferro, G., Haider, R., & Annaswamy, A. M. (2021). A proximal atomic coordination algorithm for distributed optimization. *IEEE Transactions on Automatic Control*, *67*(2), 646–661. doi:10.1109/TAC.2021.3053907

Ruan, H., Gao, H., Liu, Y., Wang, L., & Liu, J. (2020). Distributed voltage control in active distribution network considering renewable energy: A novel network partitioning method. *IEEE Transactions on Power Systems*, *35*(6), 4220–4231. doi:10.1109/TPWRS.2020.3000984

Safayet, A., Fajri, P., & Husain, I. (2017). Reactive Power Management for Overvoltage Prevention at High PV Penetration in a Low-Voltage Distribution System. *IEEE Transactions on Industry Applications*, *53*(6), 5786–5794. doi:10.1109/TIA.2017.2741925

Schwartfeger, L., Santos-Martin, D., Wood, A., Watson, N., & Miller, A., (2014, JUNE). Review of Distributed Generation Interconnection Standards. *Proceedings of Electricity Engineers' Association Conference*.

Shuvra, M. A., & Chowdhury, B. H. (2018). Distributed voltage control of active MV distribution networks in the presence of high PV penetration. *2018 IEEE Power and Energy Society Innovative Smart Grid Technologies Conference*, 1–5.

Singh, S., Pamshetti, V. B., & Singh, S. P. (2019). Time Horizon-Based Model Predictive Volt/VAR Optimization for Smart Grid Enabled CVR in the Presence of Electric Vehicle Charging Loads. *IEEE Transactions on Industry Applications*, *55*(6), 5502–5513. doi:10.1109/TIA.2019.2928490

Sun, H., Guo, Q., Qi, J., Ajjarapu, V., Bravo, R., Chow, J., Li, Z., Moghe, R., Nasr-Azadani, E., Tamrakar, U., Taranto, G. N., Tonkoski, R., Valverde, G., Wu, Q., & Yang, G. (2019). Review of Challenges and Research Opportunities for Voltage Control in Smart Grids. *IEEE Transactions on Power Systems*, *34*(4), 2790–2801. doi:10.1109/TPWRS.2019.2897948

Tanaka, K. (2009). Decentralized voltage control in distribution systems by distributed generators. *IEEE International Symposium on Industrial Electronics*, 554-559. 10.1109/ISIE.2009.5213299

Tang, Z., Hill, D. J., & Liu, T. (2021). Distributed Coordinated Reactive Power Control for Voltage Regulation in Distribution Networks. *IEEE Transactions on Smart Grid, 12*(1), 312–323. doi:10.1109/TSG.2020.3018633

Troester, E. (2009). New German grid codes for connecting PV systems to the medium voltage power grid. *2nd International workshop on concentrating photovoltaic power plants: optical design, production, grid connection*, 9-10.

Valverde, G., & Van Cutsem, T. (2013). Model predictive control of voltages in active distribution networks. *IEEE Transactions on Smart Grid, 4*(4), 2152–2161. doi:10.1109/TSG.2013.2246199

VDE. (2011). *Power generation systems connected to the low-voltage distribution network*. VDE Verlag.

Walling, R. A., Saint, R., Dugan, R. C., Burke, J., & Kojovic, L. A. (2008). Summary of distributed resources impact on power delivery systems. *IEEE Transactions on Power Delivery, 23*(3), 1636–1644. doi:10.1109/TPWRD.2007.909115

Wang, L., Yan, R., Bai, F., Saha, T., & Wang, K. (2020). A Distributed Inter-Phase Coordination Algorithm for Voltage Control with Unbalanced PV Integration in LV Systems. *IEEE Transactions on Sustainable Energy, 11*(4), 2687–2697. doi:10.1109/TSTE.2020.2970214

Wang, Y., John, T., & Xiong, B. (2019). A two-level coordinated voltage control scheme of electric vehicle chargers in low-voltage distribution networks. *Electric Power Systems Research, 168*, 218–227. doi:10.1016/j.epsr.2018.12.005

Wang, Y., Qiu, D., Strbac, G., & Gao, Z. (2023). Coordinated Electric Vehicle Active and Reactive Power Control for Active Distribution Networks. *IEEE Transactions on Industrial Informatics, 19*(2), 1611–1622. doi:10.1109/TII.2022.3169975

Yan, R., Marais, B., & Saha, T. K. (2014). Impacts of residential photovoltaic power fluctuation on on-load tap changer operation and a solution using DSTATCOM. *Electric Power Systems Research, 111*, 185–193. doi:10.1016/j.epsr.2014.02.020

Yan, R., & Saha, T. K. (2012). Investigation of voltage stability for residential customers due to high photovoltaic penetrations. *IEEE Transactions on Power Systems, 27*(2), 651–662. doi:10.1109/TPWRS.2011.2180741

Yang, Q., Wang, G., Sadeghi, A., Giannakis, G. B., & Sun, J. (2020). Two-Timescale Voltage Control in Distribution Grids Using Deep Reinforcement Learning. *IEEE Transactions on Smart Grid, 11*(3), 2313–2323. doi:10.1109/TSG.2019.2951769

Yong, C., Li, Y., Zeng, Z., Zhang, Z., Zhang, Z., & Liu, Y. (2020). Coordinated active and reactive power optimization considering load characteristics for active distribution network. *Chinese Journal of Electrical Engineering, 6*(4), 97–105. doi:10.23919/CJEE.2020.000034

Yu, L., Czarkowski, D., & De León, F. (2012). Optimal distributed voltage regulation for secondary networks with DGs. *IEEE Transactions on Smart Grid, 3*(2), 959–967. doi:10.1109/TSG.2012.2190308

Zeraati, M., Golshan, M. E. H., & Guerrero, J. M. (2017). A consensus-based cooperative control of PEV battery and PV active power curtailment for voltage regulation in distribution networks. *IEEE Transactions on Smart Grid*, *10*(1), 670–680. doi:10.1109/TSG.2017.2749623

Zhang, C., & Xu, Y. (2020). Hierarchically-Coordinated Voltage/VAR Control of Distribution Networks Using PV Inverters. *IEEE Transactions on Smart Grid*, *11*(4), 2942–2953. doi:10.1109/TSG.2020.2968394

Zhang, Z., da Silva, F. F., Guo, Y., Bak, C. L., & Chen, Z. (2021). Double-layer stochastic model predictive voltage control in active distribution networks with high penetration of renewables. *Applied Energy*, *302*, 117530. doi:10.1016/j.apenergy.2021.117530

Zhao, B., Xu, Z., Xu, C., Wang, C., & Lin, F. (2018). Network Partition-Based Zonal Voltage Control for Distribution Networks with Distributed PV Systems. *IEEE Transactions on Smart Grid*, *9*(5), 4087–4098. doi:10.1109/TSG.2017.2648779

Zhao, J., Yao, M., Yu, H., Song, G., Ji, H., & Li, P. (2020). Decentralized voltage control strategy of soft open points in active distribution networks based on sensitivity analysis. *Electronics (Basel)*, *9*(2), 295. doi:10.3390/electronics9020295

Zhao, W., Liu, D., Li, Q., & Nong, J. 2015. A novel flexible load coordination control based on distributed generation consumption in the active distribution network. *IET Conference Publications, 2015*(CP679), 7–11.

Chapter 10
Optimal Network Reconfiguration for Efficient EV Integration in the Smart Grid-Enabled Power Distribution Networks

Subhadarshini Panda
Indian Institute of Technology, Guwahati, India

Sanjib Ganguly
https://orcid.org/0000-0002-7939-7308
Indian Institute of Technology, Guwahati, India

ABSTRACT

The ever-increasing energy demand and global environmental crisis have acted like a growing push towards the adoption of electric vehicles (EVs). However, integration of EVs brings forth several challenges concerning the efficiency and reliability of power distribution networks (PDNs). These challenges must be addressed through advanced technologies and smart grid solutions. Distribution network reconfiguration (DNR) has emerged as a promising technique to enhance the reliability and efficiency of PDNs, while supporting the integration of EVs. This chapter provides a comprehensive analysis on DNR method and its application in improving the performance of smart grid enabled PDNs. The study incorporates an in-depth literature survey, highlighting the use of DNR to facilitate EV integration in PDNs. Furthermore, quantitative assessments of the literature are presented. These assessments showcase the effectiveness of DNR in achieving improved performance metrics, such as reducing losses, operational cost, enhancing voltage profile, and ensuring the reliability of EV penetrated PDNs.

DOI: 10.4018/979-8-3693-2611-4.ch010

Abbreviations

APL: Active power loss
ARSBX: Adaptive-rotation based simulated binary crossover
BPSO: Binary particle swarm optimization
CCGA: Cooperative coevolutionary genetic algorithm
DG: Distributed generation
DMS: Distribution management system
DNR: Distribution network reconfiguration
ENS: Energy not supplied
EV: Electric vehicle
GA: Genetic algorithm
IGA: Improved genetic algorithm
LDBAS: Levy flight and chaos distributed beetle antennae search
LV: Low voltage
MCOST: Minimization of cost
MGWO: Modified grey wolf optimization
MILP: Mixed integer linear programming
MINLP: Mixed integer nonlinear programming
MLBI: Minimization of load balancing index
MPL: Minimization of power loss
MIQP: Mixed integer quadratic programming
MISOCP: Mixed integer second order cone programming
MSOS: Modified symbiotic organisms search
MV: Medium voltage
MVD: Minimization of voltage deviation
MVSI: Minimization of voltage stability index
PEV: Plug-in electric vehicle
PDN: Power distribution network
SAMCSA: Self-adaptive modified crow search algorithm
SAMKHOA: Self-adaptive modified krill herd optimization algorithm
SSO: Social spider optimization
SOP: Soft-open-point
UPSO: Unified particle swarm optimization
V2G: Vehicle-to-grid

1. INTRODUCTION

In the 21st century, significant release of greenhouse gases is the primary contributor to global warming, and a major source of these emissions can be attributed to the transportation sector (García-Villalobos et al., 2014). To address the environmental threats posed by global warming, it is crucial to curb greenhouse gas emissions. In recent times, governments world-wide are competing to implement environmental-friendly alternatives to reduce the carbon foot prints. As one of the most viable alternatives to traditional gasoline-powered vehicles, EVs are increasingly viewed as a means of reducing carbon dioxide emissions (Gomez and Morcos, 2003). Hence, adoption of EVs in the transportation sector has gained global attention. Further, the modernization of the conventional grid to smart grid greatly encourages the EV integration through the utilization of vehicle-to-grid (V2G) technology (Hota et al., 2014). According to a report prepared by the authors in (IEA 2017), the number of EVs sold has already reached 12 million, with an estimated surge to 200 million by the year 2030.

However, integration of EV charging loads into the PDNs presents substantial challenges. Increased peak load demand, active power loss (APL), degradation of voltage profile, phase voltage imbalance, deterioration of power quality and compromised network reliability are some of the major consequences of uncoordinated large scale EV penetration (Dubey and Santoso, 2015). These consequences highlight the need for coordinated and efficient integration of EVs in PDNs. Moreover, the advent of the smart grid paradigm has resulted in a notable rise in distributed generation (DG) units and EV charging loads. DG units have climate-dependent outputs and hence, introduce intermittency and uncertainty into the system. Additionally, EV charging and discharging patterns can be highly unpredictable and depend on drivers' behaviour, charging infrastructure availability, and location, arrival/departure time of EVs, and battery state of charge level, etc. (Wu et al., 2017). These uncertainties can lead to additional power fluctuation and further increase the overall uncertainty level of the PDN. To ensure the efficient integration of DG units and EV charging stations, it is imperative to address these issues and develop appropriate solutions.

To combat these issues, smart grid technologies, such as distribution management system (DMS) and distribution automation can play vital roles. DMS is a computer-based system used by the electric utilities to manage and optimize the distribution of electricity to customers in a smart grid environment. It is designed to monitor and control the entire PDN in real time, from the distribution substation to the customers' meters. A DMS typically includes a number of integrated applications that help with functions such as outage management, network optimization, asset management and load management (Pilo et al., 2011). Among these various applications, network reconfiguration stands out as an effective approach to enhance system performance, reliability, and efficiency by making changes to network's layout (Vargas and Samper, 2012).

Further, automated PDNs substantially improves the operational flexibility through the utilization of remotely-controlled switches (Asrari et al., 2016). These networks rely on fast communication systems to efficiently monitor and manage the network, receiving instructions from the DMS. With flexible structure, these networks can readily adapt to changing consumer behaviour and evolving operational requirements (Rostami et al., 2015). Hence, network reconfiguration in an automated distribution network can present a promising solution to enhance the overall performance of the EV integrated PDN.

The remainder of this chapter is structured as follows. In Section 2, we offer a concise overview of the impacts of different types of EV charging on PDNs. Section 3 provides a comprehensive analysis of DNR, delving into its various objectives and the different types implemented by utility providers. Moving to Section 4, we conduct a literature survey that illuminates the crucial role of DNR in EV integrated PDNs. This section not only summarizes the key findings of various research works but also outlines the objectives of the optimization problems along with the solution strategies. Section 5 engages in a quantitative analysis of the literature, offering a comparative assessment based on performance indicators such as the impact of DNR on power loss reduction, cost reduction, enhancement of voltage profile, and reliability improvement. Finally, Section 6 brings the chapter to a close, presenting concluding remarks

2. IMPACTS OF EV CHARGING ON PDNS

The widespread adoption of EVs in PDN may create several issues, including phase voltage imbalancing, congestion and overloading of lines and transformers, node voltage deviations, power quality degradation, increased peak load demand, and power losses (Islam et al., 2019). To assess the performance of a PDN following the integration of EVs, three primary strategies have been studied. These strategies, along with their corresponding impacts, are provided in Table 1.

Table 1. Impacts of EV charging strategies (García-Villalobos et al., 2014)

Type of EV Charging	Positive Impacts	Negative Impacts
Uncontrolled charging	Easy to implement User Friendly	Increased peak load demand Overloading of feeders and transformers Increased power losses Node voltage deviation Need of network reinforcement
Off-peak charging	Easy to implement Load levelling Delay in network reinforcement Better integration of RES based DG units	Increase in demand during off peak hours due to EV charging Chances of overloading of feeders and transformers Chances of node voltage deviation Need of customer willingness to participate
Smart charging (valley filling)	Provision of ancillary services Load levelling Delay in network reinforcement Better integration of RES based DG units	Difficult to implement Need of information and communication technology Need of customer willingness to participate
Smart Charing (Peak Shaving)	Provision of ancillary services Reduction of peak load demand Optimal integration of RES based DG units Less investment in network reinforcement	Difficult to implement Need of advanced information and communication technology Premature degradation of batteries due to V2G services More energy losses in grid-battery-grid transmission

To overcome the drawbacks of uncoordinated and uncontrolled EV charging, various researchers have investigated smart charging. The concept of smart charging refers to charge the EVs by taking advantages of favourable conditions such as periods of low electricity prices, low demand, or surplus capacity. This approach ensures that EVs are charged at the most beneficial times, resulting in efficient and cost-effective charging operations. Thus, smart charging strategies are useful to manage the EV charging loads (García-Villalobos et al., 2014).

There are mainly two types of operation modes of EVs based on the direction of energy flow to/from the grid. These two modes of operation are defined as grid to vehicle and vehicle to grid. The V2G concept is an eminent extension of smart charging technology, that allows the EVs to enhance the operational efficiency of the PDN. V2G mode operation facilitates two-way flow of energy, enabling EVs to both consume and supply power to the grid. This mode offers a diverse range of services, including active power support, reactive power compensation, frequency regulation, and contribution to spinning reserve requirements. Despite these benefits, the operation of EVs in V2G mode has two significant drawbacks: battery degradation and additional energy losses (Tan et al., 2016).

3. OVERVIEW OF DISTRIBUTION NETWORK RECONFIGURATION

The PDN serves a major role within the electricity infrastructure by connecting the customers to the power grid. While the PDNs are typically structured as interconnected systems, they are operated in radial topology to leverage various benefits. These benefits include streamlined coordination of protective measures, reduced fault currents, simplified control of voltage and power flow, and ultimately, cost savings (Short, 2014). Despite evolving over the years, the main focus of PDNs has remained on providing reliable and cost-effective electric energy to customers. However, failures in PDNs account for a significant portion of customer service interruptions (Sultana et al., 2016). Therefore, evaluating reliability of PDNs has become crucial. Minimizing power losses in PDNs is also important from an economic standpoint (Badran et al., 2017).

Thus, power loss reduction and network reliability enhancement of the PDNs are two major concerns of power system researchers. Incorporation of large amount of DG units and EVs further increases these concerns. Various approaches like capacitor placement, deployment of custom power devices, DG allocation, network reconfiguration have been employed to address power loss reduction problem (Kalambe and Agnihotri, 2014). Among these methods, DNR has gained significant attention over the last four decades due to its cost-effectiveness.

DNR is a standard practice performed by utility companies to improve the operational efficiency of PDNs (Andervazh et al., 2013). This activity involves the use of controlled switches, specifically sectionalizing switches and tie-line switches, within the PDN infrastructure. Sectionalizing switches are normally closed type and designed for protection purpose. These switches isolate the faulted section from the healthy part of a feeder on the occurrence of a fault. In contrast, tie line switches are normally open type and often used in the reconfiguration process to provide an alternative service path (Roytelman et al., 1996). DNR is a method of changing the topology of the PDN by changing the status (ON/OFF) of sectionalizing and tie line switches such that radiality and operational constraints of the PDN are preserved (Das, 2006). The constraints that must be satisfied in DNR can be given as follows:

1. Every feeder should be energised.
2. The loading of feeders and transformers should be within permissible limit.
3. Voltage deviation should be within acceptable limits.
4. Radiality of the PDN must be ensured.

A simple network with several topologies is depicted in Figure 1 to illustrate both feasible and infeasible reconfiguration. The initial network prior to reconfiguration is represented by Figure 1(a). Some of the possible network reconfiguration cases are shown in Figure 1(b) to 1(g). Figure 1(b) to 1(d) are the feasible DNR solutions on the basis of radiality constraint. Because of the radial configuration of the PDN, care must be taken to prevent formation of any closed loop or isolated island following the reconfiguration (Azizivahed et al., 2020). In Figure 1(e) and 1(f), loops are formed after altering the status of the switches where as in Figure 1(g), one of the nodes remains unsupplied after the reconfiguration (indicated with a cross mark). Thus, Figure 1(e) to 1(g) represent the infeasible PDN topologies.

Figure 1. Examples of feasible and infeasible DNR

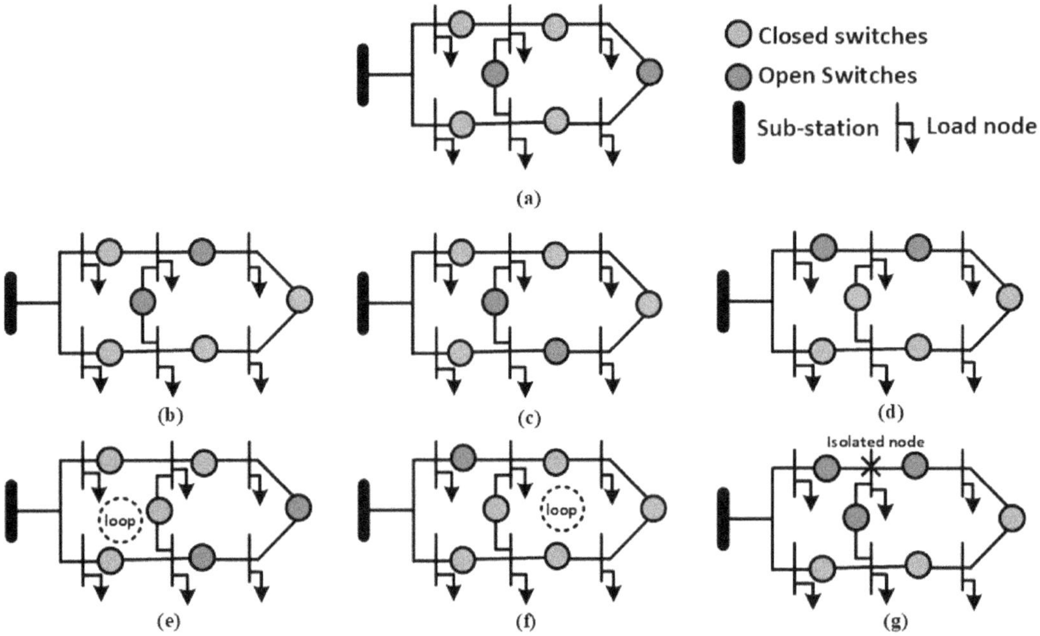

Merlin and Back first tackled the optimal DNR problem to reduce the total APL (Merlin and Back, 1975). This objective of minimizing APL has remained the primary focus of the DNR problem, persisting to this day. However, with time new DNR benefits emerge. Periodically, DNR has been used as an effective tool to achieve several objectives related to PDN operation (Ushashree and Kumar, 2023 ; Islam et al., 2019). The primary objectives of DNR can be listed as follows:

1) Minimization of active power loss
2) Improvement of voltage profile
3) Management of network congestion
4) Feeder load balancing
5) Service restoration during fault condition
6) Planning outages for maintenance

3.1. Classification of Distribution Network Reconfiguration

Based on the optimization time frame, the classification of DNR is depicted in Figure 2.

3.1.1. Static Reconfiguration

It is the process of reconfiguring a PDN during the planning stage.

- This approach seeks an improved fixed topology that is valid in terms of load and generation profiles over the entire time period under consideration (Capitanescu et al., 2015).

Figure 2. Classification of DNR approach

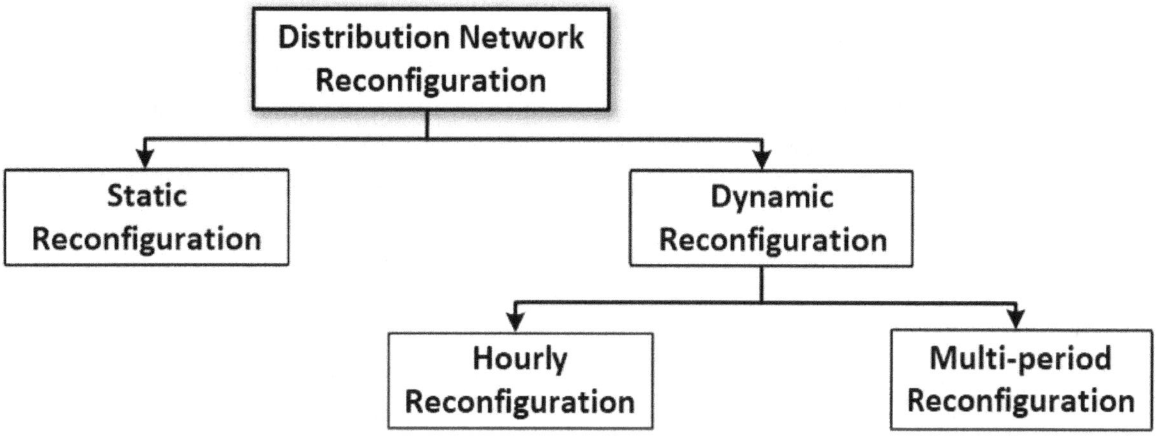

- In this type of DNR, data and constraints at a particular time period is considered. Then, using optimization methods optimal network topology is identified to meet the objective function.
- This process takes into consideration the action of both switches that are operated manually and those that are controlled remotely.
- Reconfiguration is carried for a longer time period and performed on weekly, monthly, seasonally and yearly basis (Razavi et al., 2022).

However, loads, DG units and EVs connected to the PDN have time-varying uncertain characteristics. Static DNR models can't offer reliable solutions as they are only concerned with the data and constraints of a particular time. Dynamic DNR is therefore increasingly prevalent in the present period with the development of smart grid and remote-controlled switches to assure system's high-quality operation (Pan et al., 2022). Many researchers have investigated the dynamic DNR in order to find a robust network operation topology.

3.1.2. Dynamic Reconfiguration

Dynamic reconfiguration is the process, carried out at operation stage of the network.

- This strategy allows the PDN to change its structure continually, particularly in direct response to fluctuations in the demand and sporadic resources (Sekhavatmanesh and Cherkaoui, 2020).
- In this method, reconfiguration problem is optimized with respect to time (Wang et al., 2020a).
- This process considers action of remotely-controlled switches to reconfigure the network topology (Capitanescu et al., 2015).

Dynamic DNR is further segregated into hourly reconfiguration and multi-period reconfiguration (Pan et al., 2022; Zhan et al., 2020).

- Hourly reconfiguration is carried out at each hour of the day and involves more switching actions. However, frequent switching operation is not desirable because it will affect the life of the switches and inject switching transients in the system which hampers the PDN stability (Guo et al., 2020).
- Therefore, multi-period reconfiguration is reported in which, the 24 hours of a day is divided into multiple periods depending on the load and generation profiles. Optimal network topology is then obtained for each of the period.
- This involves fewer switching actions as compared to the hourly dynamic reconfiguration (Pan et al., 2022). However, the accuracy of the optimal DNR is compromised as the uncertainties present in the network can't be modeled accurately.

In order to address the aforementioned challenges, mechanical switches can be replaced with power electronics-based switches. Integration of these smart power electronic devices, can enhance the performance and controllability of PDNs, and enable flexible system configuration during both normal and emergency situations, particularly in the event of faults (Shafik et al., 2019).

3.1.3. Network Reconfiguration Using Soft Open Points

The continuous development of power electronic control devices in active PDNs enables more frequent short-term network reconfiguration, allowing reconfigurations to be performed on an hourly timescale. One such device that has gained attention in recent years is the soft- open-point (SOP). An SOP consists of multiple power electronics converters connected to a common DC link. This device provides reactive power and transfers active power between the AC terminals. Thus, it can efficiently compensate for PDN's lack of power control capability. The involvement of SOPs is not just limited to opening and closing switches, but also extends to balancing the power flow and improving the voltage profile of PDNs (Azizi et al., 2023). It holds strong capabilities for controlling active and reactive power flow among feeders and is typically installed in place of tie-switches. Unlike the function of traditional tie-lines, SOPs possess the ability to continuously regulate and optimize active/reactive power flow in the PDN. These devices offer improved control and flexibility in the operation of active PDNs, thereby modifying the operational strategies of the system (Bai et al., 2018).

Indeed, the implementation of smart grid technology in the PDNs empowers the system with the ability to dynamically reconfigure its structure using remotely-controlled switches and power electronics-based automated switches. This advanced infrastructure enables rapid and automated adjustments to accommodate various uncertainties and evolving conditions. The network's flexibility in reconfiguration allows for significant improvements in network performance, including aspects such as reliability, power quality, and load balancing. Additionally, it facilitates operational cost minimization by optimizing the network topology based on real-time conditions and integrating distributed energy resources such as DG units and EVs. The dynamic reconfiguration capability provided by smart grid technology ensures that the PDN can effectively adapt to changing demands, maximize efficiency, and enhance overall system performance.

The progressive evolution of DNR over the last four decades is illustrated in Fig.3. This evolution has been driven by the continuous advancement of optimization techniques and the growing demand for improved power quality and reliability in PDNs. The figure highlights the significant developments and refinements that have taken place in the field of DNR, showcasing how it has become an increasingly vital aspect of power system operation and management.

Figure 3. Evolution of DNR over past four decades
(Mishra et al., 2017)

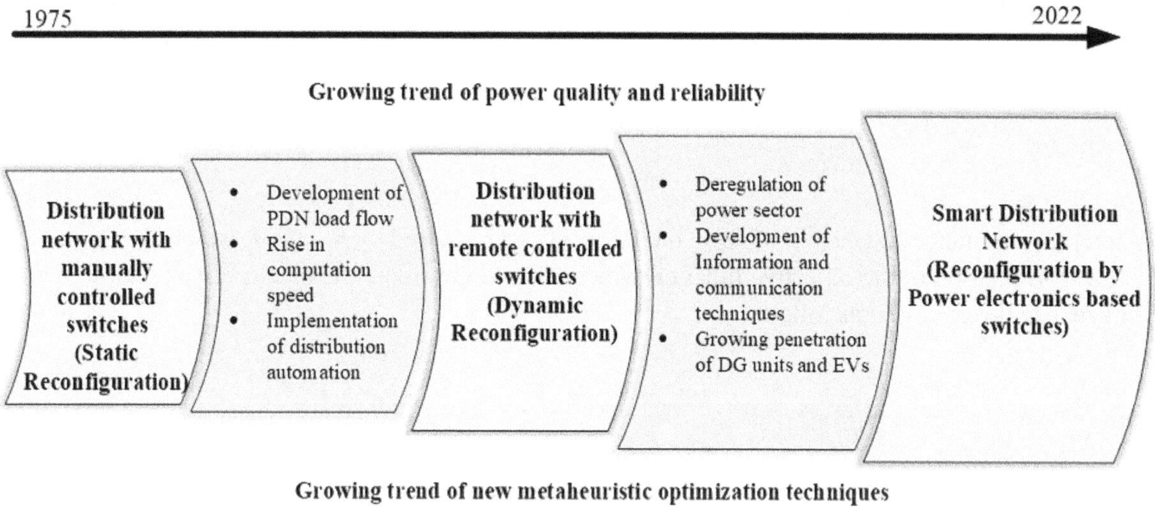

4. LITERATURE SURVEY

In this section, a comprehensive analysis has been provided on various available literatures that have focused on network reconfiguration in EV penetrated PDNs. This analysis has entailed a detailed review of the key findings and solution methods used in each study. By exploring the insights offered by these researches, a better understanding can be obtained regarding the potential benefits of DNR for efficient EV integration in PDNs.

In (Guo et al., 2020), advantages of dynamic reconfiguration are investigated considering charging demand and spatial-temporal availability of EVs. According to the authors, dynamic DNR complements the V2G provision of EVs in terms of reducing the APL and system cost. Dynamic DNR improves the network flexibility by altering the topology of the network. Thus, the V2G technology may result reduced line losses. The optimization problem includes minimization of energy loss cost (C^{EL}), switching cost (C^{swc}), and battery degradation cost (C^{deg}). The mathematical expression of the objective function is given below.

$$min\sum_{t\in T}\sum_{br\in N_{br}}(\Delta TR_{br}(P_{br}^2+Q_{br}^2)C^{EL}+x_{br,t}C^{swc})+\sum_{t\in T}\sum_{n\in N}\sum_{nv\in N_{ev}}C_{nv,t}^{deg} \tag{1}$$

Where, N_{br} represents number of branches; N_{ev} is the set of EVs at nv^{th} bus; P_{br} and Q_{br} represents active and reactive power flow in the line br; R_{br} denotes resistance of the line br; $x_{br,t}$ indicates the status of line br (connected/disconnected) at t^{th} hour; $C_{ev,t}^{deg}$ is the degradation cost of nv^{th} EV at t^{th} time interval; T is the optimization time period.

Solution strategy: The optimization approach involves formulating the problem as a MIQP model and CPLEX solver is used to determine the optimal solutions for the given problem.

In (Wang et al., 2020a), dynamic reconfiguration is carried out considering the variation of conventional loads, outputs of DG units and different EV penetration levels. The optimization problem focuses on minimization of APL, voltage deviation, and load balancing index in an EV penetrated network.

Mathematical expression of the objective to minimize the APL can be formulated by Eq.2.

$$min\sum_{t\in T}\sum_{br\in N_{br}}(x_{br}I_{br}^2R_{br}) \tag{2}$$

Where, the resistance and current flow in the branch *br* are denoted by R_{br} and I_{br} respectively.

The formulation of the objective function which aims to minimize the maximum deviation in node voltage, can be expressed as follows:

$$\min\sum_{t\in T}max\left(\frac{V_1-V_{nr}}{V_{nr}},\frac{V_2-V_{nr}}{V_{nr}}..........\frac{V_n-V_{nr}}{V_{nr}}\right) \tag{3}$$

Where, V_n, V_{nr} are the current and rated voltage of node *n* respectively.

The objective function to minimize the load balancing index can be given as follows:

$$min\sum_{t\in T}\sum_{br\in N_{br}}\left(x_{br}\frac{\sqrt{P_{br}^2+Q_{br}^2}}{S_{br}^{max}}\right) \tag{4}$$

Where, P_{br}, Q_{br} and S_{br}^{max} represent active power, reactive power and maximum complex power of branch br, respectively.

Solution strategy: Authors have used LDBAS algorithm to find optimal solution for the DNR problem.

In (Kavousi-Fard et al., 2015c), hourly dynamic DNR and V2G provision of PEVs are incorporated to improve the system reliability. Within this paper, energy not supplied (ENS) index is utilized as an indicator to demonstrate the improved reliability of the network. In addition to the reliability index, few more objectives like hourly cost of APL (C_t^{PL}), hourly cost of imported power from the substation (C_t^{ss}) and charging/discharging cost of the PEV fleet(C_t^{PEV}) are also optimized. The mathematical expression of the objective function is given by Eq. 5.

$$minf(v)=min\sum_{t\in T}(C_t^{ss}+C_t^{PL}+C_t^{PEV}+C_t^{ENS}) \tag{5}$$

Where, *v* is a vector of optimizing variables for the above optimization problem

$$v=\left[v_{switches},v_{PEV},v_{Pexc}\right]$$

Where, $v_{switches}$, v_{PEV}, v_{Pexc} are the variables to indicate status of tie line and sectionalizing switches(on/off), status of PEV fleet (charging/ discharging/idle) and amount of power exchange between substation and PEV fleet.

Solution strategy: Modified symbiotic organisms search (MSOS) algorithm is employed to determine optimal solution to DNR problem.

In (Kavousi-Fard et al., 2015b), optimal DNR strategy is investigated in the smart grid, implementing V2G provision of PEVs along with the correlated wind turbines. The objective function is formulated as an aggregated cost function consisting of minimization of operation cost and reliability cost.

$$min\sum_{t\in T}\sum_{ns\in N_{ss}} C^{ss}_{ns,t}P^{ss}_{ns,t} + \sum_{t\in T}\sum_{br\in N_{br}} x_{br}C^{PL}_{br,t}P^{PL}_{br,t} + \sum_{n\in N} A^{LD}_{n}C^{Int}_{n}\lambda_c + \sum_{t\in T}\sum_{nv\in N_{ev}} U_{nv,t}C^{PEV}_{nv,t}P^{PEV}_{nv,t} + \sum_{t\in T}\sum_{nv\in N_{ev}} C^{deg}_{ev,t} \quad (6)$$

Where, $P^{ss}_{ns,t}$ the power imported from the substation at t^{th} hour; $P^{PL}_{br,t}$ represents the power loss in branch br at t^{th} hour; A^{LD}_{n} is the average load demand at n^{th} bus; λ_c is the failure rate of cth component. $U_{nv,t}$ represents status of nv^{th} PEV connected to grid at t^{th} hour; $P^{PEV}_{nv,t}$ is the charging/ discharging capacity of nv^{th} PEV at t^{th} hour.

In (Gouin et al., 2018), a PDN planning model is reported integrating DG units and EVs. This research focuses on determining the optimal locations for line reinforcement, the location of open switches (obtained through DNR), and identifying the specific locations and quantities of active and reactive power that can be reduced using DNR techniques. These measures aim to minimize the capital expenditure of the network while ensuring efficient operation. The optimization problem to minimize the capital expenditure can be formulated as follows:

$$min\sum_{y\in Y}\sum_{br\in N_{br}} x^{rnf}_{br,y}C^{rnf}_{br}l^{rnf}_{br} \quad (7)$$

Where, Y is the total study period in year; $x^{rnf}_{br,y}$ is the binary variable indicating whether to reinforce the line br or not; C^{rnf}_{br} indicates the reinforce cost of the line br. l^{rnf}_{br} is the length of the line br to be reinforced.

Solution strategy: The problem is represented as a MILP formulation and subsequently solved utilizing the GLPK solver.

In (Antoun et al., 2022), impact of significant number of EV preconditioning actions on the PDN is investigated. To address and minimize the effects of the preconditioning load, a combined approach that incorporates both DNR and V2G technology is explored. Here, the optimization objective aims to reduce the APL.

Solution strategy: To solve the DNR problem, it is formulated as a MILP problem. The solution to this problem is obtained by utilizing the CPLEX solver.

The study presented in (Rahmani-Andebili and Fotuhi-Firuzabad, 2018), employs a stochastic model predictive control method to reconfigure the PDN in an adaptive and dynamic manner. The model also focuses on coordinated regulation of the EV charging patterns. As per the authors this approach is effective in dealing with the uncertainties introduced by the DG units and EVs. The optimization objective considers minimization of overall system operation cost, comprising the cost of energy loss, switching and charging management of EVs. The objective function can be given by Eq.8.

$$min\sum_{t\in T}(C_t^{EL}+C_t^{SWC}+C_t^{CMEV}) \quad (8)$$

Where, C_t^{EL}, C_t^{SWC}, *and* C_t^{CMEV} represent energy loss cost of branches at each hour, switching cost of hourly reconfiguration and charging management cost of EVs at each hour, respectively.

Solution strategy: In this paper, authors have used Genetic Algorithm (GA) as a solution methodology for DNR problem due to its efficiency in handling nonlinear and complex optimization problem.

In (Wu and Liu, 2023), authors have presented a DNR model to determine optimal status of switches and to develop a reactive power regulation strategy of DG units and EVs. The switching status as well as reactive power regulation outputs of DG units and EVs are taken as the variables for the DNR problem. The optimization problem aims to minimize the APL and node voltage deviation.

Solution strategy: The multi-objective DNR problem is effectively addressed through the utilization of the ARSBX algorithm for solving it.

In (Amin et al., 2020), authors have integrated DNR technology and coordinated charging scheduling of EVs to enhance the performance of the PDN. Here, both DNR and EV charging scheduling task are the complex optimization problem, aim to reduce the APL in the PDNs.

Solution strategy: Binary particle swarm optimization (BPSO) is employed for solving EV charging scheduling problem as well as the optimal DNR problem.

In (Paudel et al., 2016), hourly DNR problem is formulated to moderate the adverse impacts of EV integration in smart grid. The optimization objective is a combination of three cost components, such as, cost of power exchanged with the upstream network, cost of APL in the PDN and cost associated with operations of EVs. Mathematical expression of the objective function is given below.

$$min\sum_{t\in T}(C_t^{PL}+C_t^{PEV}+C_t^{SS}) \quad (9)$$

Solution strategy: The above-mentioned optimization problem is solved using metaheuristic BPSO technique.

In (Kothona and Bouhouras, 2022), an APL reduction approach in a medium voltage (MV) and low voltage (LV) PDN is described, incorporating DNR strategy and a two-stage EV charging scheduling respectively. The optimization process is carried out in two distinct stages. Initially, the aggregator determines the day-ahead charging schedule of EVs at a residential charging facility. In this stage, the key focus is to reduce the APL of LV PDN by preventing excessive power demand overlaps and line overloading due to uncoordinated EV charging. During last stage, the PDN operator ascertain the optimal network topology for the upcoming day, by carrying out DNR. The objective in both the stages is to minimize the APL, referring to either MV or LV network.

Solution strategy: Both DNR and EV charging scheduling methods includes continuous and discrete variables. Thus, the optimization problem is handled using a meta-heuristic technique named as UPSO.

The study presented in (Singh and Tiwari, 2020), investigates the EV scheduling strategies through cost-benefit analysis to achieve an optimal planning scheme. The paper implements two scheduling strategies that consider both the active power and reactive power dispatch from the EVs. These strategies aim to minimize the system losses by incorporating V2G operation of EVs. Here, DNR is employed in conjunction with scheduling techniques of EVs to optimize feeder investment cost and power losses of the PDN. The mathematical expression of the objective function can be given by Eq.10.

$$min\sum_{y\in Y}(C_y^{inv}+C_y^{PL}) \tag{10}$$

Where, C_y^{inv} *and* C_y^{PL} represents cost of feeder investment and cost of power loss respectively; y and Y represents the year and total planning horizon respectively.

Solution strategy: The DNR problem is solved implementing GA.

The research presented in (Rostami et al., 2015), focuses on investigating an intelligent stochastic framework for the future smart grid enabled PDNs, considering the increasing adoption of EVs. In this context, a network reconfiguration technique is implemented to mitigate the impacts of PEV charging. This study implements monte-carlo simulation to model the uncertainties related to energy price, EV loads, and conventional load. The objective function, mathematically formulated below, aims to minimize the expected cost considering these uncertainties.

$$min\sum_{t\in T}(C_t^{PL}+C_t^{SWC}+C_t^{SS}) \tag{11}$$

Solution strategy: In this paper, authors have utilized a metaheuristic optimization algorithm called SAMKHOA to obtain optimal solutions.

In (Singh and Tiwari, 2019), authors have studied DNR in an EV integrated future smart distribution grid. This study presents a two-stage optimization framework that effectively employs the DNR approach with various EV scheduling strategies to improve operational state of the PDN. Here, reduction of APL and voltage profile improvement are the primary objectives of the optimization framework.

Solution strategy: In this paper, the first stage of optimization framework determines the optimal schedule for charging/discharging of EVs. This stage focuses on minimization of load variation and operation cost and uses a GA based method to model the charging demands of EVs. In second stage, the DNR problem is optimized incorporating the EV load demand obtained from the initial stage. A modified grey wolf optimisation (MGWO) technique is utilized to obtain optimal switching status.

In (Kavousi-Fard et al., 2015a), DNR strategy is implemented to optimize the operational cost and to facilitate the EV penetration in the PDN without any extra cost. Further, the DNR technique is used to optimally manage the EV operation in a stochastic framework. The mathematical expression of the objective is given as below.

$$min\left(\sum_{t\in T}\left\{C_t^{SS}P_t^{SS}+C_t^{PL}P_t^{PL}\right\}+\sum_{t\in T}\sum_{nv\in N_{ev}}U_{nv,t}C_{nv,t}^{PEV}+\sum_{t\in T}\sum_{n\in N}\sum_{nv\in N_{ev}}C_{nv,t}^{deg}\right) \tag{12}$$

Solution strategy: Authors in this paper have modeled the stochastic DNR problem as a MINLP problem and applied a metaheuristic social spider optimization (SSO) algorithm to obtain its solution.

In (Kavousi-Fard and Khodaei, 2016), the potential of the reconfigurable microgrids in enabling PEV integration is investigated. By using remotely controlled switches, the reconfigurable microgrids facilitate significant penetration of PEVs and DG units. This integration not only helps in reducing the overall operational costs but also has the potential to enhance the reliability of the microgrid. A scenario-based approach is reported in this paper to handle various uncertainties. This stochastic framework considers multiple uncertainties such as charging/discharging pattern of PEVs, number of PEVs in each fleet, timing of daily trips for PEVs, hourly power demand, hourly generation of DG units, and hourly energy price. The optimal scheduling strategy reported, aims to minimize an aggregated cost term. The aggregated cost term involves the battery degradation cost in PEVs, cost of total power supply by DG units, cost of energy exchange with the upstream network, switching cost, and expected customer interruption cost.

Solution strategy: To find the optimal solution, researchers have employed a metaheuristic approach known as SAMCSA.

In (Sedighizadeh et al., 2019), authors have provided a stochastic model that optimizes both DNR and generation scheduling problem of microgrid. The optimization model considers the uncertainties of DG units, different charging patterns for EVs, and demand response program in a microgrid. The objective of the optimization problem is to minimize the overall operational cost of the microgrid. The total operation cost includes the active and reactive power purchasing cost from DG units and substation, demand response cost, switch operation expenses, cost of APL, and revenue generated from selling active and reactive power to the upstream network.

Solution strategy: The GAMS software is utilized to solve the optimization problem, which is formulated as a MISOCP problem.

A dynamic tariff based DNR method is provided in (Huang et al., 2016) to effectively manage the line congestion and reduce the power losses in an EV penetrated PDN. The authors, in this paper, have assumed a simplified model of EV behavior without considering the uncertainties associated with it. EVs are treated as flexible loads that can be strategically coordinated by the PDN operator to achieve optimal congestion management.

Solution strategy: The reconfiguration problem is expressed as a MILP model and solved using CPLEX.

In order to reduce the APL in a high level EV penetrated PDN, DNR along with two aggregated EV charging strategies is employed: arbitrary delay charging and peak avoiding delay charging (Cui et al., 2015). Under arbitrary delay strategy, the EV charging is delayed by some arbitrary duration while the peak avoiding delay charging strategy shifts the EV charging load to off-peak hours.

Solution strategy: Authors in (Cui et al., 2015), have modeled the optimization problem as a MISOCP problem and subsequently solved utilizing CPLEX.

In (Cheng and Li, 2019), a multi-objective DNR approach is presented considering V2G provision of EVs and renewable based DGs in the PDN. The optimization problem minimizes the node voltage deviation, APL, and voltage stability margin index.

Solution strategy: The optimization problem is solved using an improved version of GA.

The research presented in (Jangdoost et al., 2021), determines optimal network reconfiguration in an active PDN considering various charging/ discharging scenarios of EVs and power generation of DG units. The optimization problem aims to minimize the total operation cost of the PDN which includes the cost of switching, power loss, and ENS.

Solution strategy: The nonlinear optimization problem is solved using an improved version of GA(IGA).

In (Pahlavanhoseini and Sepasian, 2019), authors have provided a strategy for simultaneous planning of fast EV charging station and DNR. The framework uses a scenario-based approach to effectively handle the uncertainties related to conventional load, EV charging load and energy price. Here, DNR approach aims to optimize the aggregated initial investment cost and energy loss cost.

Solution strategy: For solving the mixed integer type nonlinear optimization problem, a metaheuristic CCGA is used.

A glimpse of different approaches and aspects of DNR in active PDNs is provided in Table 2. It involves a detailed classification of literatures based on the type of DNR used, consideration of DG units and EVs, type of uncertainty considered, type of objective function considered. Additionally, the table provides information on the objectives of optimization problem and the corresponding solution strategies employed in the cited papers.

5. QUANTITATIVE ANALYSIS OF THE LITERATURES

In order to have a comprehensive insight of DNR, it is essential to analyse the available literature quantitatively. By extensively reviewing the existing literature, this analysis seeks to provide numerical evidence and statistical insights into the impact of network reconfiguration on various performance metrics. This section offers an overview of various test networks considered in several papers. Additionally, it includes comparative assessment of literatures based on various performance indicators, exploring the impact of DNR in an EV integrated PDN.

5.1. Test Networks

Upon studying several available literatures concerning DNR techniques, it becomes evident that a review of frequently employed test networks and reported DNR results is essential. Researchers have utilized various distribution test networks over time to illustrate their own approaches and compared it with other DNR techniques. This section focuses on some of the commonly used test network and presents the optimal DNR results reported by respective authors.

5.1.1. 33 Bus Distribution Network

This test network has gained huge popularity among researchers as a means of validating the DNR methods and results. This network is composed of 33 nodes and 5 tie- line switches (Guo et al., 2020). One sectionalizing switch is considered to be equipped in each branch of the network. The initial network configuration is illustrated in Figure 4, in which switches 33, 34, 35,36 and 37 are in open state.

5.1.2. 69 Bus Distribution Network

Various studies have used this test network to implement DNR technique in a comparatively larger network(Singh and Tiwari, 2019; Wu and Liu, 2023). This test network has 69 nodes, 73 branches, and 5 tie switches. One sectionalizing switch is considered to be present in each branch of the network. In Figure 5, the single-line diagram of the test network is depicted, highlighting that switches 69, 70, 71, 72, and 73 are of the normally open type.

Table 2. Comprehensive classification of literatures

Reference	Type of Network Reconfiguration		Integration of DG/EV		Type of Uncertainty				Type of Objective Function		Objective(s)	Optimization Technique Used
	Static	Dynamic	DG	EV	Load	Price	DG	EV	Mono-Objective	Multi-Objective		
Guo et al., 2020		✓		✓			✓	✓		✓	MCOST	MIQP
Wang et al., 2020a		✓	✓	✓			✓	✓		✓	MPL, MVD, MLBI	LDBAS
Kavousi-Fard et al., 2015c		✓	✓	✓	✓	✓	✓			✓	MCOST	MSOS
Kavousi-Fard et al., 2015b		✓	✓	✓	✓	✓	✓			✓	MCOST	Not mentioned
Gouin et al., 2018	✓		✓	✓	✓		✓		✓		MCOST	MILP
Antoun et al., 2022		✓		✓	✓				✓		MPL	MILP
Rahmani-Andebili and Fotuhi-Firuzabad, 2018		✓	✓	✓			✓			✓	MCOST	GA
Wu and Liu, 2023	✓		✓	✓						✓	MPL, MVD	ARSBX
Amin et al., 2020		✓		✓	✓				✓		MPL	BPSO
Paudel et al., 2016		✓		✓	✓					✓	MCOST	BPSO
Kothona and Bouhouras, 2022		✓		✓	✓				✓		MPL	UPSO
Singh and Tiwari, 2020	✓	✓		✓						✓	MPL, MCOST	GA
Rostami et al., 2015		✓		✓	✓	✓		✓		✓	MCOST	SAMKHOA
Singh and Tiwari, 2019		✓		✓					✓		MPL	MGWO
Kavousi-Fard et al., 2015a		✓		✓	✓	✓				✓	MCOST	SSO
Kavousi-Fard and Khodaei, 2016		✓		✓	✓	✓	✓	✓		✓	MCOST	SAMCSA
Sedighizadeh et al., 2019		✓		✓	✓	✓		✓		✓	MCOST	MISOCP
Huang et al., 2016		✓		✓		✓				✓	MCOST	MILP
Cui et al., 2015		✓		✓					✓		MPL	MISOCP
Cheng and Li, 2019		✓	✓	✓	✓		✓			✓	MPL, MVD, MVSI	GA
Jangdoost et al., 2021		✓	✓	✓	✓		✓	✓		✓	MCOST	IGA
Pahlavanhoseini and Sepasian, 2019		✓			✓	✓		✓		✓	MCOST	CCGA

(Abbreviations: MPL: Minimization of power loss, MVD: Minimization of voltage deviation, MCOST: Minimization of cost, MLBI: Minimization of load balancing index, MVSI: Minimization of voltage stability index)

Figure 4. 33 bus distribution test network
(Wu & Liu, 2023)

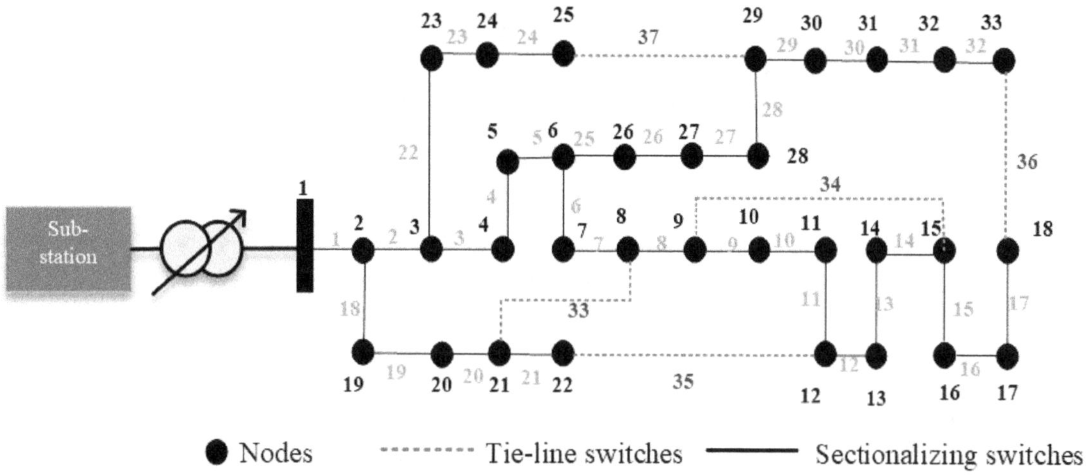

Figure 5. 69 bus distribution network
(Wu & Liu, 2023)

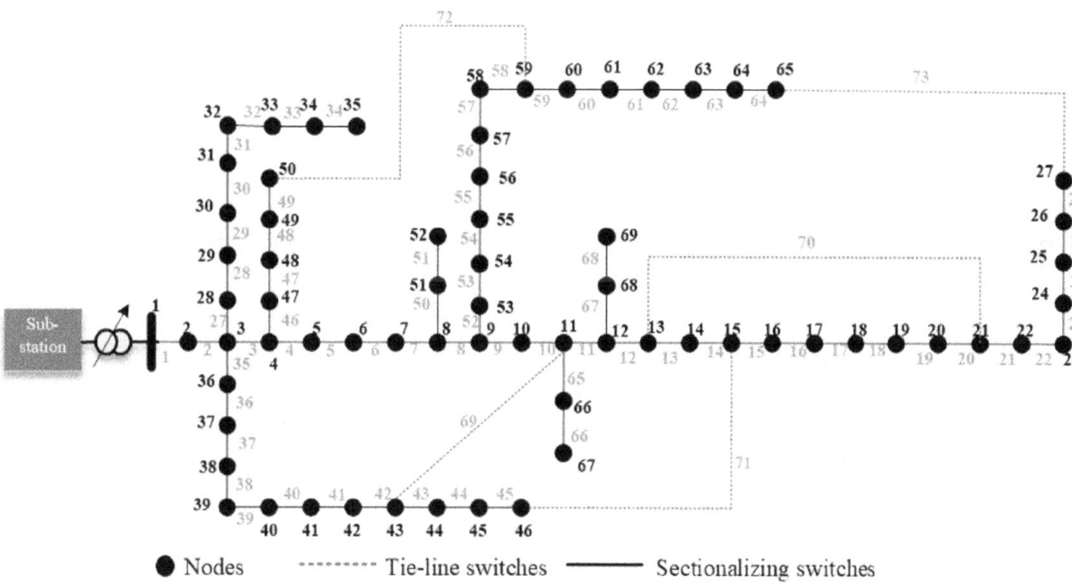

5.1.3. 118 Bus Distribution Network

This test network is widely used by researchers to validate the DNR technology on a large scale (Singh and Tiwari, 2019; Wu and Liu, 2023). It comprises a total of 118 buses, 132 branches, and 15 normally open switches. Each line of the network is equipped with a sectionalizing switch. Under normal conditions, the initial topology of the test network is given in Figure 6, with branches 118 to 132 being in an open state.

Figure 6. 118 bus distribution test network
(Singh & Tiwari, 2019)

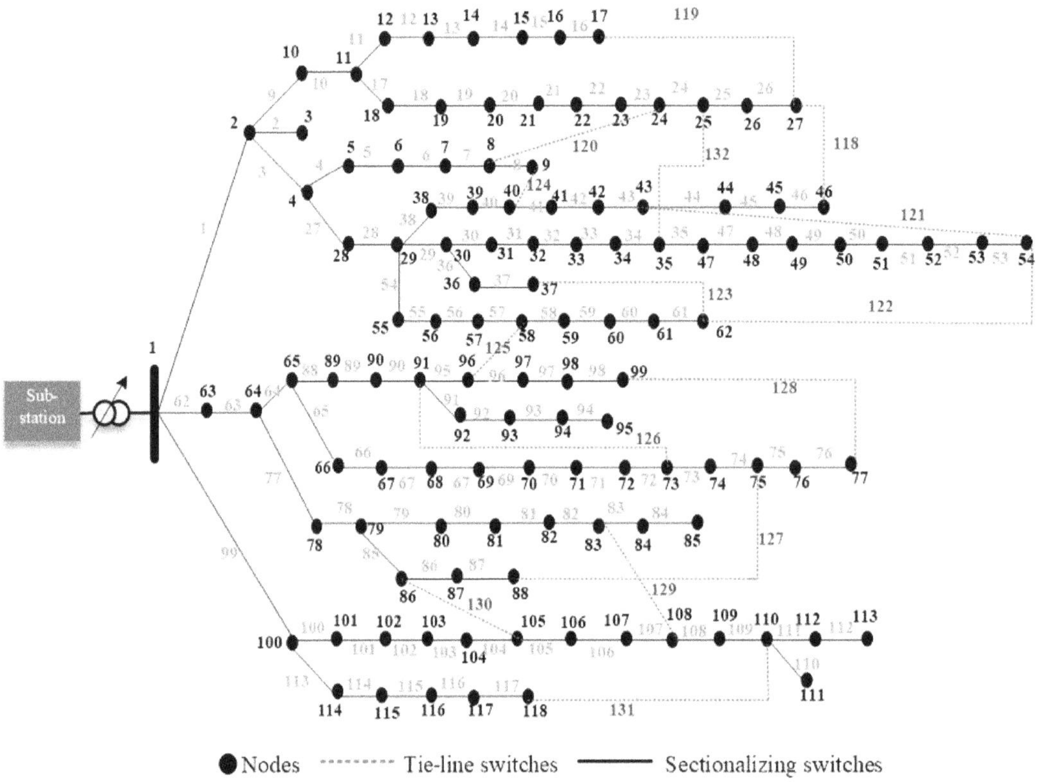

5.2. Impact of DNR on Power/Energy Loss Reduction in EV-Integrated PDNs

In power systems, losses can be generally classified into two categories: technical losses and non-technical losses. Technical losses primarily refer to APL, can greatly influence the overall efficiency of the PDNs (Sultana et al., 2016). To reduce APL in the PDNs, various methods have been reported. DNR is one of the methods that have been implemented in several literatures to minimize these losses(Amin et al., 2020; Antoun et al., 2022; Kothona and Bouhouras, 2022; Singh and Tiwari, 2020).

Moreover, with the growing popularity of EVs, the PDNs face significant challenges in managing the added load and preventing power losses. Minimizing these losses is one of the main concerns of utility sector. In this regard, DNR presents itself as a promising solution to address power losses in EV integrated PDNs.

Few researchers have investigated V2G technology to moderate the impact of EVs on the PDNs (Singh and Tiwari, 2019). Despite the valuable contributions of V2G technology in supplying energy and ancillary services to the PDN, it has been shown to be ineffective in reducing line losses due to the existence of two opposing factors (Guo et al., 2020).While V2G technology offers the advantage of peak load shaving, leading to a smoother power flow over time and subsequently reducing line losses, the charging/discharging operations of EVs under V2G technology can result in increased feeder flow. This increased flow contributes to higher overall line losses within the system.

To address this issue, DNR complements V2G technology by providing flexibility in topology reconfiguration and flow path alteration. This approach can help reduce line losses by accommodating the increased feeder flow resulting from V2G activities. Thus, by integrating DNR and V2G technologies, the power grid can effectively manage the additional load due to EVs and minimize power losses.

Table 3 provides a quantitative comparison of different literature studies, clearly illustrating the impact of DNR on power/energy loss in an EV integrated PDN.

Table 3. Quantitative comparison of DNR approaches in EV penetrated PDNs based on power/energy loss reduction

Reference	Test System	Method Used	Open Switches Before DNR	Power/Energy Loss Before DNR	Open Switches After DNR	Power/Energy Loss After DNR	% Reduction of Power/Energy Loss
Wang et al., 2020a	33 bus	LDBAS	33,34, 35,36,37	857.54kWh	7,14,10,17,28	574.7680kWh	32.97%
	IEEE 118 bus		119,120,121, 122,123,124,125, 126,127,128,129, 130,131,132,133	12485.9022kWh	45,24,6,40,47, 57,28,89,70,85,76,80,100,113,33	9851.99kWh	21.095%
Kavousi-Fard et al., 2015c	69 bus	MSOS	69.70.71,72,73	225kW	65, 71, 14,51,47	99.62kW	55.72%
Antoun et al., 2022	33 bus	MILP	33,34, 35,36,37	117.148kW	7,9,13,15,28	57.9038kW	50.57%
Wu and Liu, 2023	33 bus	ARSBX	33,34,35,36,37	208.46kW	7,9,28,17,14	64.66kW	71.26%
	69 bus		69,70,71,72,73	224.99kW	58,61,18,42,13	69.91kW	68.93%
Amin et al., 2020	33 bus	BPSO	33,34, 35,36,37	99kW	7,9,14,28,31	68kW	31.31%
Kothona and Bouhouras, 2022	33bus	UPSO	33,34,35,36,37	40.821MWh	9,14,17,25,31	14.843MWh	63.64%
Singh and Tiwari, 2019	69 bus	MGWO	69,70,71,72,73	225kW	14, 58 61, 69, 70	98.59kW	56.2%
	IEEE 118 bus		118,119,120,121, 122,123,124,125, 126,127,128,129, 130,131,132	1297.92kW	42, 25, 23, 121, 50, 58, 39, 95, 71, 74, 97, 129, 130, 109, 34	854.03kW	34.2%
Cui et al., 2015	33bus	MISOCP	Not mentioned	3680kWh	Not mentioned	3451.2kWh	6.22%
Cheng and Li, 2019	33bus	GA	33,34,35,36,37	262.8kW	7,14,30,33,37	233.8kW	11.035%

5.3. Impact of DNR on Operational/Investment Cost Reduction in EV-Integrated PDNs

In the literature, several researchers have employed DNR technique to reduce the network operational cost (Guo et al., 2020; Kavousi-Fard et al., 2015c, 2015b; Paudel et al., 2016; Rostami et al., 2015; Singh and Tiwari, 2020). The operational cost of the PDN includes, cost of power imported from the substation, cost of energy loss, switching cost, cost of energy not supplied etc. As DNR reduces the energy loss hence the cost associated with the energy loss also decreases. Furthermore, DNR effectively minimizes customer interruptions during various events such as faults or maintenance activities. This improved reliability directly translates into cost savings for both utility providers and customers, as the reduction in customer interruption cost contributes to overall operational efficiency and customer satisfaction.

In a PDN, with significant presence of EVs, the operational costs are influenced by various factors, including the additional expenses incurred from the charging/discharging activities of EVs, in addition to the different costs mentioned earlier. The lack of coordination in EV charging can lead to a substantial rise in network costs as the adoption of EVs increases within the PDNs. However, the implementation of DNR can effectively reduce the overall network cost across all levels of EV integration (Rostami et al., 2015). Further, the large integration of EVs into the system may necessitate upgrades to accommodate the growing charging load on the PDNs. However, network reconfiguration can help to reduce the need for costly network upgrades. By optimizing the network operation through reconfiguration, utilities can delay or avoid the need for costly equipment upgrades such as transformers or feeders. This can result in significant cost savings for utilities and their customers. In case of smart charging of EVs, the installation of bi-directional chargers, which are required for V2G operation, will result in additional capital costs. As per the findings in (Guo et al., 2020), the utilization of DNR technique effectively enhances the benefits of V2G services by reducing the overall system cost. Notably, this cost reduction is achieved even when considering the additional expenses associated with battery degradation during V2G operations.

A comprehensive quantitative analysis is presented in Table 4, comparing multiple literature studies. The primary focus is to investigate the effects of DNR on the reduction of operational and investment costs within PDNs integrated with EVs.

Table 4. Quantitative comparison of DNR approaches in EV penetrated PDNs based on cost reduction

Reference	Test System	Method Used	Type of Cost Considered	Total cost Before DNR (in \$)	Total Cost After DNR (in \$)	% Cost Reduction
Guo et al., 2020	33 bus	MIQP	Cost of line losses, switching, battery degradation	4.04×10^5	3.21×10^5	20.544%
Kavousi-Fard et al., 2015c	69 bus	MSOS	Cost of power exchange with upstream network, line losses, EV charging/discharging, Energy not supplied	201.33×10^5	130.39×10^5	35.24%
Kavousi-Fard et al., 2015b	69 bus	Not mentioned	Cost of expected customer interruption, power loss	189,959.90	127,721.13	32.76%
Paudel et al., 2016	69 bus	BPSO	Cost of power exchange with upstream network, line losses, EV charging/discharging	5814211.20	5611121.86	3.493%
Singh and Tiwari, 2020	33bus	GA	Cost of feeder investment	3.97×10^5	2.78×10^5	29.97%
			Cost power loss	441.29	436.11	1.17%
Rostami et al., 2015	Taiwan Power company distribution system	SAMKHOA	Cost of power exchange with upstream network, line losses, switching	7.62×10^5	7.6×10^5	0.262%
Pahlavanhoseini and Sepasian, 2019	33-bus	CCGA	Cost of fast charging station investment, energy loss	7.08×10^5	6.04×10^5	14.69%

5.4. Impact of DNR on Voltage Profile Improvement in EV-Integrated PDNs

The voltage profile optimization is essential to enhance the power quality of the network (Eldurssi and O'Connell, 2015). Improving power quality is crucial for both the financial profits of utility companies and the provision of stable and reliable delivery of electric power to customers. This objective can be accomplished by selecting a proper network configuration that exhibits the highest minimum voltage values. Several researchers have studied the impact of DNR in reducing the node voltage deviation and improving voltage profile of active PDNs (Cheng and Li, 2019; Jabr et al., 2018; Sultana et al., 2016; Wang et al., 2020b; Wu and Liu, 2023). Voltage deviation is a common phenomenon in DG and EV penetrated PDN. In situations where the power generation of the DG units surpasses the load demand, the additional power is directed towards the upstream grid. As a result, there is a chance of voltage rise at the load bus (Ismael et al., 2019). Further, injection of large number of EVs charging loads draw huge current and hence create voltage drop in the network (Deb et al., 2018). By reconfiguring the network topology, voltage deviation can be minimized, which helps to maintain stable voltage levels throughout the network. This can also help to reduce the need for expensive voltage regulation equipment to some extent.

A quantitative analysis is provided in Table 5, comparing several literatures to analyze the impact of DNR on voltage profile improvement in EV integrated PDNs.

5.5. Impact of DNR on Reliability Improvement in EV-Integrated PDNs

Reliability refers to the ability of a system to operate efficiently under planned conditions for the expected time frame (Endrenyi, 1978). In the context of PDNs, reliability evaluation focuses on the performance of each component under normal operating conditions and the impact of their performance on customer satisfaction (Shih-An Yin and Chan-Nan Lu, 2009). In recent years, there has been a considerable focus on enhancing PDNs reliability in order to ensure uninterrupted electricity supply to the consumers. The improved network reliability benefits utility companies by reducing expenses spent on post-power failure maintenance and operation of the PDNs (Sultana et al., 2016).

However, enhancing the network reliability in modern active PDN is a challenging task, mainly because of the stochastic behaviour of renewable based DG units and EVs. Several methods have been reported in the literature to overcome this challenge (Kavousi-Fard et al., 2015c, 2015b; Naguib et al., 2022). DNR is investigated as an effective method to improve network reliability without the need for additional equipment procurement (Ismael et al., 2019). DNR can be used for service restoration under faulty condition and also for planning outages for maintenance.

The impact of DNR on network reliability, considering both stationary and mobile PEV fleets, is investigated in (Kavousi-Fard et al., 2015c).The study reveals that, mobile PEVs efficiently facilitate energy transfer within the PDN, resulting in reduced operation costs and mitigation of renewable energy-based DG fluctuation. As per the findings, the ENS metric increases when the PEV fleet remains stationary, continuously connected to the same bus. Conversely, with a mobile PEV fleet and optimal DNR, the ENS decreases compared to the stationary case. Figure 7 depicts the influence of DNR on network reliability, taking into account both mobile and stationary PEV fleets.

Table 5. Quantitative comparison of DNR approaches in EV penetrated PDNs based on voltage profile improvement

Reference	Test System	Method Used	Open Switches Before DNR	Voltage Deviation Before DNR (p.u)	Open Switches After DNR	Voltage Deviation After DNR (p.u)	Minimum Node Voltage Before DNR (p.u)	Minimum Node Voltage After DNR (p.u)
Wang et al., 2020a	33 bus	LDBAS	33,34, 35,36,37	Not mentioned	7,14,10,17,28	0.5004	Not mentioned	Not mentioned
	IEEE 118 bus		119,120,121,122, 123,124,125,126, 127,128,129,130, 131,132,133	Not mentioned	45,24,6,40,47,57, 28,89,70,85,76,80,100,113,33	1.0813	Not mentioned	Not mentioned
Kavousi-Fard et al., 2015b	69 bus	Not mentioned	69,70,71,72,73	0.090559	65, 71, 14, 51, 47	0.051362	Not mentioned	Not mentioned
Kavousi-Fard et al., 2015c	69bus	MSOS	69,70,71,72,73	Not mentioned	65, 71, 14, 51, 47	Not mentioned	Not mentioned	0.94282
Antoun et al., 2022	33 bus	MILP	33,34, 35,36,37	Not mentioned	7,9,13,15,28	Not mentioned	Not mentioned	0.9632
Wu and Liu, 2023	33 bus	ARSBX	33,34,35,36,37	0.0892	7,9,28,17,14	0.0403	Not mentioned	0.9607
	69 bus		69,70,71,72,73	0.0909	58,61,18,42,13	0.0462	Not mentioned	0.9723
Amin et al., 2020	33 bus	BPSO	33,34, 35,36,37	0.06	7,9,14,28,31	0.05	0.94	0.95
Kothona and Bouhouras, 2022	33bus	UPSO	33,34,35,36,37	Not mentioned	9,14,17,25,31	Not mentioned	0.387	0.864
Singh and Tiwari, 2020	33 bus	GA	33,34,35,36,37	Not mentioned	33,21,14,37,32	Not mentioned	0.9183	0.9204
Singh and Tiwari, 2019	69 bus	MGWO	69,70,71,72,73	Not mentioned	14, 58, 61, 69, 70	Not mentioned	0.9093	0.9495.
	IEEE 118 bus		118,119,120,121, 122,123,124,125, 126,127,128,129, 130,131,132	Not mentioned	42, 25, 23, 121, 50,58, 39, 95, 71, 74, 97, 129, 130, 109, 34	Not mentioned	0.8688	0.9323
Cheng and Li, 2019	33bus	GA	33,34,35,36, 37	6.4404	7,14,30,33,37	6.3519	Not mentioned	Not mentioned

6. CONCLUSION

Smart grid enabled PDNs encourages significant integration of DG units and EVs. However, significant integration of these DG units and EVs, poses various challenges in the PDNs. In response to these challenges, several researchers have explored DNR technique as a promising solution. This chapter has complied several contributions made by the researchers to solve the optimal DNR problem to improve the overall performance of EV penetrated PDNs. Moreover, this chapter discusses the impacts of DNR on various network related issues such as APL, node voltage deviation, operational/ investment cost and reliability of the PDNs. To provide a more comprehensive understanding of the role of DNR in an

EV-integrated active distribution network, the reviewed literatures are further compared quantitatively. This quantitative analysis clearly demonstrates that implementing optimal DNR in EV integrated active PDNs leads to reduction in energy loss, improvement in voltage profile, lowered system operational costs, and enhanced network reliability.

Figure 7. Impact of optimal DNR on network reliability
(Kavousi-Fard et al., 2015c)

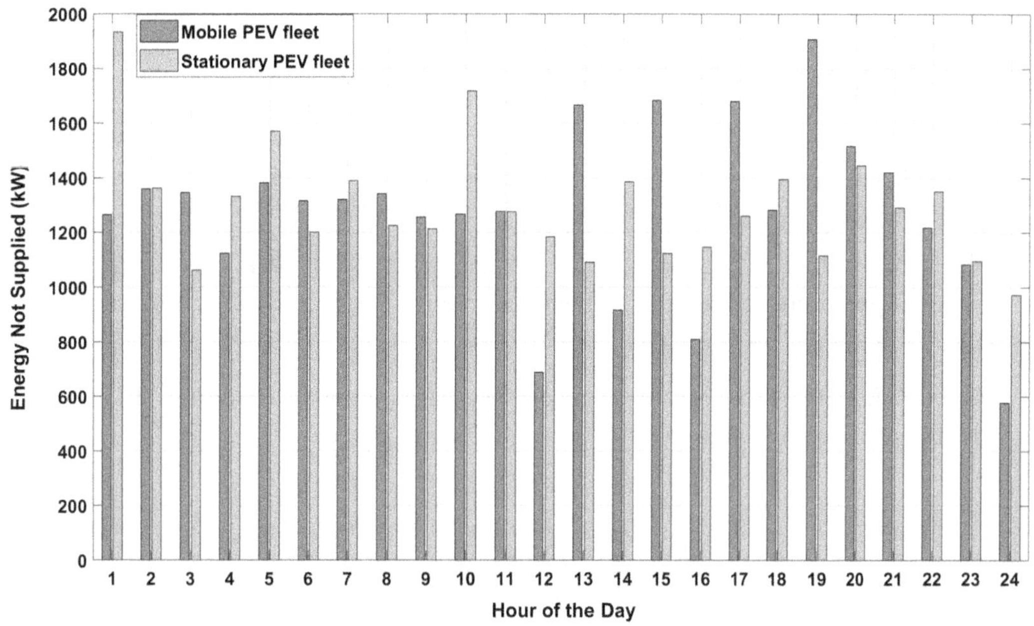

REFERENCES

Amin, A., Tareen, W. U. K., Usman, M., Memon, K. A., Horan, B., Mahmood, A., & Mekhilef, S. (2020). An Integrated Approach to Optimal Charging Scheduling of Electric Vehicles Integrated with Improved Medium-Voltage Network Reconfiguration for Power Loss Minimization. *Sustainability (Basel)*, *12*(21), 9211. doi:10.3390/su12219211

Andervazh, M., Olamaei, J., & Haghifam, M. (2013). Adaptive multi-objective distribution network reconfiguration using multi-objective discrete particles swarm optimisation algorithm and graph theory. *IET Generation, Transmission & Distribution*, *7*(12), 1367–1382. doi:10.1049/iet-gtd.2012.0712

Antoun, J., Kabir, M. E., Atallah, R., Moussa, B., Ghafouri, M., & Assi, C. (2022). Assisting Residential Distribution Grids in Overcoming Large-Scale EV Preconditioning Load. *IEEE Systems Journal*, *16*(3), 4345–4355. doi:10.1109/JSYST.2021.3104185

Asrari, A., Lotfifard, S., & Ansari, M. (2016). Reconfiguration of Smart Distribution Systems With Time Varying Loads Using Parallel Computing. *IEEE Transactions on Smart Grid*, *7*(6), 2713–2723. doi:10.1109/TSG.2016.2530713

Azizi, A., Vahidi, B., & Foroughi Nematollahi, A. (2023). Reconfiguration of Active Distribution Networks Equipped with Soft Open Points Considering Protection Constraints. *Journal of Modern Power Systems and Clean Energy*, *11*(1), 212–222. doi:10.35833/MPCE.2022.000425

Azizivahed, A., Arefi, A., Ghavidel, S., Shafie-khah, M., Li, L., Zhang, J., & Catalao, J. P. S. (2020). Energy Management Strategy in Dynamic Distribution Network Reconfiguration Considering Renewable Energy Resources and Storage. *IEEE Transactions on Sustainable Energy*, *11*(2), 662–673. doi:10.1109/TSTE.2019.2901429

Badran, O., Mekhilef, S., Mokhlis, H., & Dahalan, W. (2017). Optimal reconfiguration of distribution system connected with distributed generations: A review of different methodologies. *Renewable & Sustainable Energy Reviews*, *73*, 854–867. doi:10.1016/j.rser.2017.02.010

Bai, L., Jiang, T., Li, F., Chen, H., & Li, X. (2018). Distributed energy storage planning in soft open point based active distribution networks incorporating network reconfiguration and DG reactive power capability. *Applied Energy*, *210*, 1082–1091. doi:10.1016/j.apenergy.2017.07.004

Capitanescu, F., Ochoa, L. F., Margossian, H., & Hatziargyriou, N. D. (2015). Assessing the Potential of Network Reconfiguration to Improve Distributed Generation Hosting Capacity in Active Distribution Systems. *IEEE Transactions on Power Systems*, *30*(1), 346–356. doi:10.1109/TPWRS.2014.2320895

Cheng, S., & Li, Z. (2019). Multi-objective Network Reconfiguration Considering V2G of Electric Vehicles in Distribution System with Renewable Energy. *Energy Procedia*, *158*, 278–283. doi:10.1016/j.egypro.2019.01.089

Cui, H., Li, F., Fang, X., & Long, R. (2015). *Distribution network reconfiguration with aggregated electric vehicle charging strategy. In 2015 IEEE Power & Energy Society General Meeting*. IEEE. doi:10.1109/PESGM.2015.7285650

Das, D. (2006). A Fuzzy Multiobjective Approach for Network Reconfiguration of Distribution Systems. *IEEE Transactions on Power Delivery*, *21*(1), 202–209. doi:10.1109/TPWRD.2005.852335

Deb, S., Tammi, K., Kalita, K., & Mahanta, P. (2018). Impact of Electric Vehicle Charging Station Load on Distribution Network. *Energies*, *11*(1), 178. doi:10.3390/en11010178

Dubey, A., & Santoso, S. (2015). Electric Vehicle Charging on Residential Distribution Systems: Impacts and Mitigations. *IEEE Access : Practical Innovations, Open Solutions*, *3*, 1871–1893. doi:10.1109/ACCESS.2015.2476996

Eldurssi, A. M., & O'Connell, R. M. (2015). A Fast Nondominated Sorting Guided Genetic Algorithm for Multi-Objective Power Distribution System Reconfiguration Problem. *IEEE Transactions on Power Systems*, *30*(2), 593–601. doi:10.1109/TPWRS.2014.2332953

Endrenyi, J. (1978). *Reliability modeling in electric power systems*. Wiley.

García-Villalobos, J., Zamora, I., San Martín, J. I., Asensio, F. J., & Aperribay, V. (2014). Plug-in electric vehicles in electric distribution networks: A review of smart charging approaches. *Renewable & Sustainable Energy Reviews*, *38*, 717–731. doi:10.1016/j.rser.2014.07.040

Gomez, J. C., & Morcos, M. M. (2003). Impact of EV battery chargers on the power quality of distribution systems. *IEEE Transactions on Power Delivery*, *18*(3), 975–981. doi:10.1109/TPWRD.2003.813873

Gouin, V., Alvarez Herault, M., & Raison, B. (2018). Stochastic integration of demand response and reconfiguration in distribution network expansion planning. *IET Generation, Transmission & Distribution*, *12*(20), 4536–4545. doi:10.1049/iet-gtd.2018.5833

Guo, Z., Zhou, Z., & Zhou, Y. (2020). Impacts of Integrating Topology Reconfiguration and Vehicle-to-Grid Technologies on Distribution System Operation. *IEEE Transactions on Sustainable Energy*, *11*(2), 1023–1032. doi:10.1109/TSTE.2019.2916499

Hota, A. R., Juvvanapudi, M., & Bajpai, P. (2014). Issues and solution approaches in PHEV integration to smart grid. *Renewable & Sustainable Energy Reviews*, *30*, 217–229. doi:10.1016/j.rser.2013.10.008

Huang, S., Wu, Q., Cheng, L., & Liu, Z. (2016). Optimal Reconfiguration-Based Dynamic Tariff for Congestion Management and Line Loss Reduction in Distribution Networks. *IEEE Transactions on Smart Grid*, *7*(3), 1295–1303. doi:10.1109/TSG.2015.2419080

IEA. (2017). *Technical report*. https://www.iea.org/reports/global-ev-outlook-2017

Islam, M. R., Lu, H., Hossain, M. J., & Li, L. (2019). Mitigating unbalance using distributed network reconfiguration techniques in distributed power generation grids with services for electric vehicles: A review. *Journal of Cleaner Production*, *239*, 117932. doi:10.1016/j.jclepro.2019.117932

Ismael, S. M., Abdel Aleem, S. H. E., Abdelaziz, A. Y., & Zobaa, A. F. (2019). State-of-the-art of hosting capacity in modern power systems with distributed generation. *Renewable Energy*, *130*, 1002–1020. doi:10.1016/j.renene.2018.07.008

Jabr, R. A., Dzafic, I., & Huseinagic, I. (2018). Real Time Optimal Reconfiguration of Multiphase Active Distribution Networks. *IEEE Transactions on Smart Grid*, *9*(6), 6829–6839. doi:10.1109/TSG.2017.2724766

Jangdoost, A., Keypour, R., & Golmohamadi, H. (2021). Optimization of distribution network reconfiguration by a novel RCA integrated with genetic algorithm. *Energy Systems*, *12*(3), 801–833. doi:10.1007/s12667-020-00398-5

Kalambe, S., & Agnihotri, G. (2014). Loss minimization techniques used in distribution network: Bibliographical survey. *Renewable & Sustainable Energy Reviews*, *29*, 184–200. doi:10.1016/j.rser.2013.08.075

Kavousi-Fard, A., Abbasi, A., Rostami, M.-A., & Khosravi, A. (2015a). Optimal distribution feeder reconfiguration for increasing the penetration of plug-in electric vehicles and minimizing network costs. *Energy*, *93*, 1693–1703. doi:10.1016/j.energy.2015.10.055

Kavousi-Fard, A., & Khodaei, A. (2016). Efficient integration of plug-in electric vehicles via reconfigurable microgrids. *Energy*, *111*, 653–663. doi:10.1016/j.energy.2016.06.018

Kavousi-Fard, A., Niknam, T., & Fotuhi-Firuzabad, M. (2015b). Stochastic Reconfiguration and Optimal Coordination of V2G Plug-in Electric Vehicles Considering Correlated Wind Power Generation. *IEEE Transactions on Sustainable Energy*, *6*(3), 822–830. doi:10.1109/TSTE.2015.2409814

Kavousi-Fard, A., Rostami, M. A., & Niknam, T. (2015c). Reliability-Oriented Reconfiguration of Vehicle-to-Grid Networks. *IEEE Transactions on Industrial Informatics*, *11*(3), 682–691. doi:10.1109/TII.2015.2423093

Kothona, D., & Bouhouras, A. (2022). A Two-Stage EV Charging Planning and Network Reconfiguration Methodology towards Power Loss Minimization in Low and Medium Voltage Distribution Networks. *Energies*, *15*(10), 3808. doi:10.3390/en15103808

Merlin, A., & Back, H. (1975). Search for a minimal-loss operating spanning tree configuration in an urban power distribution system. *5th Power System Computation Conf.*, 1–18.

Mishra, S., Das, D., & Paul, S. (2017). A comprehensive review on power distribution network reconfiguration. *Energy Systems*, *8*(2), 227–284. doi:10.1007/s12667-016-0195-7

Naguib, M. A., Omran, W. E. A., & Talaat, H. (2022). Performance Enhancement of Distribution Systems via Distribution Network Reconfiguration and Distributed Generator Allocation Considering Uncertain Environment. *Journal of Modern Power Systems and Clean Energy*, *10*, 647–655. doi:10.35833/MPCE.2020.000333

Pahlavanhoseini, A., & Sepasian, M. S. (2019). Scenario-based planning of fast charging stations considering network reconfiguration using cooperative coevolutionary approach. *Journal of Energy Storage*, *23*, 544–557. doi:10.1016/j.est.2019.04.024

Pan, J.-S., Wang, H.-J., Nguyen, T.-T., Zou, F.-M., & Chu, S.-C. (2022). Dynamic reconfiguration of distribution network based on dynamic optimal period division and multi-group flight slime mould algorithm. *Electric Power Systems Research*, *208*, 107925. doi:10.1016/j.epsr.2022.107925

Paudel, A., Supingklad, W., & Ongsakul, W. (2016). Distribution network reconfiguration to support electric vehicles integration. *2016 International Conference on Cogeneration, Small Power Plants and District Energy (ICUE),* 1–6. 10.1109/COGEN.2016.7728975

Pilo, F., Pisano, G., & Soma, G. G. (2011). Optimal Coordination of Energy Resources With a Two-Stage Online Active Management. *IEEE Transactions on Industrial Electronics*, *58*(10), 4526–4537. doi:10.1109/TIE.2011.2107717

Rahmani-Andebili, M., & Fotuhi-Firuzabad, M. (2018). An Adaptive Approach for PEVs Charging Management and Reconfiguration of Electrical Distribution System Penetrated by Renewables. *IEEE Transactions on Industrial Informatics*, *14*(5), 2001–2010. doi:10.1109/TII.2017.2761336

Razavi, S.-M., Momeni, H.-R., Haghifam, M.-R., & Bolouki, S. (2022). Multi-Objective Optimization of Distribution Networks via Daily Reconfiguration. *IEEE Transactions on Power Delivery*, *37*(2), 775–785. doi:10.1109/TPWRD.2021.3070796

Rostami, M.-A., Kavousi-Fard, A., & Niknam, T. (2015). Expected Cost Minimization of Smart Grids With Plug-In Hybrid Electric Vehicles Using Optimal Distribution Feeder Reconfiguration. *IEEE Transactions on Industrial Informatics*, *11*(2), 388–397. doi:10.1109/TII.2015.2395957

Roytelman, I., Melnik, V., Lee, S. S. H., & Lugtu, R. L. (1996). Multi-objective feeder reconfiguration by distribution management system. *IEEE Transactions on Power Systems*, *11*(2), 661–667. doi:10.1109/59.496136

Sedighizadeh, M., Shaghaghi-shahr, G., Esmaili, M., & Aghamohammadi, M. R. (2019). Optimal distribution feeder reconfiguration and generation scheduling for microgrid day-ahead operation in the presence of electric vehicles considering uncertainties. *Journal of Energy Storage*, *21*, 58–71. doi:10.1016/j.est.2018.11.009

Sekhavatmanesh, H., & Cherkaoui, R. (2020). A Multi-Step Reconfiguration Model for Active Distribution Network Restoration Integrating DG Start-Up Sequences. *IEEE Transactions on Sustainable Energy*, *11*(4), 2879–2888. doi:10.1109/TSTE.2020.2980890

Shafik, M. B., Chen, H., Rashed, G. I., El-Sehiemy, R. A., Elkadeem, M. R., & Wang, S. (2019). Adequate Topology for Efficient Energy Resources Utilization of Active Distribution Networks Equipped With Soft Open Points. *IEEE Access : Practical Innovations, Open Solutions*, *7*, 99003–99016. doi:10.1109/ACCESS.2019.2930631

Short, T. A. (2014). *Electric power distribution handbook*. CRC Press.

Singh, J., & Tiwari, R. (2019). Real power loss minimisation of smart grid with electric vehicles using distribution feeder reconfiguration. *IET Generation, Transmission & Distribution*, *13*(18), 4249–4261. doi:10.1049/iet-gtd.2018.6330

Singh, J., & Tiwari, R. (2020). Cost Benefit Analysis for V2G Implementation of Electric Vehicles in Distribution System. *IEEE Transactions on Industry Applications*, *56*(5), 5963–5973. doi:10.1109/TIA.2020.2986185

Sultana, B., Mustafa, M. W., Sultana, U., & Bhatti, A. R. (2016). Review on reliability improvement and power loss reduction in distribution system via network reconfiguration. *Renewable & Sustainable Energy Reviews*, *66*, 297–310. doi:10.1016/j.rser.2016.08.011

Tan, K. M., Ramachandaramurthy, V. K., & Yong, J. Y. (2016). Integration of electric vehicles in smart grid: A review on vehicle to grid technologies and optimization techniques. *Renewable & Sustainable Energy Reviews*, *53*, 720–732. doi:10.1016/j.rser.2015.09.012

Ushashree, P., & Kumar, K. S. (2023). Power System Reconfiguration in Distribution System for Loss Minimization Using Optimization Techniques: A Review. *Wireless Personal Communications*, *128*(3), 1907–1940. doi:10.1007/s11277-022-10026-3

Vargas, A., & Samper, M. E. (2012). Real-Time Monitoring and Economic Dispatch of Smart Distribution Grids: High Performance Algorithms for DMS Applications. *IEEE Transactions on Smart Grid*, *3*(2), 866–877. doi:10.1109/TSG.2012.2187078

Wang, J., Wang, W., Wang, H., & Zuo, H. (2020a). Dynamic Reconfiguration of Multiobjective Distribution Networks Considering DG and EVs Based on a Novel LDBAS Algorithm. *IEEE Access : Practical Innovations, Open Solutions*, *8*, 216873–216893. doi:10.1109/ACCESS.2020.3041398

Wang, J., Wang, W., Yuan, Z., Wang, H., & Wu, J. (2020b). A Chaos Disturbed Beetle Antennae Search Algorithm for a Multiobjective Distribution Network Reconfiguration Considering the Variation of Load and DG. *IEEE Access : Practical Innovations, Open Solutions*, *8*, 97392–97407. doi:10.1109/ACCESS.2020.2997378

Wu, J., Wu, Z., Wu, F., & Mao, X. (2017). A power balancing method of distributed generation and electric vehicle charging for minimizing operation cost of distribution systems with uncertainties. *Energy Science & Engineering*, *5*(3), 167–179. doi:10.1002/ese3.157

Wu, R., & Liu, S. (2023). Multi-Objective Optimization for Distribution Network Reconfiguration With Reactive Power Optimization of New Energy and EVs. *IEEE Access : Practical Innovations, Open Solutions*, *11*, 10664–10674. doi:10.1109/ACCESS.2023.3241228

Yin, S.-A., & Lu, C.-N. (2009). Distribution Feeder Scheduling Considering Variable Load Profile and Outage Costs. *IEEE Transactions on Power Systems*, *24*(2), 652–660. doi:10.1109/TPWRS.2009.2016300

Zhan, J., Liu, W., Chung, C. Y., & Yang, J. (2020). Switch Opening and Exchange Method for Stochastic Distribution Network Reconfiguration. *IEEE Transactions on Smart Grid*, *11*(4), 2995–3007. doi:10.1109/TSG.2020.2974922

Chapter 11
Empowering the Future of Smart Grids:
Unveiling the Role of Electric Vehicles in V2G Integration for Sustainable Infrastructure

Aanal Sanjivbhai Raval
https://orcid.org/0000-0003-3848-0882
Gujarat Technological University, India

Arpita Pareshkumar Maheriya
https://orcid.org/0000-0001-5703-4410
Gujarat Technological University, India

Shailesh Panchal
Gujarat Technological University, India

Komal Borisagar
Gujarat Technological University, India

ABSTRACT

The mission of vehicle-to-grid (V2G) within the context of a smart grid involves leveraging electric vehicles not only as modes of transportation but also as integral components of the energy ecosystem. V2G aims to establish a bidirectional energy flow between EVs and the grid, enabling these vehicles to not only draw power for charging but also feed excess stored energy back into the grid when required. This facilitates demand response, aids in grid stability by balancing supply and demand, promotes the integration of renewable energy sources, and supports grid resilience during peak demand or emergencies. Ultimately, the mission of V2G within the smart grid framework is to enhance grid reliability, optimize energy usage, and reduce environmental impact by maximizing the potential of electric vehicles as both consumers and providers of electricity. This chapter acts as a reference and guide for the forthcoming technological development and commercialization of electric vehicles (EVs), providing insights and recommendations for the future of smart grid transportation.

DOI: 10.4018/979-8-3693-2611-4.ch011

1. INTRODUCTION

Growing concerns regarding fossil fuel consumption and environmental protection have propelled electric vehicles (EVs) to emerge as viable alternatives to traditional fossil-fuel vehicles. As a consequence, in recent years, there has been a global emphasis on advancing transportation electrification. As electric vehicles (EVs) remain stationary, their batteries can offer power to other electrical loads (V2X). Consequently, as the number of EVs surges and with an increase in the power grid's load, the Electric Vehicle-to-Grid (V2G) mode gains attention among researchers. Presently, the primary focus areas revolve around physical connectivity, charging/discharging techniques, and energy management strategies in this domain. A comprehensive analysis of the current knowledge base surrounding the prospective convergence of electric vehicles (EVs) with smart grids reveals intricate challenges facing electric grid infrastructure, communication, and control. The envisioned alignment between the growing presence of EVs and the overarching goals of smart grids necessitates a thorough exploration. The proposed integration of EVs into the grid relies on advanced research in smart metering and communication systems. Furthermore, within the domain of the vehicle-to-grid (V2G) paradigm, a meticulous examination is undertaken regarding the potentials, advantages, and limitations associated with various EV smart-charging systems.

Electric Vehicles (EVs) are positioned as a key element in achieving carbon neutrality and zero carbon emissions, garnering significant attention for their transformative potential. However, the challenges associated with EV energy supplementation have not received sufficient investigation, leading to concerns about issues such as range anxiety and the perceived inconvenience of charging services. Apart from this, V2G framework and its modeling as a business ideal including interactions among grids, load aggregators, EV users along with its spatio-temporal distributions, and adjustable capacity of EV can pave new paths to V2X integration to smart grid.

The Vehicle-to-Grid (V2G) perspective is notably aligned with the 5Ds vision, encompassing decentralization, de-carbonization, digitalization, deregulation, and democratization. The incorporation of the 5Ds in V2G implementation caters to various stakeholders actively contributing to the evolution of electric vehicles. Further at microgrid level, minimization of installation cost with optimization in grid designing, decision making, effective planning, sensitivity and uncertainty analysis of power fluctuations, optimal scheduling, analysis of effect of absence of EV and how V2G contributes to fluctuating demands along with handling low renewable resources provides comprehensive vision to escalating adoption of EVs. This study addresses these concerns and explores the dynamics surrounding EV energy supplementation, aiming to contribute valuable insights to the ongoing discourse on sustainable integration of V2X.

The objectives of this study are as follows:

- Investigation of V2G integration focusing on the bidirectional energy flow between EVs and the grid.
- Understanding sustainable energy infrastructure considering reduced carbon emissions and increased reliance on renewable energy
- Assess the influence of V2G systems on grid resilience, especially in instances of peak demand and emergencies.
- Examine how the integration of V2G can support demand response mechanisms, dynamically balancing the supply and demand of energy.

Figure 1. Flow of the chapter

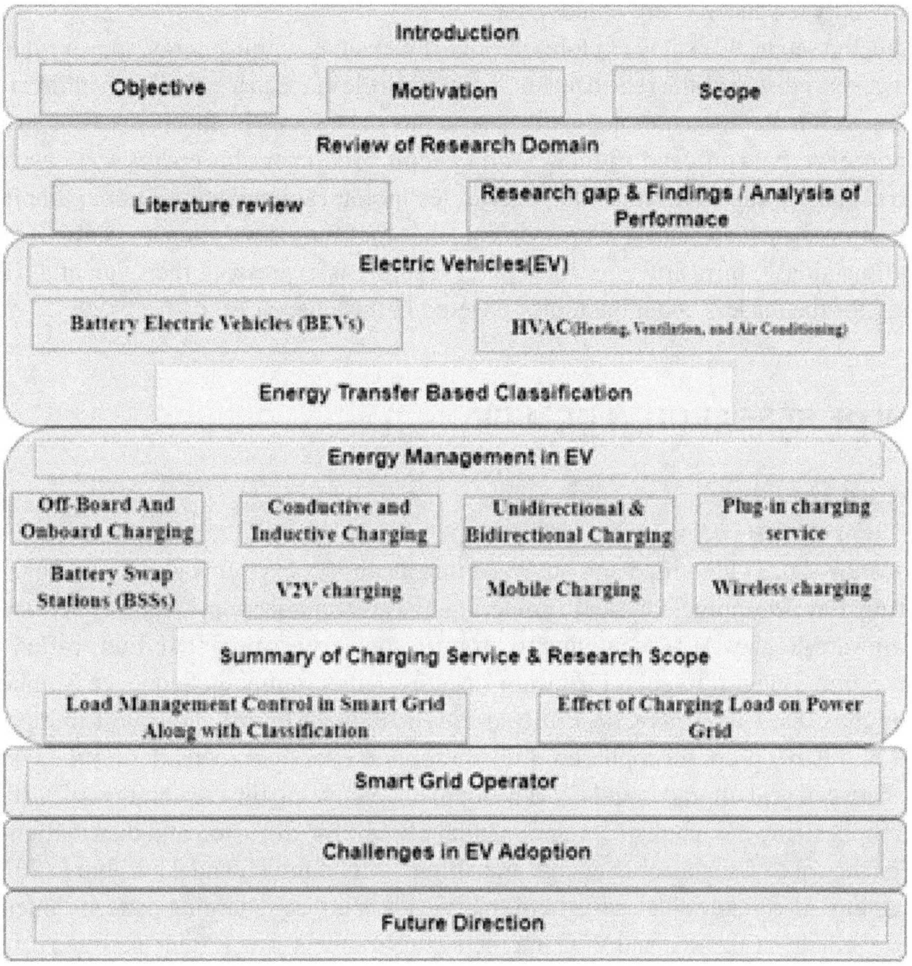

- Examine the commercial potential of V2G integration, industry trends, and business models in the context of smart grids and electric vehicles.
- Offer insights and recommendations to guide future technological developments, and commercial strategies in the fields of smart grids, electric vehicles, and sustainable infrastructure.

1.1 Motivation and Scope

The motivation for exploring "Empowering Future of Smart Grids: Unveiling the Role of Electric Vehicles in V2G Integration for Sustainable Infrastructure" is driven by the urgent need to transition to sustainable energy solutions amidst the challenges of climate change. The evolution of smart grids presents an opportunity to transform energy systems, with electric vehicles (EVs) playing a crucial role in this paradigm shift. Integrating EVs into the smart grid framework offers potential benefits such as enhanced grid resilience, dynamic demand response, and better integration with renewable energy sources. Investigating the economic and environmental advantages of V2G integration aligns with the broader goal of optimizing infrastructure and guiding future technological developments for a more sustainable and technologically advanced energy landscape.

Concerning the scope, this study explores various dimensions within energy, transportation, and sustainability. It covers advanced smart grid technologies, the role of electric vehicles in bidirectional energy flow, and the creation of sustainable energy infrastructure. The scope includes assessing grid resilience, demand response mechanisms, and aligning V2G systems with renewable energy sources. Additionally, it delves into economic and environmental benefits, technological innovations, policy implications, and commercialization trends. The aim is to provide guidance for future developments in smart grids, electric vehicles, and sustainable infrastructure. Moreover, Figure 1 is expressing the sequential flow to quick understanding for the readers about the topics covered in this chapter.

2. REVIEW OF RESEARCH IN DOMAIN

2.1 Literature Review

Novak and Ivanov (2023) Identified and analyzed threats on V2G systems like Man-in-the-Middle Attacks, Spoofing, Eavesdropping, Physical Tampering, False Data Injection, Denial of Service attacks and Malware-Firmware-Replay Attacks, aiming to propose countermeasures (like End-to-End Encryption, Intrusion Detection Systems, Regular Firmware Updates, Time-Stamping/ Sequence Numbers, Physical Security Measures, Data Integrity Checks, Role-Based Access Control and Rate Limiting) to address these. Shi et al. (2019) proposes a phased division for V2G services, considering the coevolutionary relationship between grid upgrades and EV development, analyzing the coevolution of V2G service development, emphasizing the interactive relationship between EV diffusion and distribution grid system upgrades. For the sake of sustainable energy transitions, Mohammadi and Thornburg (2023) explored connection among EVs, renewable energy and smart grids, studying charging patterns, interventions for optimization of energy utilization and V2G interactions. Ntombela et al. (2023) analysed battery in terms of power density, efficiency, dependability, fault tolerance, and cost. Focusing on challenges of energy efficiency and reducing carbon emission, Gallegos et al. (2024) studied infrastructure, its interaction and automation from perspective of data management and role of smart meters. Ramraj et al. (2023) explored EV quality of service and protocols for EV communication, data transfer, cloud services and charging infrastructure. For distributed energy management, and decentralized energy trading and secure communications, Mololoth et al. (2023) studied blockchain for optimized and reliable trusted platforms, use of ML and big data for demand management, EV charging, security and scheduling of grids as well as microgrids. Grace et al. (2023) primarily focused on electric intelligent vehicles examining their greenhouse emissions reductions, bidirectional energy flow, grid stability and optimization of energy sources enabling use of IOT devices and data analytics.

2.2 Analysis of Performance and Gap Finding

For clean energy sustainable infrastructure, use of renewable sources has increased. As this energy usage grows, requirement of distributed energy sources comes to role imposing the challenge of variable output. Though in existence of challenges, research in EV shows promise solutions in mitigating fluctuations of grid. Implementing and utilizing smart grids will contribute to achieving common objectives among grid stakeholders, fostering energy security, supporting economic growth, and aiding in the efforts to mitigate climate change. Exploratory efforts are essential to investigate natural resource utilization with their externalities, analyzing discourses as well as narratives, ultimately addressing social justice and urban resilience considerations (Mohammadi & Thornburg, 2023). Apart concerning the charging of EVs, technical considerations are no doubt explored but lacks in governmental policy regulations for the same as non technical barriers (Ntombela et al., 2023). Gallegos et al. (2024) mentioned some of parameters like Microgrid, advance metering infrastructure, EV, grid forming machine, storage, phasor measure units and flexibility in current transmission system as required for energy infrastructure and related data operations. Apart from mentioned ones, while operating with prosumer in real time, database too plays an important role. This implies a cruciality of striking a balance which enables efficient and beneficial deliverables to the end consumers. With the growing digitization of the system, there will be an expanding pool of real-time information available for resource management and auxiliary systems enhancement. This shows the essentiality of efficient operation that can prevent maintenance and other economic losses imposing electricity trading, regulations and other peer to peer business models.

3. ELECTRIC VEHICLE

Although being a small chunk of international automotive commerce, EVs are anticipated to evolve and burgeon rapidly. According to mechanism utilized for energy production, following are the types:

Battery Electric Vehicle (BEV): As the energy required by motors come solely from battery, gasoline usage isn't seen.This can be charged through level1 or level 2 or even with level 3 having commercial applications. Electric vehicles (EVs), relying solely on battery power rather than an internal combustion engine, boast longer ranges per charge compared to hybrid vehicles. DC fast charging can recharge the battery from 10% to 80% in approximately 30 minutes. It's eco friendly with only the cons of finding a charging station.

Hybrid Electric Vehicles (HEV): Hybrid Electric Vehicles (HEVs) are equipped with both an engine and a motor. The engine is fueled for power, while the motor relies on the battery as its energy source. The battery receives charges through regenerative braking and the internal combustion (IC) engine. Unlike plug-in charging, the battery in these vehicles cannot be charged externally. HEVs are renowned for their superior fuel efficiency. The vehicle initially operates using electric power upon starting. Once it reaches a certain speed, the gas engine engages. This process is managed by an onboard computer system that decides when to switch between electricity and gas for optimal performance. Regenerative braking harnesses the kinetic energy generated while stopping the car, using it to recharge the battery and assist the internal combustion engine in accelerating the vehicle. These can be either in series or parallel.

Mild hybrid electric vehicle (MHEV): MHEVs enhance fuel efficiency by employing a small 48V battery and an electric motor to optimize the performance of their internal combustion engines (ICE). Differing from hybrid electric vehicles, MHEVs primarily rely on the ICE, though it can be turned off during braking, coasting, and when stopped. Moreover, the electric motor can power auxiliary features like air conditioning or seat heaters, even when the ICE is inactive.

Fuel Cell Electric Vehicle (FCEV) (YoCharge, n.d.): These operate using hydrogen, emitting zero emissions apart from water vapor and warm air. They're the least prevalent among electric vehicles, relying on a hydrogen tank for chemical energy that is converted into electrical energy. Refueling the hydrogen tank typically takes about 10 minutes, but FCEVs come with a notably high cost. It comes up with difficulty in storing and transporting hydrogen fuel consequently being less durable.

Plug-in Hybrid Vehicle (PHEV): This vehicle type incorporates an onboard charger along with a gasoline tank. Its battery can be recharged through regenerative braking or an external plug-in. Additionally, Plug-in Hybrid Electric Vehicles (PHEVs) or Series Hybrids can seamlessly switch between electric and gas power modes. This enables the battery to accumulate sufficient power to supply the electric motor, potentially reducing gas consumption by up to 60%. PHEVs can travel up to 40 miles solely on electric power, a significant improvement compared to the limited electric range of a standard hybrid vehicle. Extended range electric vehicles (EREVs) and parallel (or blended) are its two types (Lee, n.d.), where EREVs being a series plug-in hybrids that use An electric motor that propels the vehicle, while the internal combustion engine generates electricity; when the battery is depleted, the stored electricity from the engine takes over to continue powering the vehicle. The second type relies on both its internal combustion engines and electric motors to propel itself. Table 1 shows the summary of EVs.

Table 1. Summary of EVs

	Internal Combustion	Motor	Emission	Plug-In	Charging	Power Source
BEV	✖	✓	✖	✓	Charging station	Electricity
HEV	✓	✓	✓	✖	Gas station	Electricity and gasoline
PHEV	✓	✓	✓	✓	Charging & Gas station	Electricity and gasoline
FCEV	✓	✓	✖	✖	Gas station	hydrogen
MHEV	✓(primary)	✓	✓	✖	Gas station	Electricity and gasoline

3.1. Battery Electric Vehicles

Battery Electric Vehicles (BEVs) are notably simpler in their powertrain architecture than classic internal combustion engine (ICE) vehicles, featuring components such as a high-voltage battery, an electric motor with a power electronics controller, and a single-speed gearbox. Also referred to as pure electric vehicles, they stand distinct from Hybrid Electric Vehicles (HEVs) due to their exclusive reliance on high voltage battery's electrical energy thrust. One of the primary BEV's pros lies in its being reliable, highly efficient, and electric motor's comparatively lower price. However, a notable cons include a high-voltage battery's low energy density.

3.1.1 Performance-Based Classification

Battery electric vehicles can be categorized concerning their respective acceleration performance, range and maximum speed into three types (X-enginner, n.d.):

1. Neighborhood Electric Vehicles: Neighborhood ones are small vehicles having extremely limited range, which is normally 25 km or less than it.
2. City-Electric Vehicles: Also small-sized, these offer a slightly improved range compared to neighborhood electric vehicles, usually below 50 km.

3. Battery Electric Vehicles with Full Performance: These are comparable to conventional passenger vehicles, offering a range spanning between 100 and 600 km.

3.1.2 Energy Transfer-Based Classification

The integration of Battery Electric Vehicles (BEVs) plays an important duty in the electric grid's operation. This integration relies on effectively managing the (Plug-in Electric Vehicles) PEV's charging and discharging through various control operations, techniques and mechanisms. PEVs typically serve three distinct applications: V2B, V2H, and V2G(Kiasari, 2023). Following Figure 2 shows types and concepts of EV.

Vehicle to Building (V2B): This scenario resembles Vehicle-to-Home (V2H), with the distinction that in this case, the power generated by EVs is directed towards larger structures such as hospitals or shopping centers. Additionally, the power source consists of a fleet of EVs rather than an individual vehicle.

Vehicle to Home (V2H)/Vehicle to Load (V2L): Vehicle-to-Home (V2H) refers to the connection of an EV with residential areas, serving as a power source for household loads. Its primary objective is to store and distribute energy sourced from local resources like solar and wind to power home appliances. This initiative aims to reduce energy costs, improve efficient operation along with providing backup power mode for failures. Vehicles to Home includes components like home appliances, BEV (at least one), a bidirectional charger, a home grid, small-scale distributed energy sources and a smart meter comprising the whole home energy management system. Smart metering plays a crucial role by recording system usage data and facilitating communication with utility services, potentially integrating into broader control systems in order to lessen pressure on DES (distributed energy networks).

Vehicle to Grid (V2G): The rise in EV numbers has spurred the exploration of using EV batteries' stored energy as an auxiliary power source connected to the grid. As the demand of power surpasses the base load produced by power plants, the excess energy often goes to waste. Balancing the voltage and frequency of the grid incurs considerable operational costs. To address these issues, EV batteries are being considered as a solution. Through V2G (Vehicle-to-Grid) technology, where EVs connect to the grid and offer their batteries as auxiliary power sources, these problems can be notably mitigated. Successful V2G integration relies on three key components: bidirectional power conversion, effective communication-control systems, and intelligent measurement devices. V2G presents two significant advantages: reducing power losses and potentially enabling control over power factors.

3.2 HVAC (Heating, Ventilation, and Air Conditioning) System in a Battery Electric Vehicle (BEV)

The HVAC system in Electric Vehicles (EVs) is crucial, ensuring year-round thermal comfort and regulating the High Voltage (HV) battery's operating temperature. However, this system becomes a major power consumer within EVs, comprising an electric motor-based scroll compressor for cooling and a Positive Temperature Coefficient (PTC) electric heater for warming. In extreme weather conditions like extreme heat or cold, these high-voltage components in the HVAC system substantially reduce the available driving range from the HV battery (Poal et al., 2021). To ensure proper battery sizing, estimating powertrain consumption isn't enough; auxiliary systems like HVAC play a crucial role. A variable HVAC model tailored for electric powertrain vehicles was developed (König et al., 2022), accounting for consumption based on vehicle size and weather conditions. Factoring in HVAC consumption for battery sizing, the simulations of various vehicle concepts in different weather scenarios and driving cycles, revealed the need for significant battery resizing based on weather conditions to maintain consistent driving ranges. Particularly in urban driving cycles, HVAC consumption sometimes exceeds that of the powertrain due to lower average speeds. Hence, cities, development of electric autonomous vehicles should prioritize considering HVAC and its energy demands.

Figure 2. Different concepts of BEV (Khor, n.d.)

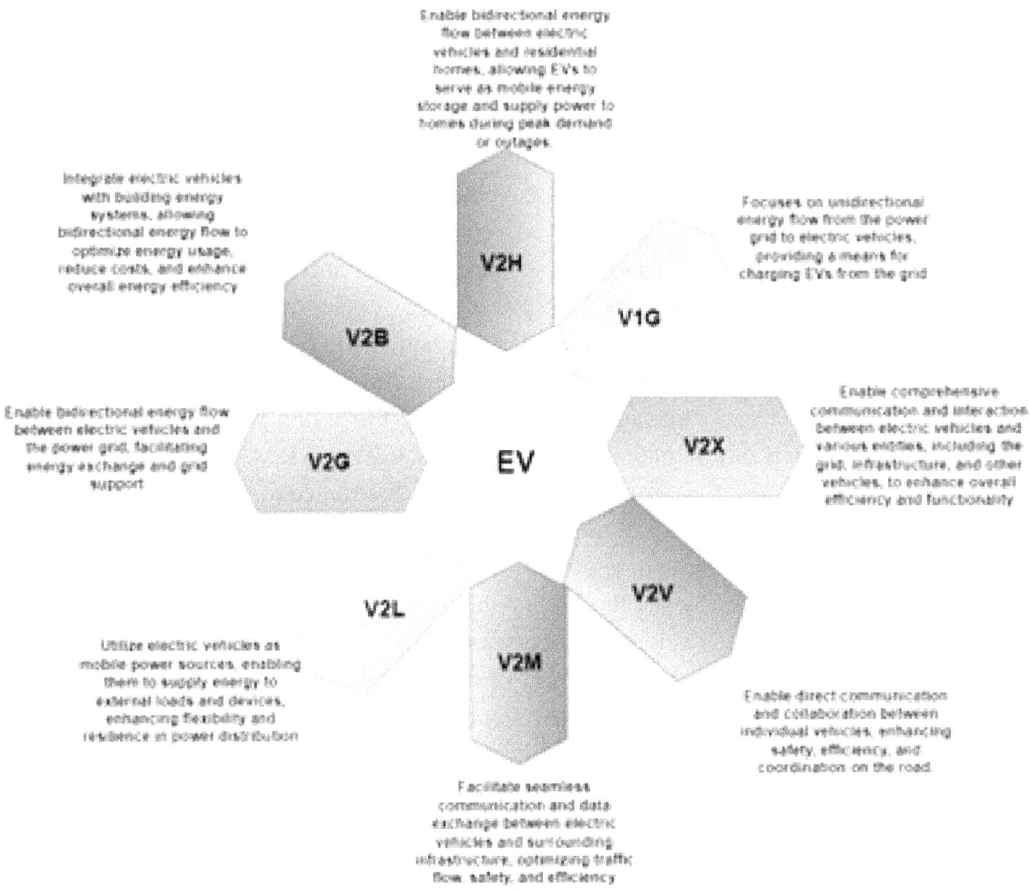

4. ENERGY MANAGEMENT

Renewable energy sources like wind and solar face challenges due to their unpredictable generation patterns, similar to other technologies designed to address daily energy needs. Despite advancements in prediction tools, the reliance on weather-dependent renewables introduces complexities for affluent countries aiming to transition to these energy sources. With the rise of Vehicle-to-Everything (V2X) technology, there's an exploration into repurposing used EV batteries as backup storage for renewables. This could enhance the storage capacity of energy, along with introducing grid complexities. Integration of EVs with renewable sources within the grid introduces variations in voltage and frequency which can destabilize the grid in addition to other complexities. Moreover, the growing number of EVs and renewable sources poses the risk of overloading the system. For successful integration of sustainable energy solutions, there is an essentiality of effective management of interplay among V2X, used batteries, and the grid.

4.1 EV Charging

In recent years, there has been a global emphasis on advancing transportation electrification as countries strive for carbon neutrality. Electric Vehicles (EVs) have emerged as a crucial pathway for achieving zero carbon emissions in the automotive sector. However, the challenges related to EV energy supplementation have not received adequate scrutiny, leading to concerns such as range anxiety and perceived inconveniences in charging services. Table 2 shows summary of Application Services of EVs Charging and Table 3 shows charging services and their respective research scope.

4.1.1 Off-Board and Onboard Charging

Onboard charging involves an AC input connected to the EV, with an internal converter within the vehicle responsible for converting AC to DC to power the battery. Conversely, off-board charging performs this conversion externally, typically at charging stations. The challenge lies in Level-1 charging, which demands power equal to the maximum capacity of the onboard charger. To reduce the onboard charger's usage, factors like weight, space, and cost are considered, and resonant circuits are utilized to mitigate adverse effects. DC stations, despite their higher cost, provide significant benefits. These advantages include decreased versatile high-power charging alternatives, faster charging rates, battery weight, improved communication (among utility firms, businesses, and commercial site owners), and refined battery control mechanisms (like managing heating concerns). This enhanced communication facilitates better contractual opportunities and billing provisions.

4.1.2 Conductive and Inductive Charging

EV batteries can be charged through conductive or inductive methods. Conductive charging involves a direct connection between the EV and the charging input, commonly seen in various AC and DC chargers available for most new cars. This method relies on external chargers, providing users flexibility in charging locations, be it at home or in public. Despite its prevalence, conductive chargers have drawbacks such as requiring user intervention and safety concerns. Inductive charging offers a solution to some of these issues by eliminating the physical connection. This process utilizes an electromagnetic field to transfer power between two objects, requiring specific charging stations designed for this purpose. While the wireless aspect is a significant advantage, aligning the coils for efficient charging has been a challenge. Recent advancements in automation, including parking assistance systems, are aiding in aligning the coils more effectively for improved efficiency (Vu et al., 2021).

4.1.3 Unidirectional and Bidirectional Charging

Unidirectional charging involves the typical flow of energy from the grid directly to EVs at charging stations. To manage the simultaneous charging of numerous EVs and prevent grid overload impacting voltage quality, an intelligent charging protocol is crucial. Disruption in the charging schedule or energy flow can harm the EV battery, accelerating its degradation. Despite its low investment cost and widespread availability in places like homes, unidirectional charging remains more popular than bidirectional charging. Bidirectional charging offers additional functionalities like load balancing, peak load management, power support, and harmonic filtering compared to unidirectional charging. Effective management of charge/discharge cycles is essential to minimize battery damage (Kiasari, 2023). Moreover, bidirectional charging demands a robust connection of EVs with the power grid, necessitating power system enhancements, substantial investments in mechanisms and foundations of V2G addressing social concerns.

4.1.4 Plug-In Charging Service

This has three typical modes where in first, to guarantee the safety of the charging process between the electrical network and the EV being charged, a control box is incorporated into the charging cable. Along with the integrated control box, this portable charger is extensively used across industrial sockets and domestic sockets, prioritizing charging performance and safety (Gonzalez Venegas et al., 2021). The second design is intended for all automatic charging systems in AC including commercial charging points and wallboxes. It's equipped to facilitate single-phase EV charging at 250V with 32A, and for three-phase it is extended to 480V with 32A, leveraging advanced equipment for these functionalities. The third charging mode is the sole method that offers DC charging. To ensure compatibility, and to transform the current from AC to DC, an external current converter is required before it passes through the charging cable to the EVs. Figure 3 shows QoE optimization and application scenarios of plug in charging services.

4.1.5 BSS

Assuming there are enough stocked batteries concerning incoming EVS, this system operates. If the inventory is insufficient, long waiting times for charging will persist until batteries become available. Additionally, due to land property considerations, Battery Swap Stations (BSSs) are typically situated in suburban areas, requiring users to travel long distances for energy refills. Furthermore, to maintain load balance, it's crucial to establish a reasonable mechanism for BSS-Selection, given service congestion exists due to limited available batteries for swapping at BSSs. The battery swapping service offers potential cost reductions in electricity through optimized battery charging. However, the construction of Battery Swap Stations (BSS) incurs considerable expenses, including those for equipment & batteries. Accounting for brand compatibility, battery size, cross-platform features and capacity is crucial. Addressing battery heterogeneity and ensuring cross-platform compatibility is essential for advancing the development of battery swapping services. Figure 4 shows QoE optimization and application scenarios of battery swapping services.

4.1.6 V2V

The V2V charging service offers flexibility both in terms of time and space. It focuses on aligning the available time of electric vehicles (EVs) with the charging station's (CSs) free time to avoid mismatches. Spatially, bypassing fixed-position CSs, it selects parking lots with available spaces. Formulation using constrained Mixed Integer Linear Programming (MILP) optimizes V2V charging.

Figure 3. QoE optimization and application scenarios of plug in charging services

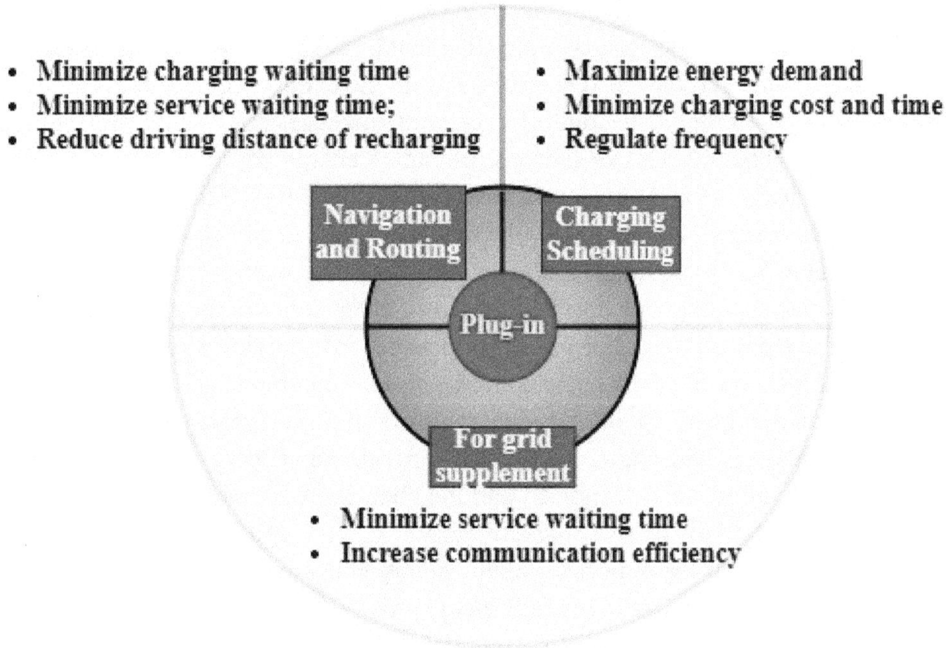

Figure 4. QoE optimization and application scenarios of battery swapping services

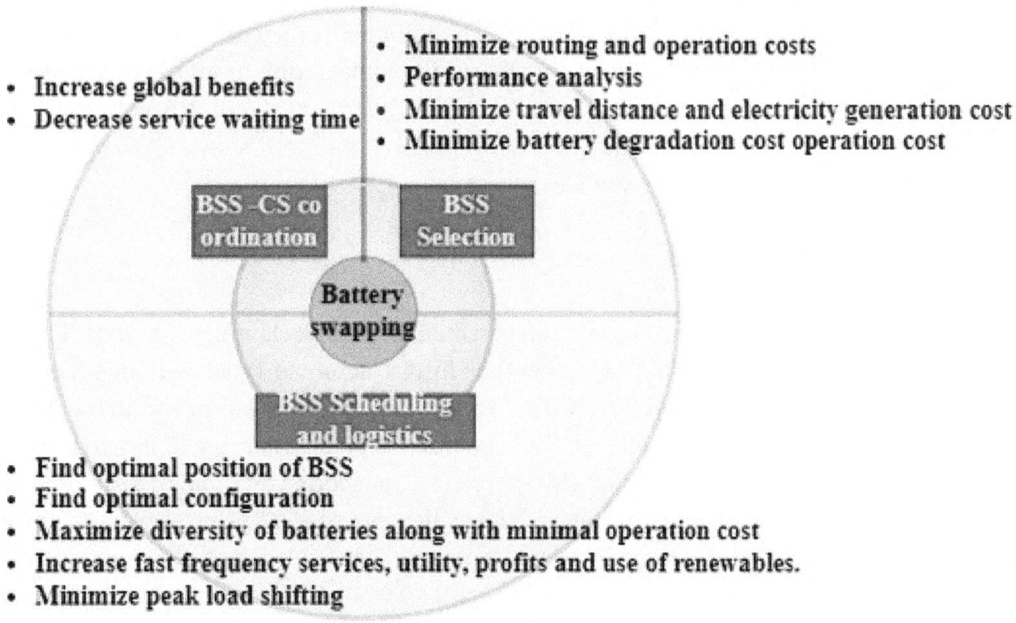

Comprehension of V2V can be divided into two categories: Asynchronous V2V Charging Optimization and Direct V2V Charging Optimization, based on their distinct service features. For direct V2V charging, an EV customer seeking charging needs to find an appropriate energy provider that maximizes charging utility while minimizing costs. On the other hand, energy providers aim to sell surplus energy at a premium price. Moreover, they also need to factor in the energy required for their travel purposes. In this first category, the GC (Global Controller) centrally manages the system, overseeing information gathering and overall planning. This includes coordinating energy consumers and providers, enabling them to access V2V charging in parking lots based on their preferences. Further division of first category includes V2V pair matching and parking lot selection. (a) V2V-Pair Matching focuses on ranking a pair of EVs for their suitability in providing P2P charging services. Two important constraints to consider are: (i) For every trip, if residuals are considered, the residual energies of the energy providers must meet the minimum requirement. (ii) Transfer of energy between EV providers results in additional energy loss during the charging process. (b) V2V-pairs must choose a common parking spot for energy transfer, focusing on factors like comfort, driving time, charging duration, and energy usage during charging.The challenge of picking the right parking spot is faced by providers and energy users on forming V2V pairs. EVs, facing parking scarcity, need optimized parking choices. Also, parking areas might lack enough chargers for V2V services. This resource scarcity affects V2V charging decisions.

In the second category of asynchronous setup, energy consumers and providers aren't required to initiate V2V charging at the same time. Instead, energy providers supply energy to a PSS (Power Storage System) which are condensers (Cao et al., 2023) equipped that store the energy. The PSS can then transfer this stored energy to energy consumers as and when required. When an EV driver requests charging, it necessitates a compatible Power Storage System (PSS) to maximize charging efficiency and minimize charging duration. As a result, the optimization problem is split into two primary categories: (i) PSS selection: Accounting essential constraints from the viewpoint of energy consumers, their priority lies in minimizing the time taken for the charging service. Additionally, they need to consider the charging cost. (ii) PSS- scheduling: Scheduling algorithms need implementation to efficiently manage the EV charging and discharging. In this context, prioritizing EVs with urgent charging needs becomes crucial, ensuring they receive high priority in the scheduling process. Figure 5 shows QoE optimization and application scenarios of V2V charging services.

4.1.7 Mobile Charging Service

Given the high mobility of EVs, there's significant potential for mobile charging solutions. This approach involves Mobile Charging Stations (MCSs) rather than infrastructure of fixed charging. It encompasses an in-motion way to charge facilitated by MCTs (Mobile Charging Trucks) or through other available private EVs. (i) Mobile Charging Station (MCS): In contrast to Fixed Charging Stations (FCSs) situated on decided places, Mobile Charging Stations (MCSs) can be transported by vehicles to different locations for charging purposes. Battery Energy Storages (BESs), the core of MCSs, are transported by or towed behind these vehicles. Further concerns in this category are MCS placement, operation optimization and scheduling. (ii) In-motion EV charging: In this scenario, when an EV identifies the immediate need for charging, it sends out a charging request. Other EVs (such as Mobile Charging Trucks or other private EVs) receiving this signal can offer the charging service. EVs engage in continuous charging via a wired or wireless approach while they are in motion. Figure 6 shows QoE optimization and application scenarios of mobile charging services.

Figure 5. QoE optimization and application scenarios of V2V charging services

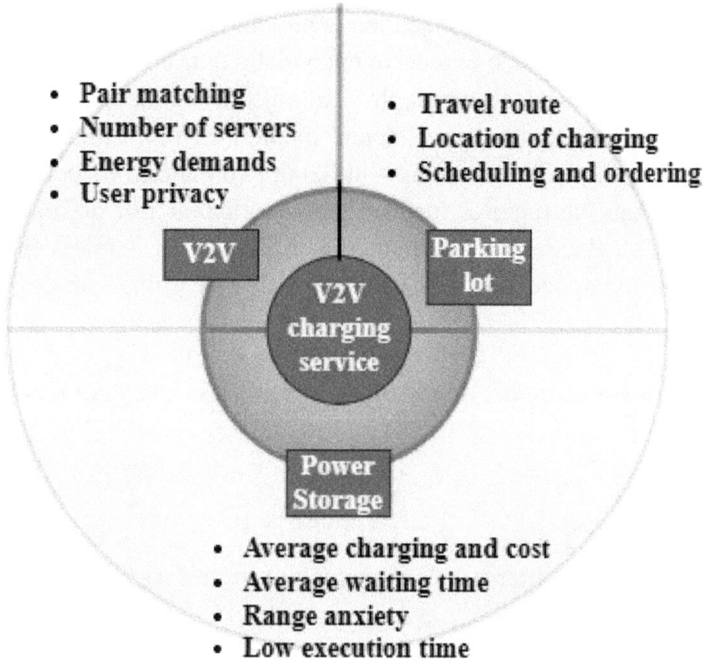

Figure 6. QoE optimization and application scenarios of mobile charging services

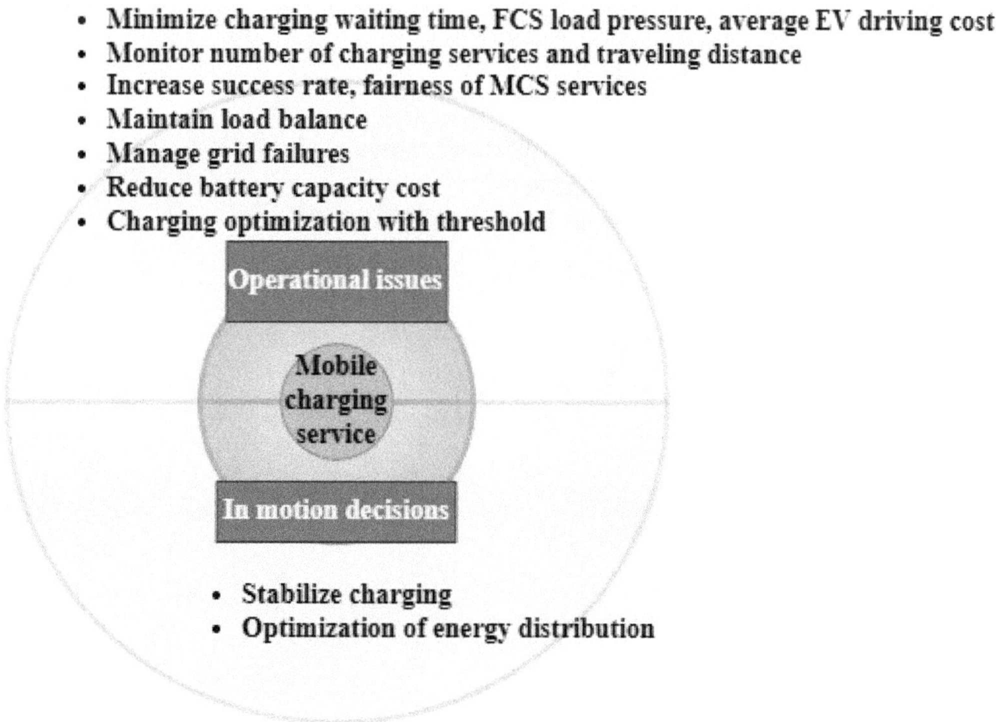

4.1.8 Wireless Charging Service

Energy transmission without a physical connection is possible here. In order to enable EV charging in absence of cables, converters are made to receive energy and convert it back into electrical power. It's a technology allowing energy transfer without physical links, using methods like capacitive and inductive charging. Capacitive charging, an experimental method, involves transferring energy wirelessly between two objects using capacitive coupling—utilizing two plates, one beneath the EV and one on the ground. Capacitive charging requires further research for practical optimization in EV charging. Inductive charging relies on electromagnetic induction principles for wireless energy transfer. Figure 7 shows QoE optimization and application scenarios of wireless charging services.

Figure 7. QoE optimization and application scenarios of wireless charging services

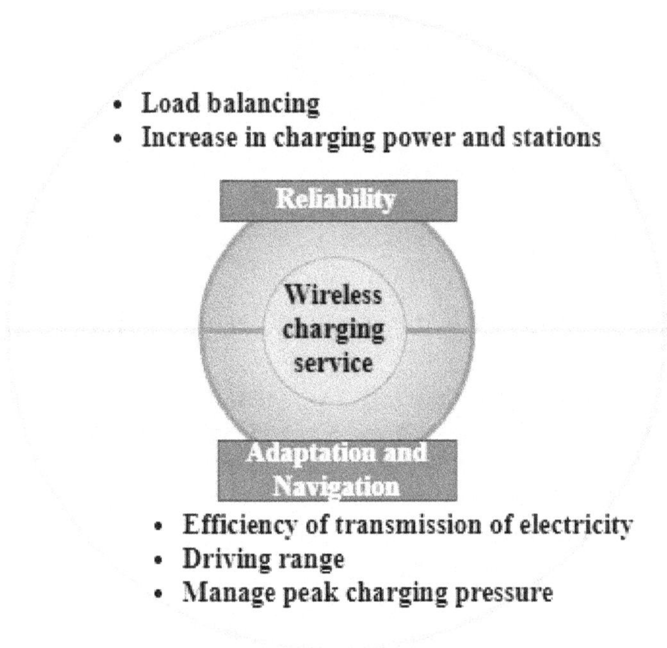

4.1.9 Summary of Charging

Table 2. Summary of application services of EVs charging (Cao et al., 2023)

Application - Charging Service (CS)	Objectives	Challenges
Plug-in	A Plug-in Charging Service refers to the infrastructure and facilities designed to charge electric vehicles (EVs) by plugging them into a power source. This service is a critical component of the electric mobility ecosystem, providing a convenient and accessible means for EV owners to recharge their vehicles.	Infrastructure Construction: need for a robust charging infrastructure, Privacy Issues: privacy and security of data collected from EVs during charging processes, Shortage of Parking Space, Weak Distribution Grid, and Uncertainty of EV Arrival
Battery Swapping	Innovative solution within the electric vehicle (EV) ecosystem that allows for the quick exchange of depleted batteries in electric vehicles with fully charged ones.	User Convenience, Initial Investment, Interchangeability- Interchangeability refers to the ability of components, devices, or elements to be easily replaced, Interchangeability, and Initial Investment
V2V CS	Vehicle-to-vehicle charging refers to a scenario where one electric vehicle (EV) transfers electrical energy directly to another EV. This can occur through physical connections or wireless technologies. Two types -Optimization of Direct V2V and Asynchronous Charging.	The Mismatch of Charging & discharging, Privacy Issues, Slow Charging Speed, Lack of Third-party Apps, Lack of Energy Providers, and Lack of Governmental Policy
Mobile CS	A service that provides on-the-go charging solutions for EV owners. Unlike traditional fixed charging stations, a mobile charging service offers flexibility by bringing the charging infrastructure directly to the user's location.	Operational Management, Privacy Protection, Range & Capacity, and Compatibility.
Wireless CS	Enables the transfer of electrical energy from a power source to the device without physical connectors or cables. Providing convenience and efficiency in EV charging as well as to other electronic devices is the primary goal of wireless charging.	Insufficient Charging, Infrastructure Deployment, Safety Concerns, Road Renovation

Table 3. Charging services and their respective research scope

Charging Services (CS)	Future Scope
plug-in	EV parking, unordered power distribution
BSS	Interchangeability among diversity, initial construction,
V2V	Third party and communication among vehicles, lack of energy providers and governmental role
MCS	Limited capacity of portable charging stations, varying charging rate and time
WCS	Required infrastructure deployment, lack of global standards for charging

4.2 Load Management Controller

Achieving rapid acceleration in electrified transportation and profitable commercialization necessitates the implementation of effective load management controllers. These controllers play a pivotal role in balancing load demand, reducing peak loads imposed on smart grid, as well as in enhancing the overall power grid stability. Various techniques, including vehicle-to-vehicle (V2V), vehicle-to-grid (V2G), vehicle-to-home (V2H), and battery swapping approaches, are available for optimizing load in the context of electrified transportation. It determines whether to charge or discharge the EV, with well-designed controllers contributing to reduced waiting times, minimized EV requests, and efficient load management.

The operation of the management network in EV charging has three main types: autonomous, centralized, and decentralized. Concern of centralized charging management is on regulation of substation's voltage throughout the sector or region, indicating the need of comprehensive communication infrastructure and specialized equipment like voltage regulators and capacitors. Whereas, the focus of decentralized charging control is to handle the regulation of voltage within a confined area, such as a parking lot. Unlike both previously discussed, autonomous charging management doesn't require a communication network and in turn can provide a more independent and self-contained approach (Gunjal et al., 2023).

Microgrids (MGs) are emerging as essential components in advanced power networks, shaped by various parameters (such as Self-sufficiency ratio, energy balance index, total energy imported among peers, maximum power load, mean electricity, cost annually) within smart-grid frameworks. Simply put, an MG represents an interconnected system consisting of wind turbines (WTs), solar panels, decentralized energy sources (DES), diesel engines, microturbines, and fuel cells. Within a small voltage distribution network, these sources are integrated with storage elements like mechanical batteries, power capacitors, and secondary rechargeable batteries. Segmentation of MG structural architecture can be viewed as stages 1, 2, and 3: three primary stages of operation. Each stage is characterized by distinct objectives and operational systems, all of which are carefully considered and implemented through the deployment of various controllers. Table 4 shows Primary Stages of MG

Table 4. Primary stages of MG

No.	Stages	Objective
1	Primary Control	The characteristics of a system that incorporates the fastest response to variations in sources or loads, eliminates circulating currents and allows for precise control of voltage magnitude and frequency tuning.
2	Supplementary controller	Managing and overseeing the operation of Microgrids (MGs). Regulating voltage and frequency in response to load variations. They are establishing connections between Distributed Generation (DG) sources and the energy storage system.
3	Tertiary level control	Control represents the ultimate and intentional stage in Microgrid (MG) operation. The focus is on identifying the optimal operation for the MG, considering both technical performance and financial considerations. From a technical perspective, if a fault occurs or a nonplanar issue arises within the MG. It aims to rectify real power from the grid.

4.2.1 Controllers Classification

Controllers can be broadly categorized into two major classes: linear controllers and nonlinear controllers. 1) Linear controllers operate based on linear mathematical relationships between inputs and outputs. 2) Nonlinear controllers incorporate intelligent actions beyond the conventional linear controllers. Generally, their operation and design become more complex due to the intricate relationships and dynamic behavior they handle. In Table 5, an overview of types of each of the linear controllers are mentioned. Table 6 shows types of Nonlinear Controllers.

Table 5. Types of linear controller

Linear Controller's Types	Description
PID controller	Classic feedback control system that uses - PID which is proportional, integral, and derivative respectively to regulate a system's output. It is widely used in industrial control systems for its simplicity and effectiveness.
Proportional resonator	It is designed to accentuate specific frequencies in a system's response. The controller amplifies the contribution of the proportional term at or near the system's resonant frequency. This allows for targeted control of oscillations or vibrations in the system.
LQR/LQG linear controller	Optimizing the performance of linear dynamic systems. LQR is used for systems without disturbances, while LQG incorporates disturbances into the control design. widely applied in areas such as aerospace, economics, and engineering
Predictive controller	Advanced control systems that use a predictive model of the process to optimize control actions over a finite time horizon. It predicts the future behavior of the system and computes optimal control actions to minimize a cost function.

Design of the predictive-based controller by adjusting the reference set current, targets minimization of error approximation ensuring precise tracking. Objective lies in reaching the stage that has minimal distortion, with minimum internal harmonic disturbance and in achieving high accuracy in controlling current. Notable examples within this category include the Deadbeat Controller (DB) and Model Predictive Controller (MPC).

Model Predictive Control (MPC) distinguishes itself by focusing on adjusting the current period, in contrast to classical PID controllers that operate based on upcoming timeslots stored in memory. Its primary goal is to improve the performance index by forecasting the trajectory of future manipulations through control parameters. Finite control set MPC also incorporates nonlinearity aspects and relies solely on the network dynamics, distinguishing it as a class with the potential to surpass conventional PI controllers in terms of adaptability and performance. However, It involves careful consideration of various parameters, model dynamics, and tuning to achieve robust performance across different operating conditions and uncertainties. As a result, expertise and practical experience play a crucial role in effectively implementing robust MPC controllers for real-world applications.

The generalized classical feedback Deadbeat (DB) controller is known for its rapid transient response and minimal Total Harmonic Distortion (THD). It finds extensive use in active power filters due to its enhanced dynamic effects in diverse fields. In DB control, the assignment involves determining what input signal should be applied to a system to achieve a steady-state output in the shortest possible time. These characteristics make the DB controller particularly well-suited for applications where achieving a rapid and accurate response with minimal distortion is crucial, such as in active power filters. The selection of a specific controller hinges on factors such as the characteristics of the controlled system, desired performance criteria, and specific control objectives.

Smart control achieves automation by emulating biological intelligence. Additionally, it excels at drawing insights from biological structures, identifying issues, and utilizing them to address control challenges.

Table 6. Types of nonlinear controllers (Arfeen et al., 2019)

Nonlinear Controller's Types	Description
Hysteresis current controller	Hysteresis current control is a non-linear control technique commonly used in power electronic systems. It involves switching between control states based on the comparison of the actual and reference currents with predefined hysteresis bands. Widely used in applications like motor drives and power converters.
Sliding mode control	It is a robust control strategy that aims to drive the system state onto a predefined sliding surface. It involves the creation of a sliding mode where the system dynamics become simplified and easier to control. Suitable for systems with uncertainties and disturbances.
Partial feedback linearization	Partial feedback linearization is a control strategy that aims to linearize only a subset of the system's states, leaving other states nonlinear. This technique simplifies the control design process by achieving linearization for specific variables. Often applied in systems with mixed linear and nonlinear dynamics.
Robust controller	Robust control techniques are designed to provide stable performance despite uncertainties and variations in the system. These controllers can handle parameter variations and external disturbances. Commonly used in systems where uncertainties are present, such as aerospace and automotive control. Both of its types, H-infinite controllers and μ-synthesis controllers are powerful tools in control theory, providing robustness and performance guarantees even in the presence of uncertainties and disturbances.
Smart controllers	Smart controllers encompass a range of control strategies that incorporate intelligent algorithms, machine learning, or adaptive mechanisms to enhance system performance. These controllers adapt to changing conditions and learn from the system's behavior. Applied in various fields, including autonomous systems, robotics, and smart grids.

It is classified into the following types:

1) The Repetitive Response Controller (RC) is implemented based on the principles of an Iteration Learning Algorithm (ILA). It oversees the operation of a utility power converter by adhering to defined reference periodic values, especially in scenarios with disturbances. The RC controller exhibits a null definite dynamic response, while its dynamic behavior is specifically tailored for handling periodic nonlinear loads. To address certain issues, the RC controller can be complemented by highly dynamic response controllers, either in a shunt or series configuration.
2) The inherent self-learned characteristics of the Artificial Neural Network (ANN) adaptation rule allow for a versatile and straightforward design, accommodating various grid disturbances and operating conditions. These features contribute to a more robust control function compared to traditional classical controls. Moreover, it introduces a novel control strategy, such as line voltage/current control without the need for sensors and system synchronization. The ANN can assess alternative function mappings. A proposed reactive power controller utilizes a Probabilistic Wavelet Fuzzy Neural Network (PWFNN) for managing reactive and active power control. This controller demonstrates the ability to facilitate low-voltage ride-through (LVRT) processes with power balance even under various grid fault situations.
3) The Adaptive Neuro-Fuzzy Inference System (ANFIS) is a member of the neural network family that is based on the Takagi–Sugeno–Kang (TSK) fuzzy inference system. ANFIS establishes mappings between inputs and outputs using membership functions and their associated parameters. The membership functions and related parameters of the Input-Output (IO) system dynamically adjust with acquired knowledge. This adaptive controller not only possesses the capability to address voltage profile issues but is also acknowledged for its effectiveness in mitigating the charging-discharging cycle.

4) Rule-based control approaches can be categorized into two main types: Deterministic rule-based algorithms and fuzzy rule-based algorithms constitute two major categories in control strategy methodologies. In deterministic rule-based approaches, experts in control strategy strive to formulate deterministic procedures, often implemented through lookup transition tables and flowcharts. These algorithms heavily depend on human expertise and network practices to establish rules that facilitate effective communication among energy resources and nonlinear loads, such as electric vehicles. They excel in online power management, utilizing their computational proficiency to provide accurate results tailored to specific output scenarios. The Fuzzy Logic Controller comprises four major components: the fuzzification interface, rule-based decision-making logic, inference engine, and defuzzification interface. In contrast to binary logic, fuzzy logic operates on a many-valued level, where variables have truth table values extending in diverse degrees, ranging between null and one—commonly referred to as their membership grade. Fuzzy logic manages randomness in the network through a straightforward if-then rule-based approach, eliminating the need for complex mathematical modeling of the system. This characteristic renders it particularly advantageous for intricate systems where a comprehensive mathematical design representation may not be feasible.

Measuring the performance of a controller can be accomplished through various techniques, with four key error-integrating functions particularly relevant for performance assessment (Arfeen et al., 2019). Table 7 shows performance Measurement of Control System

Table 7. Performance measurement of control system (Arfeen et al., 2019)

Parameter	Description
IAE- Integral Absolute Error	t combines the absolute error with respect to time, providing a measure of the error in any system response irrespective of its weight.
ITSE- Integral of Time-Squared Error	It exhibits lower sensitivity and requires less computational effort
ISE- Integral of Squared Error	It imposes harsher penalties on substantial errors compared to minor ones, emphasizing that the square of a larger error holds much more significance. Control systems designed for minimizing ISE and ITSE are effective not only in swiftly correcting major errors but also in permitting minor errors to persist over an extended period.
The performance index integral of time-weighted absolute error	Calculates integral of the absolute error, with the contribution of each error value weighted by its corresponding time duration.

4.3 Model Charging Load of EV

EV charging behaviors exhibit complexity often characterized by spatial and temporal randomness. Accurate modeling of charging load characteristics greatly relies on identifying and analyzing various factors. Therefore, establishing the precision of EV charging load models necessitates considering numerous relevant factors (Xiang et al., 2019). Figure 8 shows factors influencing charging load

Figure 8. Factors influencing charging load

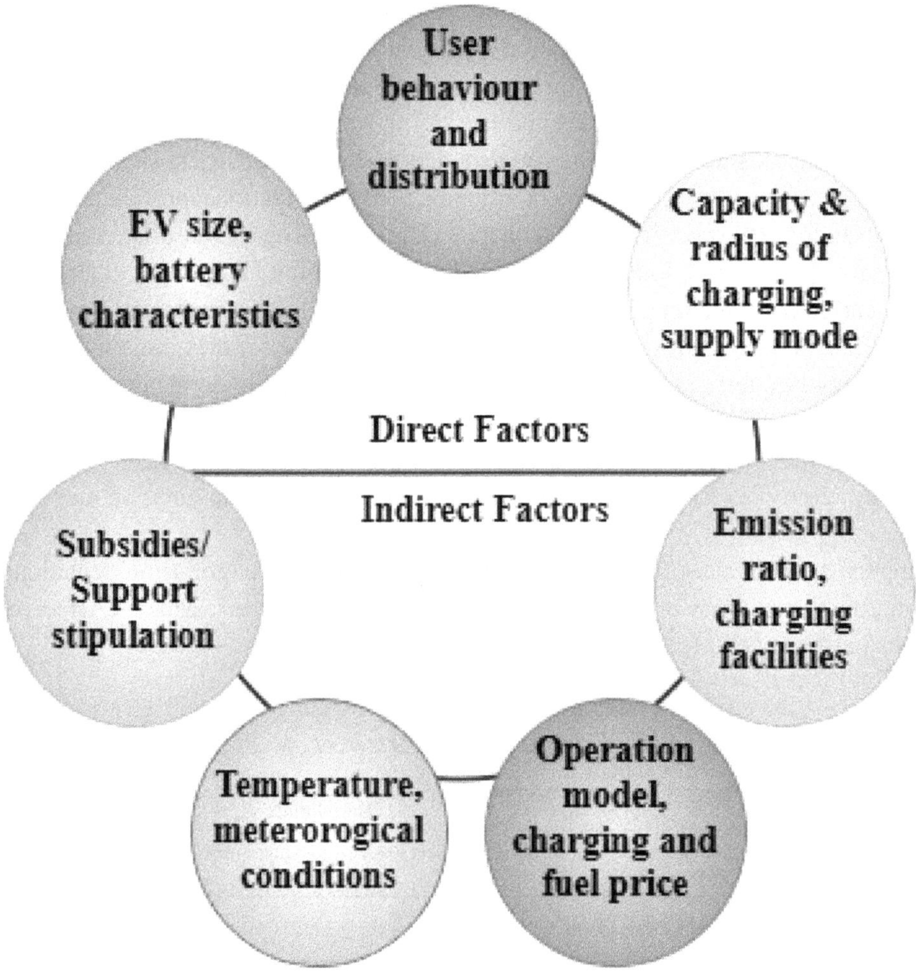

4.3.1 EV Charging Load Modeling for Temporal Dimension

The Monte-Carlo method is employed to determine the charging duration and power for Electric Vehicles (EVs) by utilizing the probability distribution of a set of Charging Vehicles (CVs) or a small sample of EVs. This method involves simulating both the driving and charging processes of these vehicles to derive relevant parameters. To determine driving process parameters such as driving time and distance, it is often assumed that the start time, distance covered, and end time by a driver are independent. Probability density functions are then derived by fitting traffic statistics data, utilizing Monte Carlo sampling methods for separate generation

Analyzing statistical data reveals a correlation between driving time and distance, which is addressed through two methods:

a. Relevant driving data is generated by utilizing the conditional probability function, taking into account the correlation.

b. A multidimensional joint probability distribution function involving driving start time, end time, and mileage is created using the copula function. These methods generate a coupled random number sequence that accurately simulates a user's driving behavior.

Markov-chain: The way EVs charge depends on their mobility patterns and the availability of charging stations. Predicting this behavior is tricky, but random methods using factors like state, dwell time, and transition probabilities help describe what to expect. Markov processes specifically capture how EVs move between starting points, destinations, and the time spent at different locations, revealing insights into their travel patterns and residence times.

Fuzzy logic System: In typical driving scenarios, EV users struggle to precisely assess battery charge and estimate residence time for charging decisions. Instead, they rely on broad categories like 'low', 'medium', or 'high' for charging status and 'short', 'middle', or 'long' for expected residence time. These vague guidelines form the basis for users to establish rules guiding their charging behavior during journeys.

4.3.2 EV Charging Load Modeling for Spatial Dimension

Traffic flow dynamism and simulation: In a power distribution network (PDN), the charging load of EVs varies based on the spatial distribution of vehicles. The mobility of EVs inherently ties their widespread access to the grid, transportation network, and charging service infrastructure. Predicting the charging load at a station within a transportation network involves utilizing dynamic traffic flow models from the traffic field. These models help estimate the rate of EV arrivals at charging stations. To forecast charging loads accurately, Monte Carlo simulation, queuing theory, and various other techniques are combined to perform charging load predictions.

Additionally, each vehicle can be portrayed as an intelligent agent, following a day-wise agenda and making decisions guided by a utility function. This function influences choices such as route selection and the identification of additional fuel points tailored to specific vehicle types.

Parking: The approach encompasses the segmentation of the Electric Vehicle (EV) charging area into discrete regions, followed by a thorough analysis of the utilization of different land types within each zone, taking into account specific parking characteristics. This comprehensive assessment facilitates the estimation of anticipated parking demand across the entire area, considering its spatial distribution.

Demography: The primary aim of Electric Vehicles (EVs) is to facilitate transportation across various locations, and the number of vehicles in a specific area is contingent on user count and the area's attributes. Moreover, vehicle presence varies throughout the day; residential zones may witness fewer vehicles during working hours, with a peak in demand during evenings. Conversely, industrial and commercial areas exhibit an opposite trend. Therefore, a demographic model is essential for accurately modeling EV charging loads, considering the dynamic patterns influenced by the area's characteristics and human activities throughout the day. Dynamic traffic flow and simulation are pivotal in comprehending the charging load dynamics of EVs in a power distribution network. The spatial distribution of vehicles profoundly impacts the charging load, and the mobility of EVs is intricately linked to grid access, transportation networks, and charging infrastructure. To precisely forecast charging loads at stations within a transportation network, dynamic traffic flow models are utilized, drawing insights from the field of traffic engineering. Techniques such as queuing theory, Monte Carlo simulation, and other methodologies are integrated to forecast charging loads effectively.

4.3.3 Hybrid Dimension

The trip-chain theory delineates the entire journey of an individual from the initial point to the ultimate destination, traversing through multiple transit points in a specific chronologically. Methods rooted in this theory combine temporal and spatial dimensions, synthesizing these aspects for comprehensive analysis. By developing interactive trajectories of EVs, these temporal-spatial models effectively construct behavioral patterns in travel.

4.4 Effect of Charging Load on Power Grid

These methods aim to smoothen enhance the power grid's load factor, charging profiles, minimize network losses, and guarantee safe and cost-efficient grid operation. Rapid advancements in energy storage technologies have led to the emergence of self-storage charging stations (Steinschaden & Baptista, 2020). These stations incorporate on-site energy storage devices capable of storing energy for subsequent vehicle charging. At times, these storage systems incorporate renewable energy sources or join forces with other energy storage devices, known as hybrid energy storage systems, to enhance the reliability of the charging system (Sabzi et al., 2019). As the number of EVs requiring electric energy grows, the need for charging stations becomes more evident. The expansion of the charging infrastructure not only satisfies this demand but also promotes the broader acceptance of EVs (Energy Efficiency and Renewable Energy, n.d.). Various EU projects, such as those detailed in Charging, (n.d.), "Electrified L-Category Vehicles Integrated Into Transport and Electricity Networks" (n.d.), "Novel, automated charging infrastructure for electric vehicles" (n.d.), and ZAPINAMO (n.d.), focus on enhancing charging infrastructures to accommodate the influx of new EVs joining the grid. For instance, Austrian projects Charging (n.d.) and "Novel, automated charging infrastructure for electric vehicles" (n.d.) introduce Matrix charging stations, designed without external moving parts, showcasing impressive efficiency rates of 99% transmission and offering a 30% cost reduction compared to competitors. Additionally, British company ZPN proposes easily deployable, stored-energy-based fast-charging systems resembling larger and quicker versions of power banks for mobile phones (ZAPINAMO, n.d.). The flexibility for EV owners to charge their vehicles at their convenience may require utilities to enhance charging capacity and expand the number of charging stations (Cannon, 2020). In Csonka (2020), researchers designed a centralized strategy linking EVs to the distribution network, treating EVs as mobile energy storage to stabilize load fluctuations. The aim was to align flexible charging needs with varying renewable energy availability, maximizing renewable energy utilization. Charging EVs during the day using solar energy showed potential to eliminate additional generation when a portion of flexible load was engaged. In Yu et al. (2019), a real-time smart charging schedule was employed to curb grid peak demand. Using dynamic programming for optimization, they managed a 24% reduction in annual grid demand but saw no significant change in hourly demand patterns.

5. OPERATOR

Grid operators, also referred to as system operators, hold a pivotal role in the energy sector, tasked with upholding the dependable and secure functioning of the electrical grid. Their responsibilities include continuous monitoring of the grid's performance, ensuring it operates efficiently within its capacity, and consistently meeting electricity demands. Additionally, these operators manage balancing and ancillary service markets, aligning demand with supply to ensure system security and stability. In Europe, grid operators are categorized into two main groups: Transmission System Operators (TSOs) and Distribution System Operators (DSOs) (Operator, n.d.). A Transmission System Operator (TSO) is responsible for the efficient and reliable transmission of electricity from power generation plants through the power grid to regional or local electricity distribution operators. On the other hand, a Distribution System Operator (DSO) operates, manages, and sometimes owns the local and regional energy distribution networks responsible for delivering electricity to end-users. These distribution grids include both low-voltage networks (250-400 V) and medium-voltage networks (6-50 kV).

6. CHALLENGES IN EV ADOPTION

Range anxiety: Range anxiety, as defined, represents the mental distress stemming from fearing an abrupt power loss during EV driving (Cao et al., 2023). The driving range of Electric Vehicles (EVs) is significantly reduced in winter conditions due to battery characteristics, and the use of air conditioners further affects this range.

Lack of infrastructures: The lack of infrastructures for V2V charging is due to varying charging protocols among EVs, potentially causing failures in the charging process. This shortage highlights the necessity for a DC-DC converter capable of accommodating multiple charging protocols. The charging habits and compatibility of EVs pose a significant challenge to power grid load & stability management. Given the power system's restricted load capacity, the installation of charging infrastructure requires meticulous planning. Additionally, continual enhancements in grid serviceability are crucial to meet the rapidly increasing demands for EV charging.

Battery constraints: The effectiveness, durability, and charging speed of batteries greatly impact EV adoption. Progress in battery technology, encompassing increased energy density, quicker charging capabilities, and enhanced durability, is essential to address consumer concerns and elevate the allure of EVs.

Pricing constraints: The considerable initial expense of Electric vehicles in contrast to traditional internal combustion engine vehicles remains a major hurdle for numerous potential buyers (Ragavendran & Kumar, 2023). This price difference is mainly influenced by the battery costs, which constitute a substantial portion of an EV's overall price.

Unawareness of role: Numerous consumers are yet to grasp the advantages of EVs, like reduced carbon emissions, lower operational expenses, and a quieter, smoother driving encounter. Informative campaigns, focused marketing tactics, and demonstration initiatives can significantly enhance awareness and correct misunderstandings about EVs. These efforts play a critical role in generating consumer interest and demand for EVs.

Uncertain parameters: Many problems assume that accurate data has been known in advance. In contrast, considering the practical scenario, power generation and demand, charging data like plug-out and plug-in times (Nimalsiri et al., 2019) as well as its price aren't known precisely.

Instability of ancillary services: V2G (Vehicle-to-Grid) services are prone to instability compared to conventional ancillary services (like demand side management (Raval et al., n.d.), frequency response and reserve services]) provided by generators that typically prefer stable interfirm relationships. Electric vehicles (EVs) primarily serve as transportation tools, hence, the charging & dis-charging patterns of Electric vehicle owners can be significantly disrupted by the stochastic alterations in their driving schedules or other random factors.

7. FUTURE DIRECTIONS

Through the strategic application of technology, smart homes, smart healthcare (Kaiser et al., 2023; Maheriya & Panchal, 2023) and smart lifestyle can enhance efficiency, sustainability, and overall quality of living. This transformative approach holds the promise of improving daily lives and fostering a future that is both prosperous and environmentally friendly. Efforts are underway to transform the use of EVs to augment the reliability and sustainability of the transportation system. The increasing adoption of EVs has presented challenges such as the need for charging infrastructure development and the prediction of peak loads, making management a complex task.

Machine Learning: In recent times, the growing adoption of EVs has prompted interest in utilizing EV batteries as a mobile energy storage solution for the electricity grid through V2G technology. However, the smooth integration of EVs into the grid presents various challenges. A notable issue revolves around power quality, particularly arising from fluctuations in battery power output. Traditional droop control techniques have been proposed to govern power flow in V2G systems. However, these methods may face limitations in effectively regulating battery power output during periods of high demand, potentially leading to grid instability. This research proposes the implementation of a droop-Artificial Neural Network model. The suggested approach involves incorporating an Artificial Neural Network model into the droop control system, aiming to provide accurate predictions for both the charger's voltage and frequency (Aurangzeb et al., 2023).

The increase in electric vehicle (EV) charging, especially on electricity grids relying on fossil fuels, raises concerns about grid safety, stability, and heightened greenhouse gas emissions. The uncontrolled integration of EVs into the grid is anticipated to grow significantly, leading to challenges such as new peaks, transformer limitations, and substation capacity violations (Raval et al., n.d.). To address the impact on energy consumption costs and reduce greenhouse gas emissions, a robust and dynamic multi-level optimization approach is essential. A machine learning (ML)-based charge management system considers various charging technologies, aiming to guide EVs efficiently while minimizing expenses, managing voltages, handling load fluctuations, and reducing power loss (Mazhar, 2023). Table 8 presents spotlights on certain control methodologies that can be deployed to solve the problem for EV charging management .

Figure 9 Provides an overview of the methods and techniques explored, accompanied by the objectives pursued. It summarizes various control methods employed by researchers for implementing Electric Vehicle (EV) charging management within the use of each state of art method and it's objective . Furthermore, each sector size and color represent individual methods and it extend table 8 by showing the objective of method to solve the problem.

Table 8. Spotlights certain control ML techniques chosen by researchers for deploying EV charging management analong with problem solved with it

Techniques	Issue Addressed
Ant-Based Optimization	Uncontrolled Charging, High Energy Consumption, Charging Station Congestion
Artificial Bee Colony	Grid Imbalance, Voltage Fluctuation, Uncontrolled Charging Schedule
Genetic Algorithm	EV Routing Problem,Grid Load Variance,High EV Charging Demand, Poor Charging Infrastructure Location, Uncoordinated Charging, Transformer Overload Thermal Line Limit, Voltage Limit
Particle Swarm Optimization	Demand Side Management, Uncontrolled Charging and Discharging, Grid Overload, High Energy Consumption
Neural Network	EV Charging Impact on the Grid, High Energy Consumption, Grid Voltage Fluctuation, Grid Power Losses, Transformer Overload, EV Charging Load Variation
Social Spider Optimization	Grid Congestion, Grid Voltage Drop on Feeder Lines, Economic Load Dispatch Problem, High Energy Consumption
Reinforcement Learning	EV Energy Consumption, Uncontrolled Charging of EVs, High Cost of Electricity, EV Scheduling Problem Congestion Control Problem, High Peak Load
Load Modelling Techniques	EV Charging Traffic Flow Problem, High Energy Demand by EVs, High Voltage Variation on the Grid, Uncontrolled Charging of EVs
ADMM Technique	Grid Load Variation, High Energy Demand, Energy Management Problem Random Large-Scale EV Charging, Charging Station Congestion, Long Charging Duration
Monte Carlo Simulation	EV High Energy Consumption, Increased Peak Load
Hybrid Optimization: PSO, GA, Dynamic Programming, Monte Carlo, Hybrid Crow Search	Stochastic EV Charging Process, High Energy Consumption, Grid Instability Uncontrolled Charging, Charging Infrastructure Locations, Grid Power Losses Charging Station Location, Increased Peak Load, Grid Load Variance Frequency Control Problem

Cyber Security and Blockchain: A consortium blockchain network is established to evaluate and monitor the degradation cost of Electric Vehicle (EV) batteries. This blockchain involves all relevant stakeholders, including the EV user, Aggregator, EV Charging Service Operator (CSO), and EV Management Service Provider -MSP, aiming to optimize the use of parked EVs. The blockchain records crucial data for accurate estimation of EV degradation costs and tracks cash flow among the involved parties, particularly for Vehicle-to-Grid (V2G) applications (Raval et al., n.d.). Compared to regular databases, blockchains offer enhanced security, and their consensus mechanisms increase resistance to hacking attempts. Once added to the blockchain, past transactions cannot be modified, providing an additional layer of security, which is particularly valuable in addressing cybersecurity concerns in EV charging networks (Gowda, 2021). Due to the open structure of the Vehicle-to-Grid network, there exists a significant risk of exposing users' personal information. While several solutions aim to tackle this issue, many fall short of creating a fully distributed environment, leaving room for a potential single point of failure. To bolster security, blockchain technology, with a focus on Ethereum and Ganache software, can be utilized to secure transactions between Electric Vehicles and microgrids. Additionally, there is a future prospect of implementing a Hyperledger Fabric-based blockchain to closely monitor transactions on an hourly basis (Ramasamy et al., 2023).

As electric vehicle (EV) technology progresses, EVs have become essential elements in the prospective smart grid. The surge in EVs on roads and the adoption of vehicle-to-grid (V2G) technology, facilitating bidirectional electrical flows between EVs and the power grid, has garnered considerable attention. Prior to availing a charging service, verifying an individual's identity is crucial to safeguard their privacy from potential breaches. This study introduces a privacy protection strategy for electric cars, employing a blind signature-based distributed approach. The system is designed to operate within a fully distributed environment, mitigating concerns related to central failure points and concentrated attacks, all the while ensuring user privacy. The blockchain-based technique, as analyzed, exhibits reduced computational overhead without compromising security (Wang et al., 2023). Nevertheless, creating a fair and confidential electricity exchange system has emerged as a significant challenge. The introduction of blockchain technology offers a fresh perspective to tackle this issue (Zhang et al., 2022). Furthermore, it presents a privacy-preserving blockchain-powered electricity auction strategy designed for Vehicle-to-Grid (V2G) networks in the smart grid. The approach employs PS group signatures to protect the privacy of Electric Vehicles (EVs) or charging stations and utilizes blockchain for automated auction execution.

Decentralized Energy Systems: In the traditional power setup, electricity flows one way, originating from large power plants and reaching consumers via transmission and distribution networks. But the shift towards cleaner, more distributed energy systems is reshaping grid operator roles. This shift demands closer collaboration between Distribution System Operators - DSOs and Transmission System Operators - TSOs, especially in systems with intermittent renewables and distributed energy resources (DERs). Peaks in electricity demand, such as when EVs are plugged in after work or high solar input at midday, strain the grid. TSOs often lack comprehensive DER insights, leading to inaccurate load and generation forecasts, risking supply-demand balance. In this evolving DER-powered grid, DSOs shoulder extra duties, managing peak loads, providing support to TSOs, and ensuring voltage stability. Addressing these decentralized system complexities requires innovative collaboration among stakeholders.

Figure 9. Control methodologies implemented for electric vehicle (EV) charging within the grid (Kene & Olwal, 2023)

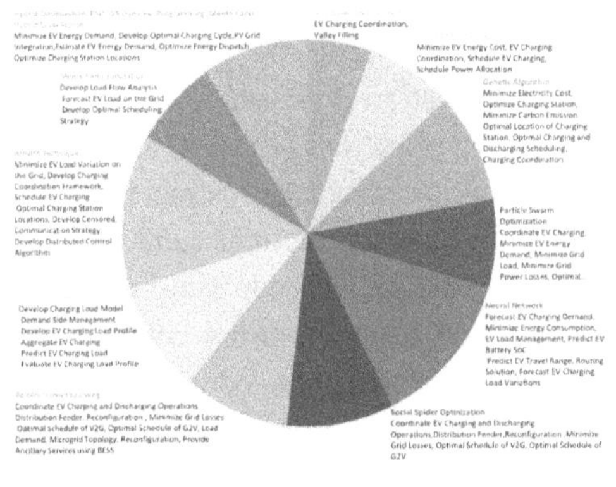

Peer-to-peer: P2P involves trading of energy for flexible exchange of surplus energy from small-scale Distributed Energy Resources -DERs among local consumers. In practical terms, this system employs sophisticated metering infrastructure (smart meters), digital platforms, and automated systems to track the real-time generation and consumption of energy. People with surplus energy, such as from solar panels or stored in batteries, have the opportunity to directly sell it to those in need. Pricing strategies are crafted to incentivize this energy exchange, creating a more financially attractive proposition for local energy trading compared to buying from or selling back to the grid.

Vehicle as platforms: Vehicle Computing encompasses the technologies enabling computational tasks to be carried out within Connected Vehicles (CVs) (Lu & Shi, 2023; Maheriya & Panchal, 2023; Raval & Borisagar, n.d.). This functionality transforms these vehicles into mobile computing units, facilitating computation, communication, energy usage, sensing, and data storage for diverse edge-enabled (Maheriya & Panchal, 2023; Raval & Borisagar, 2022; Raval & Borisagar, n.d.) services. It underscores that CVs serve as robust mobile computing platforms capable of analyzing data from onboard sensors and connected devices, even while parked or undergoing charging processes.

8. CONCLUSION

The rapid adoption of low carbon and renewable energy sources (RESs) is emerging as a crucial defendable solution to meet the rising global demand for electricity while mitigating carbon emissions. In light of severe challenges related to energy and environment, global attention is progressively directed towards E-vehicles (EVs) and associated emerging technologies. These technologies hold immense potential to expedite the progress of intelligent and defendable transportation, contributing to the establishment of future smart cities. This chapter conducts a comprehensive review of the existing literature on the current state of Electric Vehicles (EVs) and their interface technology with the power grid. The primary insights and statistical information are extracted from recent publications, shedding light on the latest technological advancements, constraints, and potential future market developments. Furthermore, electric vehicle technology holds significant potential for technological innovation in the energy market, envisioning vehicles serving as both a transportation means and a dynamic energy vector interfacing with the grid (V2G), buildings (V2B), and other applications (V2X).

REFERENCES

Arfeen, Z. A., Khairuddin, A. B., Larik, R. M., & Saeed, M. S. (2019). Control of distributed generation systems for microgrid applications: A technological review. *International Transactions on Electrical Energy Systems*, *29*(9), e12072. doi:10.1002/2050-7038.12072

Arpita, M., & Panchal, S. (2023). Smart Health and Cybersecurity in the Era of Artificial Intelligence. In M. S. Kaiser, J. Xie, & V. S. Rathore (Eds.), *Information and Communication Technology for Competitive Strategies (ICTCS 2021). Lecture Notes in Networks and Systems* (Vol. 401). Springer. doi:10.1007/978-981-19-0098-3_5

Aurangzeb, M., Xin, A., Iqbal, S., Afzal, M. Z., Kotb, H., AboRas, K. M., Ghadi, Y. Y., & Ngoussandou, B. P. (2023). A Novel Hybrid Approach for Power Quality Improvement in a Vehicle-to-Grid Setup Using Droop-ANN Model. *International Journal of Energy Research*, *2023*, 2023. doi:10.1155/2023/7786928

Cannon, P. (2020). How a Lack of Public EV Charging Stations Is Creating Big Business Opportunities. *ABB*, (Jun), 15.

Cao, Y., Cui, J., Liu, S., Li, X., Zhou, Q., Hu, C, & Liu, Z. (2023). A Holistic Review on E-Mobility Service Optimization: Challenges, Recent Progress and Future Directions. *IEEE Transactions on Transportation Electrification*.

Charging, M. (n.d.). *Novel, automated charging infrastructure for electric vehicles*. Available at: https://cordis.europa.eu/project/id/830239

Csonka, B. (2020, June). Centralized charging power distribution method for electric vehicles. In *2020 Smart City Symposium Prague (SCSP)* (pp. 1-6). IEEE. 10.1109/SCSP49987.2020.9134002

Electrified L-Category Vehicles Integrated Into Transport and Electricity Networks. (n.d.). https://cordis.europa.eu/project/id/769926

Energy Efficiency and Renewable Energy, Developing Infrastructure to Charge Plug-In Electric Vehicles. (n.d.). https://afdc.energy.gov/fuels/electricity_infrastructure.html

Gallegos, J., Arévalo, P., Montaleza, C., & Jurado, F. (2024). Sustainable Electrification—Advances and Challenges in Electrical-Distribution Networks: A Review. *Sustainability (Basel)*, *16*(2), 698. doi:10.3390/su16020698

Gonzalez Venegas, Petit, & Perez. (2021). Plug-in behavior of electric vehicles users: Insights from a large-scale trial and impacts for grid integration studies. *eTransportation, 10*.

Gowda, S. N. (2021). *Assessment and tracking electric vehicle battery degradation cost using blockchain. IEEE Power & Energy Society Innovative Smart Grid Technologies Conference*.

Grace, O., Iqbal, K., & Rabbi, F. (2023). Creating Sustainable Urban Environments: The Vital Link between Development, Health, and Smart Cities. *International Journal of Sustainable Infrastructure for Cities and Societies*, *8*(1), 53–72.

Gunjal, M., Mathew, L., & Sudha Letha, S. (2023). Electric vehicle controllers for sustainable energy management: Recommendations and challenges. *Environment, Development and Sustainability*, 1–31. doi:10.1007/s10668-023-03561-x

Kene & Olwal. (2023). Energy Management and Optimization of Large-Scale Electric Vehicle Charging on the Grid. *World Electric Vehicle Journal, 14*(4), 95.

Khor. (n.d.). *Battery Electric Vehicle (BEV).* https://medium.com/@ackhor/battery-electric-vehicle-bev-d69f8683b141

Kiasari, M. (2023). *Comparative Analysis of Machine Learning Techniques for an Hour-Ahead Forecasting of Electric Vehicle States*. Academic Press.

König, A., Mayer, S., Nicoletti, L., Tumphart, S., & Lienkamp, M. (2022). The Impact of HVAC on the Development of Autonomous and Electric Vehicle Concepts. *Energies*, *15*(2), 441. doi:10.3390/en15020441

Lee, C. (n.d.). *What is EV, BEV, HEV, PHEV? Here's your guide to types of electric cars*. www.marketwatch.com/story/what-is-ev-bev-hev-phev-heres-your-guide-to-types-of-electric-cars-11617986782tch

Lu, S., & Shi, W. (2023). Vehicle as a Mobile Computing Platform: Opportunities and Challenges. *IEEE Network*.

Maheriya, A., & Panchal, S. (2023). A Neoteric Approach to Improvise Privacy Shielding of Sensitive Healthcare Information & Prediction of Disease. *International Journal of Computing and Digital Systems*, *13*(1), 1–12.

Mazhar, T. (2023). Electric Vehicle Charging System in the Smart Grid Using Different Machine Learning Methods. *Sustainability, 15*(3), 2603.

Mohammadi, M., & Thornburg, J. (2023). Strategizing EV Charging and Renewable Integration in Texas. *arXiv preprint arXiv:2310.17056*.

Mololoth, V. K., Saguna, S., & Åhlund, C. (2023). Blockchain and machine° learning for future smart grids: A review. *Energies*, *16*(1), 528. doi:10.3390/en16010528

Nimalsiri, N. I., Mediwaththe, C. P., Ratnam, E. L., Shaw, M., Smith, D. B., & Halgamuge, S. K. (2019). A survey of algorithms for distributed charging control of electric vehicles in smart grid. *IEEE Transactions on Intelligent Transportation Systems*, *21*(11), 4497–4515. doi:10.1109/TITS.2019.2943620

Novak, A., & Ivanov, A. (2023). Network Security Vulnerabilities in Smart Vehicle-to-Grid Systems Identifying Threats and Proposing Robust Countermeasures. *Journal of Artificial Intelligence and Machine Learning in Management*, *7*(1), 48–80.

Novel, automated charging infrastructure for electric vehicles. (n.d.). https://cordis.europa.eu/project/id/781780

Ntombela, M., Musasa, K., & Moloi, K. (2023). A Comprehensive Review for Battery Electric Vehicles (BEV) Drive Circuits Technology, Operations, and Challenges. *World Electric Vehicle Journal*, *14*(7), 195. doi:10.3390/wevj14070195

Operator, G. (n.d.). *gridX*. www.gridx.ai/knowledge/what-is-a-grid-operatorned – gridX

Poal, V., Panchare, D., & Mehta, B. (2021). *Balancing of HVAC System Energy Consumption in Electric Vehicles*. SAE Technical Paper 2021-28-0124. doi:10.4271/2021-28-0124

Ragavendran, S., & Kumar, P. (2023). Exploring Barriers and Challenges of Electric Vehicles in India and Vehicle-to-Grid Optimization: A Comprehensive Review. *International Journal Of Novel Research In Engineering Sciences,* 16-19.

Ramasamy, S., Ganesan, K., & Koodalsamy, B. (2023). Realistic choice-based decision-making mechanism in electric vehicles and secure energy transaction using ethereum blockchain. *Electrical Engineering*, 1–21. doi:10.1007/s00202-023-02074-0

Ramraj, R., Pashajavid, E., Alahakoon, S., & Jayasinghe, S. (2023). Quality of Service and Associated Communication Infrastructure for Electric Vehicles. *Energies*, *16*(20), 7170. doi:10.3390/en16207170

Raval, A., & Borisagar, K. (2022, March). A Survey on Techniques and Methods of Recommender System. In *International Conference on Computational Intelligence in Data Science* (pp. 97-114). Cham: Springer International Publishing. 10.1007/978-3-031-16364-7_8

Raval, A., Lohia, A., & Bhatt, U. Y. AI Business Model is Emerging Energy Market and Smart Grid. In *Applications of Big Data and Artificial Intelligence in Smart Energy Systems* (pp. 169–192). River Publishers. doi:10.1201/9781003440864-8

Raval, A. S., & Borisagar, K. R. (n.d.). *Document Recommendation: An anatomization of neoteric advancements, problems and potential scope*. Academic Press.

Sabzi, Asadi, & Moghbeli. (2019). Regenerative energy management of electric drive based on Lya-punov stability theorem. *J. Mod. Power Syst. Clean Energy, 7*(2), 321–328. doi:10.1007/s40565-018-0497-y

Shi, L., Lv, T., & Wang, Y. (2019). Vehicle-to-grid service development logic and management formulation. *Journal of Modern Power Systems and Clean Energy*, *7*(4), 935–947. doi:10.1007/s40565-018-0464-7

Steinschaden, S., & Baptista, J. (2020, January). Development of an Efficient Tool for Solar Charging Station Management for Electric Vehicles. *Energies*, *13*(11), 11. Advance online publication. doi:10.3390/en13112979

Vu, V.-B., Gonzalez-Gonzalez, J. M., Pickert, V., Dahidah, M., & Trivino, A. (2021). A Hybrid Charger of Conductive and Inductive Modes for Electric Vehicles. IEEE Trans. Ind. Electron., 68(12), 12021–12033. doi:10.1109/TIE.2020.3042162

Wang, B., Gou, Y., & Zhang, S. (2023). V2G Blind Signature Privacy Protection Scheme Based on Blockchain. *Journal of Electronics and Information Science*, *8*(1), 36–41. doi:10.1016/j.ins.2023.01.021

X-enginner. (n.d.). *What is a battery electric vehicle (BEV).* https://x-engineer.org/battery-electric-vehicle-bev/

Xiang, Y., Hu, S., Liu, Y., Zhang, X., & Liu, J. (2019). Electric vehicles in smart grid: A survey on charging load modeling. *IET Smart Grid*, *2*(1), 25–33. doi:10.1049/iet-stg.2018.0053

YoCharge. (n.d.). *The Different Types of Electric Vehicles.* https://yocharge.com/ev/types/

Yu, D., Adhikari, M. P., Guiral, A., Fung, A. S., Mohammadi, F., & Raahemifar, K. (2019, May). The impact of charging battery electric vehicles on the load profile in the presence of renewable energy. In *2019 IEEE Canadian Conference of Electrical and Computer Engineering (CCECE)* (pp. 1-4). IEEE. 10.1109/CCECE.2019.8861730

ZAPINAMO. (n.d.). *Affordable, Future-proof Rapid Charging Infrastructure For Electric Vehicles From stored [clean and economical] energy.* https://cordis.europa.eu/project/id/790471

Zhang, W., Yang, W., Chen, C., Li, N., Bao, Z., & Luo, M. (2022). Toward Privacy-Preserving Blockchain-Based Electricity Auction for V2G Networks in the Smart Grid. *Security and Communication Networks*, *2022*. doi:10.1155/2022/6911463

Chapter 12
A Comprehensive Review of Electric Vehicles:
A Smart Choice for Environmental Sustainability and Energy Conservation

Sanketh C. Naik
Sahyadri College of Engineering and Management, India

Mohammed Hyder Ali A. H.
Sahyadri College of Engineering and Management, India

Arnel Ralf Lobo
Sahyadri College of Engineering and Management, India

Sona S. S.
Sahyadri College of Engineering and Management, India

Manjunatha Badiger
Sahyadri College of Engineering and Management, India

ABSTRACT

Electric vehicles (EVs) are crucial for long-term resource management, combating climate change, and enhancing energy efficiency. They significantly reduce harmful emissions, addressing challenges like greenhouse gas escalation and fossil fuel depletion. Exploring EVs involves analyzing their ecological footprint, energy efficiency, and addressing challenges like range anxiety and charging infrastructure. Beyond environmental benefits, EVs offer reduced emissions and potential integration with renewables, promoting overall sustainability. Advances in battery technology and EV design address concerns, ensuring continuous evolution. The chapter expands to social and economic implications, including job creation, lowered healthcare costs, and improved air quality. Global policy initiatives and case studies highlight growing global acceptance, emphasizing EVs' transformative potential in reshaping transportation for environmental sustainability and global energy conservation.

DOI: 10.4018/979-8-3693-2611-4.ch012

1. INTRODUCTION

Electric vehicles (EVs) are becoming a very attractive alternative to conventional internal combustion engine vehicles in the light of urgent environmental concerns and the continuous search for sustainable transportation options. Conventional vehicles burn fossil fuels and release pollutants including hydrocarbons, nitrogen oxides, and carbon oxides (Obebe et al., 2021), which are major contributors to environmental problems and are getting worse. This even leads to scarcity of energy resources. The conventional vehicles is having very low energy efficiency. As an intermediate step we have hybrid electric vehicles, which are powered by both combustion engine and also one or more electric motors, which use energy stored in batteries. However, HEVs are only a transitory phase in the transition from internal combustion engines (ICEVs) to pure electric vehicles (Shengxi Bai & Chunhua Liu, 2021), and thus cannot fully resolve the problem. Pure electric vehicles technology therefore emerged since vehicle exhaust emissions are becoming the main cause of air pollution in densely populated urban areas.

The primary cause is the current escalation in the severity of environmental problems. Currently, vehicle exhaust emissions stand out as the primary culprits behind air pollution, particularly in densely populated regions. Air pollution is a serious environmental concern to human health that occurs at dangerously high levels in many regions of the world (Manisalidis et al., 2020). Notably, vehicle emissions rank among the top three contributors to outdoor air pollution. However, the transportation industry is one of the biggest contributors to greenhouse gas (GHG) emissions, which are responsible for the effects of global warming on people's lives and health. GHG emissions are mostly generated by CO2 (90%) (Kelvin O. Yoro & Michael O. Daramola, 2020). Global warming serve as a stark warning for humanity. The intensifying global climate crisis, characterized by surging temperatures, unprecedented extreme weather events, and a surge in air pollution, has underscored the critical need for swift and effective action to address environmental challenges.

In this context, electric vehicles have gained significant traction as a viable means to curtail greenhouse gas emissions, offering a tangible solution to mitigate the environmental repercussions linked to traditional fossil fuel-powered vehicles. Thus, it has evolved into a prevailing trend within the automobile industry (Muratori et al., 2021). Nations worldwide, including economic powerhouses such as India, China, and Korea, have recognized the urgency of this shift and instituted policies to propel the technological evolution towards electric vehicles (Sida Feng & Christopher L. Magee, 2020). The effect of these initiatives is evident in the increasing rate of electric vehicle adoption across the globe (Zhang et al., 2014). This collective effort, materializing through meticulous planning and strategic policies, underscores a shared commitment among nations to expedite the widespread integration of electric vehicles. This coordinated effort is seen as a crucial step toward creating a cleaner, more sustainable transportation system, which is in line with the necessity of tackling the pressing environmental issues at hand.

2. LITERATURE SURVEY

In their study, (Barman et al., 2023) investigate the strategic approach to the energy crisis and environmental concerns of integrating renewable energy (RE) into electric vehicle (EV) charging networks. It emphasizes EVs' 60% energy efficiency compared to traditional vehicles but highlights emissions' reliance on EVs' power source. Four core EV charging elements are explored: global RE sources' advantages and limitations, energy storage tech, charging systems, and smart grid integration for RE incorporation.

Additionally, industry-driven smart charging approaches leveraging RE in line with global EV energy trends are scrutinized. Challenges and opportunities in sustainable RE integration into EV charging infrastructure, including grid integration, standardization, and network security, are also addressed. The study's overall goal was to improve the understanding and adding to the progress of sustainable electric vehicle (EV) charging technology within the discussion.

In the study conducted by (Zhihan Lv & Wenlong Shang, 2023), focuses on exploring environmentally conscious alternatives in response to growing concerns about the environmental impact stemming from the widespread use of fossil fuels in industrial processes and transportation. It focuses on Hybrid Electric Vehicle (HEV) technology, driven by the need for better fuel economy, reduced emissions, and adaptation to escalating fuel costs. HEVs integrate HEVs integrate both internal combustion engines and electric motors powered by rechargeable battery packs, charged via onboard generators and regenerative braking. The key challenge lies in devising an effective Energy Management Strategy (EMS) that fulfills objectives without compromising vehicle performance. The paper revisits and reviews recent EMS developments for HEVs, shedding light on Plug-in HEVs from an EMS perspective. It also explores a new perspective on Through-the-Road (TtR) HEVs employing in-wheel motors, a novel concept requiring fewer hardware components and offering potential cost reduction. Evaluating recent research, the paper envisions the hypothetical realization of TtR HEVs. It thoroughly discusses the concept's pros and cons, compares its performance with conventional HEVs, and outlines future directions for this innovative approach.

The study carried out by (Iqbal et al., 2021), explores the crucial point where smart grid technology and sustainability converge. Emphasizing the significance of load scheduling and demand-side management, the study underscores their pivotal roles in enhancing load factor efficiency and optimizing energy consumption patterns. Distributed Generation (DG) technologies, such as renewables (solar and wind), gas turbines, microturbines, and fuel cells, play a significant role in decentralized grids and provide economic benefits for cogeneration and peak-shaving. To achieve emission-free generating and sustainable development, it is imperative to integrate renewable energy sources, especially solar and wind. The paper articulates India's ambitious vision for its renewable energy sector, as it aspires to reach a significant milestone of 175 gigawatts (GW) in renewable electricity capacity by the year 2022. Smart home technologies, including LED lights, street lights, and water pumps powered by solar photovoltaic systems, contribute to energy efficiency and sustainability. India has already implemented a significant number of smart home lights, street lights, and water pumps. One important component of the smart grid is smart home energy management, which focuses on managing and optimizing household appliances to lower electricity use, increase energy efficiency, and preserve a clean and green environment.

In the scholarly work led by (Mobarak et al., 2020), explores a pioneering concept centered on the integration of low-cost, flexible thin film photovoltaic (PV) cells onto the steel body panels of electric vehicles (EVs). The primary objective of this innovative approach is to address challenges related to solar installation cost and storage issues. The proposed solar EV (SEV) option aims to reduce grid charging needs and extend driving range by harnessing solar power during parking or driving outside. The most optimal option for PV integration is thought to be copper indium gallium selenide (CIGS) cells because of their high efficiency, versatility, and shown ability to integrate onto steel. The paper highlights the challenges of integrating PV cells onto automotive steel panels, optimizing series and parallel PV cell configurations, designing MPPT tracking controllers, and developing efficient power electronic converters. Based on the study, many people interested in electric vehicles (EVs) may discover that affordable mass-produced models, such as standard electric vehicles (SEVs), fit their budget. They can easily manage the additional cost through established car financing options.

According to (Xue et al., 2020), control algorithms and energy management strategies (EMS) have a significant impact on the energy efficiency, overall dependability, and efficiency of hybrid electric vehicles (HEVs). Addressing specific powertrain configurations, the challenge lies in crafting efficient EMS and suitable control algorithms that meet diverse development objectives without compromising vehicle performance. The study introduces a detailed multi-level classification system for HEVs, considering aspects such as hybridization degree, motor placement, powertrain components, and external charging. It systematically summarizes and reviews the principles and research status of EMSs for each HEV type. Moreover, the research conducts a comparative analysis of EMSs and control algorithms, examining their characteristics, applications, real-time capabilities, and historical evolution. It concludes with discussions on potential future research directions and challenges in advancing the energy management systems of HEVs. This review aims to contribute significantly to the development of intelligent, efficient, and advanced EMSs crucial for enhancing future HEV energy management systems.

In the review led by (Rajeev Ranjan Kumar & Kumar Alok, 2020), the primary objective is to offer a thorough and comprehensive understanding of the dynamics surrounding the adoption of electric vehicles (EVs) and the ensuing prospects for sustainability. The review identifies variables in five categories: antecedents, mediators, moderators, consequences, and socio-demographics. It maps out the nomological network of EV adoption, integrating key constructs from antecedents to consequences. The review also explores the impact of EV adoption on the three dimensions of sustainability: economic, environmental, and social. Some of the key findings include the importance of factors such as dealership experience, charging infrastructure resilience, and marketing strategies in influencing EV adoption. The review also emphasizes well-researched subjects including purchase-based incentive programs, total cost of ownership, and the development of charging infrastructure. Mediators and moderators that affect EV adoption are identified, providing a better understanding of the adoption process. The study fills a gap in the literature by analyzing all sustainable consequence variables collectively. The findings have implications for researchers and policymakers, providing insights into the complexities of EV adoption and offering recommendations for manufacturers, policymakers, governments, and academicians.

In the research conducted by (Yang et al., 2020), is dedicated to the meticulous optimization of Hybrid Electric Vehicle/Plug-in Hybrid Electric Vehicle (HEV/PHEV) operation, recognizing their pivotal contribution to transportation electrification for enhanced fuel efficiency. The research highlights the role of Energy Management Strategy (EMS) which plays a critical role in governing energy flow between fuel and electric storage, vital during HEV/PHEV design. As intelligent connected vehicle technology evolves, developing an efficient EMS utilizing Vehicle-to-Infrastructure/Vehicle-to-Vehicle (V2I/V2V) data remains a significant challenge. The research offers a review of diverse EMS approaches for conventional HEV/PHEV and those integrating V2I/V2V information. It provides a comprehensive survey of methodologies used in various EMS designs. Specifically, it delves into EMS for single-vehicle and multi-vehicle scenarios within an intelligent transport system framework. Finally, the study outlines future research challenges, identifying areas for further investigation. As a comprehensive reference, this research aims to assist researchers in the HEV/PHEV field by offering insights into EMS methodologies and outlining potential avenues for future advancements.

In the scholarly work by (Li et al., 2019), conducts an in-depth and comprehensive analysis of the technologies employed in pure electric vehicles (EVs).The paper covers topics such as energy storage, energy generation systems, energy management strategies, and charging technologies. It explains the advantages of pure electric vehicles over hybrid electric vehicles and explains the many energy storage system types found in fully electric cars. The paper also evaluates the environmental impacts of pure electric vehicles and discusses the challenges faced by these vehicles, along with corresponding solutions. The latest developments in pure electric vehicles are also presented. Overall, the paper serves as a helpful guide for anyone interested in learning about the key technologies for pure electric vehicles.

The study conducted by (Krishna Veer Singh et al., 2019) provides an extensive and comprehensive assessment that summarizes the current state of research on hybrid electric vehicles (HEVs). The authors begin by discussing the motivation for developing HEVs, including the need to reduce petroleum consumption and toxic emissions. They then provide an overview of the different architectures of HEVs, including series, parallel, and series-parallel configurations. For each architecture, the authors discuss the advantages and disadvantages, as well as the key components, such as the electric motor, battery, and power electronics. The authors also review the literature on the control strategies for HEVs, including rule-based, optimization-based, and model-based approaches. They discuss the challenges of designing effective control strategies, such as the need to balance conflicting objectives, such as fuel economy and drivability. In addition, the authors review the literature on the design and optimization of HEV components, such as the electric motor and battery. They discuss the trade-offs between performance, cost, and weight, and highlight the importance of considering the entire system when designing individual components. Overall, the literature survey in this paper provides a comprehensive overview of the current state of research on HEVs, highlighting the key challenges and opportunities for future research.

(Khan et al., 2018) research work covers a wide range of subjects that are closely related to electric vehicles (EVs) and the infrastructure that supports them. The focus is on employing renewable energy sources, specifically solar photovoltaic, for power generation in island systems like the Lakshadweep islands. It also mentions initiatives like the International Renewable Energy Agency's SIDS Lighthouses Initiatives and the Global Renewable Energy Islands Network. The survey further highlights the economic and environmental challenges of transitioning from fossil fuel-based systems to renewable energy systems. It looks at various methods of charging, placement strategies, and the effects of electric cars. Additionally, it touches upon topics like charging infrastructure design, control systems, safety considerations, and social acceptance of renewable energy innovation. Overall, the literature survey provides insights into the current state of electric vehicle charging infrastructure, renewable energy integration, and the challenges and opportunities associated with the transition to cleaner transportation systems.

The research project by (Un-Noor et al., 2017) offers a thorough analysis of electric vehicle (EV) technology, including a range of topics including EV configurations, electrical machines, battery energy sources, charging methods, optimization strategies, impacts, trends, and upcoming advancements. The study emphasizes the potential of EVs to significantly impact the environment, power systems, and related sectors, while also acknowledging the challenges associated with EV penetration. The authors review the state-of-the-art technologies related to EVs and their impacts, while also addressing the obstacles that EVs must overcome to replace internal combustion engine vehicles. They conclude by highlighting the potential of EVs to become the future of transport, offering a viable alternative to conventional vehicles dependent on diminishing fossil fuel reserves. In summary, the paper functions as a valuable reference for comprehending the present status and prospective future advancements in electric vehicle (EV) technology, addressing the need for further research in this rapidly evolving sector.

The study conducted by (Wisdom Enang & Chris Bannister, 2017) provides a comprehensive and in-depth analysis of engine emissions control. It highlights the importance of reducing emissions from internal combustion engines due to their impact on air quality and human health. The survey covers the different types of pollutants emitted by engines, including NOx, CO, and HC, and the various methods and technologies that have been developed to reduce engine emissions, such as catalytic converters, EGR, and SCR. The survey also discusses the challenges and limitations of emissions control, including the trade-offs between reducing emissions and maintaining engine performance and efficiency. It emphasizes the need for effective control strategies that can balance emissions reduction with other factors such as fuel economy and engine durability. Finally, the survey underscores the significance of research and development in the realm of engine emissions control, including the growing adoption of alternative fuels and the necessity for more stringent emissions standards.

The study by (Sabri et al., 2016), carries out the pursuit of cleaner solutions amid environmental concerns triggered by fossil fuel usage in industries and transportation. It focuses on Hybrid Electric Vehicle (HEV) technology, driven by the need for better fuel economy, reduced emissions, and adaptation to escalating fuel costs. HEVs integrate electric motors and internal combustion engines powered by rechargeable battery packs, charged via onboard generators and regenerative braking. The key challenge lies in devising an effective Energy Management Strategy (EMS) that fulfills objectives without compromising vehicle performance. The paper revisits and reviews recent EMS developments for HEVs, shedding light on Plug-in HEVs from an EMS perspective. It also explores a new perspective on Through-the-Road (TtR) HEVs employing in-wheel motors, a novel concept requiring fewer hardware components and offering potential cost reduction. Evaluating recent research, the paper envisions the hypothetical realization of TtR HEVs. It thoroughly discusses the concept's pros and cons, compares its performance with conventional HEVs, and outlines future directions for this innovative approach.

The survey conducted by (Jaguemont et al., 2016), provides a comprehensive review of the effects of cold temperatures on lithium-ion batteries used in hybrid and electric vehicles. It discusses the challenges faced by these batteries in cold weather, including performance loss and degradation, and proposes various thermal management strategies to address these issues. The paper also covers different types of battery models, such as electrochemical, empirical, and electrical models, and how they can be used to understand internal processes and degradation effects. Additionally, the paper explores the impact of low temperatures on aging mechanisms and proposes methods for addressing these effects. Overall, this paper offers valuable insights into the challenges and potential solutions for using lithium-ion batteries in cold weather conditions

In the paper by (Andreas Poullikkas, 2015), a comprehensive overview of electric vehicle (EV) technologies and associated charging methods is provided. This encompasses considerations such as fuel economy, CO2 emissions, vehicle-to-grid and grid-to-vehicle architectures, as well as a detailed exploration of the technical characteristics of different types of EVs. It goes into detail about the distinctions between plug-in electric vehicles, hybrid electric vehicles, and fully electric vehicles, emphasising the latter two's capacity for external battery recharging. The authors stress the importance of taking into account all system impacts when highlighting the need for standards and regulations to facilitate the shift to a market that includes EVs. A vast array of studies and research articles about EV technologies, charging infrastructure, battery design, driving habits, and the economic and environmental analysis of EVs are also cited in the work.

The study carried out by (Hannan et al., 2014) delves further into a variety of alternative energy sources that are seen to be feasible alternatives to petroleum in automobiles. The following renewable energy sources are being researched for potential application in Hybrid Electric Vehicles (HEVs) to meet future transportation needs: solar cells, batteries, fuel cells, and supercapacitors. The study comprehensively explores various facets of HEV technology, including Energy Management Systems (EMS), power conditioning, propulsion systems, DC machines, and vehicle systems, encompassing models, algorithms, simulations, and experimental derivations.It synthesizes the performance outcomes of different combinations of HEV systems, encapsulating them in a comprehensive table with corresponding references. This survey paper offers an in-depth overview of HEVs, their energy source combinations, model variations, and EMS development by diverse researchers. Despite current technologies demonstrating satisfactory HEV performance, the paper highlights shortcomings in reliability and intelligent system advancements. Through this extensive review, the study unveils multifaceted factors, challenges, and issues concerning sustainable next-generation hybrid vehicles.

The study conducted by (Aishwarya Panday & Hari Om Bansal, 2014) explores the significance of optimal power distribution between alternative energy sources and Internal Combustion Engines (ICE) in Hybrid Electric Vehicles (HEVs), emphasizing their superior fuel efficiency over conventional vehicles. Energy Management Strategies (EMS) play a pivotal role in determining the power allocation between the engine and motor, enhancing fuel economy and optimizing HEV performance. The study comprehensively reviews various EMS detailed in existing literature, particularly focusing on battery-powered hybrid vehicles and extending insights into Plug-in Hybrid Electric Vehicles (PHEVs) . It highlights extensive research efforts dedicated to energy optimization, consolidating numerous methods and control strategies proposed in literature. The paper offers a structured overview of these strategies within a coherent framework, aiming to be a valuable reference for researchers delving into hybrid vehicle energy optimization, providing insights and facilitating further advancements in this field.

The thorough study by (Bayindir et al., 2011) examines the field of Hybrid Electric Vehicles (HEVs), offering a detailed analysis of their technology, design, and functionality. It explores the integration of multiple power sources, typically Internal Combustion Engines (ICE) and electric motors, focusing on their combined efficiency and reduced environmental impact compared to conventional vehicles. The study encompasses various aspects, including energy management strategies, power distribution, and control systems within HEVs. It discusses the advancements in battery technology and the incorporation of regenerative braking systems to enhance energy efficiency. Furthermore, the paper evaluates the performance metrics, advantages, and limitations of HEVs, serving as a holistic reference for researchers and enthusiasts seeking a comprehensive understanding of these eco-friendly vehicles.

(Shen et al., 2011) conducted a comprehensive study that explores Hybrid Electric Vehicles (HEVs) as a viable alternative to traditional fuel-based energy sources, driven by considerations of resource scarcity, environmental consciousness, and evolving emission regulations. It highlights HEV's internal composition, comprising internal combustion engines, electric machines, and power electronic equipment. The overview focuses on HEV configurations, energy management strategies, and electronic control units, detailing the advantages and drawbacks of each configuration. The paper categorizes and thoroughly describes existing powertrain control techniques tailored for HEVs while elaborating on the electronic control units utilized within HEV setups. Additionally, it discusses the latest trends and forthcoming technological challenges expected to impact HEVs in the near future, providing insights into the evolving landscape of HEV technology and the hurdles it faces.

3. CLASSIFICATION OF ELECTRIC VEHICLES

There are different kinds of electric vehicles (EVs), identified by their system architectures, each offering unique advantages in the realm of sustainable transportation. The first type is Battery Electric

Vehicles (BEVs), which exclusively use electric batteries for power and do not have traditional internal combustion engines (JongRoul Woo & Christopher L. Magee, 2020). Another category is Plug-in Hybrid Electric Vehicles (PHEVs), which combine an electric motor with an internal combustion engine. PHEVs can operate on electric power alone for shorter trips and switch to the combustion engine for longer journeys, providing flexibility and reduced reliance on fossil fuels (Himanshu Singh et al., 2021). Fuel Cell Electric Vehicles (FCEVs) constitute the third type, utilizing hydrogen fuel cells to generate electricity (Anchasa Pramuanjaroenkij & Sadık Kakaç, 2023). Lastly, Hybrid Electric Vehicles (HEVs) integrate a regular engine with an electric motor, working together to improve fuel efficiency and lower emissions (Nirendra Nath Mustafi, 2022). These diverse electric vehicles cater to specific needs and help move us towards more sustainable and environmental friendly ways of getting from one place to another.

3.1. Battery Electric Vehicle (BEV)

Battery Electric Vehicles (BEVs) mark a revolutionary shift in the automotive landscape, representing a clean and sustainable alternative to traditional internal combustion engine vehicles. These vehicles only use electricity to operate. These are more efficient than HEVs. An alternative name for them is All-Electric Vehicles (AEV). BEV-equipped electric vehicles are completely battery-powered and require no external power source. The vehicle's power is stored in a substantial battery pack, which can be charged by connecting to the electrical grid. One or more electric motors are then powered by the charged battery pack, enabling the electric automobile to be driven by them. With advancements in battery technology, BEVs boast impressive ranges, making them increasingly practical for daily use (Lebrouhi et al., 2021). As nations globally intensify efforts to reduce greenhouse gas emissions, Battery Electric Vehicles stand out as a cornerstone in the quest for a greener and more sustainable future in the realm of transportation. Electric motor, inverter, battery, control module, and drive train are the major parts of a battery-electric vehicle. Examples for BEV include TATA Nexon, TATA Tigor. Figure 1 illustrates the comprehensive architecture of a Battery-Electric Vehicle (BEV).

Figure 1. Architecture of BEV

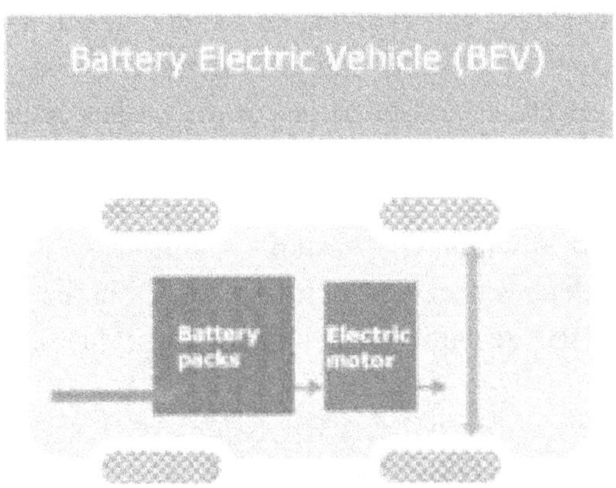

3.2. Hybrid Electric Vehicle (HEV)

Series hybrid and parallel hybrid are other names for HEVs. Engine and electric motor are both features of HEVs. Unlike traditional vehicles relying solely on fossil fuels, HEVs leverage a combination of an internal combustion engine and one or more electric motors powered by an onboard battery. This dynamic integration enables HEVs to optimize fuel efficiency, reduce emissions, and enhance overall performance (Wisdom Enang & Chris Bannister, 2017). The concurrent rotation of the transmission by both the engine and electric motors propels the wheels into motion. The engine, electric motor, battery pack with controller and inverter, gasoline tank, and control module are the essential parts of a HEVs. The engine of a HEV receives energy from the fuel tank just like in a conventional car. The electric motor powers the batteries. The transmission can be turned simultaneously by the engine and the electric motor. This dual-power approach allows for improved fuel economy and reduced environmental impact compared to traditional counterparts. Examples of HEVs include Toyota Prius and the Honda Insight. Figure 2 depicts the intricacies of the Hybrid Electric Vehicle (HEV) architecture.

Figure 2. Architecture of HEV

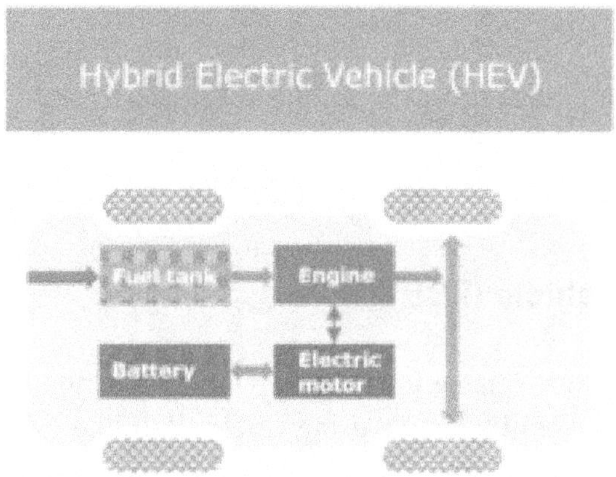

3.3. Plug-in Hybrid Electric Vehicle (PHEV)

PHEVs are also known as series hybrids. Combining elements of both traditional internal combustion engines and electric propulsion, PHEVs offer a dual power source through a rechargeable battery and an internal combustion engine. There is an option for us to select between traditional fuels like gasoline and alternative fuels like biodiesel. Rechargeable batteries can also serve as a power source for it. It's possible to charge the battery externally. Operating in two distinct modes, PHEVs showcase their adaptability: the All-electric mode utilizes the motor and battery for energy, while the Hybrid mode seamlessly integrates both electricity and petrol/diesel for energy generation. Battery, inverter, control module, gasoline tank, engine, electric motor, and battery charger are the key parts of a plug-in hybrid electric vehicle. PHEVs operate on the basis of using energy only until their battery pack runs out. They switch to full electric mode and start working. When the battery runs out and the engine takes over, the car runs like a conventional, plug-in hybrid. Charging options for PHEVs are diverse, ranging from external electric power sources and the engine to regenerative braking (Himanshu Singh et al., 2021). Notably, the electric motor supplements the engine's power, allowing for the use of smaller, more efficient engines without compromising performance. Examples for PHEVs include BMW i8, Hyundai Sonata. Figure 3 elucidates the architectural intricacies of the Plug-in Hybrid Electric Vehicle (PHEV).

Figure 3. Architecture of PHEV

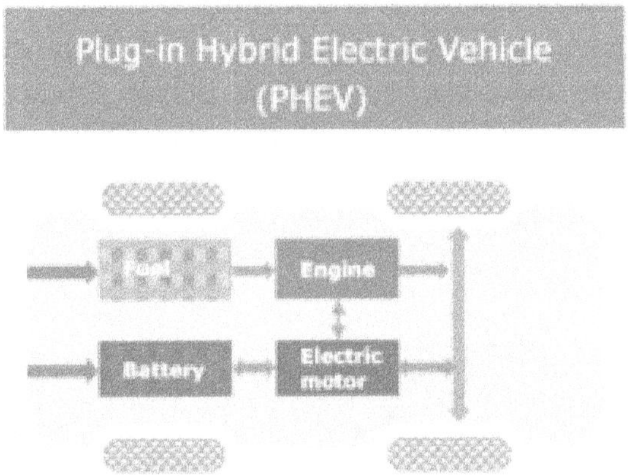

3.4. Fuel Cell Electric Vehicle (FCEV)

Zero-Emission vehicles are another name for FCEVs. Unlike automobiles with conventional internal combustion engines that run on fossil fuels. By using "fuel cell technology," which produces only water vapor as a byproduct of a chemical reaction between hydrogen and oxygen, they are able to produce the electricity needed to power the car. Electric energy directly generated from the fuel's chemical energy. Because of their efficiency and emission-free operation, the adoption of FCEVs is a viable way to fight air pollution and cut greenhouse gas emissions (Usman Asif & Klaus Schmidt, 2021). The essential parts of FCEVs are a fuel cell stack, electric motor, hydrogen storage tank, battery with converter, and controller. The way FCEVs operate is that the car itself produces the power needed to drive it. While challenges such as infrastructure development and hydrogen production still need addressing, the advent of Fuel Cell Electric Vehicles marks a significant stride toward a cleaner, more sustainable future in the realm of automotive technology. Examples for FCEVs include Toyota Mirai, Hyundai Nexo. Figure 4 below illustrates the intricate architecture of a Fuel Cell Electric Vehicle (FCEV).

Figure 4. Architecture of FCEV

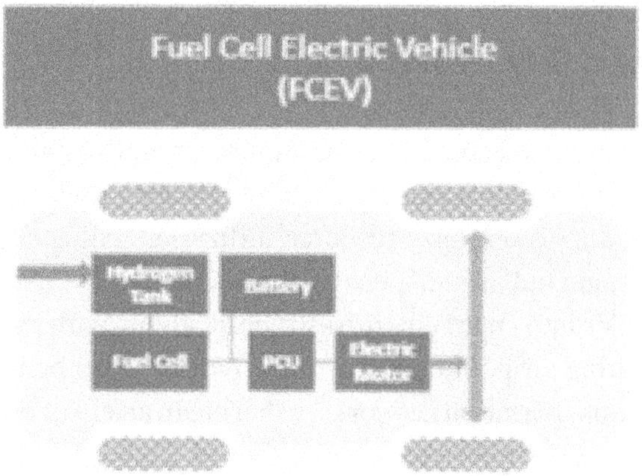

4. HARNESSING RENEWABLE ENERGY FOR ELECTRIC VEHICLES

The utilization of renewable energy sources in Electric Vehicles (EVs) represents a pivotal step towards fostering sustainable and environmentally friendly transportation. EVs can drastically lower their carbon footprint and aid in the global fight to address climate change by utilizing energy from renewable sources like solar, wind, and hydropower. Integrating renewable energy into EV charging infrastructure not only diminishes the dependence on fossil fuels but also enhances the overall eco-friendliness of electric transportation. This synergy between renewable energy and EVs exemplifies a progressive approach to transportation, aligning with the broader goal of creating a cleaner, greener future (Wei-Jen Lee & Tanya Shiramagond, 2018). As technology continues to advance, the seamless integration of renewable energy in EVs holds promise for a more sustainable and resilient transportation system, ultimately benefitting both the environment and society at large.

Electric vehicles equipped with solar panels integrated into their bodies or roofs harness solar energy to supplement their power needs. The solar panels capture sunlight and convert it into electricity, directly charging the vehicle's battery (Mobarak et al., 2020). The solar panels play a crucial role in extending the car's range. Mitigating the energy demand from the battery for auxiliary functions, such as air conditioners and interior electronics, through the integration of solar panels represents a notable advancement in the energy efficiency of electric vehicles. This enhances the vehicle's range and signifies a noteworthy advancement in the evolution of environmentally conscious and sustainable transportation alternatives. The incorporation of solar technology not only reduces the reliance on the primary power source but also underscores the potential for innovative energy solutions in the automotive sector. This dual benefit of increased energy efficiency and a reduced environmental footprint underscores the pivotal role that solar panels play in shaping the future landscape of eco-conscious and sustainable transportation.

The electricity generated by wind turbines makes a significant contribution to the broader electrical grid. It offers a renewable and environmentally friendly source that enhances the grid's sustainability. The energy contained in batteries for charging electric vehicles originates from a variety of grid-fed sources, such as wind power. In essence, the utilization of wind energy represents a pivotal step toward fostering a more resilient, eco-friendly, and interconnected framework for the electrification of transportation.

4.1 Role of EVs in Smart City Development

Electric Vehicles (EVs) play a crucial role in the development and sustainability of smart cities by actively contributing to the enhancement of cleaner and more efficient urban transportation systems (Casini et al., 2017). As smart cities aim to optimize resource utilization, enhance connectivity, and minimize environmental impact, EVs stand out as indispensable elements crucial to realizing these smart city objectives. The integration of EVs into smart city infrastructure aligns with the overarching goal curtailing carbon emissions and alleviating air pollution. Beyond environmental benefits, EVs can be seamlessly incorporated into intelligent transportation networks, ushering in an era of enhanced traffic management, diminished congestion, and overall improved urban mobility. The use of EVs in smart cities promotes energy efficiency, leveraging advancements in technology to optimize charging infrastructure and grid management (Pandiyan et al., 2023). By strategically implementing Electric Vehicles (EVs), smart cities can cultivate an urban environment that is sustainable and resilient, fostering a future of urban mobility that is both cleaner and more intelligent.

4.2 Policies to Improve Adoption of EVs

Countries around the world are implementing a range of policies to accelerate the adoption of Electric Vehicles (EVs) as part of their broader commitment to sustainable and low-carbon transportation. To promote wider accessibility of electric vehicles (EVs) to the general public, many countries have introduced financial incentives, such as tax credits, rebates, and subsidies (Zhang et al., 2014). Additionally, some governments are investing in the development of robust charging infrastructure networks to alleviate range anxiety and facilitate widespread EV usage. Stringent emission standards and regulations are being imposed to encourage automakers to produce more electric and hybrid vehicles. Even more aggressive goals are being set by several nations to completely phase out internal combustion engine cars (Morfeldt et al., 2021). Furthermore, initiatives to promote research and development in battery technology and electric drivetrains are aimed at overcoming technological barriers and improving the overall performance of EVs. These multifaceted policies reflect a global commitment to fostering a sustainable and electrified future for transportation, encouraging the widespread adoption of Electric Vehicles.

5. CASE STUDIES AND RECENT ADVANCEMENTS IN EV'S

5.1 Case Studies on EV's for Sustainable Energy Conservation

Numerous businesses and organizations have successfully integrated electric vehicles (EVs) to champion environmental sustainability and conserve energy. FedEx, a prominent global logistics company, has been a leader in environmental sustainability (David M. Herold & Ki-Hoon Lee, 2018). They strategically deployed electric delivery vans in urban areas, resulting in a significant reduction in greenhouse gas emissions and fuel costs. FedEx remains steadfast in expanding its electric vehicle fleet, a commitment that aligns with the company's environmental goals and serves as a testament to the tangible benefits achievable through the integration of electric vehicles in logistics operations. Similarly, United Parcel Service (UPS), a global powerhouse in package delivery and supply chain management, has proactively responded to the imperative shift towards sustainability. Recognizing the significance of addressing environmental concerns, UPS has tactically integrated electric vehicles into its extensive fleet, exemplifying a clear commitment to reducing its carbon footprint and embracing eco-friendly practices. Google, committed to carbon neutrality, has invested in electric vehicle (EV) infrastructure. As a global tech giant, Google acknowledges the pivotal role it plays in minimizing its environmental impact. IKEA, a global furniture retailer, has embraced last-mile sustainability with electric vans and bikes, aligning with its commitment to environmental responsibility (Lisa Johansson & Matilda Holmqvist, 2021). Beyond the environmental impact, these initiatives have garnered positive feedback from environmentally conscious customers, underscoring the importance of sustainable practices in the retail sector. These cases highlight the positive impact of integrating EVs into various operational facets, emphasizing the potential for businesses to achieve both environmental and economic benefits through sustainable transportation practices.

5.2 Recent Advancements in EVs

Tackling the challenges associated with electric vehicles (EVs), especially concerns like range anxiety and charging infrastructure, requires innovative solutions to drive the widespread adoption of sustainable transportation. To mitigate range anxiety, advancements in battery technology play a crucial role (Rainieri et al., 2023). With ongoing research focused on delivering batteries that boast extended ranges and faster charging capabilities. This not only mitigates concerns about the limitations of EV travel but instills confidence in users regarding the reliability and convenience of their electric vehicles. Complementing this, the rise of intelligent charging infrastructure emerges as a pivotal solution. Enabling fast DC charging stands out as a practical solution to reduce recharging time and extend the range for intercity travel along highways (Xie et al., 2018). Recognizing the diverse energy and recharge requirements associated with various driving styles, planners of electric vehicle (EV) infrastructure should take these factors into careful consideration. A strategic and dynamic approach to building EV recharging infrastructure becomes crucial in mitigating the prevalent issue of range anxiety. Simultaneously, wireless charging technologies are stepping into the spotlight (Katharina Csillak & Mariana Moreno Kuhnke, 2023), aiming to streamline the recharging process and enhance user convenience. The integration of these technological advancements not only addresses the practical aspects of EV usage but also contributes to a more seamless and user-friendly electric vehicle experience, fostering a transition towards sustainable and eco-friendly mobility solutions.

6. CONCLUSION

In conclusion, the comprehensive review of electric vehicles (EVs) presented herein establishes them as a smart and strategic choice for advancing both environmental sustainability and energy conservation in the realm of transportation. Electric vehicles (EVs) have emerged as a forefront solution, offering a range of benefits from reduced greenhouse gas emissions to enhanced energy efficiency. The evolution from conventional internal combustion engine vehicles to HEVs and, ultimately, to pure EVs is a testament to the collective commitment to address pressing environmental concerns. The four different kinds of electric cars offer a range of choices to suit different requirements and tastes. BEVs, powered solely by electricity, stand out as a clean and sustainable alternative. Renewable energy integration in EVs, particularly through solar panels and wind energy, further enhances the eco-friendliness of electric transportation (Irfan et al., 2023). Using these sustainable energy sources not only helps the world's efforts to battle climate change, but also reduces dependence on fossil fuels. The confluence of financial incentives, strategic investment in charging infrastructure, rigorous emission standards, and ambitious targets to phase out traditional vehicles serves as a resounding global declaration of commitment toward cultivating a sustainable and electrified future for transportation (Yoomin Lee & Muneeza Mehmood Alam, 2021). The synchronized implementation of these initiatives signifies a shared vision among nations to accelerate the electrification of transport and embrace a future where sustainable mobility is not just an aspiration but a global reality. In essence, the journey from conventional vehicles to electric mobility is a dynamic and promising evolution. As technology advances and global collaboration intensifies, electric vehicles are positioned as a pivotal element in creating a cleaner, greener transportation landscape, ultimately benefiting both the environment and society at large.

REFERENCES

Alam, M. M., & Lee, Y. (2021). *Cleaner Vehicles and Charging Infrastructure: Greening Passenger Fleets for Sustainable Mobility*. Academic Press.

Asif, U., & Schmidt, K. (2021). Fuel cell electric vehicles (Fcev): Policy advances to enhance commercial success. *Sustainability (Basel)*, *13*(9), 5149. doi:10.3390/su13095149

Bai, S., & Liu, C. (2021). Overview of energy harvesting and emission reduction technologies in hybrid electric vehicles. *Renewable & Sustainable Energy Reviews*, *147*, 111188. doi:10.1016/j.rser.2021.111188

Barman, P., Dutta, L., Bordoloi, S., Kalita, A., Buragohain, P., Bharali, S., & Azzopardi, B. (2023). Renewable energy integration with electric vehicle technology: A review of the existing smart charging approaches. *Renewable & Sustainable Energy Reviews*, *183*, 113518. doi:10.1016/j.rser.2023.113518

Bayindir, K. Ç., Gözüküçük, M. A., & Teke, A. (2011). A comprehensive overview of hybrid electric vehicle: Powertrain configurations, powertrain control techniques and electronic control units. *Energy Conversion and Management*, *52*(2), 1305–1313. doi:10.1016/j.enconman.2010.09.028

Casini, M. (2017, August). Green technology for smart cities. *IOP Conference Series. Earth and Environmental Science*, *83*(1), 012014. doi:10.1088/1755-1315/83/1/012014

Csillak, K., & Kuhnke, M. M. (2023). Interoperability Framework for Electromobility (INFRA): The main results from the USER-CHI-framework in a new spotlight. *Transportation Research Procedia*, *72*, 1926–1933. doi:10.1016/j.trpro.2023.11.672

Enang, W., & Bannister, C. (2017). Modelling and control of hybrid electric vehicles (A comprehensive review). *Renewable & Sustainable Energy Reviews*, *74*, 1210–1239. doi:10.1016/j.rser.2017.01.075

Feng, S., & Magee, C. L. (2020). Technological development of key domains in electric vehicles: Improvement rates, technology trajectories and key assignees. *Applied Energy*, *260*, 114264. doi:10.1016/j.apenergy.2019.114264

Hannan, M. A., Azidin, F. A., & Mohamed, A. (2014). Hybrid electric vehicles and their challenges: A review. *Renewable & Sustainable Energy Reviews*, *29*, 135–150. doi:10.1016/j.rser.2013.08.097

Herold, D. M., & Lee, K. H. (2018). Carbon disclosure strategies in the global logistics industry: Similarities and differences in carbon measurement and reporting. *Pathways to a Sustainable Economy: Bridging the Gap between Paris Climate Change Commitments and Net Zero Emissions*, 87-101.

Holmqvist, M., & Johansson, L. (2021). *A self-service design concept for last mile delivery: The solution for retail stores in urban environments*. Academic Press.

Iqbal, S., Sarfraz, M., Ayyub, M., Tariq, M., Chakrabortty, R. K., Ryan, M. J., & Alamri, B. (2021). A comprehensive review on residential demand side management strategies in smart grid environment. *Sustainability (Basel)*, *13*(13), 7170. doi:10.3390/su13137170

Irfan, M., Deilami, S., Huang, S., & Veettil, B. P. (2023). Rooftop Solar and Electric Vehicle Integration for Smart, Sustainable Homes: A Comprehensive Review. *Energies*, *16*(21), 7248. doi:10.3390/en16217248

Jaguemont, J., Boulon, L., & Dubé, Y. (2016). A comprehensive review of lithium-ion batteries used in hybrid and electric vehicles at cold temperatures. *Applied Energy*, *164*, 99–114. doi:10.1016/j.apenergy.2015.11.034

Khan, S., Ahmad, A., Ahmad, F., Shafaati Shemami, M., Saad Alam, M., & Khateeb, S. (2018). A comprehensive review on solar powered electric vehicle charging system. *Smart Science*, *6*(1), 54–79. doi:10.1080/23080477.2017.1419054

Kumar, R. R., & Alok, K. (2020). Adoption of electric vehicle: A literature review and prospects for sustainability. *Journal of Cleaner Production*, *253*, 119911. doi:10.1016/j.jclepro.2019.119911

Lebrouhi, B. E., Khattari, Y., Lamrani, B., Maaroufi, M., Zeraouli, Y., & Kousksou, T. (2021). Key challenges for a large-scale development of battery electric vehicles: A comprehensive review. *Journal of Energy Storage*, *44*, 103273. doi:10.1016/j.est.2021.103273

Li, Z., Khajepour, A., & Song, J. (2019). A comprehensive review of the key technologies for pure electric vehicles. *Energy*, *182*, 824–839. doi:10.1016/j.energy.2019.06.077

Lv, Z., & Shang, W. (2023). Impacts of intelligent transportation systems on energy conservation and emission reduction of transport systems: A comprehensive review. *Green Technologies and Sustainability*, *1*(1), 100002. doi:10.1016/j.grets.2022.100002

Manisalidis, I., Stavropoulou, E., Stavropoulos, A., & Bezirtzoglou, E. (2020). Environmental and health impacts of air pollution: A review. *Frontiers in Public Health*, *8*, 14. doi:10.3389/fpubh.2020.00014 PMID:32154200

Mobarak, M. H., Kleiman, R. N., & Bauman, J. (2020). Solar-charged electric vehicles: A comprehensive analysis of grid, driver, and environmental benefits. *IEEE Transactions on Transportation Electrification*, *7*(2), 579–603. doi:10.1109/TTE.2020.2996363

Morfeldt, J., Kurland, S. D., & Johansson, D. J. (2021). Carbon footprint impacts of banning cars with internal combustion engines. *Transportation Research Part D, Transport and Environment*, *95*, 102807. doi:10.1016/j.trd.2021.102807

Muratori, M., Alexander, M., Arent, D., Bazilian, M., Cazzola, P., Dede, E. M., Farrell, J., Gearhart, C., Greene, D., Jenn, A., Keyser, M., Lipman, T., Narumanchi, S., Pesaran, A., Sioshansi, R., Suomalainen, E., Tal, G., Walkowicz, K., & Ward, J. (2021). The rise of electric vehicles—2020 status and future expectations. *Progress in Energy*, *3*(2), 022002. doi:10.1088/2516-1083/abe0ad

Mustafi, N. N. (2022). An overview of hybrid electric vehicle technology. *Engines and Fuels for Future Transport*, 73-102.

Obebe, S. B., Kolo, A., & Yusuf, P. P. (2021). Automotive emissions: Sources, environmental effects, and control measures. *Int. J. Innov. Sci. Res. Technol.*, *6*(4), 288–293.

Panday, A., & Bansal, H. O. (2014). A review of optimal energy management strategies for hybrid electric vehicle. *International Journal of Vehicular Technology*, *2014*. doi:10.1155/2014/160510

Pandiyan, P., Saravanan, S., Usha, K., Kannadasan, R., Alsharif, M. H., & Kim, M. K. (2023). Technological advancements toward smart energy management in smart cities. *Energy Reports*, *10*, 648–677. doi:10.1016/j.egyr.2023.07.021

Poullikkas, A. (2015). Sustainable options for electric vehicle technologies. *Renewable & Sustainable Energy Reviews*, *41*, 1277–1287. doi:10.1016/j.rser.2014.09.016

Pramuanjaroenkij, A., & Kakaç, S. (2023). The fuel cell electric vehicles: The highlight review. *International Journal of Hydrogen Energy*, *48*(25), 9401–9425. doi:10.1016/j.ijhydene.2022.11.103

Rainieri, G., Buizza, C., & Ghilardi, A. (2023). The psychological, human factors and socio-technical contribution: A systematic review towards range anxiety of battery electric vehicles' drivers. *Transportation Research Part F: Traffic Psychology and Behaviour*, *99*, 52–70. doi:10.1016/j.trf.2023.10.001

Sabri, M., Danapalasingam, K. A., & Rahmat, M. F. (2016). A review on hybrid electric vehicles architecture and energy management strategies. *Renewable & Sustainable Energy Reviews*, *53*, 1433–1442. doi:10.1016/j.rser.2015.09.036

Shen, C., Shan, P., & Gao, T. (2011). A comprehensive overview of hybrid electric vehicles. *International Journal of Vehicular Technology*, *2011*, 2011. doi:10.1155/2011/571683

Shiramagond, T., & Lee, W. J. (2018, September). Integration of renewable energy into electric vehicle charging infrastructure. In *2018 IEEE International Smart Cities Conference (ISC2)* (pp. 1-7). IEEE. 10.1109/ISC2.2018.8656981

Singh, H., Ambikapathy, A., Logavani, K., Arun Prasad, G., & Thangavel, S. (2021). Plug-in hybrid electric vehicles (PHEVs). *Electric Vehicles: Modern Technologies and Trends*, 53-72.

Singh, K. V., Bansal, H. O., & Singh, D. (2019). A comprehensive review on hybrid electric vehicles: Architectures and components. *Journal of Modern Transportation*, *27*(2), 77–107. doi:10.1007/s40534-019-0184-3

Un-Noor, F., Padmanaban, S., Mihet-Popa, L., Mollah, M. N., & Hossain, E. (2017). A comprehensive study of key electric vehicle (EV) components, technologies, challenges, impacts, and future direction of development. *Energies*, *10*(8), 1217. doi:10.3390/en10081217

Woo, J., & Magee, C. L. (2020). Forecasting the value of battery electric vehicles compared to internal combustion engine vehicles: The influence of driving range and battery technology. *International Journal of Energy Research*, *44*(8), 6483–6501. doi:10.1002/er.5382

Xie, F., Liu, C., Li, S., Lin, Z., & Huang, Y. (2018). Long-term strategic planning of inter-city fast charging infrastructure for battery electric vehicles. *Transportation Research Part E, Logistics and Transportation Review*, *109*, 261–276. doi:10.1016/j.tre.2017.11.014

Xue, Q., Zhang, X., Teng, T., Zhang, J., Feng, Z., & Lv, Q. (2020). A comprehensive review on classification, energy management strategy, and control algorithm for hybrid electric vehicles. *Energies*, *13*(20), 5355. doi:10.3390/en13205355

Yang, C., Zha, M., Wang, W., Liu, K., & Xiang, C. (2020). Efficient energy management strategy for hybrid electric vehicles/plug-in hybrid electric vehicles: Review and recent advances under intelligent transportation system. *IET Intelligent Transport Systems*, *14*(7), 702–711. doi:10.1049/iet-its.2019.0606

Yoro, K. O., & Daramola, M. O. (2020). CO2 emission sources, greenhouse gases, and the global warming effect. In *Advances in carbon capture* (pp. 3–28). Woodhead Publishing. doi:10.1016/B978-0-12-819657-1.00001-3

Zhang, X., Xie, J., Rao, R., & Liang, Y. (2014). Policy incentives for the adoption of electric vehicles across countries. *Sustainability (Basel)*, *6*(11), 8056–8078. doi:10.3390/su6118056

Chapter 13
A Comprehensive Review on Artificial Intelligence and Blockchain Technologies for Electric Vehicle Charging Ecosystems

Poras Khetarpal
Department of Information Technology, Bharati Vidyapeeth College of Engineering, India

Mahesh Kumar
Department of Information Technology, Bharati Vidyapeeth College of Engineering, India

Kumar Lakshay
Maharaja Agrasen Institute of Technology, India

Neelu Nagpal
https://orcid.org/0000-0003-1923-2760
Maharaja Agrasen Institute of Technology, India

Pierluigi Siano
Department of Management and Innovation Systems, University of Salerno, Italy

ABSTRACT

Many attempts have been taken to offer the consumer/user an exceptional service quality (SQ) due to this opportunity which is so profitable and expanding for both the researchers and financial investors. Some significant issues still need to be resolved given the current procedures and establishment of charging infrastructure for EV. In this chapter, two major challenges that EV development has to face are discussed: (1) the charging infrastructure for EVs is quite weak and (2) charging schedules for EVs are not structured or random. Considering the above problems, the integration of EV charging infrastructure with blockchain and AI is reviewed in this chapter. The current blockchain and AI charging solutions for EV charging are discussed in this chapter. In this chapter, a combination of both the technologies are reviewed for developing EV charging ecosystem that is secure, effective, and decentralised. Stakeholders and policymakers can use this chapter as a starting point to determine prospective paths and managing charging infrastructure for EVs.

DOI: 10.4018/979-8-3693-2611-4.ch013

Nomenclature

AI Artificial Intelligence
CMS Central Management System
CS Charging Stations
EV Electric Vehicle
EVCS Electric Vehicles Charging Stations
IoT Internet of Things
ITS Intelligent Transportation System
SQ Service Quality
SG Smart Grid
IEA International Energy Agency
TNE 1 Taiwan National Expressway 1
OBU On-Board Unit
GPS Global Positioning Systems
GCS Global Charging System
DC Direct Current
RCE Remote Code Execution
DoS Denial-of-Service
EVSE Electric Vehicles Supply Equipment
LAN Local Area Network
WAN Wide Area Network
OCPP Open Charge Point Protocol
STLF Short Term Load Forecasting
MTLF Medium Term Load Forecasting
MILP Mixed Integer Linear Programming
ML Machine Learning
LTLF Long Term Load Forecasting
WNN Wavelet Neural Network
PKI Public Key Infrastructure
P2P Peer-To-Peer
PoW Proof of Work
PoS Proof of Stake
ZKP Zero-Knowledge Proofs
MAS Multi-Agent System
MFA Multi-Factor Authentication
MIP Mixed integer programming
TLS Transport Layer Security
SSL Secure Sockets Layer

1. INTRODUCTION

IoT enables the integration of sensors, actuators, and internet connectivity into everyday objects, transforming them into intelligent entities. This integration fosters a diverse range of unique applications and services, such as smart cities, development of smart transportation, smart healthcare services, industrial automation, precision agriculture, and many more (Chavhan et al., 2020). Given the substantial impact of IoT technology on the development of smart applications and services, it is anticipated that this market will generate approximately US$6.2 trillion in revenue annually by the year 2025 (Fuqaha et al., 2015). ITS and smart grids are two crucial and intriguing IoT-based services. An ITS includes a spectrum of services such as traffic monitoring, street surveillance, and self-driving cars, and many more. These services can be coupled to form an intelligent infrastructure of vehicles incorporating elements such as intelligent road-side units and traffic signals to give customers a smooth driving experience (Xia et al., 2016). On the other hand, smart meters and controllers are used in a smart grid (SG) to observe and regulate energy consumption (Relan et al., 2023). While both ITS and smart grid system are specialised services, different new opportunities result from their intersection, one such opportunity is the EV industry and the required infrastructure supporting its charging (Sun et al., 2023). Significant developments in automotive technology have made EVs an emerging market for business as well as research (Chakrapani et al., 2023). The International Energy Agency (IEA) estimates that there were more than 5.1 million EVs sold worldwide in 2018. Additionally, it is predicted that by 2030, there will be up to 250 million EVs on road, which will reduce the need for oil-based products by nearly 130 tonnes (Bunsen et al., 2019). The requirement for acquiring a charging infrastructure that can meet the demands of EV drivers is essential due to the rapid penetration rate of EVs. As a result, numerous businesses, including Siemens, Shell, and others, have decided to start installing EV Charging Stations (EVCS). The market is being led by young companies like ChargePoint and Enel. Providing EV users with a hassle-free and comfortable driving experience is what all these businesses have in common. The number of EV users in the market will increase if EVCS localization is done, and easy payment options like bespoke payments and membership cards are available. The infrastructure for EV is facing two main issues which includes the 1) timing/scheduling of EV charging 2) security and privacy issues. The scheduling of EV charging creates a dilemma. An excess of available EV Charging Stations (EVCS) used simultaneously can lead to disruptions on the power grid, while an insufficient number may inconvenience EV drivers. Enhancing the power of charging stations and incorporating multiple charging ports in certain EV models contribute to faster charging times. Optimization strategies, such as aligning charging periods with lower electricity demand, especially during overnight charging, prove effective. Moreover, advancements in battery technology, particularly higher energy densities, significantly accelerate charging processes (Mastoi et al., 2023). Despite these benefits, challenges may arise, including the necessity for substantial infrastructure upgrades to accommodate higher-power charging stations and potential constraints in battery technology development (Danese et al., 2022). Striking a balance among these factors is crucial for achieving optimal and efficient charging solutions for EVs.

The critical aspect of time and scheduling considerations for electric vehicles (EVs) traveling on Taiwan National Expressway 1 (TNE1) has been examined in a study (Yang et al., 2013). The primary goal is to reduce the waiting duration for vehicles at EVCSs during long-distance trips, recognizing the importance of efficient charging for energy-efficient transport and reduced carbon emissions. The study has introduced an Electric Vehicle Charging Station (EVCS)-selection problem, emphasizing the selection of stations with shorter waiting times, especially when fast-charging poles at charging stations

(CSs) are occupied. To evaluate the effectiveness of CS-selection algorithms in minimizing waiting times, local information (OBU data) and GPS information (real-time queue lengths maintained by a server) have been combined. The study has compared algorithm performance concerning waiting times at CSs, emphasizing the essential role of global information for making informed decisions based on real-time charging infrastructure availability. Additionally, EV charging behavior has been examined through a coordinated charging model aimed at minimizing total charging time (An et al., 2023). The introduction of an earliest finish charging scheduling algorithm and two practical swarm-optimization-based algorithms has addressed the combinatorial nature of the problem, providing practical solutions for large numbers of EVs. The study has been exemplified through a real-life case study, showcasing the effectiveness of the proposed approaches in optimizing EV charging schedules. While these reported case studies have contributed to improving the scheduling and timing of EV charging, addressing limitations such as real-time data accuracy and communication latency has been crucial for broader algorithm applicability.

Further, the random selection of EVCSs can significantly disrupt the overall efficiency of charging systems and pose challenges to the grid. The examination of the factors influencing EV users' preferences of charging stations has been done to understand their charging selection behavior (Hanemann et al., 2017). This analysis is crucial to allocate the charging resources to meet the diverse requirements of EVs in optimum manner. Strategic resource allocation is critical for enhancing charging service efficiency and addressing concerns related to mileage anxiety. Consequently, researchers are increasingly exploring topics such as optimal charging resource placement, time-of-use pricing mechanisms, and the impact of charging on the power grid (El-Bayeh et al., 2021). While extensive work has been conducted on the coordinated charging of EVs, primary factors like predicting demand during charging, assessing criteria for charging resources, monitoring electricity prices, and evaluating the status of the grid have been considered (Ubaid, 2022). While these investigations have played a role in lowering expenses for EV users, alleviating the challenges faced by charging station operators, and mitigating risks to the power grid, there remains a necessity for additional progress. Furthermore, given the evolution of the e-mobility industry, relying exclusively on historical charging data for enhancements in EV performance, as well as adjustments in travel habits, may lead to significant delays in optimizing charging resource stations. Recognizing the potential intelligence and interconnectivity of EVs, the utilization of live traffic information can be employed to optimize the charging of EVs. This involves obtaining real-time state information, including the positions, travel speeds, charging requirements, the arrangement of charging stations, and the conditions of electric vehicles at these stations (Li et al., 2018). Further, reducing charging time of EV involves leveraging various strategies including faster-charging technologies, such as DC fast charging, stand out, enabling an 80% charge in a few minutes [Kumar et al., 2023]. EV charging scheduling and timing challenges have been reported with the applications of various optimization algorithms including dynamic and mixed integer linear programming (MILP) including adaptive features (Sassi et al., 2017, Siddique et al., 2023), meta-heuristic algorithms (Zaidi et al., 2023). The objective of these algorithms is to achieve an optimal pairing of EVs and their EVCSs, ensuring the efficient distribution of charging resources to EVs while adhering to specified constraints. MILP, when applied to large-scale problems, becomes computationally expensive and lacks guaranteed global optimality, facing challenges with scalability and handling nonlinearities. Metaheuristics, while versatile, may converge to local optima, are sensitive to parameter tuning, and may struggle with complex constraints. In response, leveraging AI, such as ML and deep learning, offers a potential solution. Integrating AI into metaheuristics enhances performance by addressing convergence issues, adapting parameters dynamically, and learning from experiences (Khetarpal et al. 2023). Hybrid approaches that merge metaheuristics with AI present a potential avenue for more efficient and robust solutions to intricate optimization problems like EV charging scheduling.

From the above-mentioned discussion, it is inferred that the challenges in the EV charging ecosystem are multifold. The rapid penetration of EVs raises concerns about the timing/scheduling of EV charging, creating a delicate balance to avoid disruptions to the power grid while ensuring convenience for EV drivers. Security and privacy issues also loom large, with cyber threats potentially compromising energy theft and customer data. Existing research, such as the study by (Ubaid et al., 2022) on EV charging considerations on Taiwan National Expressway 1, addresses aspects of the scheduling problem, aiming to minimize waiting times at charging stations during long-distance trips. Moreover, the random selection of EV Charging Stations (EVCS) can disrupt overall efficiency, emphasizing the need to analyse factors influencing EV users' choices systematically. Current research trends include optimizing the placement of charging resources, exploring time-of-use pricing mechanisms, and understanding the influence of charging on the power grid. While efforts have been made in coordinated charging of EVs, there's a call for advancements that consider real-time traffic conditions and leverage the intelligence and interconnectivity of EVs (Li et al., 2018). Numerous research projects have concentrated on utilising various AI-based technologies to deliver efficient charging schemes that satisfy the power grid as well as EV drivers (Islam et al., 2023) have been proposed.

The growing integration of sensors in Electric Vehicles (EVs) has led to emerging cybersecurity challenges and privacy concerns, primarily attributed to the sensitive nature of the gathered and stored data. These sensors capture diverse information, encompassing driver and passenger identification, location data, and vehicle performance metrics, posing a potential risk for malicious exploitation [Bharathidasan et al., 2023]. Addressing security and privacy concerns in EV infrastructure becomes imperative due to the involvement of diverse stakeholders and varied interests. Inadequate cybersecurity measures for Electric Vehicle Supply Equipment (EVSE), EVs, or grid operator communication systems pose a significant risk, impacting adoption and industry perceptions (Johnson et al., 2023). Presently, the lack of a comprehensive EVSE cybersecurity approach and limited industry best practices highlight the need for thorough research to establish comprehensive recommendations that ensure the safeguarding of EV charging systems. EVCSs are susceptible to a variety of cyber-threats like remote code execution (RCE) and denial-of-service (DoS). Such flaws could result in the theft of energy and customer data as well as pose a threat to the entire charging infrastructure, all of which would influence the uptake of EVs (Pourmirza and Walker, 2021). So, it becomes imperative to address the paramount concerns of security and privacy. A review report on associated risks and challenges, emphasizing the increased vulnerability to cyberattacks has been conducted (Euceda, 2023). This proposes a research framework to guide academics and practitioners in addressing gaps in the implementation of EVs within transportation networks, with the aim of enhancing cybersecurity measures for the evolving fleet of connected vehicles. Also, a study addressing the critical cybersecurity threat posed by malware to the power grid and electric vehicle (EV) infrastructure has been conducted (Mousavian, 2015). The focus was on the worm propagation in EV to EVSE networks, and a probabilistic model is proposed to assess threat levels. Here, a MILP protection scheme is introduced, strategically isolating infected nodes with the aim of optimizing the response until patches are released for effective cyber-attack prevention. Although a study has been conducted addressing cybersecurity and privacy threats to EVs, evaluating the impact on sustainability and proposing specific security controls (Muhammad et al. 2023), our proposed study delves into the integration of AI and blockchain to enhance security in e-mobility ecosystems as discussed in the following sections. Notably, privacy and security remain persistent challenges in the IoT paradigm, prompting substantial research endeavors to leverage technologies such as blockchain and AI, along with lightweight security methods, to effectively mitigate these vulnerabilities.

It is inferred that the above-mentioned difficulties could be handled with the combination of AI and blockchain technologies together using hashing, consensus processes, and smart contracts. Blockchain is a decentralised, distributed, and secure database of transactions that enables transactions amongst different entities in a trust less, secure and decentralised, environment (Wang et al. 2019). However, AI has excelled at identifying previously unknown patterns and making judgments in accordance with them. AI and blockchain technologies play crucial roles in enhancing the inter-operational efficiency of the EV charging system, smart system's security, and transparency of the EV charging ecosystem to the stakeholders.

AI influences in three different activities such as, 1) Intelligent Charging Infrastructure Management, 2) Intelligent Charging Recommendations, and 3) Grid Load Balancing. It is used to optimize the charging infrastructure by analyzing data on charging patterns, energy demand, and grid conditions. It can dynamically allocate resources, predict charging needs, and manage charging stations effectively to reduce congestion and ensure optimal charging experiences for EV users. AI algorithms can analyze various factors like driving patterns, battery conditions, and user preferences to provide personalized charging recommendations. This can help EV owners make informed decisions about when and where to charge their vehicles, considering factors such as cost, availability, and environmental impact. AI can assist in balancing the load on the power grid by intelligently managing the charging process. It can optimize charging schedules based on grid conditions, time-of-use tariffs, and renewable energy availability. By ensuring a more even distribution of charging load, AI can help prevent grid overloads and reduce electricity costs.

Blockchain enhances the security systems of the EV charging system in the following aspects 1) Payment and Billing Systems, 2) Data Security and Privacy, 3) Authentication and Authorization, and 4) Supply Chain Management. Blockchain technology can provide secure, transparent, and decentralized payment and billing systems for EV charging. By using smart contracts, blockchain can automate transactions, enable peer-to-peer energy trading, and ensure accurate and tamper-proof billing. This eliminates the need for intermediaries, reduces costs, and enhances trust between charging station operators and EV users. Blockchain can enhance data security and privacy in the EV charging ecosystem. It can securely store charging data, identity information, and transaction records, ensuring that sensitive data remains protected from unauthorized access. Blockchain's decentralized nature and cryptographic techniques provide robust security measures. Blockchain can be used to establish a decentralized and tamper-proof system for authenticating and authorizing EV charging sessions. This ensures that only authorized users can access charging stations and prevents fraud or misuse of charging services. Blockchain can improve transparency and traceability in the supply chain for EV charging infrastructure components, such as charging stations and batteries. It can track the origin, quality, and maintenance history of these components, reducing counterfeiting risks and enhancing overall reliability.

Principal contributions of this paper are:

- Comprehensive overview of the present situation and standards for EV charging and highlighting the major difficulties related to planning and security of the EV infrastructure.
- Analysis of AI and blockchain technologies for dealing with the issues related to planning and security of EV charging ecosystem. Examination of various academic and industrial solutions that

use these two technologies independently to uncover gaps and understand unresolved problems related to the EV charging ecosystem.
- Examining the potential for the two technologies to be combined to create a more intelligent and robust EV charging ecosystem.

In this paper existing protocols and difficulties encountered for the infrastructure of EV charging are discussed. Then, the potential applications of AI and blockchain to address these issues and to provide user optimal charging service are discussed along with potential research directions.

2. COMPONENTS OF EV CHARGING INFRASTRUCTURE

The infrastructure for EV charging is made up of various parts that communicate with one another through a variety of protocols. Figure 1 shows the components of EV charging infrastructure. Major components for EV charging infrastructure could be summarized as:

- Energy Supplier: Any organisation that can provide enough energy to run the charging stations qualifies as an energy supplier.
- Electric Vehicle Charging Station (EVCS): The EVCS is the place where EV drivers charge their cars with power provided by the power grid. EVCS consists of EVSE which are the real physical objects that are connected to EVs through connectors and plugs. According to their power level, EVSEs are majorly categorised into three levels: Level 1, Level 2, and Level 3. Level 3 is the highest power level among the three levels (Kaur et al. 2021). EVSEs could be public or private set-ups. EVSEs are often connected to the gateway station through a local area network (LAN), the gateway station then establishes a wide area network (WAN) connection with the central management system.
- Central Management System (CMS): The CMS is the brain or central programme which schedules EVs, records their transactions, compiles user databases, etc., for the charging stations.
- Electric Vehicles: EVs are the customers that get the power from EVSE through direct physical connection.

A wide variety of communication protocols are available in the EV charging ecosystem. One such communication protocol which is open source and is functioning between EVCS and CMS is Open Charge Point Protocol (OCPP) (Härer et al. 2022). This protocol oversees EVCS registration and controls their operation remotely. Several businesses like Schneider Electric, Siemens, and others, use their own communication protocols between EVCS and the CMS. Additionally, a series of communication protocols are also used to control the flow of power from EVSE to EVs. The infrastructure and charging ecology for the EV market is flourishing from the past 10 years. There are two major difficulties regarding the charging ecosystem of EV which are discussed below:

Figure 1. Abstraction of the components and present deployment of the EV charging infrastructure

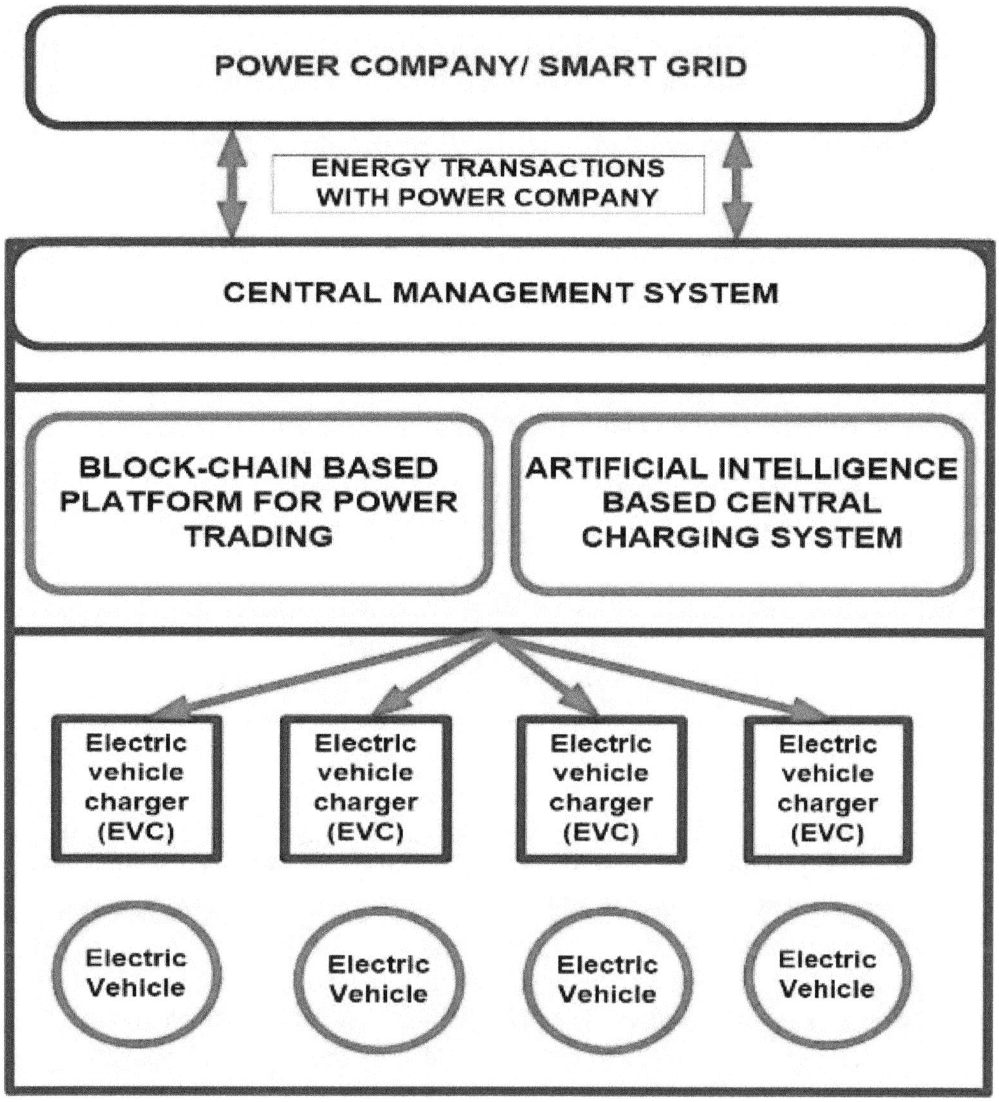

2.1 Security and Privacy Issues for EV Charging Ecosystem

Whenever a new service is launched, security and privacy are always a matter of concern. These concerns are also there for the infrastructure for EV charging. For the current EV charging system, two key issues regarding security and privacy are there. First key issue is using multiple protocols. Distinct sets of weaknesses are introduced using different protocols imposing vulnerability on the system. With multiple sets of protocols being used together for a charging ecosystem the total system becomes vulnerable to attacks. Second key issue is having numerous points of entrance into the system, for example multiple charging points. Hence, the charging infrastructure's entities and protocols all have vulnerabilities that could be abused and compromised. For instance, it was discovered that many EVCSs used by significant operators were susceptible to RCE and buffer overflow attacks. As a result, an enemy may use the most vulnerable entities to attack the more important ones for e.g. Power Grid. With the capacity to steal a user's vital data, and launch a DoS attack, an attacker can get a significant advantage (Xu et al. 2021). Given the high power ratings of these EVCSs, any compromise or concurrent schedule alterations could exacerbate the problem, potentially leading to grid disturbances due to an unforeseen surge in demand.

2.2 Optimal Charging Schedules for EV Charging Ecosystem

The absence of ideal scheduling plans is the second biggest issue facing the EV charging environment. As per Clean Technica, 40% of 3000 drivers are of the opinion that Electric Vehicle Charging Stations (EVCSs) are adequately situated for their needs. Furthermore, close to 50% of these drivers express the view that the existing infrastructure is only somewhat suitable for long-distance travel. From the viewpoint of the grid operator, these projections might be used to plan the deployment locations of EVCSs. So, the function of AI may be summed up as follows (Bae et al., 2022):

1. Prediction of the load profile for EVCS deployment.
2. Predicting EV drivers' behaviour for more effective scheduling plans

AI has been used by a few businesses and startups to enhance the EV charging infrastructure since, with the help of AI load profile scheduling and charging behaviour of the EV and EVCS can be predicted. As an example, upon Oracle's acquisition of Opower Inc. in 2016, it gained access to a vast dataset containing billions of data points related to the energy consumption of households from 60 million customers spanning 100 utilities. Utilizing this expansive dataset, predictions for EV loads were made, allowing power utilities to receive alerts and adjust their power generation accordingly to manage the load efficiently. This was done by leveraging Oracle's deep learning technology. Another intriguing application of AI was shown in CES 2019 by the startup GBatteries with a patent. By gathering health indications from the EV battery, the company employed AI to expedite the charging process. The battery's ability to be charged at maximum power was decided after careful consideration of these indicators. The trade-off in this situation was that charging at maximum power speeds up the process but depletes the battery in the process. Therefore, an AI-driven decision was made regarding the allocation of power to the vehicle battery. Essentially, AI's strength lies in its capacity to identify intricate patterns, offering predictions and assessments that can be utilized by power utilities for improved power generation management or by EV drivers to expedite their charging times.

The AI based predictions for EVs are broadly divided into two categories (Zhang et al., 2021): (i) load forecasting for charging, (ii) charging pattern forecasting. The load profile for charging of EV is majorly divided into three classes:

1. Short term load forecasting (STLF): To predict the load demand of EVs for charging ranging from hours to a week STLF is done (Lakshmi et al., 2023). With STLF stability of the power system is maintained and it helps in optimizing the charging schedules and location for EVCS.
2. Medium term load forecasting (MTLF): For load forecasting of EV charging ranging for more than one week to upto a year MTLF is used. MTLF planning helps in power system planning.
3. Long term load forecasting (LTLF): For load forecasting of EV charging ranging from one year and upto many years LTLF is used. LTLF is majorly required for planning of infrastructure for EV installation.

The data for forecasting EV load is majorly collected from surveys collected from EVCS through smart meters installations. These smart meters are installed in public or private. EVCS give extensive data for EV charging behaviour. In (Lopez .et al., 2019), AI based unsupervised learning algorithm is used for LTLF forecasting of EV charging. The GPS data collected from 17 EVs are used for load forecasting. SLTF method is proposed for EV charging where the proposed algorithm is a supervised learning algorithm and is used for scheduling as well as forecasting of EV charging (Gao et al. 2018). In (Lei Z. et al. 2014), MTLF method of EV bus charging is proposed. This model is based on the Wavelet neural network (WNN) and is based on data obtained from historical EV loads. Table 1 gives AI based models used in literature for load forecasting for EV charging.

Table 1. AI-based models used in literature for load forecasting for EV charging

Ref.	Supervised/ Unsupervised	Forecasting Type	Model	Objective	Data Collected	Finding	Region
[13]	Unsupervised	LTLF	Markov based decision process	Forecasting	GPS data from different EV	Average savings improved up to 88%	Winnip city, Canada
[14]	Supervised	STLF	WNN	Forecasting & Scheduling	Survey data (actual)	Accuracy of prediction improved	Baoding in China
[15]	Supervised	MTLF	WNN	Forecasting & Scheduling	14 electric bus and 14 ECVS for 35 days	A scientific based proposed for charging electric buses in a city	Guangdong in China

3. BLOCKCHAIN TECHNOLOGY IN EV CHARGING ECOSYSTEM

Blockchain refers to an evolving technique which enables the organisations to efficiently handle transactions within a setting that is both distributed and decentralized, ensuring security. Following are the three types of environments (Liu et al., 2022):

1. Distributed environment: all network nodes receive copies of the transaction history.
2. Decentralised environment: no centralised body is present which is in-charge of the transactions.
3. Secure environment: by utilising public key infrastructure (PKI) which involves a set of policies, rules and regulations secure environment is maintained among network nodes.

Blockchain technology lets various nodes and entities connect with one another in a secure environment without depending on a central authority to monitor these transactions. Table 2 gives differences between a centralised transaction system and decentralised blockchain based system. Blockchain's utility has evolved from its initial role as a distributed ledger to encompass the exchange of digital assets, including but not limited to energy and money. Blockchain was used for the first time in 2008 by Santoshi Nakamoto for money trade by the creation of the Bitcoin network. The Bitcoin network the money trade is done between multiple sources with the use of bank or central authority. Later, Ethereum was created, enabling the development of Decentralized Apps, which made blockchain technology to realise its full potential. The idea of smart contracts was first introduced by Ethereum in which a blockchain network was used along with a software algorithm that automatically decides the trade between various decentralized entities when the conditions are fulfilled without dealing with any central authority. Smart contracts use blockchain technology to develop decentralised economies for trading of different assets (Muneeb et al., 2022). Blockchain offers an advantage in developing decentralised economies for exchanging nearly any asset thanks to smart contracts. Three basic layers make up a blockchain network abstraction (Wang et al., 2019):

1. Network Layer: This layer in blockchain network is a representation of the protocol which is used for communications among network nodes. Typically, Peer-to-Peer (P2P) protocols are used by the blockchain network to function.
2. Consensus Layer: This layer of the blockchain network is in charge of assuring the accuracy of the data being shared. This layer ensures that a legitimate transaction is made when the transaction is initiated and broadcasted over the network. Various consensus methods have been suggested for this purpose, including, for instance, Proof of Work (PoW), Proof of Stake (PoS), and others. Miners/validators within the network are responsible for executing the consensus processes. Based on how the network is implemented, these validators and miners can receive cryptocurrency as compensation for their services.
3. Transaction layer: It is the top layer in the stack and is the in-charge of managing data transactions inside the network. Smart contracts are defined and invoked properly by this layer.

Table 2. Differences between a centralised transaction system and decentralised blockchain based system

Attributes	Centralised Transaction Systems	Blockchain-Based Decentralised Systems
Third party is involved	Yes	No
Safety	Unsafe Data and high risk of hacking.	Safe data with low risk of hacking.
Data management	Data transfer between clients and the server.	Data with clients only.
Stability	Low	Very high
Execution time	High	Low as compared to centralised system
Complex system	No	Yes

3.1 Blockchain Network Types

Blockchain based networks could be defined as follows (Wang et al., 2021):

- Public or permission less network: Public networks allow anybody to join and take part in the network's consensus and validation processes.
- Private or permissioned network: Only a select group of authorised nodes are allowed to administer the network and update it. Only these authorised nodes validates, decides entry of other nodes in the network and oversees network maintenance.

The above said blockchain network implementation has certain advantages and disadvantages. Blockchain establishes a trustless and decentralized environment for networks. In a public blockchain network, the functionality relies on the collaboration of all participating nodes. The participation of network nodes influences the processing speed. On the other hand, in private blockchain networks, only a few nodes are authorized, hence are faster as compared to public blockchain networks and are also scalable.

Table 3 gives comparison of different blockchain platforms used for the EV ecosystem. With the help of Blockchain technology, the privacy and security issues regarding charging of EV vehicles with present infrastructure can be solved. As a result, there have been initiatives to use blockchain networks with EV charging from both the research and industry sectors. Earlier, P2P energy trading was used by EV charging systems to digitise energy trade between the EV and EVCS but nowadays blockchain technology is on rise for EV charging ecosystem taking into consideration the trade possibility between the smart grid, EVCS and EV. Several research initiatives have been taken on developing blockchain networks to safely handle the energy trade between energy traders/generators/distributors, EVCS and EV (Jiang et al. 2018). In this type of blockchain based energy trading network for EV charging, the prices would be lower than other energy trading networks for EVs since, there would be no middlemen and hence no role of bad elements would be there in the network. With the help of blockchain technology, normal homes (with renewable energy generation for e.g. solar panels) can sell their generated energy to EVs. This would promote business making normal homes public EVCS. Also, various businesses have launched projects to create EV charging networks that are supported by blockchain in the context of P2P energy trade. For example, the ISO-15118 EV charging protocols have been extended through the efforts of the Oxygen Initiative, with the additional proposal of integrating a blockchain network. This network enables EVCS to furnish costing information and grid parameters that are already in place and has suggested a blockchain network that would allow EVCS to provide pricing as well as grid conditions for vehicles.

Therefore, the primary function of blockchain is to enable EV owners and homeowners to securely exchange energy without the need for middlemen. Since, home-owners can sell their energy to the grid and EVs can charge through EVCS available removing one to one interaction between the EV and home-owners, which reduces the number of EVCS and price of energy is increased.

4. BLOCKCHAIN AND AI COMBINED TOGETHER FOR EV CHARGING ECOSYSTEM

AI can help in managing EV charging schedules and in coordinating their deployments by forecasting load profiles and the predictions of their charging behaviour. Also, AI based energy trading systems depend upon a central organization like power utility for overall control. This centralized trust base has security concerns associated with it e.g. some harmful elements can easily get access to the EVCS and can disrupt or harm the working of power grid through this access. Also, sensitive user information could be stolen by this type of access. In contrast, blockchain aspires to offer a decentralised, secure, and trust less energy trading system. As a result, this method might eliminate the need for intermediaries and centralised bodies, saving money on superfluous operating expenses.

Table 3. Comparison of different blockchain platforms used for EV ecosystem

Type of Blockchain	Documentation/Support Available	Development	Programming Language	Additional Services	Transaction per Second Rate (TPS)	Incentive	Consensus Mechanism
IBM	GitHub repository, FAQ, Technical support available	Hyperledger platform started with Linux	Javascript/Java, Chaincode	Blockchain cloud service	Near to reach 100,000	No financial incentive	User can create its own algorithm for consensus
Etherium	Website by Etherium has proper documentation	Developed in 2014. Useful for real time applications.	Solidity	Very much flexible and cloud services available	1000-4000 With Etherium 2: 10,000	Financial rewards for blocks.	PoW and PoS
NEM	Technical documentation by NEM	Both public and private blockchain.	Java, python, c++	Integration with mobile API	4000	Financial incentive as reward.	Proof of importance
Intel	Documentation on website and GitHub repository.	private	Javascript, API, Go, Rust, C++	Integration with IoT	1300	No incentive	PoET
Oracle	Enterprise edition	Fabric platform	Python, C++, Golang	Integration with cloud.	4000 transaction per hour	No financial reward	RAFT
Bitcoin	Documentation on website	public	C++	Financial applications and transactions	9000	Financial reward given	PoW

The blockchain based charging network can make the selling and purchase of energy through bidding between ECVS owners, power utilities, other energy providers etc. Through which the availability of the energy and prices could be broadcasted over the network and the EVs can purchase the energy. The availability of energy and prices are broadcasted so that EVs can reach the best EVCS and charge themselves in a hassle-free manner. However, the smart contract implementation through blockchain technology could not be implemented properly without the idea of EV driver's dynamic behaviour, charging habits and load profile of the EVCS for which AI based analysis and load prediction is required. Consequently, AI and blockchain should work together to create a full-fledged EV charging system. The work done in (Liu C. et al., 2018) is one such example of how AI and blockchain could function together. The authors suggested an EV charging system that would be supported by blockchain, in which:

1. EVs could be charged through power utility or grid directly, through privately owned EVCS or through publicly owned EVCS.
2. EVs could discharge back into the system again, relieving the strain on the power grid. An adaptive algorithm is proposed which optimizes the charging and discharging schedules of the EV taking into consideration the load requirements by forecasting techniques.

Another example of AI and blockchain working together can be seen in (Liu et al., 2019), where the authors put more emphasis on the blockchain network's scalability than on the EVs' scheduling. Based on EV data, the authors have used deep learning in particular to boost transactional throughput while maintaining the system's decentralisation, latency, and security. This is especially intriguing because the suggested system may adjust its operational parameters like block dynamics, block size, etc. in response to dynamic environmental requirements.

The Porsche Digital Lab team from the Odyssey Hackathon have suggested a blockchain-AI enabled EV charging environment. This model reduced the stress on the grid, boosted customer satisfaction, and maintained EV security. The team suggested a system of energy trading in which EV drivers could resell their energy and local governments could lease their EVCSs. An AI based agent was also developed to manage EV schedules by forecasting the costs, supply, and driving habits of the EV drivers.

The collaboration of AI and blockchain for EV charging and its infrastructure is not just about an open trading market for digital assets but many other problems are solved by their combined functioning. Figure 2 gives blockchain and AI based EV charging systems. With PKI and consensus-based procedures, blockchain technology offers immutable and secure transactions. Blockchain decentralizes and disperses the infrastructure providing multiple charging points making EV charging ecosystem strong. Hence, a decentralised system makes it unnecessary for CMS and the electrical grid to serve as managing entities. Instead, they serve as facilitators, which reduces the risk of attacks and privacy issues in the charging ecosystem. Additionally, AI has maintained its dominance in both forecasting and anomaly detection.

Table 4 gives combined benefits of advanced AI and blockchain technologies used together. Some of the emerging trends in AI such as digital twins (Mihai et al., 2022), hybrid models of learning (Yang et al., 2020), explainable AI (Heimerl et al., 2022) and automated ML (Manduchi et al. 2022) are listed with extended benefits provided by blockchain technology. As shown in Table 4 with the combination of both the technologies, a better, secure and trusted EV charging ecosystem could be developed. Table 5 gives the essential features which are obtained by integration of both the technologies.

Figure 2. Blockchain and AI-based EV charging system

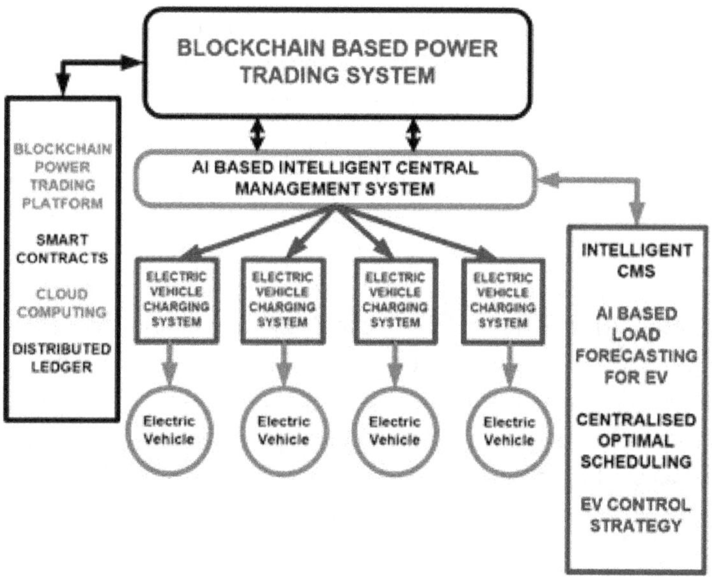

Table 4. Combined benefits of advanced AI and blockchain technologies used together

AI Trending Technologies	Objective of the Technology	EV Application	Blockchain Additional Benefits
Digital twins	Conversion of complex systems into digital simulations.	Useful for EV based simulations. Simulations of renewable systems for smart grid applications.	Makes the system trustworthy and reliable.
Hybrid models of learning	Combination of different AI and ML models to achieve better decision-making power.	Real time decision making for autonomous vehicles and EVs for charging decisions.	Making the system decentralized and trustworthy.
Explainable AI	AI based model which is interpretable regarding decision making.	Military based EV, autonomous EV.	Making the system trustworthy and secure.
Augmented data learning	Applying transfer learning among various AI models.	Critical forecasting situations for EV.	Reliable, trustworthy and secure system.
Automated ML	Automating the complete process of AI and ML based models.	Big data analytics	Immutable system

Table 5. The essential features which are obtained by integration of both the technologies

AI	Blockchain	Benefits of Integration
Centralised	Decentralised	Data security and privacy
Dynamic decisions	Deterministic decisions	Decisions are more robust.
Volatile	Data integrity	Intelligence becomes decentralised.
Probabilistic approach	Immutable	Decision making improved collectively.
Data dependent and centric	Attack resilient	Improved efficiency

As a result, combining AI and blockchain paves the door for the development of self-sustaining, self-improving charging ecosystems. AI can be included into the blockchain network at several layers, each of which would have a different capability. AI can be utilised at the network layer as an additional security layer to foresee network problems and take appropriate action for e.g. with the help of AI malicious nodes could be isolated. Furthermore, AI can be utilized within the consensus layer as a scalability gauge to predict system dynamics, including network size and data frequency. It can then adjust consensus methods such as block size and block generation time accordingly. In the transaction layer, AI may also be utilised to forecast system dynamics and detect abnormalities, which can all be used to improve EV scheduling processes in smart contracts. Theoretically, a self-maintained EV charging network can be developed using the integration of blockchain and AI technologies. But, there is a need for more advanced and practical research even though the theoretical findings from the study are encouraging. It is crucial to thoroughly evaluate and comprehend the scalability and dependability of blockchain technology in EV charging networks. With AI and Blockchain technology new consensus based algorithms could be developed that can decrease the overall network overhead while considering security issues of the EV charging ecosystem. For privacy issues regarding EV charging, research is going into the direction of Zero-Knowledge Proofs (ZKP) and homomorphic encryption making the Blockchain transactions transparent, visible, and accessible. Furthermore, AI based learning embedded within the blockchain network is an intriguing study area that might help create a completely decentralised network having dispersed AI nodes. Alternative techniques for decentralized Electric Vehicle Charging Stations (EVCSs) networks involve the use of a Multi-agent system, wherein a subset of agents in the blockchain network is assigned tasks such as data gathering, AI computations, block validation, and more. Each agent can thus carry out its assigned mission on their own.

To adequately assess the efficacy of such proposed systems, real-world implementation is required. The influence of AI and Blockchain technology in the Electric Vehicle (EV) charging ecosystem can be extended to encompass the IoT networks integrated with the grid. Also, this research could be developed to improve current protocols for improved energy management of EV charging systems.

5. PRACTICAL RECOMMENDATIONS FOR SECURITY IMPROVEMENT AND PRIVACY

This section presents a set of practical recommendations that focus on the security and privacy preservation of the EVCS utilizing the intricate landscape of AI and blockchain technologies. Key measures involve a judicious integration of encryption, authentication protocols, and secure communication channels including robust end-to-end encryption for secure data transmission. The integration of robust end-to-end encryption stands as a foundational measure to secure data transmission between EVCSs, EVs, and central servers. Through the implementation of state-of-the-art cryptographic algorithms the confidentiality and integrity of sensitive data can be ensured Lakshmi et al., 2023, Ji et al., 2017). A secure access framework through Multi-Factor Authentication (MFA) is suggested for both EV owners and EVCS operators through a combination of passwords, biometrics, and token-based verification enhances access control, mitigating the risk of unauthorized access. In addition, employing established secure communication protocols, such as Transport Layer Security (TLS) or Secure Sockets Layer (SSL), is essential to create secure connections between EVCSs, and backend servers. This safeguards against potential eavesdropping and man-in-the-middle attacks, fortifying overall communication channels. The incorporation of blockchain technology serves to establish an immutable and transparent ledger of transactions within the EV charging ecosystem. This decentralized approach enhances data integrity, providing a secure foundation for recording and verifying charging transactions. Conducting periodic security audits and penetration testing is vital to identify and address potential vulnerabilities within the EV charging infrastructure. Proactive measures, including vulnerability assessments, contribute to the creation of a resilient and secure system. Embracing data minimization practices to collect only essential information and prioritize the anonymization of personally identifiable information. This dual approach minimizes the risk of data breaches and safeguards user privacy within the EV charging ecosystem. Staying abreast of and complying with evolving regulatory standards in the EV charging sector is crucial. Aligning the ecosystem with industry-specific regulations ensures the integration of the latest security requirements and best practices. Development and implementation of the educational initiatives to enhance awareness among EV owners, CS operators, and other stakeholders regarding security best practices as educated users contribute to a more security-conscious and resilient EV charging ecosystem. The judicious incorporation of these measures involving stakeholders builds a secure, reliable, and privacy-centric foundation for integrating AI and blockchain technologies in EV charging.

CONCLUSION

This paper investigated the infrastructure, protocols, and current deployments within the EV charging ecosystem, with a focus on the trade-off between grid operations and user satisfaction, and concerns related to security and privacy. The chapter explored the potential of AI to coordinate EV schedules and facilitate EVCSs provisioning. The key considerations involved enhancing security in the electric vehicle (EV) charging infrastructure and implementing a trustless transaction mechanism for energy exchange among EVs, individually owned EV charging stations (EVCS), and energy providers. The findings highlighted the notable contributions of Blockchain technology, such as incentivizing EV owners to sell energy back to the grid for peak shaving, enhancing EV owners' satisfaction through the sale of energy by individually owned EV charging station (EVCS) operators and local communities, and reducing overall expenses by eliminating inefficiencies. Recognizing the distinct advantages of AI and Blockchain independently, this study proposes their collaborative integration to establish a robust and efficient environment for EV charging. While recognizing the individual merits of AI and Blockchain, our research proposed the necessity of their combined integration to create a more resilient and efficient EV charging environment. Various use cases were explored to demonstrate how AI and Blockchain, working in tandem, can elevate the overall charging ecosystem. In conclusion, this study advocates for a holistic approach that leverages the complementary strengths of AI and Blockchain for a comprehensive and robust EV charging solution.

REFERENCES

Al-Fuqaha, A., Guizani, M., Mohammadi, M., Aledhari, M., & Ayyash, M. (2015). Internet of Things: A Survey on Enabling Technologies, Protocols, and Applications. *IEEE Communications Surveys and Tutorials*, *17*(4), 2347–2376. doi:10.1109/COMST.2015.2444095

An, Y., Gao, Y., Wu, N., Zhu, J., Li, H., & Yang, J. (2023). Optimal scheduling of electric vehicle charging operations considering real-time traffic condition and travel distance. *Expert Systems with Applications, 213*(B). doi:10.1016/j.eswa.2022.118941

Bae, S., Gros, S., & Kulcsár, B. (2022). Can AI Abuse Personal Information in an EV Fast-Charging Market? *IEEE Transactions on Intelligent Transportation Systems*, *23*(7), 8759–8769. doi:10.1109/TITS.2021.3086006

Bharathidasan, M., Indragandhi, V., Suresh, V., Jasiński, M., & Leonowicz, Z. (2022). A review on electric vehicle: Technologies, energy trading, and cyber security. *Energy Reports, 8*, 9662-9685. doi:10.1016/j.egyr.2022.07.145

Bunsen, T. (2019). *Global EV Outlook (2018).* Towards Cross-Modal Electrification.

Chakrapani, B., Chakrapani, K., Kavitha, T., Iqubal, S. M., & Kempanna, M. (2023). Sustainable Smart Energy Systems and Energy Preservation Strategies in Intelligent Transportation Sectors. In Applications of Big Data and Artificial Intelligence in Smart Energy Systems: Volume 2 - Energy Planning, Operations, Control and Market Perspectives. River Publishers.

Chavhan, S., Gupta, D., Chandana, B. N., Khanna, A., & Rodrigues, J. J. P. C. (2020). IoT-Based Context-Aware Intelligent Public Transport System in a Metropolitan Area. *IEEE Internet of Things Journal, 7*(7), 6023–6034. doi:10.1109/JIOT.2019.2955102

Danese, A., Torsæter, B. N., Sumper, A., & Garau, M. (2022, March 22). Planning of High-Power Charging Stations for Electric Vehicles: A Review. *Applied Sciences (Basel, Switzerland), 12*(7), 3214. doi:10.3390/app12073214

El-Bayeh, C. Z., Alzaareer, K., Aldaoudeyeh, A.-M. I., Brahmi, B., & Zellagui, M. (2021). Charging and Discharging Strategies of Electric Vehicles: A Survey. *World Electric Vehicle Journal, 12*(1), 11. doi:10.3390/wevj12010011

Euceda, R., Akundi, A., & Luna, S. (2023). Cybersecurity Challenges in Electric Vehicles: An initial literature review and research agenda. *IEEE International Systems Conference (SysCon),* 1-8. 10.1109/SysCon53073.2023.10131069

Gao, Y., Guo, S., Ren, J., Zhao, Z., Ehsan, A., & Zheng, Y. (2018). An electric bus power consumption model and optimization of charging scheduling concerning multi-external factors. *Energies, 11*(8), 2060. doi:10.3390/en11082060

Hanemann, P., Behnert, M., & Bruckner, T. (2017). Effects of electric vehicle charging strategies on the German power system. *Applied Energy, 203*, 608–622. doi:10.1016/j.apenergy.2017.06.039

Härer, F., & Fill, H.-G. (2022). Decentralized Attestation and Distribution of Information Using Blockchains and Multi-Protocol Storage. *IEEE Access : Practical Innovations, Open Solutions, 10*, 18035–18054. doi:10.1109/ACCESS.2022.3150356

Heimerl, A., Weitz, K., Baur, T., & André, E. (2022). Unraveling ML Models of Emotion With NOVA: Multi-Level Explainable AI for Non-Experts. *IEEE Transactions on Affective Computing, 13*(3), 1155-1167. . doi:10.1109/TAFFC.2020.3043603

Islam, S., Badsha, S., Sengupta, S., Khalil, I., & Atiquzzaman, M. (2023). An Intelligent Privacy Preservation Scheme for EV Charging Infrastructure. *IEEE Transactions on Industrial Informatics, 19*(2), 1238-1247. . doi:10.1109/TII.2022.3203707

Jiang, T., Fang, H., & Wang, H. (2018). Blockchain-Based Internet of Vehicles: Distributed Network Architecture and Performance Analysis. *IEEE Internet of Things Journal.*

Johnson, J., Anderson, B., Wright, B., Quiroz, J., Berg, T., Graves, R., Daley, J., Phan, K., & Kunz, K. (2022). *Cybersecurity for Electric Vehicle Charging Infrastructure.* Sandia Report. https://www.osti.gov/servlets/purl/1877784

Kaur, K., Kaddoum, G., & Zeadally, S. (2021, August). Blockchain-Based Cyber-Physical Security for Electrical Vehicle Aided Smart Grid Ecosystem. *IEEE Transactions on Intelligent Transportation Systems, 22*(8), 5178–5189. doi:10.1109/TITS.2021.3068092

Khetarpal, P., Nagpal, N., Al-Numay, M. S., Siano, P., Arya, Y., & Kassarwani, N. (2023). Power Quality Disturbances Detection and Classification Based on Deep Convolution Auto-Encoder Networks. *IEEE Access : Practical Innovations, Open Solutions, 11*, 46026–46038. doi:10.1109/ACCESS.2023.3274732

Kumar, M., Panda, K. P., Naayagi, R. T., Thakur, R., & Panda, G. (2023). Comprehensive Review of Electric Vehicle Technology and Its Impacts: Detailed Investigation of Charging Infrastructure, Power Management, and Control Techniques. *Applied Sciences (Basel, Switzerland)*, *13*(15), 8919. doi:10.3390/app13158919

Lakshmi, D., Nagpal, N., Chandrasekaran, S., & Hemanth, D. J. (2023). A quantum-based approach for offensive security against cyber-attacks in electrical infrastructure. *Applied Soft Computing, 136*. doi:10.1016/j.asoc.2023.110071

Lakshmi, D., Tiwari, R. S., Nagpal, N., Kassarwani, N., Vishnuvarthanan, G., & Srivastava, A. (2023). Comparative Study on Forecasting of Schedule Generation in Delhi Region for the Resilient Power Grid Using Machine Learning. *IEEE Transactions on Industry Applications*. Advance online publication. doi:10.1109/TIA.2023.3316646

Lei, Z., Chun, H., & Haoming, Y. (2014). Study of charging station short-term load forecast based on wavelet neural networks for electric buses. *Proc. Chin. Conf. Pattern Recognit.*, 555-564. 10.1007/978-3-662-45643-9_59

Li, H., Yang, J., & Zhu, J. (2021). An Y. Electric vehicle charging scheduling algorithm based on traffic condition. *Computer System & Applications*, *30*(7), 178–183.

Liu, C., Chai, K. K., Zhang, X., Lau, E. T., & Chen, Y. (2018). Adaptive Blockchain-Based Electric Vehicle Participation Scheme in Smart Grid Platform. *IEEE Access : Practical Innovations, Open Solutions*, *6*, 25,657–665. doi:10.1109/ACCESS.2018.2835309

Liu, C., Guo, H., Xu, M., Wang, S., Yu, D., Yu, J., & Cheng, X. (2022). Extending On-Chain Trust to Off-Chain – Trustworthy Blockchain Data Collection Using Trusted Execution Environment (TEE). *IEEE Transactions on Computers*, *71*(12), 3268–3280. doi:10.1109/TC.2022.3148379

Liu, M. (2018). *Deep Reinforcement Learning Based Performance Optimization in Blockchain-Enabled Internet of Vehicle, ICC 2019-2019 IEEE Int'l. Conf. Commun.* ICC.

Lopez, K. L., Gagne, C., & Gardner, M. A. (2019). Demand-side management using deep learning for smart charging of electric vehicles. *IEEE Transactions on Smart Grid*, *10*(3), 2683–2691. doi:10.1109/TSG.2018.2808247

Manduchi, E., Le, T.T., Fu, W., & Moore, J.H. (2022). Genetic Analysis of Coronary Artery Disease Using Tree-Based Automated Machine Learning Informed By Biology-Based Feature Selection. *IEEE/ACM Transactions on Computational Biology and Bioinformatics, 19*(3), 1379-1386. . doi:10.1109/TCBB.2021.3099068

Mastoi, M. S., Zhuang, S., Munir, S., Malik, H., Hassan, M., Alqarni, M., & Alamri, B. (2023). A study of charging-dispatch strategies and vehicle-to-grid technologies for electric vehicles in distribution networks. *Energy Reports, 9*, 1777-1806. doi:10.1016/j.egyr.2022.12.139

Mihai, S. (2022). Digital Twins: A Survey on Enabling Technologies, Challenges, Trends and Future Prospects. *IEEE Communications Surveys & Tutorials, 24*(4), 2255-2291. Advance online publication. doi:10.1109/COMST.2022.3208773

Mousavian, S., Erol, K., Melike, & Ortmeyer, T. (2015). *Cyber Attack Protection for a Resilient Electric Vehicle Infrastructure*. . doi:10.1109/GLOCOMW.2015.7414174

Muhammad, Z., Anwar, Z., Saleem, B., & Shahid, J. (2023). Emerging Cybersecurity and Privacy Threats to Electric Vehicles and Their Impact on Human and Environmental Sustainability. *Energies*, *16*(3), 1113. doi:10.3390/en16031113

Muneeb, M., Raza, Z., Haq, I. U., & Shafiq, O. (2022). SmartCon: A Blockchain-Based Framework for Smart Contracts and Transaction Management. *IEEE Access, 10*, 23687-23699. . doi:10.1109/ACCESS.2021.3135562

Ons, S., & Ammar, O. (2017). Electric vehicle scheduling and optimal charging problem: Complexity, exact and heuristic approaches. *International Journal of Production Research*, *5*(2), 519–535. doi:10.1080/00207543.2016.1192695

Pourmirza, Z., & Walker, S. (2021). Electric Vehicle Charging Station: Cyber Security Challenges and Perspective. *IEEE 9th International Conference on Smart Energy Grid Engineering (SEGE),* 111-116. 10.1109/SEGE52446.2021.9535052

Relan, D., Khatter, K., & Nagpal, N. (2023)/ Blockchain based Solution for Electricity Supply Chain in Smart Grids. In *Applications of Big Data and Artificial Intelligence in Smart Energy Systems: Volume 2 - Energy Planning, Operations, Control and Market Perspectives.* River Publishers.

Siddique, A. B., & Gabbar, H. A. (2023). Adaptive Mixed-Integer Linear Programming-Based Energy Management System of Fast Charging Station with Nuclear–Renewable Hybrid Energy System. *Energies*, *16*(2), 685. doi:10.3390/en16020685

Sun, Y., Hu, Y., Zhang, H., Chen, H., & Wang, F.Y. (2023). A Parallel Emission Regulatory Framework for Intelligent Transportation Systems and Smart Cities. *IEEE Transactions on Intelligent Vehicles, 8*(2), 1017-1020. . doi:10.1109/TIV.2023.3246045

ur Rehman. (2022). A robust vehicle to grid aggregation framework for electric vehicles charging cost minimization and for smart grid regulation. *International Journal of Electrical Power & Energy Systems, 140*. doi:10.1016/j.ijepes.2022.108090

Wang, S., Ouyang, L., Yuan, Y., Ni, X., Han, X., & Wang, F.-Y. (2019). Blockchain-Enabled Smart Contracts: Architecture, Applications, and Future Trends. *IEEE Transactions on Systems, Man, and Cybernetics. Systems*, *49*(11), 2266–2277. doi:10.1109/TSMC.2019.2895123

Wang, T., Zhao, C., Yang, Q., Zhang, S., & Liew, S.C. (2021). Ethna: Analyzing the Underlying Peer-to-Peer Network of Ethereum Blockchain. *IEEE Transactions on Network Science and Engineering, 8*(3), 2131-2146. . doi:10.1109/TNSE.2021.3078181

Xia, Y., Zhang, L., & Liu, Y. (2016). Special issue on big data driven Intelligent Transportation Systems. *Neurocomputing, 181*, 1-3. doi:10.1016/j.neucom.2015.10.121

Xu, S., Chen, X., & He, Y. (2021). EVchain: An Anonymous Blockchain-Based System for Charging-Connected Electric Vehicles. *Tsinghua Science and Technology*, *26*(6), 845–856. doi:10.26599/TST.2020.9010043

Yang, S., Cheng, W., Hsu, Y., Gan, C., & Lin, Y. (2013). Charge scheduling of electric vehicles in highways. *Mathematical and Computer Modelling, 57*(11–12), 2873-2882. doi:10.1016/j.mcm.2011.11.054

Yang, W., Joo, M., Kim, Y., Kim, S. H., & Chung, J.-M. (2020). Hybrid Machine Learning Scheme for Classification of BECTS and TLE Patients Using EEG Brain Signals. *IEEE Access : Practical Innovations, Open Solutions*, *8*, 218924–218935. doi:10.1109/ACCESS.2020.3038948

Zaidi, I., Oulamara, A., Idoumghar, L., & Basset, M. (2023). Electric Vehicle Charging Scheduling Problem: Heuristics and Metaheuristic Approaches. *SN Computer Science*, *4*(3), 283. Advance online publication. doi:10.1007/s42979-023-01708-1

Zhang, X., Chan, K. W., Li, H., Wang, H., Qiu, J., & Wang, G. (2021, June). Deep-Learning-Based Probabilistic Forecasting of Electric Vehicle Charging Load With a Novel Queuing Model. *IEEE Transactions on Cybernetics*, *51*(6), 3157–3170. doi:10.1109/TCYB.2020.2975134 PMID:32248136

Chapter 14
Modelling and Simulation of Electric Vehicle for Performance Analysis:
BEV and HEV Electrical Vehicle Implementation Using Simulink for E-Mobility Ecosystems

Kutubuddin Kazi
https://orcid.org/0000-0001-5623-9211
BMIT Institute, Solapur, India

ABSTRACT

The mathematical model of the BEV is created in this instance using the MATLAB/Simulink environment and evaluated using several standard drive cycles in order to evaluate the performance. To ensure that the battery continues to operate in a safe manner, fuzzy logic control is employed to determine the regenerative braking component. The implemented EV system's performance is checked by the authors with regard to RES, regenerative braking, battery performance analysis. This is accomplished using artificial intelligence (AI) techniques. The sensors detect the external physical parameters. The overall performance is monitored using a two-way communication system implemented using cloud computing techniques and wireless technology.

INTRODUCTION

Transportation is an important factor in daily life. This is accomplished through the use of diesel and gasoline vehicles, which contribute to various forms of pollution in the environment. Electric vehicles are being implemented to address the shortcomings of existing vehicles by Karale (2020). The existence and introduction of electric vehicles came in 19th century. The advancement of power electronics combined with microelectronics leads to the enhancement and development of electric vehicles by Hosmani (2022).

DOI: 10.4018/979-8-3693-2611-4.ch014

Electric vehicles (EV) have been gaining tremendous attention in recent years. The growing popularity of EVs is largely due to the environmental benefits they offer. EVs are seen as an important component of a sustainable transport system as they emit less pollution and require no gasoline or diesel fuel. With the development of innovative technologies and battery management systems, the range and performance of electric vehicles continues to improve. In order to ensure that EVs continue to be an effective and reliable form of transportation, modelling and simulation of their performance is essential. This article will discuss the use of Simulink for modelling and simulation of electric vehicles, with special focus on Battery Electric Vehicles(BEV) and Hybrid Electric Vehicles(HEV).

The development of electric vehicles are rapidly increased due to the eco-friendly nature by Kazi k (2022). The rise of electric vehicles are due to the reduced amount of fossil fuels that also destructs the environment with various climatic change by wale(2019). The process of attaining clean energy through the electric vehicles is also an important parameter for switching over the electric vehicles. The rise in the petrol and diesel price is also a considering parameter for the switching of electric vehicles than the conventional vehicle by Halli(2023). The emissions from the petrol and diesel vehicle cause larger emission of toxic compounds in the environment which adversely affects the health of living organisms by K S L (2022). To overcome the constraints of commercial vehicles, the electric vehicles are introduced by Kazi K (2022). This is highly recommended due to the ease of use and coziness level. The electric vehicle is defined as the automotive vehicle that operates through the electrical energy. It adopts the electric motor for propulsion by Priya (2023).

There are four kinds of electric vehicles based upon the performance and manufacturing. The hybrid electric vehicles by Kazi (2022) play a prominent role in saving the usage of electric energy through adopting renewable energy source by Mrunal (2022) & Kazi (2022). This hybrid electric vehicles includes adopting energy from non-conventional energy source such as solar, tidal and wind energy. This helps in the enhancement of renewable energy source by Nikita(2022). In hybrid electric vehicles, the artificial intelligence by K K (2022) with cloud computing techniques by K K (2023) are implemented to enhance the control of power and to maintain the overall performance of the vehicles. This helps to maintain the battery's State of Charge (SOC) by Halli (2023). This helps to maintain the life span of battery that adversely promotes the performance efficiency of the electric vehicles. Thus they are attained through the optimization techniques that are integrated with wireless technology by Kazi (2023). In the hybrid electric vehicles, the electrical energy obtained from the grid can be stored when there is larger availability for further usages. This can also help by sending them back to the electrical grid by Mrunal (2022) when there is a demand in the supply system. This helps to improve the storage of energy capacity in the electric vehicle by Hosmani(2022). The overall performance is monitored and automated through two way communication system with sensors by Kamuni (2022). The electric vehicles are the efficient alternative for the fossil fuel vehicles. The EVs has zero tailpipe discharges. The absence of tailpipe emission leads to the reduction of carbon footprints in the environment. The electric vehicles have lesser maintenance cost due to the absence of various movable parts inside the vehicles which also reduces the running cost of the vehicle by K S L (2022).

Thus the electric vehicles provides higher efficiency with better performance. This is done through the use of lithium ion batteries for storage of electrical power by Shweta (2022). The electric vehicles are much convenient and easier to operate because it does have any complicated system. They provide lesser noise pollution when compared to the traditional vehicles. This is due to the absence of combustion engines in the electric vehicles. They also provide easier and convenient in charging at any place without any rush at the petrol pump as similar to the conventional vehicle by K S L (2022).

Figure 1. Electric vehicle (EV)

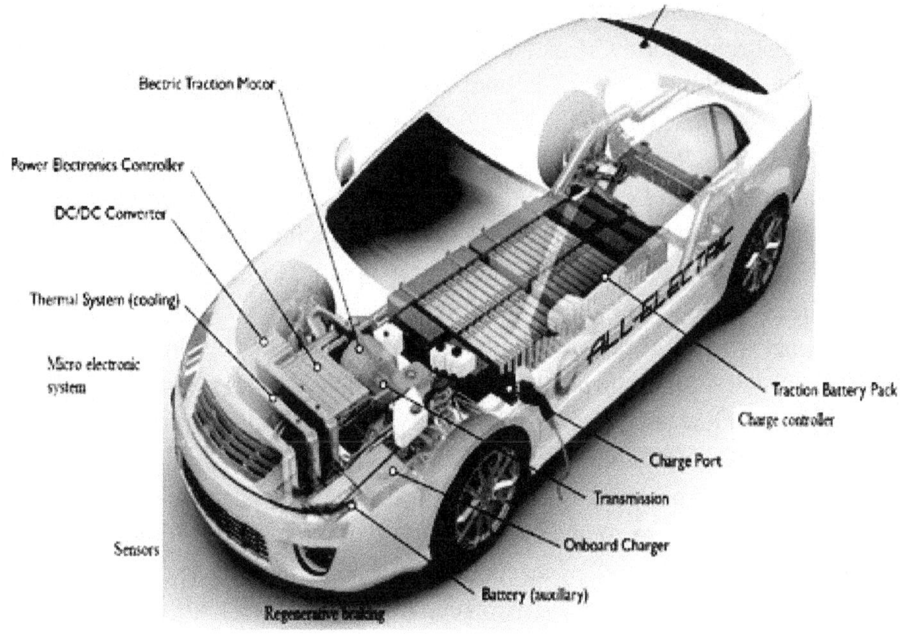

Figure 1 demonstrates the model of electric vehicles. The advancement of electric vehicles are implemented thru artificial intelligence with cloud computing techniques. Artificial Intelligence (AI) by Vahida (2023) techniques helps in versatile forms with various task scheduling and source distribution by Mulani 2019) & Gouse (2018). This helps in numerous automation and sophistications. These electric vehicles are integrated with wireless technology with sensors to adopt automation in the systems. The real time information processing are enhanced through the AI and IoT by Sultana (2023).

Various mechanism are implemented to control the loss of energy and to improve efficiency of a system. It is done with optimization techniques by Yogita (2026) with algorithms with the adoption of AI by Pradeepa (2022). This helps in the management of congestion control. To improve the functioning of electric vehicles, the integration of smart grid by Xu (2021) with electrical vehicles plays prominent role. This helps in optimization of charging and discharging of electrical vehicles based upon the rush and non-rush hours. It helps to maintain demand side management by Liyakat (2023). Here loads are scheduled through the optimization techniques adopting AI which helps to schedule the loads based upon priority of consumers by K S L (2022).

The Government has launched the Electric Vehicle Initiative (EVI) to enhance the development of EVs to reduce the usage of fossil fuels and develop a sustainable greener ecosystem. In this initiative sixteen countries are participating including India. There are various financial advantages in the electric vehicles ranging from lower cost of registration fees and road tax when compared to the purchase of the petrol or diesel vehicles. There are numerous policies by Ravi (2022) and benefits offered by the Government. Thus the electric vehicles are the way forward in the transportation by Vinay(2022) sector that helps in versatile benefits. The electric vehicle industry also helps to enhance various employment opportunities as reported by the Ministry of Skill Dev. and Entre. The adoption of electric vehicles leads to the improvement of green energy with economic benefits in the upcoming decades by Radanliv(2022).

When talking about electric vehicles(EVs), 3- main types are HEV-hybrid electric vehicles, PHEV-plug-in hybrid electric vehicles, and BEV-battery electric vehicles. The automotive industry has advanced into one of the most important worldwide sectors, both financially and in terms of R&D Departments by priyanka (2020). To develop the safety of both travelers and striders, more high-tech features are frequently added to cars. Furthermore, Rane (2022) there are more cars on the road, allowing for more comfortable and quick travel. Nevertheless, this has led to a notable increase in the concentration of air pollutants in urban areas, such as particulate matter (PM), carbon monoxide (CO), nitrogen oxides (NOX), and sulfur dioxide (SO2). Furthermore, a study conducted indicates that the transport sector is responsible for nearly 28% of the total emissions of carbon dioxide (CO2), with road transport contributing to over 70% of these emissions by Sultana (2023).

Thus, the governments of most developed countries are encouraging the use of electric vehicles (EVs) to lower the concentration of air pollutants, such as CO2 and other greenhouse gases [4]. More precisely, they promote efficient and sustainable mobility through a range of initiatives, mainly through tax benefits, financial assistance, or other special measures like free usage of public transport or highways.

A vehicle's electric engine is powered by a physical process developed in the latter part of 19th century. To do this, a current must be used to create a magnetic field at the machine's stationary "stator," whose displacement drives the rotating "rotor" by sweta (2026). The technology sector has grown throughout the world in recent years. It is evolving swiftly to keep up with the demands of the modern world. For humans, technology has proven to be a reliable and loyal ally. We are constantly astonished and surprised by new inventions. There is a petrol crisis sweeping the globe. Fuel is not always readily available and is relatively expensive. The system of electric cars is friendly to the environment. In by Mrunal (20220, it can be powered by energy sources that can be rented to produce electricity. If the system is inexpensive to use, build, and maintain, the general public will find it appealing. Vehicles with proper design will function and provide more mileage with the specified battery specifications. The designer will find this portal useful for estimating static parameters by Turjaman (2020). It provides the vehicle's range before the construction of an electric car by Karale(2019).

To increase the vehicle's performance and range, the right parameters can be chosen. The vehicle designer can begin experimenting with electric vehicles by selecting the desired parameters. We will introduce several electrical vehicle design parameters in this paper. Well-built cars offer excellent performance for a slightly higher price to the consumer. Vehicle range is increased as a result of advancements in battery technology. The more individuals who participate in the build and usage of electric vehicles, the stronger the movement is to adopt them by Rathi (2021).

Vehicles fueled by petroleum and diesel predominated the automation sector until the negative effects on the environment were clearly apparent. BEVs, in particular, have emerged as a possible substitute by Maheshwari (2020). This is entirely battery-powered, as the name would imply, and has no internal combustion engines whatsoever. Although this vehicle has benefits like no emissions, is environmentally friendly, and is powered by a clean sources of energy by Waghmode(2022), several drawbacks like its limited range and high cost function as roadblocks to the advancement of this technology by Satpute (2022). The paper provides examples of BEV modelling, simulation, and analysis.

Unlike internal combustion(IC) technology, which uses pressure and combustion to move a vehicle, electric vehicles, or just EVs, are powered by electromagnetism. These cars' electric motors run on electricity, which is typically stored in batteries by Tadalgi (2022). The most important component of an electric vehicle is the battery. Depending on the type of EV, the battery technology will vary and its modelled dynamics play an important role in the overall performance of the electric vehicle. Simulink is a modelling and simulation tool commonly used to simulate the design and performance of BEV and HEV. Simulink models are used to analyse the dynamic behaviour of the battery, motor and drivetrain of an EV. The models can be used to simulate the vehicle performance in terms of energy efficiency, recharge time, range and architecture. The models can also be used to analyse the thermal management behaviour of the vehicle and the influence of the climate conditions on the vehicle performance.

The models used for simulation are developed based on the vehicle specifications such as battery capacity, motor type and design, vehicle size, etc. The Simulink model of a BEV or HEV consists of a dynamic battery model and a motor model. The parameters of the battery model and the motor model are tuned to accurately represent the real-world behaviour of the battery and motor. This ensures the accuracy of the simulation results and the ability to replicate the actual performance of the vehicle. The battery model is used to simulate the charging and discharging behaviour of the battery, while the motor model is used to simulate the torque and speed output of the motor.

Advantages of Simulink in Modelling and Simulation of Electric Vehicle-

Simulink provides a powerful platform to simulate the performance of electric vehicles. It allows for the development of complex models, efficient simulation and rapid prototyping of EV. The models allow for an accurate estimation of electric vehicle performance in various conditions and can be tuned to reflect the real-world performance of the vehicle.

The models can also be used to optimise the design of an electric vehicle with respect to energy efficiency, range, acceleration and overall performance. Furthermore, the ability to simulate various driving cycles allows for comprehensive testing and analysis of EV and can lead to improved design and increased efficiency.

PROPOSED SYSTEM

The e-vehicle or the electric vehicle is the combination of vehicle with electric propulsion system with battery storage and energy management system. The electric vehicles are differentiated into three categories. They are composed of microelectronics with chemical and automotive applications. The electric vehicles adopts batteries by Waghmare (2022) to store the energy for functioning. These rechargable batteries helps in proper charging and discharging operations that leads to improve the life span. This is also done by employing the regenerative braking. The electric vehicle can able to extract the kinetic energy at standstill conditions . This also helps to improve the performance of the battery. The traction batteries are the important part in the electric vehicles. This helps to enhance the energy storage process in the vehicles.

Figure 2 demonstrates the electric vehicle with their various functioning parameters. This includes the AI techniques to provide automation with improve performance in the system. The components of electric vehicle includes power supply, electric gear box, inverter, charger, battery, sensors, servo-steering and servo-brake system. The charging of battery are classified into three types namely the current, voltage and pulse charging. These charging properties of batteries helps in the development of electric vehicles. Thus, energy management system shows a prominent part in the functioning of the electric vehicles. The hybrid electric vehicle includes the extraction of energy for renewable energy source by Divya (2022). This includes the solar panel at the roof of the electric vehicle that extracts energy at the day time and stores in the battery for further usages. The source of power for the electric vehicles are adopted through the battery by Kutubuddin(2022). This helps in the improvement of energy management system in the electric vehicles.

Figure 2. Components of electric vehicle (EV)

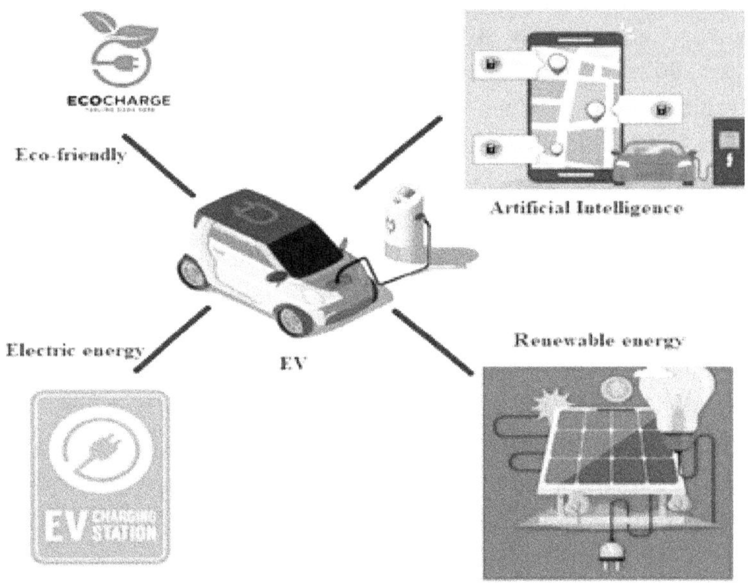

IMPLEMENTATION OF E-VEHICLES

The e-vehicles are functioned using the electric energy from the grid that are stored in the battery by Aragonda(2018) . The control management of the energy in the electric vehicle are maintained and regulated through the batteries inside the electrical vehicles. To adopt faster energy charging of batterie, the technology of multi-energy system is employed. This helps to charge the batteries in a short span of time. Due to the absence of various ignition engines and motors inside the electric vehicle leads to saves the interior space. The body structure of the electric vehicle is composed of aluminum with plastic over coating to enhance light weight structure. The important parameter is the electric propulsion system that consists of single motor and multi motor drive.

Figure 3 represents the classification of electric vehicles. Ther are differentiated based upon the usage of conventional engine and the type of battery charging and discharging techniques.

1. HEV-HYBRID ELECTRIC VEHICLE

The HEV is vehicle whuch integrates with the Internal Combustion(IC) engine with an electric propulsion system. The Figure 4 represents the functioning of hybrid electric vehicles. It is composed of internal combustion engine, generator and electric motors. The energy storage system includes batteries and supercapacitors . The numerous categories of batteries are lead acid battery, Li battery and Ni-metal hydride batteries. Fuel economization is attained through the hybrid electric vehicles. This enhances various supportable modes of mobility in the system. This invloves various techniques such as the regenerative braking system to improve an efficiency of system.

Figure 3. Classification of EV

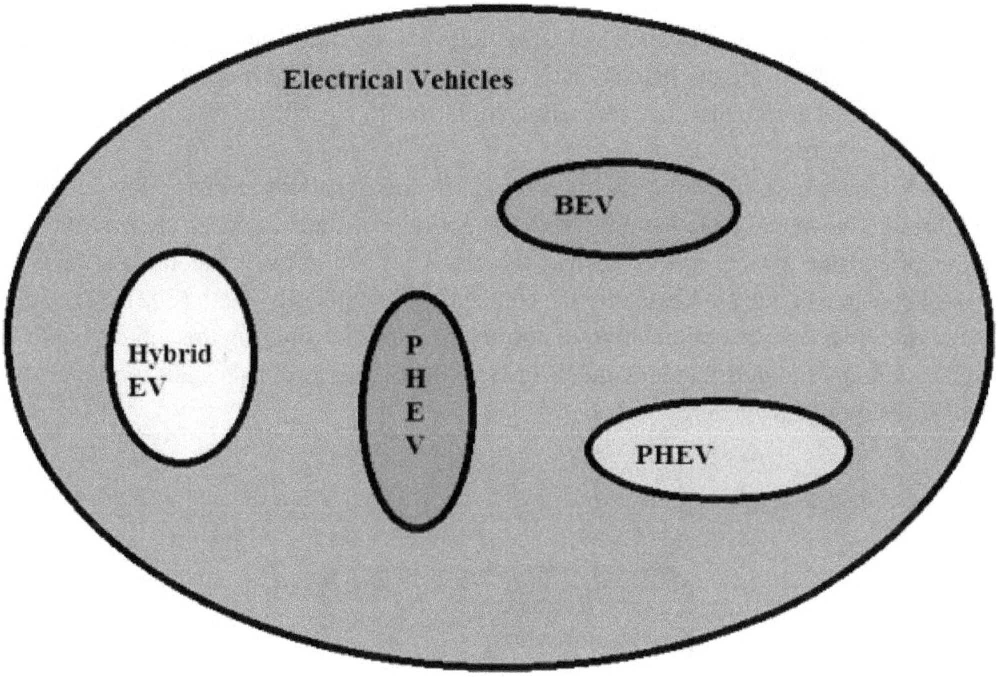

Figure 4. Hybrid electric vehicle

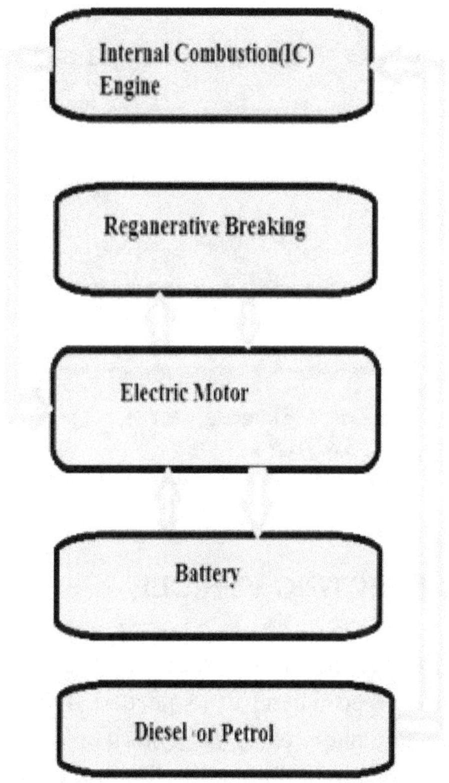

2. BEV-BATTERY ELECTRIC VEHICLE

The BEV is represented as the complete electric vehicle that are operated through the chaemical reactions that are takes place inside the battery. This can be charged and discharged based upon the needs of the users. They replaces the internal combustion engines with electric motors and various controllers for functioning the vehicle.

The Figure 5 demonstrates the components and functioning of battery electric vehicle. Thus they are formulated through the rechargable batteries with the absence of fuels in operating the vehicles. There is no another propulsion system in the electric vehicle. They are much efficient in reducing various environmental constraints that are caused by the conventional vehicles by Neto (2021). They are inbuilt with AI techniques to adopt the optimization of power This helps in charging the vehicle based upon the rush and off rush hours and dependent on the priority of the consumers. This leads to reduce the energy demand and saves the electric energy.

Figure 5. Battery electric vehicle (BEV)

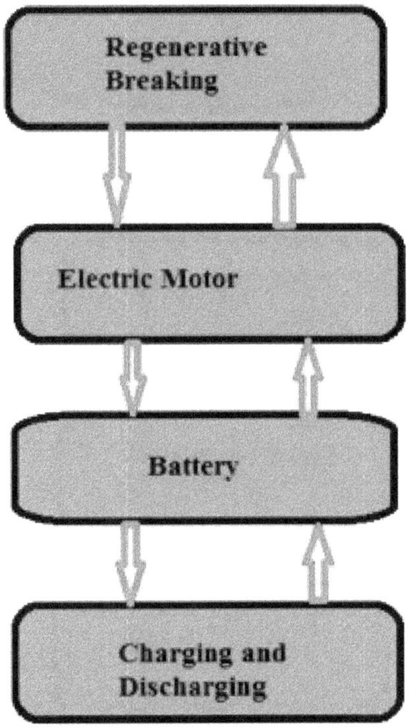

3. PHEV-PLUG IN HYBRID ELECTRIC VEHICLE

The PHEV is defined as the vehicles which adopts two source for the internal combustion engine in which the battery can be charged and discharged whenever its needed. This can be done through plugging a charge cable with an external source for battery charging. Thus it uses charging elements and regenerative braking system. This includes both the usage of electric power and fossil fuels based upon availability.

Figure 6 demonstrates the PHEV components and functioning. It is composed of internal combustion engine, generator and electric motors. They are formulated through the rechargeable batteries with the absence of fuels in operating the vehicles.

Figure 6. Plug in hybrid electric vehicle

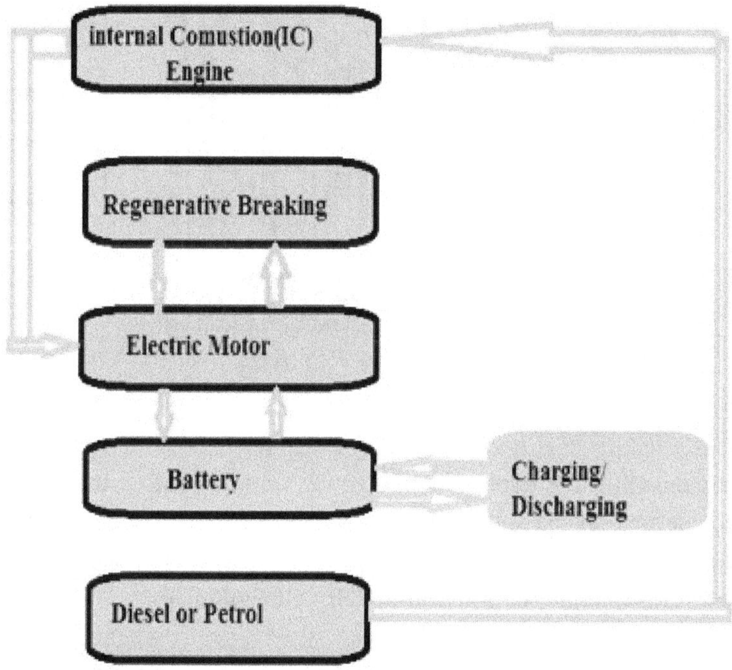

4. INTEGRATION OF E-VEHICLES WITH ARTIFICIAL INTELLIGENCE

The integration of e-vehicles with AI gets rise with the vehicle to grid technology (V2G). This involves the process of charging and discharging techniques. This helps in implementation of two way communication system interacting with the grid and electric vehicle by Wanaka (2021). This includes various control strategies with sensors and hardware implementation in the electric vehicles. This helps to improve the efficiency through attaining stability in the system.

With the introduction of AI in the charging and discharging system, it helps in attaining intelligent trends in upcoming decades. The load balancing is attained through the introduction of optimization techniques. They are done through deep learning techniques. This helps in the combination of renewable energy for charging and leads in the reduction of greenhouse gas effects and its constraints. This helps in charging the battery through extracting solar energy from solar panels that converts heat energy in electrical energy accompanied with charge controllers. Then it is stored in battery for further usages. They help in management of congestion and promotes voltage regulation by Kazi K (2022), & Sultana(2023).

Figure 7. Evolution of electric vehicle

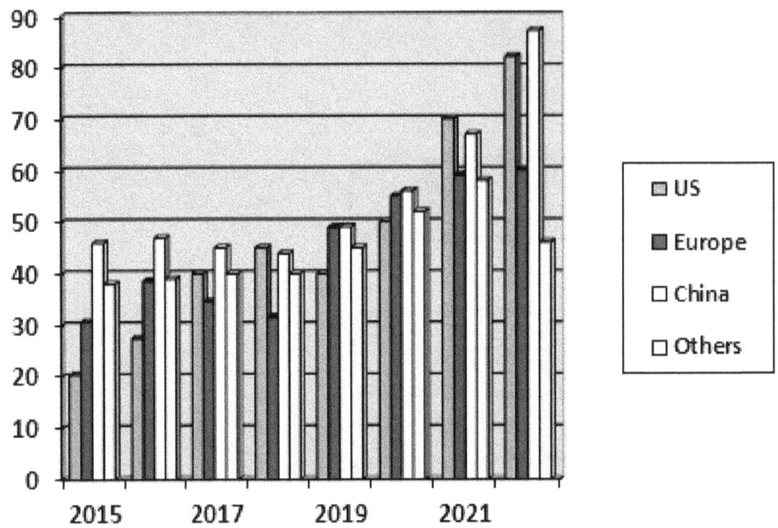

The Figure 7 demonstrates the evolution of electric vehicles across the Globe till date. The deep learning technique is defined as the subcategory of machine learning. It is consists of numerous layers that helps to process the data to produce the output. This includes extraction of optimum results with numerous layers that process them through optimization techniques.

Figure 8. Deep learning

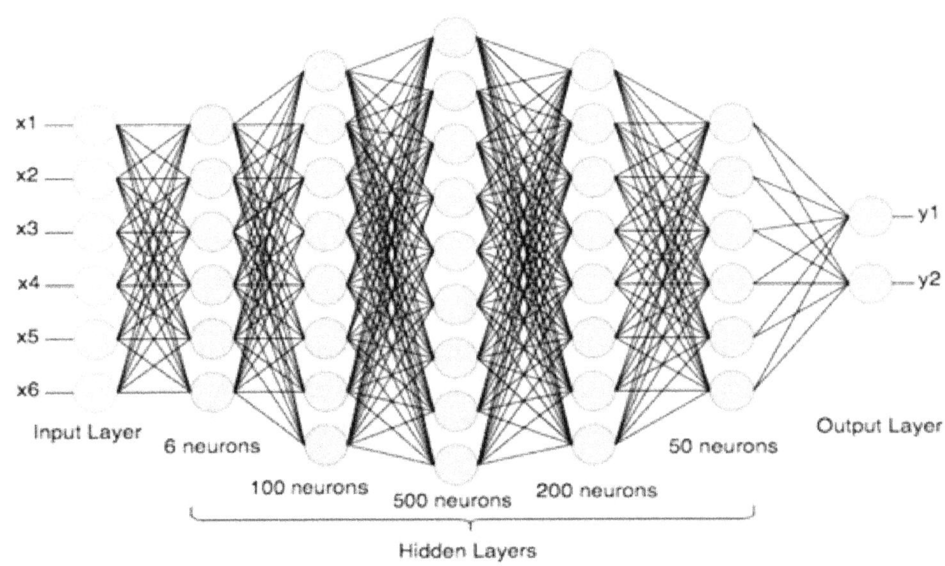

The Figure 8 represents the deep learning with numerous input and hidden layers that process the output. This is implemented to reduce the human interference in the functioning. This is largely used language and speech processing to improve automation with accurate results in the electrical vehicles. The scheduling is done through the optimization techniques. It is composed of input layer, hidden layer and the output layers. This works based upon the functioning of human intelligence. It resembles the neural network of human brain. It includes various data and algorithm to test and train the prototype.

The constraints are solved through the genetic algorithm. The genetic algorithm is a heuristic approach is solve the problems to achieve the optimum results. This is enhanced through the techniques of natural selection.

Figure 9 demonstrates Genetic Algorithm process in obtaining the optimum outcome.

Figure 9. GA-Genetic algorithm

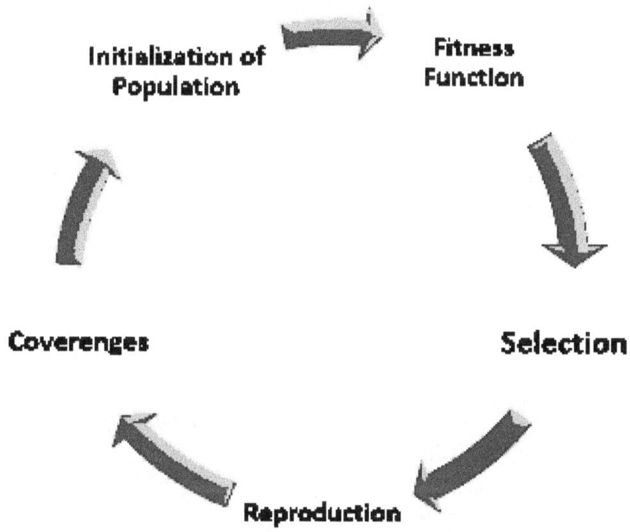

Figure 10 represents priority scheduling of loads for charging the electric vehicles to maintain demand side management.

GA is used as an optimization technique to schedule the loads based upon the priority. The electric vehicle's charging is based upon the rush and off rush hours. This reduces the demand of electric power from the grid. This also helps to maintain the balance of power for the domestic and industrial usages.

Figure 10. Priority scheduling

```
function_inputs = [4,-2,3.5,5,-11,-4.7]
desired_output = 44

def fitness_func(solution, solution_idx):
    output = numpy.sum(solution*function_inputs)
    fitness = 1.0 / (numpy.abs(output - desired_output) + 0.000001)
    return fitness

fitness_function = fitness_func

def on_start(ga_instance):
    print("on_start()")

def on_fitness(ga_instance, population_fitness):
    print("on_fitness()")

def on_parents(ga_instance, selected_parents):
    print("on_parents()")

def on_crossover(ga_instance, offspring_crossover):
    print("on_crossover()")
```

Figure 11. Model of BEV

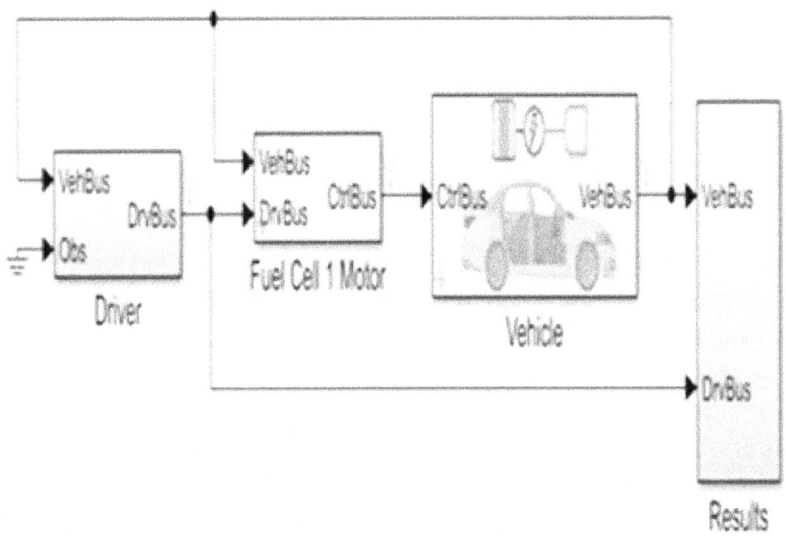

SIMULATION IMPLEMENTATION AND OUTCOME

1. Implementation of BEV

The logic utilised for BEV modelling is shown in Fig. 11, where the set drive cycle is given as input to the PID controller for generating the necessary acceleration or braking force after being compared to the actual speed of the vehicle. The driving cycle is a common pattern that displays speed in millimetres per second (m/s) in relation to time. Regeneration is subject to several restrictions based on the battery SOC and driving circumstances, such as the required braking effort and vehicle speed. The regenerative brake, whose braking force is divided by a fuzzy logic controller, is assisted by the mechanical brake.

The result of the propelling and regenerative torques is the motor output torque. Losses in motor power are also computed. The input power of an induction motor is the result of adding the motor's output power and motor losses. The tractive effort needed for speed estimation is calculated by the driveline model as shown in Figure. The torque loss component within the vehicle is also calculated by the model. Figure depicts the speed estimation model. The total tractive effort is calculated by summing the mechanical braking force and the net torque applied to the wheels by the engine. It determines the acceleration to determine the vehicle's speed. The model assumes that the battery is made of lithium iron phosphate (LiFePO4).

The BEV model is thoroughly analysed for many conventional drive cycles, including (Figure 12):

Figure 12. Proposed power and break of EV system

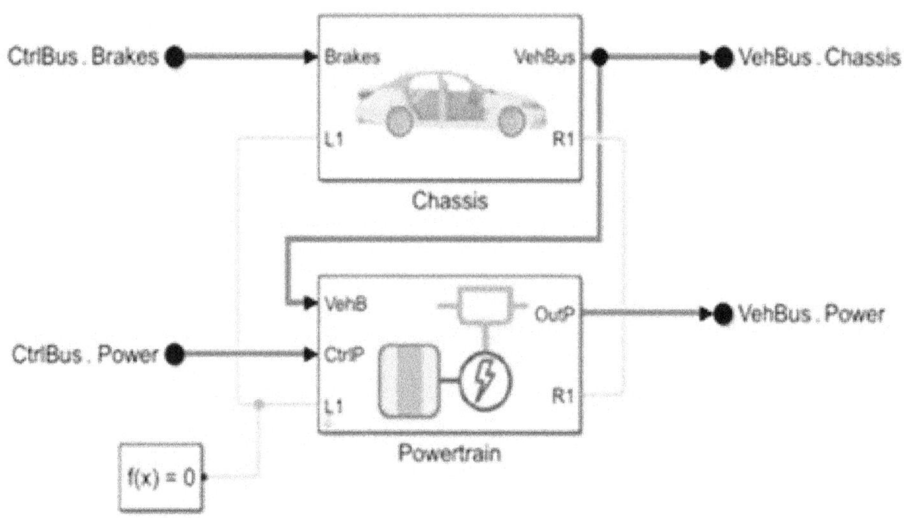

The engine emulate is shown in Figure 13 below, where the torque envelope for motoring and regeneration is developed using the motor speed, regenerative torque, and accelerator as inputs. The sum of the regenerative and propelling torques in a motor is its output torque. Additionally, the motor power losses are computed. The sum of the motor's output power and motor losses is the induction motor's input power. The tractive effort needed to estimate the speed is calculated by the driveline model, as seen in

Figure The torque loss component in the car is also computed by the model. Figure displays the speed estimation approach. The overall torque of the motor applied to the wheels plus the mechanical braking force equals the total tractive effort. It determines the acceleration in order to estimate the vehicle's speed.

The battery design that determines the battery's SOC and current is shown in Figure 14. Lithium iron phosphate (LiFePO4) is presumed as the battery within the design. The necessary input power determines the battery's Ah capacity.

An H-LDC, an L-LDC, and a nonisolated OBC make up the suggested system. To boost efficiency and power density while operating at a broad input and output voltage range, each LDC uses an LLC resonant network; a selecting switch is added to keep the H-LDC's resonant tank apart from the OBC by Yogita (2014).

Figure 13. Motor model

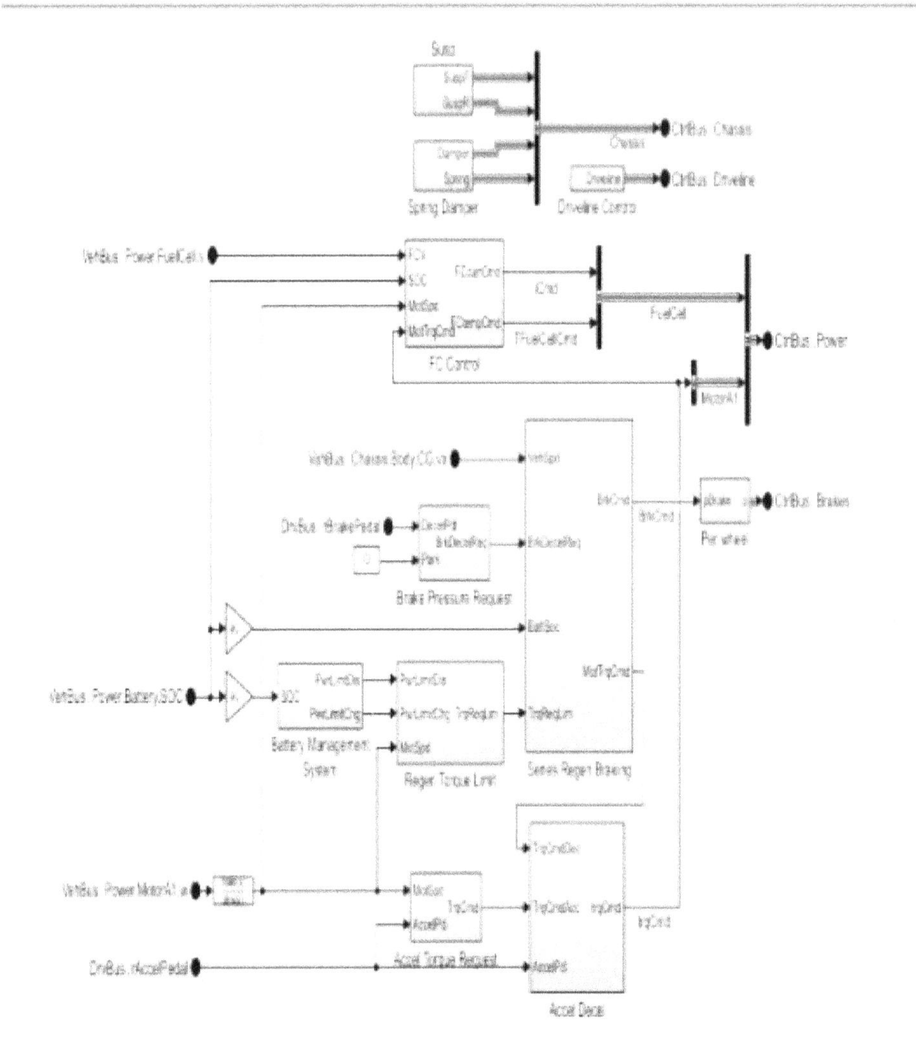

Figure 14. Battery model

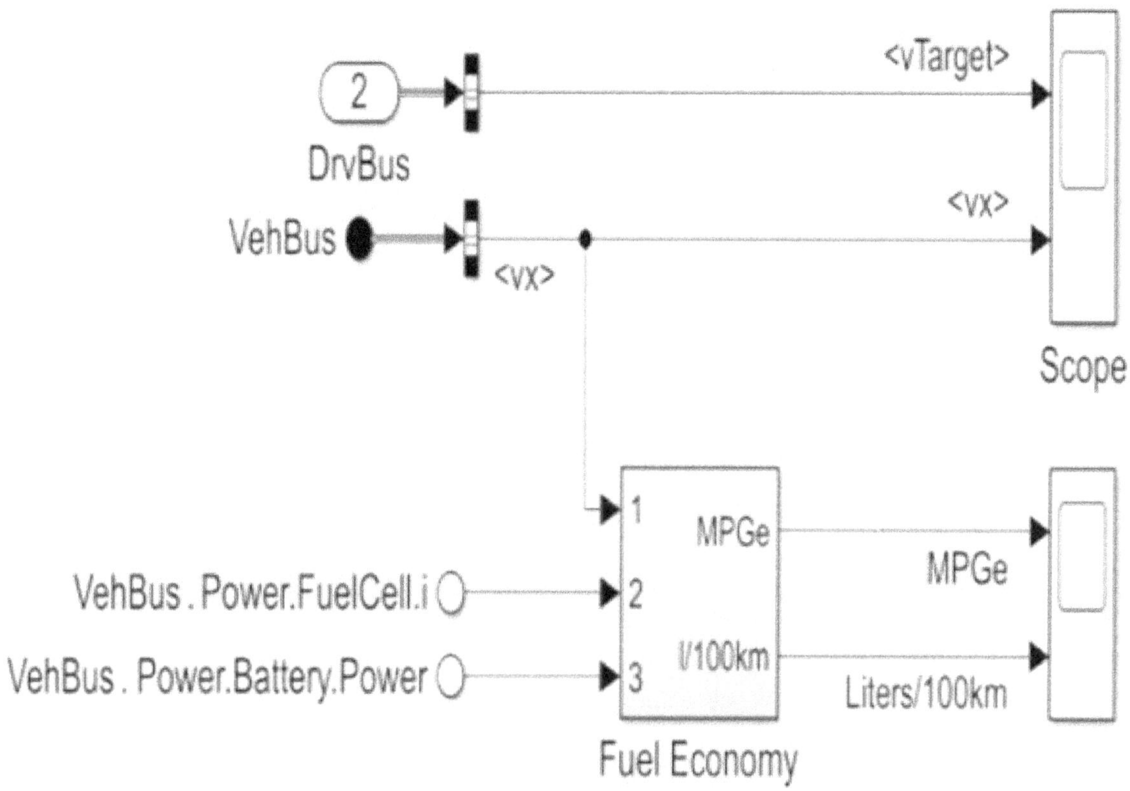

The EV battery, the on-board charger(OBC) (Figure 15,16), and the grid make up the vehicle's three primary parts. The charger has a 230 V, 50 Hz, 1 kW rating. A 60V, 50Ah, 3kWh Li-ion battery powers the EV. 400V is maintained for the intermediate capacitor, and 16.67A is maintained for the battery charging current.

In this example, an isolated full bridge converter is used to provide galvanic isolation between the battery and the grid. 400 V and 6 A is the rating of the high-frequency inverter that is attached to the transformer's primary side. The transformer's secondary side is connected to diodes with a 150 V and 20 A rating.

The braking system depicted in Figure 17 determines the necessary amount of regenerative force to be applied to the motor during non-stop operation. One computes an regenerative force. A gearbox with a G ratio connects the motor to the wheel. The symbol for the wheel's radius is r.

In electric vehicles (EVs), regenerative braking allows the vehicle to recycle its internal power and charge the battery (Figure 18). However, the battery's lifespan is impacted if power is applied to it without supervision. Many methods and strategies have been in use to prolong the battery's life. Regenerative braking is a technique used in electric vehicles (EVs) to recharge the battery and recycle electricity. However, the battery's life is impacted if it receives electricity without being in control. Various methods and approaches are already used to extend the battery's life. The graphs of the vehicle speed (m/s) are shown in Figure 19 below.

Figure 15. OBC model

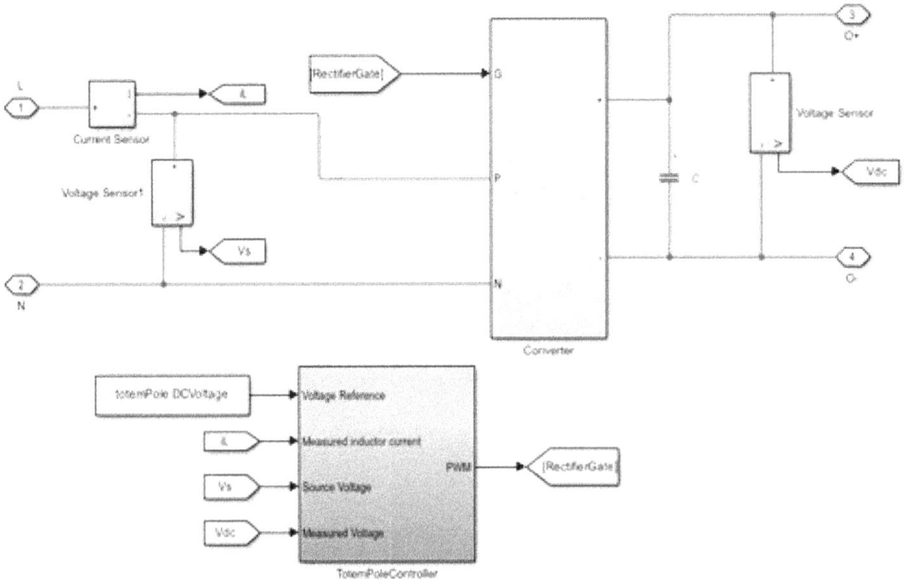

Figure 16. Configuration of the proposed battery charger

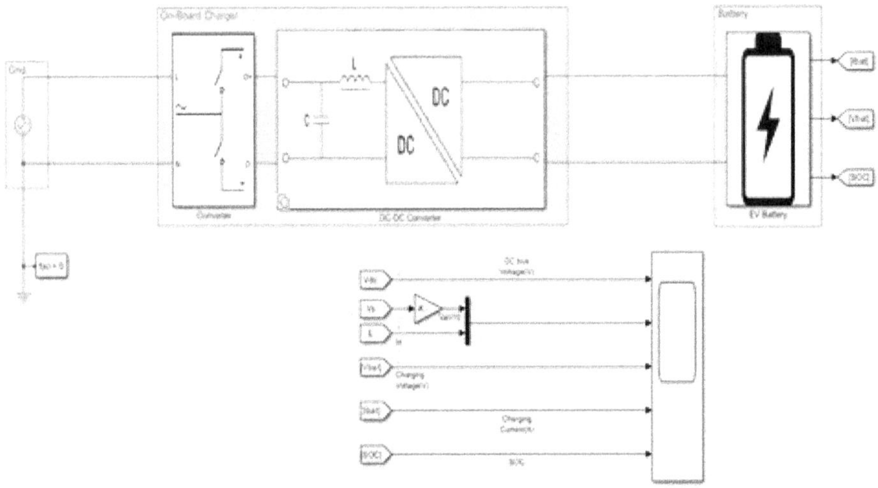

Figure 20 below illustrates the battery's power and state of charge for an improved Indian drive cycle, taking into account the urban and highway drive cycles independently. When the results of the ADVISOR software and MATLAB/SIMULINK have been contrasted, it is discovered that the results are coordinated. Rather than 90 km/h, the maximum EV speed in the Modified Indian drive cycle is 80 km/h.

Figure 17. Break model of vehicle

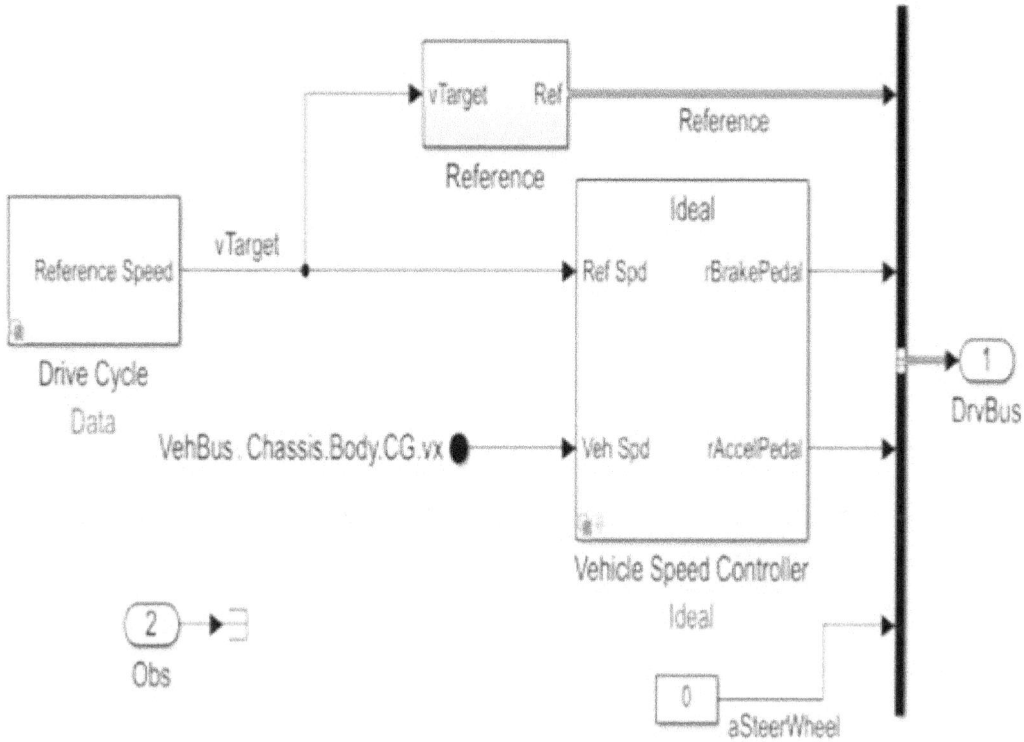

Figure 18. Power and break of EV system

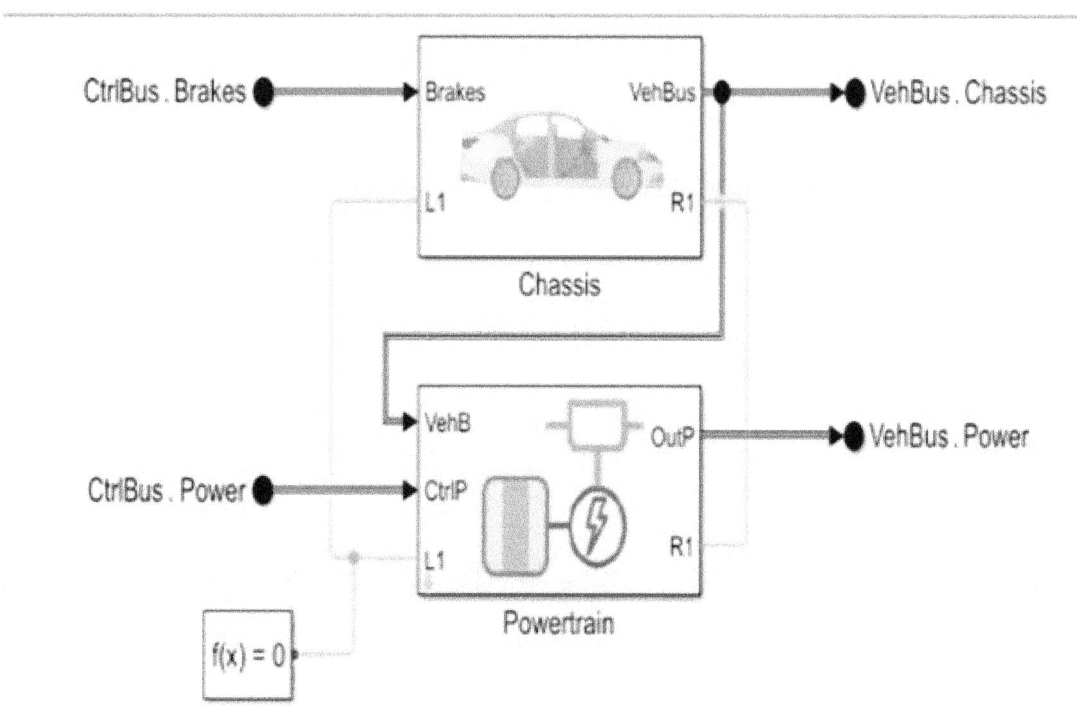

Figure 19. Speed vs. time

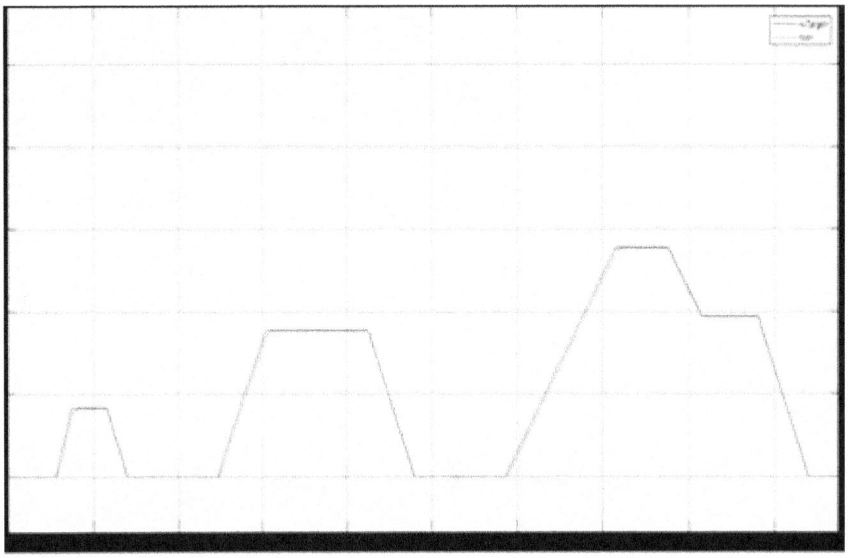

Figure 20. Battery power vs. time

Figure 21 shows the results of the simulation for battery charge current, batteries SoC, batteries terminal voltage, grid voltage and current, and DV voltage. IBC topologies were tested using hardware experiments and simulations to validate the developed machine model and its control. Because of the selected current references, the charger operates at unity power factor discharging, 0.9 power factor charging, and unity power factor charging.

The plot below (Figure 21) shows the DC bus voltage, grid current, battery terminal voltage, and charging current for both the ideal switching totem pole converter and the typical DC-DC converter. The intermediate capacitor voltage stabilises at 400V, the grid voltage and current are in phase, and the 16.67A battery charging rate is constant.

Figure 21. Simulated results

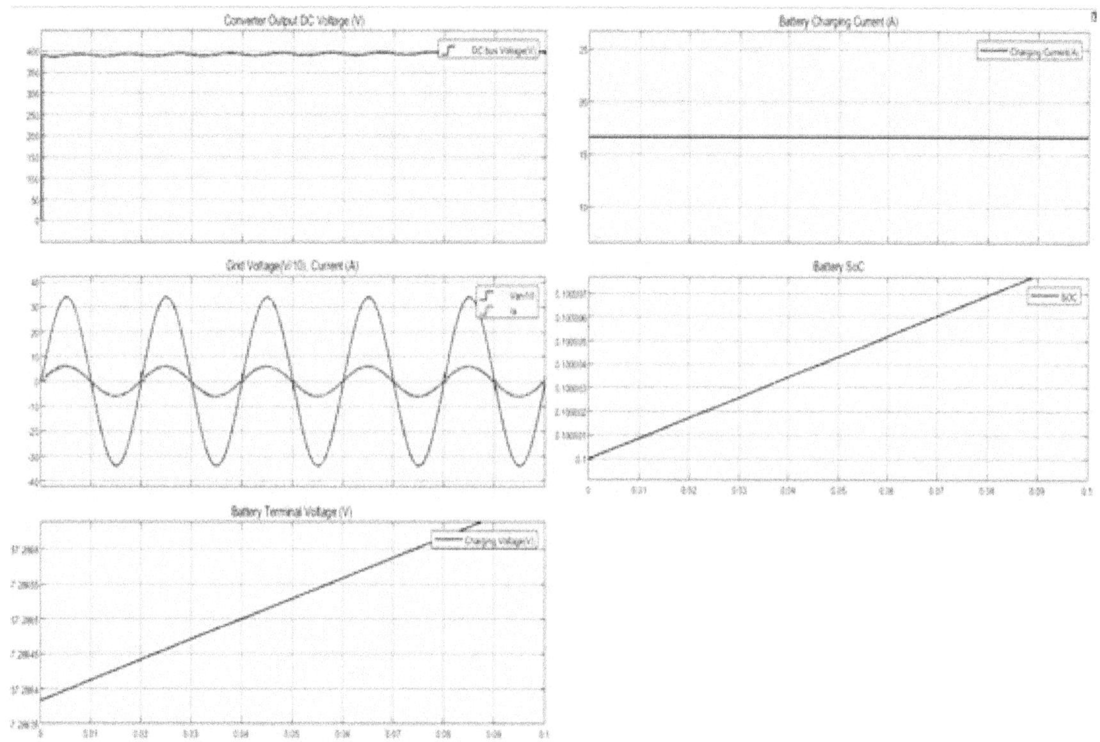

2. Implementation of HEV

The implementation of e-vehicles with AI and cloud computing techniques with sensors are evaluated and established in matlab Simulink. This helps to obtain the performance efficiency and its outcome.

Figure 22 demonstrates the modelling of e-vehicles in matlab Simulink. This includes various sensors with regenerative braking, motors and batteries.

Figure 23 represents the modelling of vehicle body with sensors and storage systems.

Figure 22. Modelling of e-vehicles

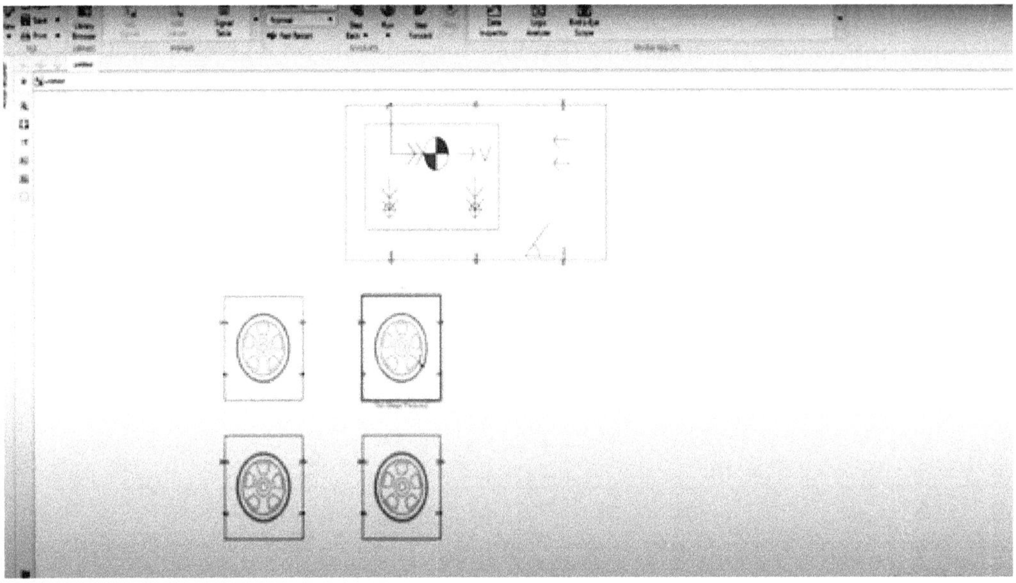

Figure 23. Vehicle body

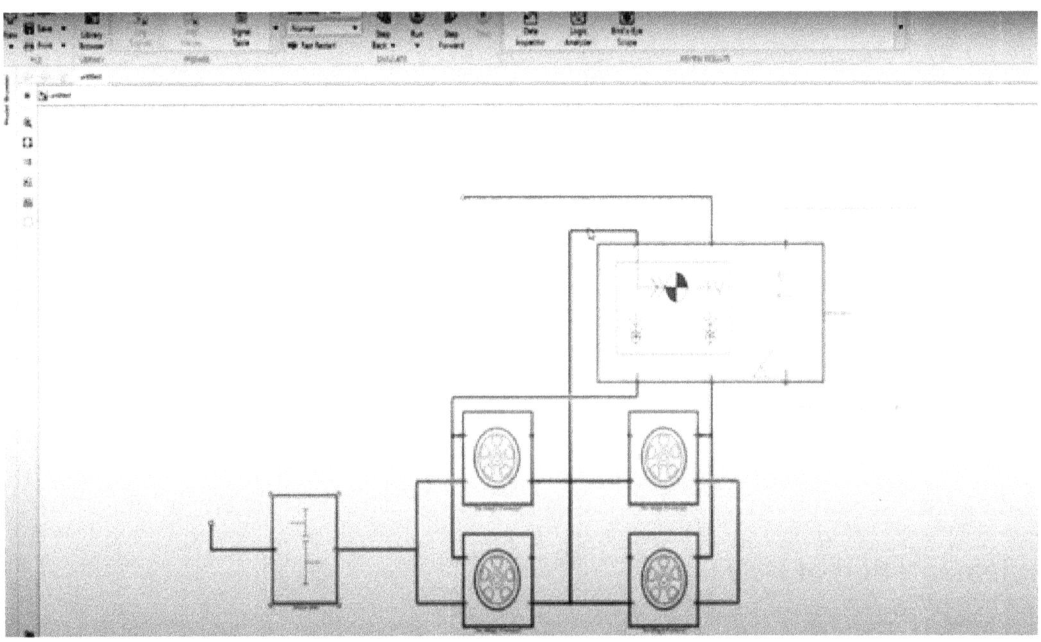

Figure 24 represents the implementation of electric vehicles with renewable energy systems. The renewable energy here utilised is the solar energy source. The solar panel helps in the conversion of energy and saves in the battery for further usages.

Figure 25 represents the implementation of regenerative braking schemes in the electric vehicles.

Figure 26 demonstrates the battery's performance parameter analysis. This helps in the calculation of charging and discharging of battery to enhance its life cycle.

Figure 27 demonstrates the performance efficiency of the electrical vehicles. This is calculated through various parameters involving battery's state of charge to regenerative braking.

Figure 24. Electric vehicles with RES

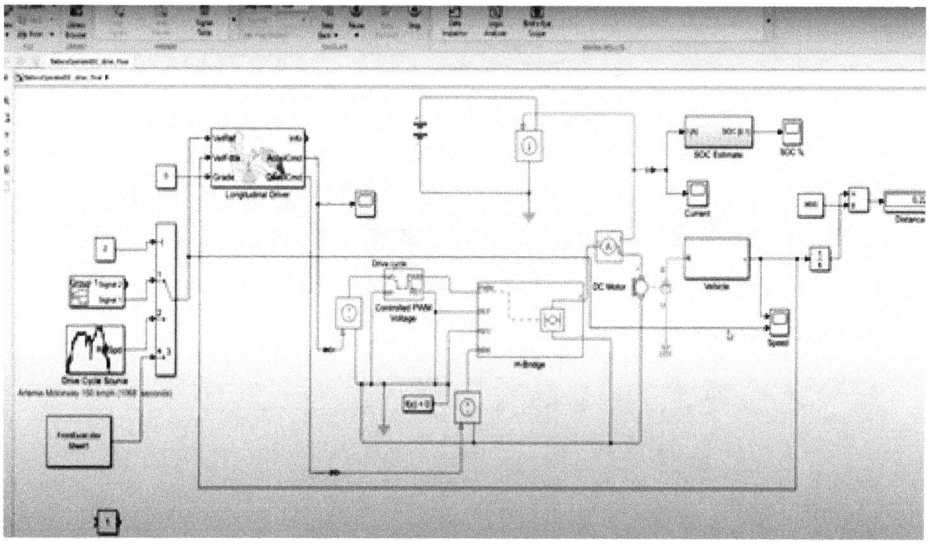

Figure 25. Regenerative braking

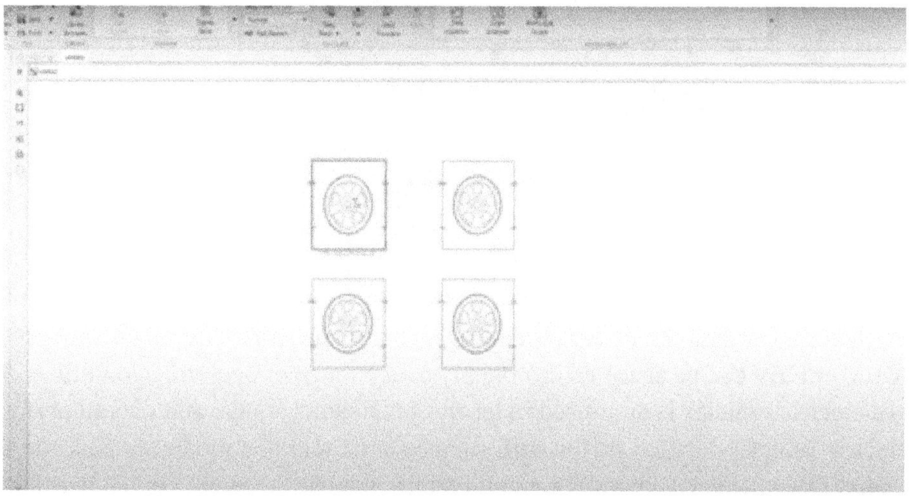

Figure 26. Battery's performance analysis

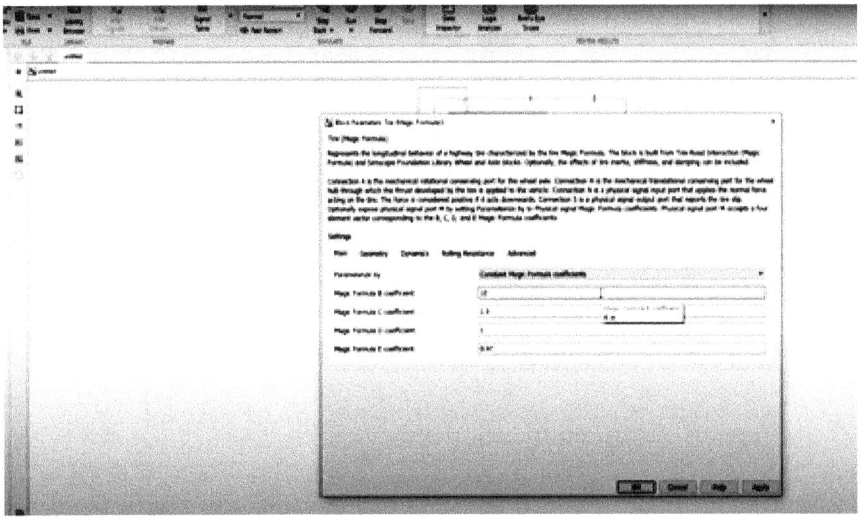

Figure 27. Performance efficiency

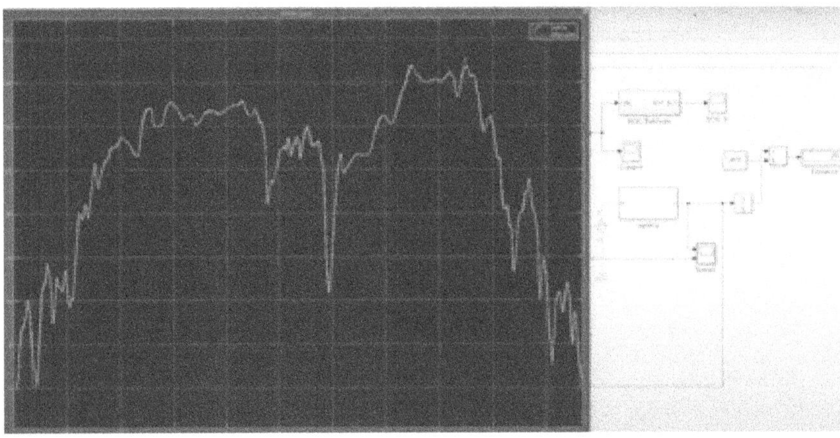

6. CONCLUSION

The simulation of electric vehicles has become increasingly important in the development of efficient and reliable vehicles. Through the use of modelling and simulation tools such as Simulink, the performance of a BEV or HEV can be analysed thoroughly over a range of operating conditions. This ensures that the use of electric vehicles is optimised in terms of efficiency, range and overall performance. The success of electric vehicles depends on the performance of battery and motor systems and simulation is an essential part of their design and development. Electric vehicle development has accelerated over the last few decades. These are due to their environmentally friendly nature and decreased use of fossil fuels. This contributes to a greener environment by preventing the release of various toxic pollutants into the environment. One of the things keeping consumers from purchasing EVs is the high price tag. In order to combat this, government has promoted the use of EVs more widely by offering financial incentives for business vehicles. But foreign batteries are the primary cause of the continued at least 30% premium for electric cars. Under the Center's Faster Adoption and Manufacturing of Hybrid and Electric Vehicles (FAME) plan, electric commercial vehicles began receiving subsidies in 2015. Experts say that a dearth of appropriate charging stations and the EV industry's reliance on imported batteries and parts are its two main issues. But by 2020, things might be different. The discovery of lithium reserves in Bangalore will surely result in a drop in the cost of battery imports. During the last quarter, manufacturers unveiled several new EV models, some of which promise a significantly longer range than the 80–90 km that EVs currently offer. The EVs are implemented with the help of AI to improve overall implementation of vehicles. This improves the operation of electric vehicles by enabling two-way communication via cloud computing techniques. We checked the performance of EV on Electric vehicles with RES, Regenerative braking, Battery's Performance analysis basis. To eliminate demand, the genetic algorithm is used to optimise the use of electric energy from the grid based on priority and peak hours. Thus the use of electrical vehicles with AI techniques is the way forward measures to enhance various benefits with reducing the constraints caused by the conventional vehicles.

Fast charging is made possible by the integrated battery chargers (IBC) that are suggested in this work. In contrast to exterior chargers that are designed only for battery charging, the IBC solution is smaller and lighter. Due to the additional sources of losses (contactors and rotor magnets), the efficiency of IBCs may differ from that of dedicated battery chargers; however, these losses can be minimised by optimising the drive during the design stage. The integrated battery chargers' efficiencies were not formally examined in this thesis because the lab prototypes do not accurately reflect the intended system. Finite element analysis was used to examine the effects on the six-phase synchronous machine when it was an element from an integrated battery charger. The effects that were examined were the distribution of force and flux densities in the machine airgap during the torque cancellation charging operation. The analysis demonstrates that the charging operation's current limit is determined by the flux density distribution that results in stator saturation. It was demonstrated that the charging current ought to be lower than the device's rated current. Additionally, it was demonstrated that for a six-phase synchronous motor with the same rating, a full pitch winding is preferable to a short pitch (by one slot) winding in order to prevent saturation. There were also a number of design suggestions offered, based on the motor windings' inductance.

REFERENCES

Aavula, & Deshmukh, & Mane. (2022). Design and Implementation of sensor and IoT based Remembrance system for closed one. *Telematique*, *21*(1), 2769–2778.

Akansha. (2022). Email Security. *Journal of Image Processing and Intelligent Remote Sensing*, *2*(6).

Al-Turjman, F., & Alturjman, S. (2020). 5G/IoT-enabled UAVs for multimedia delivery in industry-oriented applications. *Multimedia Tools and Applications*, *79*(13–14), 8627–8648. doi:10.1007/s11042-018-6288-7

Alhayani, B. (2022). 5G standards for the Industry 4.0 enabled communication systems using artificial intelligence: Perspective of smart healthcare system. *Applied Nanoscience*, (0123456789). Advance online publication. doi:10.1007/s13204-021-02152-4 PMID:35096498

Da Xu, L. (2021). Emerging Enabling Technologies for Industry 4.0 and Beyond. *Information Systems Frontiers, 2022*. Advance online publication. doi:10.1007/s10796-021-10213-w

de Oliveira Neto, G. C. (2022). *How can Industry 4.0 technologies and circular economy help companies and researchers collaborate and accelerate the transition to strong sustainability? A bibliometric review and a systematic literature review, no. 0123456789*. Springer Berlin Heidelberg.

Gil, S., Zapata-Madrigal, G. D., García-Sierra, R., & Cruz Salazar, L. A. (2022). Converging IoT protocols for the data integration of automation systems in the electrical industry. *Journal of Electrical Systems and Information Technology*, *9*(1), 1–21. doi:10.1186/s43067-022-00043-4

Gouse Mohiuddin Kosgiker. (2018). Machine Learning- Based System, Food Quality Inspection and Grading in Food industry. *International Journal of Food and Nutritional Sciences*, *11*(10), 723–730.

Halli, U. M. (2022a). Nanotechnology in E-Vehicle Batteries. *International Journal of Nanomaterials and Nanostructures.*, *8*(2), 22–27.

Halli, U. M. (2022b). Voltage Sag Mitigation Using DVR and Ultra Capacitor. *Journal of Semiconductor Devices and Circuits.*, *9*(3), 21–31p.

Hiekata, K., Wanaka, S., Mitsuyuki, T., Ueno, R., Wada, R., & Moser, B. (2021). Systems analysis for deployment of internet of things (IoT) in the maritime industry. *Journal of Marine Science and Technology*, *26*(2), 459–469. doi:10.1007/s00773-020-00750-5

Hosmani, S. M. (2022). Implementation of Electric Vehicle system. *Gradiva Review Journal*, *8*(12), 444–449.

Kazi, K. (2022a). Hybrid optimum model development to determine the Break. *Journal of Multimedia Technology & Recent Advancements*, *9*(2), 24–32.

Kazi, K. (2022b). Big data and HR Analytics in Talent Management: A Study. *Recent Trends in Parallel Computing*, *9*(3), 16–26.

Kazi, K. S. (2022c). IoT-Based Healthcare Monitoring for COVID-19 Home Quarantined Patients. *Recent Trends in Sensor Research & Technology*, *9*(3), 26–32.

Kazi, S. S. L. (2023). Integrating IoT and Mechanical Systems in Mechanical Engineering Applications. *Journal of Mechanisms and Robotics*, *8*(3), 1–6.

Kazi, V. (2023). Deep Learning, YOLO and RFID based smart Billing Handcart. *Journal of Communication Engineering & Systems*, *13*(1), 1–8.

Kutubuddin, K. (2022). Smart Grid energy saving technique using Machine Learning. *Journal of Instrumentation Technology and Innovations*, *12*(3), 1–10.

Kutubuddin, K. (2022). Detection of Malicious Nodes in IoT Networks based on packet loss using ML. *Journal of Mobile Computing, Communication & Mobile Networks*, *9*(3), 9–16.

Liyakat, K. K. S. (2023). Machine Learning Approach Using Artificial Neural Networks to Detect Malicious Nodes in IoT Networks. In P. K. Shukla, H. Mittal, & A. Engelbrecht (Eds.), *Computer Vision and Robotics. CVR 2023. Algorithms for Intelligent Systems*. Springer. doi:10.1007/978-981-99-4577-1_3

Liyakat, K. K. S. (2024). Machine Learning Approach Using Artificial Neural Networks to Detect Malicious Nodes in IoT Networks. In S. K. Udgata, S. Sethi, & X. Z. Gao (Eds.), *Intelligent Systems. ICMIB 2023. Lecture Notes in Networks and Systems* (Vol. 728). Springer. doi:10.1007/978-981-99-3932-9_12

Liyakat, S. (2022). *Multiple object detection and classification based on Pruning using YOLO. Lambart Publications.*

Maheswari, C., Priyanka, E. B., Thangavel, S., Vignesh, S. V. R., & Poongodi, C. (2020). Multiple regression analysis for the prediction of extraction efficiency in mining industry with industrial IoT. *Production Engineering*, *14*(4), 457–471. doi:10.1007/s11740-020-00970-z

Maqbool, R., Saiba, M. R., & Ashfaq, S. (2022). Emerging industry 4.0 and Internet of Things (IoT) technologies in the Ghanaian construction industry: Sustainability, implementation challenges, and benefits. *Environmental Science and Pollution Research International*, *30*(0123456789), 37076–37091. Advance online publication. doi:10.1007/s11356-022-24764-1 PMID:36565423

Mulani, A. O. (2019). Effect of Rotation and Projection on Real time Hand Gesture Recognition system for Human Computer Interaction. *Journal of The Gujrat Research Society*, *21*(16), 3710–3718.

Nerkar, P. M., & Shinde, S. S. (2023). Monitoring Fresh Fruit and Food Using Iot and Machine Learning to Improve Food Safety and Quality. *Tuijin Jishu/Journal of Propulsion Technology, 44*(3), 2927 – 2931. https://propulsiontechjournal.com/index.php/journal/article/view/914

Nikita, K., & Supriya, J. (2020). Design of Vehicle system using CAN Protocol. *International Journal for Research in Applied Science and Engineering Technology*, *8*(V), 1978–1983. doi:10.22214/ijraset.2020.5321

Pankaj, & Kulkarni. (2019). Implementation of Low Power and area efficient carry select Adder. *International Journal of Research in Engineering, Science and Management*, *2*(4), 183–184.

Peng, Y., & Wu, I. C. (2021). A cloud-based monitoring system for performance analysis in IoT industry. *The Journal of Supercomputing*, *77*(8), 9266–9289. doi:10.1007/s11227-021-03640-8

Pradeepa, M. (2022). Student Health Detection using a Machine Learning Approach and IoT. *2022 IEEE 2nd Mysore sub section International Conference (MysuruCon).*

Radanliev, P., De Roure, D., Nicolescu, R., Huth, M., & Santos, O. (2022). Digital twins: Artificial intelligence and the IoT cyber-physical systems in Industry 4.0. *International Journal of Intelligent Robotics and Applications*, *6*(1), 171–185. doi:10.1007/s41315-021-00180-5

Rahman, A., Chakraborty, C., Anwar, A., Karim, M. R., Islam, M. J., Kundu, D., Rahman, Z., & Band, S. S. (2022). SDN–IoT empowered intelligent framework for industry 4.0 applications during COVID-19 pandemic. *Cluster Computing*, *25*(4), 2351–2368. doi:10.1007/s10586-021-03367-4 PMID:34341656

Rane, S. B., & Narvel, Y. A. M. (2022). Data-driven decision making with Blockchain-IoT integrated architecture: A project resource management agility perspective of industry 4.0. *International Journal of System Assurance Engineering and Management*, *13*(2), 1005–1023. doi:10.1007/s13198-021-01377-4

Rathee, G., Balasaraswathi, M., Chandran, K. P., Gupta, S. D., & Boopathi, C. S. (2021). A secure IoT sensors communication in industry 4.0 using blockchain technology. *Journal of Ambient Intelligence and Humanized Computing*, *12*(1), 533–545. doi:10.1007/s12652-020-02017-8

Rosati, R., Romeo, L., Cecchini, G., Tonetto, F., Viti, P., Mancini, A., & Frontoni, E. (2022). From knowledge-based to big data analytic model: A novel IoT and machine learning based decision support system for predictive maintenance in Industry 4.0. *Journal of Intelligent Manufacturing*, *34*(1), 107–121. doi:10.1007/s10845-022-01960-x

Salunke. (2022). Announcement system in Bus. *Journal of Image Processing and Intelligent Remote Sensing, 2*(6).

Sha, L., Xiao, F., Chen, W., & Sun, J. (2018). IIoT-SIDefender: Detecting and defense against the sensitive information leakage in industry IoT. *World Wide Web (Bussum)*, *21*(1), 59–88. doi:10.1007/s11280-017-0459-8

Shreya & Afrin. (2022). 3D E-Commers using AR. *International Journal of Information Technology and Computer Engineering, 2*(6), 18-27.

Shweta Nagare. (2014). Different Segmentation Techniques for brain tumor detection: A Survey. *MM-International Society for Green*, *Sustainable Engineering and Management*, *1*(14), 29–35.

Swami. (2022). Sending notification to someone missing you through smart watch. *International Journal of Information Technology and Computer Engineering, 2*(8), 19 – 24.

Vaijnath, S. P., & Prajakta, M. (2022). Smart safety device for women. *International Journal of Aquatic Science*, *13*(1), 556–560.

Waghmare. (2022). Smart watch system. *International Journal of Information Technology and Computer Engineering, 2*(6), 1 - 9.

Waghmode. (2022). *Voltage Sag mitigation in DVR based on Ultra capacitor. Lambart Publications.*

Wale, A. D., & Dipali, R. (2019). Smart Agriculture System using IoT. *International Journal of Innovative Research In Technology*, *5*(10), 493–497.

Yogita Shirdale,. (2014). Analysis and design of Capacitive coupled wideband Microstrip antenna in C and X band: A Survey. *Journal GSD-International Society for Green*, *Sustainable Engineering and Management*, *1*(15), 1–7.

Chapter 15
Technological Trends in Electric Passenger Car Segment in India and Medium-Term Demand for Magnet-Based Rare Earth Materials

Arpita Pandey
https://orcid.org/0009-0000-3678-0908
The Energy and Resources Institute, New Delhi, India

Rudrodip Majumdar
https://orcid.org/0000-0001-9362-8945
Energy, Environment, and Climate Change Programme, National Institute of Advanced Studies, Bengaluru, India

ABSTRACT

The green mobility aspirations of India have necessitated substituting conventional vehicles with electric motor-driven counterparts. The rare earth permanent magnet-based motor is the dominant solution for electric vehicle motor architectures as it ensures energy efficiency with reduced kerb weight and increased vehicle range. This study estimates the magnet-based rare earth demand for the realistic aspirational growth trajectories of electric four-wheelers up to 2030. During the first phase (2021-25), a high year-on-year growth rate for the e-car segment is assumed, which creates the base for electric vehicle penetration, whilst the second phase (2026-30) envisages a rapid product substitution at the cumulative level by 2030. The estimated total demand for rare earth materials during 2021-30 is between 1434 and 3584 tonnes. Analysis indicates that India could leverage its strong position in the global automotive market and geopolitical proximity to the favourable global powers for collaboration in research and development, technology transfer, and securing raw material supply.

DOI: 10.4018/979-8-3693-2611-4.ch015

Nomenclature

GDP	Gross Domestic Product
HCVs	Heavy and Commercial Vehicles
R&D	Research and Development
CAGR	Compound Annual Growth Rate
ICE	Internal Combustion Engine
EVs	Electric Vehicles
PLI	Production Linked Incentive
FAME	Faster Adoption and Manufacturing of Hybrid and Electric Vehicles
NEM	National Electric Mobility
2W	Two-Wheeler
3W	Three-Wheeler
NdFeB	Neodymium Iron Boron
REE	Rare Earth Elements
E4W	Electric Four-Wheeler
SUVs	Sport Utility Vehicles
MPVs	Multi-Purpose Vehicles
Y-o-Y	Year-on-Year
PMSM	Permanent Magnet Synchronous Motor
LREE	Light Rare Earth Elements
HREE	Heavy Rare Earth Elements
SIAM	Society of Indian Automotive Industry
SMEV	Society of Electric Vehicles
IGP	Initial Growth Phase
RSP	Rapid Substitution Phase
kW	Kilo Watt
Nd	Neodymium
Dy	Dysprosium
Pr	Praseodymium
BSM	Beach Sand Mineral
QUAD	Quadrilateral Security Dialogue
ASEAN	Association of Southeast Asian Nations

1. INTRODUCTION

The Indian automotive industry contributes to 7.1% of the country's overall GDP, and it also contributes 49% to the manufacturing GDP. The total annual automotive production stood at 22.7 million units in FY 2020-21. India exports about 18% of the total automotive production (SIAM, 1998-2020; Invest India, 2022; Press Trust of India, 2021). The industry comprises passenger vehicles, heavy and commercial vehicles (HCVs), three-wheelers, and two-wheelers. India's automotive sector has been strengthened by a few key factors, such as the availability of a skilled engineering labour force at a lower cost, and tie-ups with leading foreign companies and R&D institutes. The conducive arrangements have led to large manufacturing and assembling capabilities in the country (Miglani, 2019, pp 439-463). Among the leading global automotive manufacturing countries, India's labour market offers a lower cost of production for the automotive industry in terms of minimum hourly wage (US$ 0.9) as compared to that in Germany (US$ 46.4), Japan (US$ 35.4), and China (US$ 2.8) (Gonsalves, 2019). Moreover, the large unsaturated passenger car market accompanied by a high middle-class population makes India one of the fastest-growing automotive markets (Miglani, 2019, pp 439-463). Currently, the passenger car segment accounts for 13% of the total production in India's automotive industry. Globally, India stands 4th in the production of passenger cars, preceded by China, Japan, and Germany (SIAM, 1998-2020; SESEI, 2018).

In FY2019, the country produced a total of 30.91 million units of automotive vehicles, in which the passenger segment exhibited moderate growth. The shortage in the supplies of essential raw materials and critical automotive parts during the last two quarters (Q3 and Q4) of FY2019-20 negatively affected the sector. The nationwide lockdown from 23rd March 2020 onwards, prompted by the COVID-19 pandemic, further added to the de-growth. However, the social distancing norms led to the preference for personal mobility, which in turn resulted in a rather quick recovery of the passenger car market in India. The passenger vehicles' retail increased by 43.45% in June 2021, compared to the level observed during June 2020 (HT Auto Desk, 2021). The passenger car segment in India is envisaged to reach a high penetration level of 175 vehicles per 1000 population in 2040 from the current level of 22 vehicles per 1000 population with the growth in the economy (Abbas, 2018). It indicates that the industry will continue to witness growth in the near-to-medium term. Assuming recovery from the pandemic effects, it is expected to grow at a faster pace (CAGR of about 9%).

The Indian transport sector contributes to 13.2% of the country's total carbon dioxide emissions (Saurabh and Majumdar, 2022; Saurabh and Majumdar, 2023a). Amidst the global concerns around climate change, the current challenges of policymakers include measures to ensure stable economic growth trajectories along with reduced emissions (Kamath et al., 2022; Pujari et al., 2023; Saurabh and Majumdar, 2023b). The published literature pertinent to automotive research suggests two pathways to reduce vehicular emissions, a) enforcement of stringent emission norms by installing improved and advanced catalytic convertors for ICE-based conventional vehicles, and b) transition towards electric propulsion drive systems through the fully electric or hybrid category of vehicles (Chan, 2016; Vidhi and Shrivastava, 2018; Dey and Mehta, 2020; Mukherjee et al., 2016).

The e-mobility offers a substantial environmental benefit since it is devoid of local pollutant emissions. Electric Vehicles (EVs) would also help the country reduce dependence on imported Petro-based fuels. Although EVs have been part of the automotive industry for over a decade, they have gained prominence since 2017. The global initiatives for reducing carbon emissions and the aspirations of the Government of India to achieve a predominantly green energy-driven economy are likely to boost the production of EVs. Electric vehicles are also expected to serve as temporary energy storage facilities that can supply electricity to the grid in case of local emergencies, which would also augur well with the larger framework of futuristic vehicle grid integration (Khwaja et al., 2023).

Through the implementation of different schemes such as the Production Linked Incentive (PLI), FAME - I and FAME – II, as well as the state-wise EV policies, the Government of India (GOI) is pushing towards quicker adoption of e-mobility and local manufacturing of electric vehicles including key automotive parts & components. These schemes are mainly aimed at promoting the manufacturing of automotive parts, and batteries and the creation of charging infrastructure through the initial investment support and tax exemptions. A direct monetary benefit to the consumer amounting to about US$ 2,000 (assuming 1 USD = Rupees 75) has been declared by the Government of India for the purchase of every electric passenger car. The purchase of electric cars is also being encouraged through relaxed loan norms (NITI Aayog and RMI India, 2022; BEE, 2020). The National Electric Mobility (NEM) Mission envisages an EV penetration of 30% by 2030. This has emphasised the substitution of all types of vehicles ranging from short-range vehicles to the ones used in the last mile connectivity in the near to medium term.

India's domestic automotive market is exhibiting a transition towards e-mobility largely through the ongoing rapid substitution of ICE-based two-wheeler (2W) and three-wheeler (3W) products by their electric counterparts. The passenger car segment is likely to follow a similar trend soon.

The motor architecture adopted in electric cars provides a basis for the estimation of the future demand for magnetic raw materials. The two widely used motor architectures adopted in e-car powertrains are induction motors and permanent magnet-based electric motors. Recent trends show that vehicle manufacturers prefer to use NdFeB (Neodymium Iron Boron) rare earth permanent magnets in motor architectures.

The present study pivots around the estimation of rare earth materials (neodymium, praseodymium, and dysprosium) demand between 2021 and 2030 for a realistic aspirational growth trajectory of the electric passenger car segment in India. Along with the estimated RE demand, this study also aims to delineate the possible challenges and bottlenecks that the E4W segment is likely to witness in case of a supply shortage.

The indigenous supply chain of essential components such as motors and permanent magnets will help reduce import dependency and production costs. The Indian automotive industry employs about 35 million people. It is therefore expected that the economic benefits of the planned manufacturing ecosystem will trickle down to the other sectors as well. The issues related to raw-material supply risks, price volatility, technological upgradation, etc., could be addressed by adopting suitable business strategies. Such a strategy should inherently delineate the pathways to promote collaborations with advanced industrial powers for R&D, technology transfer, and supply of raw materials. The infrastructural and technological preparedness in the country has been evaluated for the envisaged green mobility transition. Further, the key bottlenecks and the possible mitigation measures are also highlighted in this work.

1.1 Status of the Total Four-Wheeler Industry in India

For the period 2008-2018 (pre-covid years), passenger car production in India grew at a CAGR of about 13% and considering some slowdown even during the pre-COVID period, it was well poised to grow at a CAGR of 9 -10% CAGR. However, the COVID-19 pandemic has slowed down the growth substantially. Over the past three years (2019-21), the production of passenger cars has been observed to grow at a low CAGR of about 4% (SIAM, 1998-2020).

1.2 The Product-type Distribution of the Passenger Car Market

In the ICE-based conventional passenger car segment, the industry has six major manufacturers accounting for more than 89% of the market share and has a huge production line comprising different vehicle product types. A scrutiny of different product types prevalent over the recent years indicated a gradual substitution of hatchbacks and lower-range sedan cars by SUVs and MPVs (Sport Utility Vehicles and Multi-Purpose Vehicles) (Statista, 2021a). This shift in market dominance is primarily attributed to the better range, baggage space and seating capacity offered by SUVs and MPVs alike (Sharma, 2021). Although hatchbacks comprise more than 50% of the market, a decrease in market share is expected due to the shift towards higher-range vehicles.

1.3 Status of the Electric Four-Wheeler Segment in India

The E4Ws gained prominence in India in 2018. Since then, there has been a continuous Y-o-Y growth in the sales of e-passenger cars in India. The cumulative sales of E4Ws in India stood at 19577 units at the end of 2021.

The shift towards low-emission vehicles has resulted in an increased demand for E4Ws. The annual production of E4Ws in India reached 13600 units in 2021. Currently, the penetration of e-cars in the total 4W market in India at the cumulative level is negligible (~0.04%). The penetration of electric cars in the total annual production of 4Ws is also very small (~0.3%).

In 2021, Tata Motors and Morris Garage (MG) Motors accounted for 90% of the electric passenger car market in India. Both these companies use rare earth permanent magnet (PM) based electric motor architecture in their drivetrains.

A closer scrutiny of the currently available e-car models exhibits the dominance of the SUV category. SUVs comprise about 85% of the electric cars sold in India. Presently, sedans account for about 15% of the market, and hatchbacks are absent in the e-car segment. The Luxury cars which use rare earth PM-based motors have a negligible share in the market.

The organization of the rest of this chapter is as follows: **Section 2** of this chapter delves into a nuanced assessment of technological trends in electric passenger car motor architecture. The section also discusses the use of rare earth (RE) permanent magnets in electric passenger car powertrains, considering realistic RE compositions used in the EV motors of different power ranges. **Section 3** explains the approach and methodology adopted in this study for the future projections of EV penetration in the passenger car segment. It also provides the basis for assessing the total rare-earth material requirement. **Section 4** presents the scenario-based analysis carried out to estimate the total demand for magnet-based rare earth materials till 2030 emanating from the electric passenger car segment in India. **Section 5** discusses the current market situation of rare earth materials and suggests plausible pathways to deal with the supply-demand mismatch that could be expected in the foreseeable future. The chapter concludes **in Section 6** that a robust indigenous manufacturing ecosystem accompanied by a prudent resource-sourcing strategy would be key to realizing the aspirational "*green mobility transition*" in a sustainable manner.

2. LITERATURE REVIEW ON THE TECHNOLOGICAL TRENDS

2.1. Motor Architecture in Electric Passenger Cars

The powertrain architectures of the electric cars are more flexible and compact as compared to that of the ICE-based counterparts (Thakur, 2021; Argueta, 2010). In a typical electric passenger car powertrain, three main components are crucial for the vehicle's performance, i.e., electric motor, battery pack and charge controller. These subsystems can be configured in several different ways as per the choice of the manufacturer, and the vehicle design aspects can be adjusted accordingly (Un-Noor et al., 2017).

The electric propulsion system of the vehicle is supported by the electric motor drive, transmission device, and wheels. The electric motor drive comprises an electric motor, a power converter, and an electronic controller. The electric motor consists of two essential parts, namely the stator or the fixed part, and the rotor (which moves with the rotation of the shaft). The motor utilises the electrical energy from the battery and provides mechanical energy to the wheels. The fundamental working principle of the electric motor pivots around magnetism. The inverter module in the propulsion system converts the direct current produced by the battery pack into an alternate current. The alternate current helps in reversing the polarity of the magnetic field, which facilitates the rotor movement. The motor propulsion system determines the key performance parameters, such as the torque to inertia ratio, torque density, speed range, noise, maintenance, control feasibility, and cost (Argueta, 2010; Un-Noor et al., 2017; Chan, 2002; Hudda and Kumar, 2014). In case the brake is applied to slow down the vehicle, some motors provide the regenerative function that results in recharging the battery pack by a limited amount. This energy transformation mechanism, which converts motor-rotor mechanical energy into electrical energy, provides 25% higher driving efficiency (Juan, 2017; Lakshmi et al., 2017). The lesser vibration in the e-cars reduces the noise, and the powertrain being devoid of ignition and combustion of fuel makes the driving of electric vehicles emission-free.

In the electric passenger car segment, there are currently two predominant types of electric motor architectures, and these are expected to prevail in upcoming years. These motor architectures are asynchronous induction motors and the permanent magnet synchronous motor (PMSM). Asynchronous induction motors or AC induction motors are driven by electromagnets, and rotor current facilitates torque production. The absence of brushes, commutators and slip rings makes the induction motor a low-maintenance and cost-effective option for electric vehicles (Tahir, Galadima and Maruf, 2019). On the other hand, the PMSM architecture houses a permanent magnet in its rotor to create the necessary magnetic field. The quality of the permanent magnet has a critical role in ensuring torque density, efficiency, operational reliability, and durability. The active weight of the motor (magnets are a crucial part of the active weight) affects the vehicle's kerb weight or zero load weight. In turn, vehicle kerb weight directly affects the vehicle range for a particular motor power rating and the battery capacity. Therefore, there has been a continuous effort towards improving the material efficiencies in the permanent magnets used in the EV motors, aimed at reducing the vehicle kerb weight and improving functionality. Further, the reduced weight of the magnet implies a reduced level of raw material requirement in the manufacturing of permanent magnets.

2.2 Rare Earth Permanent Magnet Application in Electric Cars

Neodymium-Iron-Boron (NdFeB) RE permanent magnets offer substantial weight reduction (of up to 60%) and higher torque per unit mass as compared to their ferrite-based counterparts (Simsir and Ertan, 1999). This results in improved performance for complex automotive designs in terms of the riding range. Reduced motor weight, improved riding range, performance reliability, and durability offered by the RE permanent magnets have resulted in a trend that indicates the dominance of RE permanent-based PMSM motor architecture in the global electric passenger car market.

In a NdFeB permanent magnet-based motor, two Light Rare Earth Elements (LREE), viz. neodymium (Nd) and praseodymium (Pr), and one Heavy Rare Earth Element (HREE), namely dysprosium (Dy), are needed. The previous studies indicate that, on average, the RE elements comprise about 35% of magnet weight (Elwert et al., 2016). In the NdFeB magnets, the magnetic properties offered by Nd and Pr are very similar, and their percentages in the material mix are altered based on the availability and price fluctuations of the raw materials. Dysprosium is required to reduce the chances of circuit failure due to overheating, and it improves the resistance to demagnetisation (Fears, 2021).

The literature pertinent to motor designs based on permanent magnets and company-level specifications bring forth two key criteria for assessing the quantum of permanent magnets in the active weight of the motor and the share of the RE materials in the magnet composition, viz., the technology used in manufacturing the magnets (i.e., *compression injection moulded or sintered*) and the motor power density.

Based on the available literature, the motors with power ratings of 150kW or less predominantly use permanent magnets containing 27% Nd, 1% Pr, and 7.5% Dy by weight (Elwert et al., 2016). The higher motor powers (> 150 kW) aimed at achieving improved vehicle range increase the internal operating temperature of the e-cars up to 200 °C (Elwert et al., 2016). Consequently, to ensure higher durability and maintain coercivity and heat retentivity, the dysprosium content needs to be higher. It is assumed that permanent magnets used in electric vehicles with higher motor powers (>150kW) contain 23% of Nd, 3% of Pr and 9% of Dy by weight (Elwert et al., 2016).

Along with the motor's internal temperature constraints, the material costing is another reason for content modification in the permanent magnets. There has been a conscious effort towards reducing the share of neodymium in the NdFeB magnets used in high-end cars. The primary reason that can be attributed to this trend is the high market price of neodymium (Tahanian, Aliahmadi and Faiz, 2020). Since the magnetic properties are similar, the share of Nd is reduced from 27% to 23%, and that of Pr is increased from 1% to 3% so that the overall share of the RE materials in the magnet weight is maintained at around 35%.

The *compression injection moulding/bonded technology* of NdFeB permanent magnet manufacturing is prevalent in the market due to lower prices. These magnets have a lesser material density which affects the magnetic performance, and therefore, the magnetic quantity required in a motor is more. Literature related to magnet quantity in motor architecture provides the range of permanent magnet requirement per vehicle as 2-6 kg for motor powers ranging between 80 and 400 kW (Fears, 2021).

Another NdFeB permanent magnet manufacturing technology, named *sintered magnetisation,* uses a powder metallurgical process (Davies, Mottram and Harris, 2001). This process provides superior magnetic characteristics with higher heat resistance (Sugimoto, 2011). The application of sintered magnet in motor parts reduces the magnet requirement to 1 kg in lower power ranges (35-150 kW) and to approximately 4 kg for vehicles with higher motor power ratings (> 150 kW). With improved magnet grade and lesser weight, magnets produced via this process are particularly beneficial for vehicles with higher motor power ratings as well as high riding ranges. The R&D efforts towards increasing the vehicle riding range would necessitate better performance of the motor and reduction in the vehicle kerb weight. Both objectives can be achieved by using magnets with a higher motor power density and a lesser physical weight.

3. METHODS FOR PROJECTION OF VEHICLE STOCK AND MAGNET REQUIREMENT

3.1 Datasets Used

The datasets utilised for the analysis are obtained from SIAM (Society of Indian Automotive Industry) Annual Reports 1994-2020. The annual reports of the major manufacturing companies are studied to estimate market shares, product portfolios, and technological choices (including predominant motor architecture) of the individual players in the passenger cars segment. The Society of Manufacturers of Electric Vehicles (SMEV) provides a few statistics related to EV adoption and product availability. Several research papers, international reports (IEA, McKinsey), market analytics reports (JMK research), and other open-source information are used to understand market trends, dominant technological solutions, and product substitution. Published literature pertinent to the composition of permanent magnets used in the traction motor architecture and the quantity of rare earth materials in those magnets formed the basis for the projection of future demand levels (Tahir, Galadima and Maruf, 2019; Elwert et al., 2016, Fears, 2021; Li et al., 2010).

3.2 Future Projection of Four-Wheeler Industry and Electric Cars

Considering the pandemic-induced slowdown and observed market recovery in the last year, a CAGR of 7.5% is used for estimating 4W production in India for the period 2022-2030. The annual production of 4Ws in India stood at the 4.03-million mark in 2021, and the cumulative production during the period 1994-2021 is estimated to be about 57.81 million vehicles.

Similarly, for the electric car segment, the annual and cumulative production levels in 2021, which stand at 13600 units and 21800 units, respectively, are used as the baseline for the projection of E4W production for the period 2022-30.

Some analytics firms have anticipated rapid growth of the E4W market in India. Such projections indicate a year-on-year (Y-o-Y) growth rate of 150% or more, which would lead to a substitution rate of about 12% at the annual level by 2025 (Gulia and Jadhav, 2021). Although these high growth rates are aspirational, considering the current state of affairs, such growth rates are unlikely to be realized.

In this study, it is assumed that the growth of the E4W in India is going to happen in two phases. Temporally these two phases are described as the *Initial Growth Phase (IGP)*, which comprises the time segment between 2021 and 2025, and the *Rapid Substitution Phase (RSP)*, which comprises the time segment from 2026 to 2030.

During the *Initial Growth Phase*, a realistic and adequately high Y-o-Y growth in the production of e-cars is likely to be witnessed, which would create a base for a faster substitution of ICE-based conventional vehicles by the electric motor-driven counterparts at the cumulative production level during the *Rapid Substitution Phase (RSP)*.

The current and upcoming E4W vehicle models were scrutinised to understand the future technological trends in this segment. Company-level information suggests that the global trend is moving towards higher energy efficiency of motors. The global trend is also increasingly moving towards vehicles with higher power ratings, higher riding ranges, and lower vehicle kerb weights.

There can be many growth trajectories for a growing industry segment. However, an aspirational trajectory exhibits a continuously growing trend without any abrupt discontinuities. This study considers two aspirational growth scenarios for the E4W segment. The scenarios are described in **Table 1**.

Table 1. Different growth scenarios for the E4W segment in India

Scenario	Y-o-Y Growth Rate During IGP (2021-25)	Penetration of E4Ws at the Cumulative Level in the 4W Market by 2030 (End of RSP)
A	70%	2%
B	100%	5%

For the estimation of rare earth material requirements from E4Ws by 2030, we have considered only those e-car models (currently available and coming up by 2022-23) which use permanent magnet-based motor architecture in the respective powertrains. At present, induction motor-based E4Ws comprise only 2% of the market. However, with the growing demand for short-range passenger cars in urban households as well as from the ride-sharing transport service providers (e.g., Ola, Uber, etc.), newer E4W models in the hatchback and sedan classes equipped with induction-based motor architecture are expected to enter the market in the near to medium term. This study assumes that the market share of induction motor-based E4Ws will increase from the current level to reach 15% by 2030.

Using company-level data available from the open-source information provided by the automotive manufacturers (including vehicle, motor, and magnet) (Campbell, 2008; TATA Motors, 2022; Mahindra, 2022; MG Motors, 2022; Hyundai, 2022; Audi, 2022; Mercedes – Benz, 2022; Jaguar, 2022), the current market trends and prevalent dominant magnet compositions are ascertained. Therefore, a pragmatic approach will assume an increased demand for the RE materials used in the NdFeB magnets in conjunction with the growth trajectory of the electric four-wheeler segment.

3.3 Estimation of Magnet Requirements in Electric Cars

The two criteria of magnet application, as mentioned in **section 2.2**, were considered for the magnet quantity estimation in future e-car production.

Since the E4W segment is in a very nascent stage and a lot of technological options are still on a trial basis, the manufacturers are more prone to using the bonded magnets in the existing models. The market information reveals that the current e-car models are in the motor power range of 50 kW to 300 kW, with maximum models in the lower motor power density category. Compression-moulded or bonded magnets offer a lower price option, leading to affordability for the current level of e-car production. Whereas the sintered magnet application in electric car motor architecture is limited in the current production scale. The increasing demand for high-end vehicles with reduced vehicle kerb weight would drive the technological shift in magnet application.

The motor's active weight (including permanent magnets) is evaluated by dividing the average motor power (kW) by the minimum allowable motor power-to-active mass ratio of 2.88 kW/kg. For the compression moulded permanent magnets, the ratio of the magnet weight to that of the active component is about 0.0545 (Campbell, 2008). It is assumed that the usage of such magnets in the PM-based traction motor architecture will prevail till 2025.

For the sintered magnets, the ratio of the magnet weight to the active weight comes to about 0.034 (Campbell, 2008). It is assumed that the sintered magnets will gain a major market share in future e-car models, especially in the post-2026 period.

3.3.1 Initial Growth Phase (2021-25)

The market shares of different motor power ranges currently available in the PM-based vehicle category are used to calculate the weighted average magnet quantity per vehicle for the near term. Assuming the current dominance of bonded technology in permanent magnet manufacturing and a larger share of lower motor power vehicles in the market, the estimated average magnet requirement works out to be about 1.87 kg per E4W till 2025. The corresponding requirements of Nd, Dy and Pr are 0.5, 0.142, and 0.021 kg, respectively. Table 2 exhibits the details of magnet and RE requirements per E4W for the initial growth phase (2021-25).

Table 2. Magnet and rare earth material requirements per E4W during 2021-25

Motor Power Range (kW)	Median Power Considered (kW) (Based on Currently Available Models)	Market Share	Magnet Requirement per Vehicle (kg)	Nd (kg)	Dy (kg)	Pr (kg)
50-80	65	13%	1.23	0.332	0.0923	0.012
95-110	100	85%	1.89	0.511	0.142	0.019
230-300	280	2%	5.29	1.220	0.476	0.159
Weighted Average			1.872	0.501	0.142	0.021

3.3.2 Rapid Substitution Phase (2026-30)

The diverse product mix of electric cars comprising a wide range of motor powers is likely to influence the magnet requirement for the PM-based E4Ws. It is assumed that the sintered magnets will be the dominant technological solution for magnet-making post-2026 (i.e., during the rapid substitution phase in India). The market share of different motor power ranges is derived from the vehicle models that are likely to become available in the country in the upcoming years. For the period 2026-30, the weighted average magnet quantity per vehicle is estimated at 2.10 kg. Correspondingly, Nd, Dy, and Pr requirements are 0.513, 0.178, and 0.048 kg, respectively. **Table 3** provides the details of the estimation of magnet and RE quantities.

Table 3. Magnet and rare Earth material requirements per E4W during 2026-30

Motor Power Range (kW)	Median Power Considered (kW) (Based on Available and Upcoming Models)	Market Share	Magnet Requirement per Vehicle (kg)	Nd (kg)	Dy (kg)	Pr (kg)
<50	35	5%	0.413	0.112	0.031	0.004
50-100	85	35%	1.003	0.271	0.075	0.010
100-200	150	22%	1.770	0.478	0.133	0.0177
200-300	280	28%	3.305	0.760	0.297	0.099
300-475	350	10%	4.132	0.950	0.372	0.124
Weighted Average			2.10	0.513	0.178	0.048

4. RESULTS

Assuming the 7.5% CAGR, the total passenger car production level in the country is estimated to reach the 110.83-million mark by 2030. The current annual production level of 40.33 lakh units (as of 2021) is expected to increase about 1.9 times by 2030, to reach 77.32 lakh units in 2030. Figure 1 exhibits the projected annual production of passenger cars in India during the period 2022-30.

The study is focused on estimating the future demand for rare earth materials; considering the growth of PM-based E4W segment by 2030. As per the assumption mentioned in Section 3.2, the annual E4W production of 13600 units in 2021 would increase to 7.71 lakh units (Scenario A) in 2030. In Scenario B, the annual production of E4Ws in India is estimated to reach 22.205 lakhs in 2030. The assumption indicates that 15% of the total E4W produced in 2030 (1.15 lakh units in Scenario A and 3.33 lakh units in Scenario B) would use induction motors in their powertrain architecture.

Figure 1. The base and projected annual production of total four-wheeler vehicles

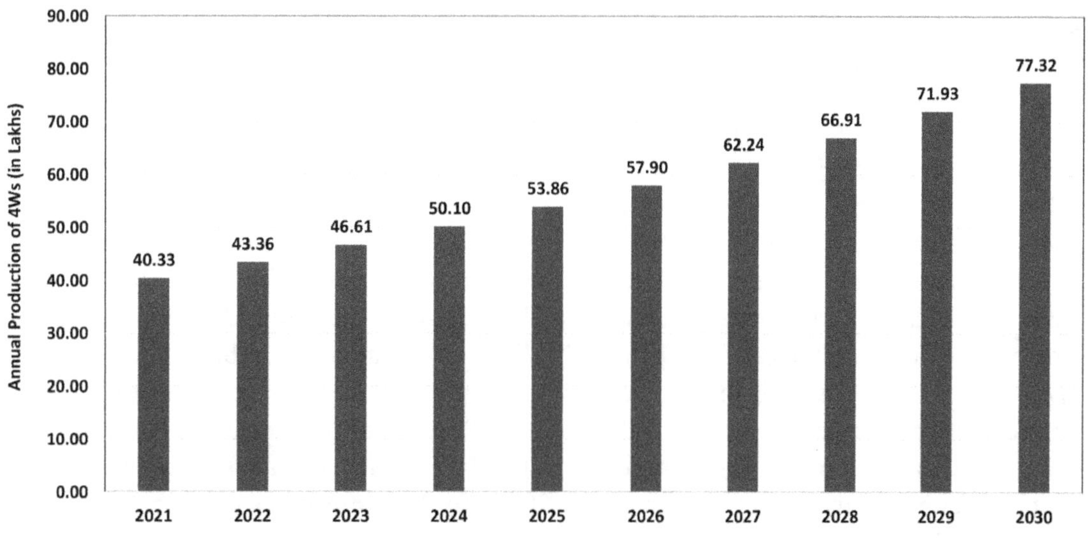

According to Scenario A (see Table 1), the annual production level of the permanent magnet motor-based E4Ws is estimated to reach 6.55 lakh units in 2030, and the total production between 2021 and 2030 is estimated at 19.66 lakh units.

According to Scenario B (see Table 1), the annual production level of PM-based E4Ws will reach 18.87 lakh units by 2030. The total production between 2021 and 2030 is estimated at 48.92 lakh units in Scenario B.

Figure 2 shows the year-wise annual production levels of PM-based E4Ws in India during the period 2021-30.

Figure 2. The growth projection of the PM-based E4Ws segment in Scenarios A and B

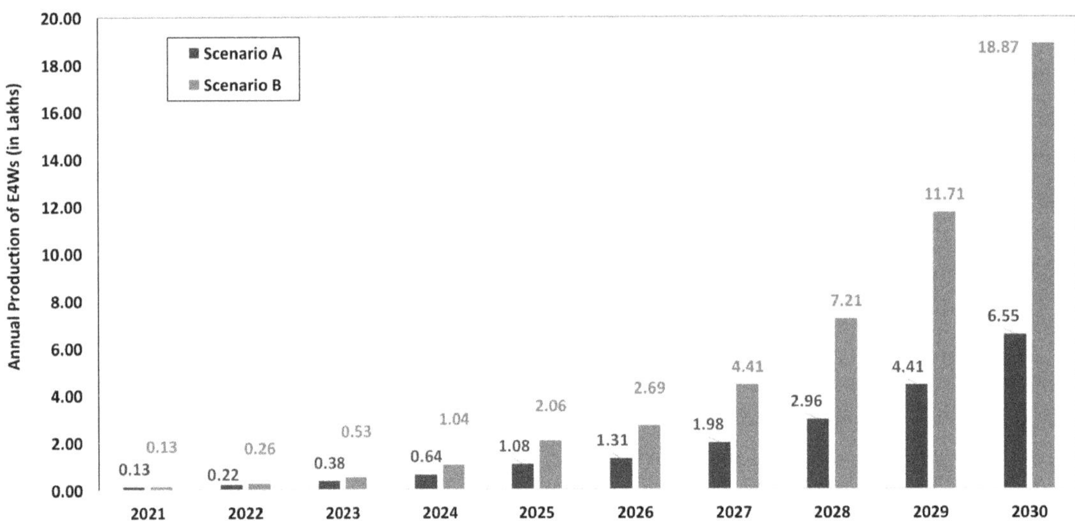

Using the values presented in Tables 2 and 3, the total quantity of the magnet required for the PM-based electric passenger cars between 2021 and 2030 is 4072 tonnes (Scenario A) and 10180 tonnes (Scenario B), respectively. Figures 3(a) and 3(b) show the year-wise demand for Neodymium, Dysprosium and Praseodymium for PM-based passenger cars in India for the period 2021-30. From the figures, it is evident that the demand is going to increase sharply after 2027, which is attributed to the rapid substitution of ICE-based passenger cars with their electric counterparts.

The production level of E4Ws in 2021 led to a combined annual RE demand for magnet materials of about 8.8 tonnes. The combined annual RE demands are estimated to reach 484 tonnes (Scenario A) and 1394 tonnes (Scenario B), respectively, in 2030. The year-wise total demand for rare earth materials and combined demand for the projected period (2021-30) are mentioned in Table 4.

The assumed abrupt change in magnet requirement attributed to the envisaged shift in magnet manufacture technology (from compression moulded to sintered) and diversification of the motor power range of E4Ws would impact the demand for rare earth materials. Consequently, a drastic change is observed during the latter half of the period of interest (i.e., 2026-30). The estimations indicate that the increase in the demand for dysprosium and praseodymium will be substantial post-2026. The demand growth for neodymium (Nd) will slow down after 2026, largely because of the product mix that is going to be available, accompanied by the envisaged shift in the magnet composition.

Table 4. Projected year-wise E4W growth and magnetic material requirement

Year	2021	2022	2023	2024	2025	2026	2027	2028	2029	2030	Grand Total
Total Passenger 4Ws in India (Lakhs)	40.33	43.36	46.61	50.10	53.86	57.90	62.24	66.91	71.93	77.32	570.57
Scenario A: 70% Y-o-Y Growth till 2025, 2% Substitution at Cumulative Level by 2030											
PM-based Passenger E4Ws (Lakhs)	0.13	0.22	0.38	0.64	1.08	1.31	1.98	2.96	4.41	6.55	19.66
Magnet (tonnes)	24.9	42.1	71.0	119.8	201.6	275.6	414.7	621.3	925.2	1376.1	4072.2
Rare Earth Materials Demand (Scenario A)											
Nd (tonnes)	6.7	11.3	19.0	32.1	54.0	67.4	101.4	151.9	226.2	336.5	1006.6
Dy (tonnes)	1.9	3.2	5.4	9.1	15.3	23.3	35.1	52.5	78.2	116.3	340.4
Pr (tonnes)	0.3	0.5	0.8	1.3	2.2	6.3	9.4	14.1	21.0	31.3	87.2
Total RE (tonnes)	8.8	14.9	25.2	42.6	71.6	96.9	145.9	218.6	325.5	484.1	1434.1
Scenario B: 100% Y-o-Y Growth till 2025, 5% Substitution at Cumulative Level by 2030											
PM-based Passenger E4Ws (Lakhs)	0.13	0.26	0.53	1.04	2.06	2.69	4.41	7.21	11.71	18.87	48.92
Magnet (tonnes)	24.9	49.5	98.3	195.1	386.2	564.5	926.6	1513.9	2458.9	3962.7	10180.6
Rare Earth Materials Demand (Scenario B)											
Nd (tonnes)	6.7	13.3	26.4	52.3	103.5	138.0	226.6	370.2	601.3	968.9	2507.1
Dy (tonnes)	1.9	3.8	7.5	14.8	29.3	47.7	78.3	128.0	207.9	335.0	854.3
Pr (tonnes)	0.3	0.6	1.1	2.2	4.3	12.8	21.1	34.4	55.9	90.1	222.6
Total RE (tonnes)	8.8	17.6	34.9	69.3	137.2	198.6	326.0	532.6	865.0	1394.0	3584.0

5. DISCUSSION

5.1 Current Status of Raw Material Availability

India's aspirational target of green mobility relies on the rapid deployment of a large fleet of electric vehicles in the near to medium-term future. The RE permanent magnet-based motors are the dominant solution for EV powertrains owing to the higher efficiencies offered by them. As of 2019, 95% of the EVs produced worldwide are reported to use RE permanent magnet-based traction motors (Gauß et al., 2021). However, the aspirational growth trajectories for green mobility should also be anchored on the availability of critical raw materials (e.g., RE materials for NdFeB magnets, materials used in EV batteries, etc.). Therefore, it is necessary to take stock of the demand and supply of RE magnet materials in the Indian context.

Figure 3. Projected annual magnet and RE material requirements during the period 2021-30 for (a) Scenario A and (b) Scenario B

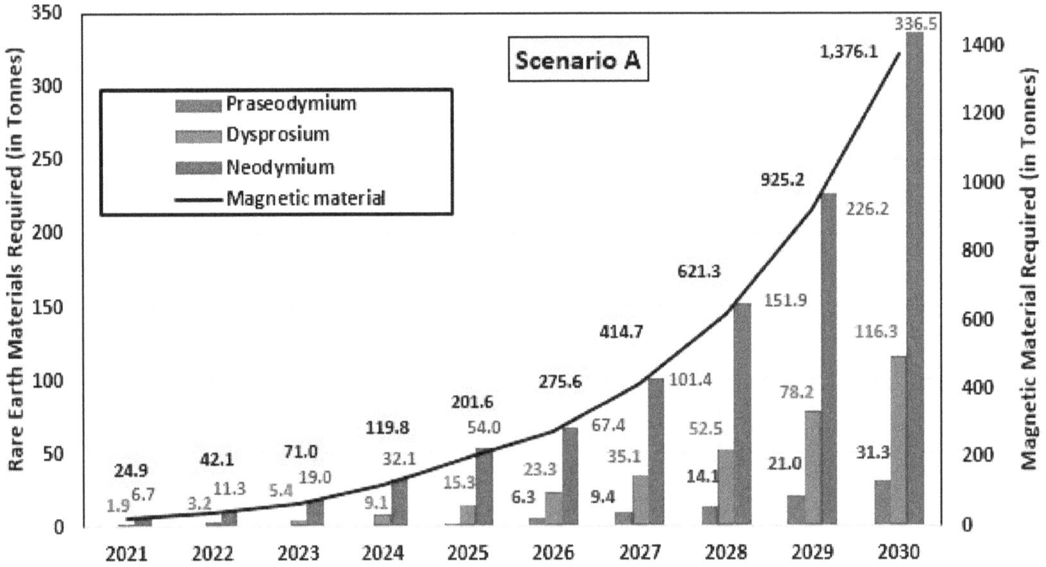

(a)

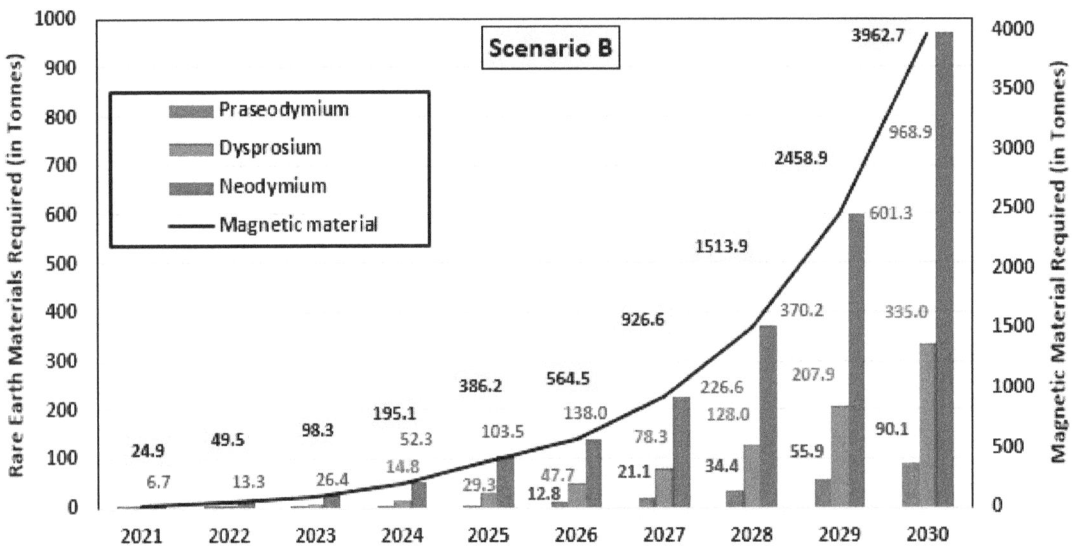

(b)

India possesses about 6% of global rare earth reserves and stands in the fifth position, following China (36.6%), Vietnam (18.3%), Brazil (17.5%), and Russia (10%) (Statista, 2021b; ECFR, 2021). China has been the largest producer of RE materials for quite some time now. In 2021, 60.6% of the RE material supply came from China, whereas the USA, Myanmar, and Australia accounted for 15.5%, 9.4%, and 7.9% of the global supply, respectively. As per the open-source information, India's total production of rare earth oxide equivalent was about 3000 metric tonnes, amounting to 1.05% of the global supply (ECFR, 2021; Statista, 2021c).

India's rare earth material extraction potential lies in Monazite Beach Sand Mineral (BSM). BSM contains light rare earth elements (LREE), such as neodymium (Nd) and praseodymium (Pr). However, the presence of heavy rare earth elements (HREE), such as dysprosium, is negligible (Indian Bureau of Mines, 2020). The global trend shows an increasing demand for rare earth elements for the manufacturing of permanent magnets (Gauß et al., 2021; Trench and Sykes, 2020). The RE permanent magnets are used in a wide array of end-use applications, such as wind turbine generators, energy-efficient consumer durables, industrial automation, loudspeakers and household electronics, water pumps, etc. In 2008, the manufacturing of permanent magnets consumed 20% of global RE oxides, and the volumetric consumption of RE raw materials for magnet-making increased to 38% by 2019 (Constantinides, 2012; Serpell, Paren and Chu, 2021).

Currently, India does not have any major manufacturer of RE permanent magnets. There is a small nucleus of magnet manufacturers in the country. However, the Government of India needs to incentivise small-scale manufacturers to transform them into larger producers of RE permanent magnets.

5.2 Sustainable Growth of Electric Four-Wheeler Segment

India has adequate technological capabilities associated with the mining and processing of RE raw materials. However, the constraints on the feed availability for the processing units, accompanied by the lack of demand for the domestically produced RE materials (most of RE demand in India is embedded in end products, of which a large fraction is imported as completely built units, systems, components, and sub-systems) render the exercise unviable for the RE producers. Direct support from the government is needed to change this current state of affairs. Further, an RE industry ecosystem needs to be formed, which will connect the complex network of the upstream, midstream and downstream parts of the RE value chain. This integration, along with proper incentives, will encourage the industry to scale up indigenous manufacturing and strengthen the production lines of the end products to achieve economies of scale.

The current projection of demand for REEs indicates an increase in RE demand (by 50 to 150 times) by 2030 compared to the current levels for the chosen conservative and realistic growth scenarios of the electric passenger car segment. It is to be noted that the NEM (National Electric Mobility) Mission envisages '*vehicle penetration targets*' much higher than those presented in this study. Therefore, to meet the aspirational green energy goals, it is important to take stock of the potential raw material supply risks that might prove to be a critical bottleneck in the foreseeable future.

The current level of preparedness (in terms of the existence of an industrial nucleus and the availability of raw materials) indicates that substantial additional logistic support will be needed to fulfil the country's domestic demands. If the country chooses to bolster the indigenous manufacturing of RE-based products, the resource base expansion needs to be taken up on a priority basis at the national level, which would encompass exploration, mining, processing, and production of RE metals. This would ensure smooth and rapid production of alloys and powders and, finally, production of RE magnets for different end-use applications, including the EV traction motors.

India is a member of QUAD, along with the USA, Japan, and Australia. India also has an adequately strong relationship with multiple ASEAN countries. Using the geopolitical proximity with QUAD, ASEAN countries and the Western Alliance (involving the UK and European Countries), navigate through various geopolitical equations to ensure a favourable ambience for the technology transfer and R& D collaborations. It can also secure raw material supplies from friendly countries.

USA's recent plans for market development for critical minerals pivot around building a low-cost supply chain for magnetic materials (USGS, 2022). India could explore trade agreements to purchase neodymium-praseodymium materials for the indigenous manufacture of permanent magnets. Japan has advanced technologies for the extraction and refinement of critical materials such as dysprosium. Therefore, knowledge sharing, collaborative R&D efforts, and technology transfer would facilitate the creation of the industrial ecosystem in India.

Recent estimation pertinent to the rare earth reserves indicates abundant availability of rare earth minerals in multiple countries in Africa (e.g., Republic of Burundi, Tanzania), Asia (e.g., Myanmar and Vietnam) and South America (e.g., Brazil) (ECFR, 2021). India's favourable relations with member countries of such regions could be leveraged in future. The multi-national association for raw material supply chain would provide impetus to the Indian economy for rapid indigenous development of the electric automotive industry. The reuse and recycling of critical components such as motors and permanent magnets might be a viable solution to deal with raw material supply issues. The literature suggests that the life of an electric car is around 15-20 years, and the traction motors used in e-cars also have a similar life expectancy (Jerew, 2021). Therefore, this study does not cover the reuse and recycling of magnet material or the built motor architecture, as the chosen period is only ten years. It is likely that a mix of virgin and recycled rare earth magnets in the electric motor architecture would prove to be a viable solution in the long term (Pandey and Majumdar, 2023). However, the techno-economic viability of these advanced technologies may deter India from adopting them in the near term.

Figure 4 depicts a possible sustainability approach that might prove to be effective in realising India's green mobility aspirations within the timeline of 2030 that the country is aiming for.

6. CONCLUSION

India is looking for the progressive substitution of the prevalent ICE-based vehicles with their electric motor-driven counterparts in all segments. Currently, the penetration of electric vehicles in the overall automotive market is very small, and the EV segment in the country is in the nascent stage. The national aspirations towards electric vehicle deployment are high to reduce the CO_2 and other harmful emissions that are attributable to the road transport sector. Further, a shift towards electricity-based mobility would also reduce dependency on imported petro-based fuels (petrol and diesel). Consumers' preference for personal vehicles would drive the demand for electric passenger cars in the near to medium-term future. The financial support provided by the central and state governments has an immense role in driving the growth of EVs. Government support will also be crucial in transforming ICE-based vehicle manufacturers to become larger producers of electric passenger cars. Additionally, the growing share of the Indian automotive industry globally provides ample opportunity to capture a major market share in the EV supply chain.

Figure 4. India's E4W growth scenarios and sustainable pathways to achieve aspired targets

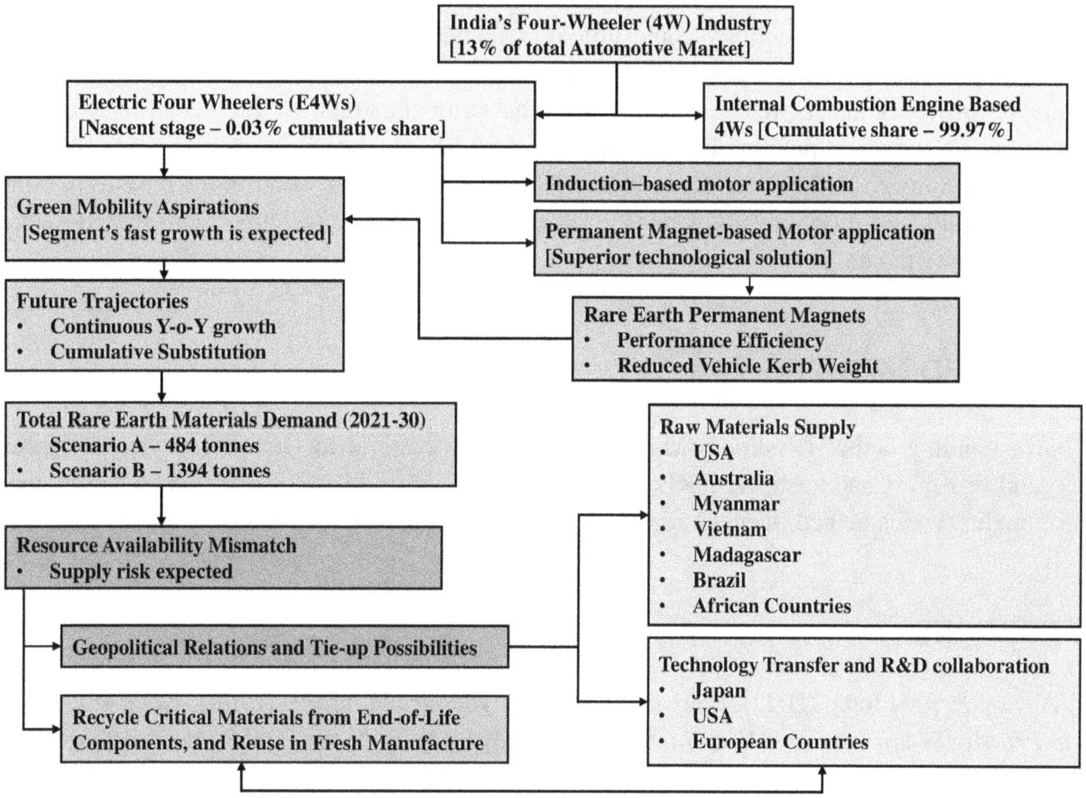

A survey of global EV and traction motor manufacturers suggests that the NdFeB permanent magnet-based motor architecture is a preferred solution. The Neodymium Iron Boron (NIB) permanent magnets need three key rare earth materials, viz. neodymium, praseodymium, and dysprosium. The study projects the technological trends in the electric four-wheeler (E4W) segment till 2030, and accordingly, the demand for rare earth materials is also estimated. The improvement in material efficiencies has also been considered in the projections.

The study considered two scenarios, where E4Ws reach up to 2% and 5% cumulative penetration, respectively, in the overall passenger car segment by 2030. The realistic growth ambitions for PMSM electric cars would increase the annual production size to 6.55 lakh and 18.87 lakh units by 2030, in Scenario A and Scenario B, respectively. The total RE requirements are estimated as 1434 tonnes (Scenario A) and 3584 tonnes (Scenario B) for the period 2021-30.

The demand for RE materials for the E4W segment and the current level of availability of indigenously produced raw materials indicate a supply-demand mismatch if the country wants to go for a rapid product substitution. A suitable geopolitical collaboration strategy is suggested to avoid potential raw material supply risk and the associated bottlenecks.

The study proposes exploring geopolitical strategies to ensure access to sustainable global supply chains and strengthen indigenous capabilities. The international relations with major RE producers in the global market could provide advanced technology for resource extraction and refining. In the long term, reusing essential parts and recycling critical materials could be a viable way towards sustainable production and consumption.

The shift towards green mobility could unfold tremendous trading opportunities for the country that would encompass rare earth raw materials, rare earth oxides and other intermediaries, alloys and powders, permanent magnets, traction motors and other critical automotive parts, and fully assembled vehicles. India's dominant position in the competitive global automotive market presents the country with an opportunity to build a rare earth industrial ecosystem which would have long-term benefits concerning both socio-economic and strategic dimensions, leading to resource security and resilience of this developing nation.

ACKNOWLEDGMENT

The corresponding author (Rudrodip Majumdar) is highly grateful to Dr Shailesh Nayak (Director, NIAS) and Prof. S. Chandrashekar (JRD Tata Visiting Professor, NIAS) for their invaluable insights and comments, which helped in enhancing the quality of the chapter.

REFERENCES

NITI Aayog, & RMI India (2022). *Banking on Electric Vehicles in India – A Blueprint for inclusion of EVs in Priority Sector Lending Guidelines*. A Joint Study by NITI Aayog, and RMI India. https://www.niti.gov.in/sites/default/files/2022-01/Banking-on-EV_web_2.0a.pdf

Abbas, M. (2018, Dec.12). India has 22 cars per 1,000 individuals: Amitabh Kant. *Economic Times (ET) Auto*. https://auto.economictimes.indiatimes.com/news/passenger-vehicle/cars/india-has-22-cars-per-1000-individuals-amitabh-kant/67059021

Argueta, R. (2010). *A Technical Research Report: The Electric Vehicle*. University of California. Retrieved from https://www.ehcar.net/library/rapport/rapport044.pdf

Audi (2022). *E-Tron Technical Specifications and Data*. Audi Media Centre. https://audimediacenter-a.akamaihd.net/system/production/car_motorizations/984/file_en/d933a6f0a1f3a3b-c0915035a491b370621af532d/eTechnische_Daten_Audi_e-tron_50_quattro_230kW.pdf

HT Auto Desk (2021). *Auto Industry in Recovery Mode, Buoyed by Demand for Passenger Vehicles: FADA*. Hindustan Times (HT) Auto Desk. https://auto.hindustantimes.com/auto/news/auto-industry-in-recovery-mode-buoyed-by-demand-for-passenger-vehicles-fada-41625722389929.html

BEE. (2020). *E-Mobility*. Bureau of Energy Efficiency (BEE). https://beeindia.gov.in/content/e-mobility

Campbell, P. (2008). *System Cost Analysis for an Interior Permanent Magnet Motor*. AMES Laboratory, US Department of Energy. https://www.osti.gov/servlets/purl/940187

Chan, C. C. (2002). The state of the art of electric and hybrid vehicles. *Proceedings of the IEEE, 90*(2), 247–275. doi:10.1109/5.989873

Chan, C. C. (2016). An overview of electric vehicle technology. *Proceedings of the IEEE, 81*(9), 1202-1213. https://ieeexplore.ieee.org/abstract/document/237530 doi:10.1109/5.237530

Constantinides, S. (2012, Sept. 30 – Oct. 3). *The Demand for Rare Earth Materials in Permanent Magnets*. COM 2012, Niagara Falls. https://www.arnoldmagnetics.com/wp-content/uploads/2017/10/Demand-for-rare-earth-materials-in-permanent-magnets-Constantinides-COM-2012-psn-hi-res.pdf

Davies, B. E., Mottram, R. S., & Harris, I. R. (2001). Recent developments in the sintering of NdFeB. *Materials Chemistry and Physics*, *67*(1-3), 1–3. doi:10.1016/S0254-0584(00)00450-8

Dey, S., & Mehta, N. S. (2020). Automobile pollution control using catalysis. *Resource. Environmental Sustainability*, *2*, 100006. Advance online publication. doi:10.1016/j.resenv.2020.100006

ECFR. (2021). *The Power Atlas – Seven battlegrounds of a networked world.* Published Jointly by the European Council of Foreign Relations (ECFR) and Stiftung Mercator. https://ecfr.eu/wp-content/uploads/power-atlas.pdf

Elwert, T., Goldmann, D., Roemer, F., & Schwarz, S. (2016). Recycling of NdFeB Magnets from Electric Drive Motors of (Hybrid) Electric Vehicles. *Journal of Sustainable Metallurgy*, *3*(1), 108–121. Advance online publication. doi:10.1007/s40831-016-0085-1

Fears, P. (2021, Jan. 5). Rare Earth Magnets in Electric Vehicle Motors. *Bunting - Berkhamsted.* https://www.bunting-berkhamsted.com/rare-earth-magnets-in-electric-vehicle-motors/

Gauß, R., Burkhardt, C., Carencotte, F., Gasparon, M., Gutfleisch, O., Higgins, I., Karajić, M., Klossek, A., Mäkinen, M., Schäfer, B., Schindler, R., & Veluri, B. (2021). *Rare Earth Magnets and Motors: A European Call for Action*. A report by the Rare Earth Magnets and Motors Cluster of the European Raw Materials Alliance. Berlin. https://eitrawmaterials.eu/wp-content/uploads/2021/09/ERMA-Action-Plan-2021-A-European-Call-for-Action.pdf

Gonsalves, O. (2019, Jan. 17). The Labor Market in India: Structure and Costs. *India Briefing*. https://www.india-briefing.com/news/labor-market-india-structure-costs-18264.html/

Gulia, J., & Jadhav, A. (2021). *E-Cars Market in India- October 2021*. JMK Research & Analytics. https://jmkresearch.com/wp-content/uploads/2021/10/E-Cars-Market-in-India_JMKResearch_Oct-2021.pdf

Hudda, V., & Kumar, A. (2014). *A Summary of Electric Vehicle Propulsion Technologies*. GRIN Verlag, Munich. Document no. 277317. https://www.grin.com/document/277317

Hyundai. (2022). *Kona Electric Specifications*. Hyundai. https://www.hyundai.com/in/en/find-a-car/kona-electric/specification

Invest India (2022). *Sector Automobile – National Investment Promotion and Facilitation Agency*. https://www.investindia.gov.in/sector/automobile

Indian Bureau of Mines. (2020). *Indian Minerals Yearbook 2019 (Part- III: Mineral Reviews), 58th Edition, Rare Earths (Advance Release).* https://ibm.gov.in/writereaddata/files/10012020172151RareEarth_2019_AR.pdf

Jaguar. (2022). *E – Pace Specifications.* Jaguar. https://www.jaguar.com/jaguar-range/e-pace/specifications/index.html

Jerew, B. (2021). *EV Lifespan: Do They Last as Long as Gasoline Cars?* Lifewire – Tech for Humans. https://www.lifewire.com/do-evs-last-as-long-as-gasoline-cars-5202392

Juan, L. P. (2017). *Regenerative Braking*. Stanford University. http://large.stanford.edu/courses/2017/ph240/leis-pretto1/

Kamath, H. G., Majumdar, R., Krishnan, A. V., & Srikanth, R. (2022). Cost and environmental benefits of coal-concentrated solar power (CSP) hybridization in India. *Energy, 240*, 122805. Advance online publication. doi:10.1016/j.energy.2021.122805

Khawaja, Y., Shankar, N., Qiqieh, I., Alzubi, J., Alzubi, O., Nallakaruppan, M. K., & Padmanaban, S. (2023). Battery management solutions for li-ion batteries based on artificial intelligence. *Ain Shams Engineering Journal, 14*(12), 102213. doi:10.1016/j.asej.2023.102213

Lakshmi, D. N., Sandhyaa, R., Kanwar, P. D., & Priya, L. S. (2017). Energy Efficient Electric Vehicle Using Regenerative Braking System. *International Journal of Advance Research, Ideas, and Innovations in Technology, 2*(3). http://large.stanford.edu/courses/2017/ph240/leis-pretto1/docs/lakshmi.pdf

Li, Y., Gerling, D., Ma, J., Liu, J., & Yu, Q. (2010). The Comparison of Control Strategies for the Interior PMSM Drive used in the Electric Vehicle. *World Electric Vehicle Journal.* https://www.researchgate.net/publication/279899500_The_Comparison_of_Control_Strategies_for_the_Interior_PMSM_Drive_used_in_the_Electric_Vehicle

Mahindra. (2022). *E-Verito Electric Car. Features and Specifications*. Mahindra Electric. https://www.mahindraelectric.com/vehicles/everito/

Mercedes – Benz. (2022). *Mercedes – Benz EQC Technical Specification*. https://www.mercedes-benz.co.in/passengercars/mercedes-benz-cars/models/eqc/specifications/aerodynamics.module.html

Miglani, S. (2019). Innovation, Economic Development, and Intellectual Property in India and China. ARCIALA Series on Intellectual Assets and Law in Asia. In *The Growth of the Indian Automobile Industry: Analysis of the Roles of Government Policy and Other Enabling Factors* (pp. 439-463). https://link.springer.com/chapter/10.1007/978-981-13-8102-7_19

MG Motors (2022). *MG ZS EV Specifications and Features.* MG Motors. https://s7ap1.scene7.com/is/content/mgmotor/mgmotor/documents/mg-dc-pdf-0186.pdf

Tata Motors (2022). *Specifications and Features of TATA Nexon EV.* TATA Motors. https://nexonev.tatamotors.com/features/

Mukherjee, A., Bagchi, J., Roy, K., & Mondal, K. (2016). Catalytic Converter in Automobile Exhaust Emission. *Journal for Research., 02*(10), 29–33. http://www.journal4research.org/articles/J4RV2I10020.pdf

Pandey, A., & Majumdar, R. (2023). Anthropogenic Environmental Hazards: Compensation and Mitigation. In P. Pathak, R. R. Srivastava, & S. Ilyas (Eds.), *Assessing End-of-Life Room Air Conditioner Recycling Potential for Sustainable Resource Utilization in India: A Case Study for Reducing Environmental Burden'* (pp. 41–69). Springer Nature Switzerland. Retrieved from https://link.springer.com/chapter/10.1007/978-3-031-41013-0_3

Press Trust of India. (2021). *'Government aims to raise auto sector contribution to GDP, job creation: Gadkari'*. Business Standard. https://www.business-standard.com/article/automobile/govt-aims-to-raise-auto-sector-contribution-to-gdp-job-creation-gadkari-121082501375_1.html

Pujari, A. S., Majumdar, R., Saha, S. K., Saha, S. K., & Subramaniam, C. (2023). Annular vertical cylindrical thermochemical storage system with innovative flow arrangements for improved heat dispatch towards space heating requirements. *Renewable Energy*, *217*, 119168. Advance online publication. doi:10.1016/j.renene.2023.119168

Saurabh, K., & Majumdar, R. (2022). Clean Fuels for Mobility. Energy, Environment, and Sustainability. In G. Di Blasio, A. K. Agarwal, G. Belgiorno, & P. C. Shukla (Eds.), *Fuels for Sustainable Transport in India* (pp. 27–55). Springer. doi:10.1007/978-981-16-8747-1_3

Saurabh, K., & Majumdar, R. (2023a). Renewable Fuels for Sustainable Mobility. Energy, Environment, and Sustainability. In P. C. Shukla, G. Belgiorno, G. D. Blasio, & A. K. Agarwal (Eds.), *Functional Use-Based Positioning of Conventional Vehicles in Conjunction with Alternate Low-Emission Fuels* (pp. 97–159). Springer. doi:10.1007/978-981-99-1392-3_5

Saurabh, K., & Majumdar, R. (2023b, December). Sub-segment-based approach for assessing growth and energy transition trajectories of Indian two-wheelers. *Journal of Cleaner Production*, *430*(10), 139559. doi:10.1016/j.jclepro.2023.139559

Serpell, O., Paren, B., & Chu, W.-Y. (2021). *Rare Earth Elements – A Resource Constraint of the Energy Transition.* Kleinman Center for Energy Policy. https://kleinmanenergy.upenn.edu/wp-content/uploads/2021/05/KCEP-Rare-Earth-Elements.pdf

SESEI. (2018). *Indian Automobile Industry*. Seconded European Standardization Expert in India (SESEI). https://www.sesei.eu/wp-content/uploads/2018/12/Automotive-Sector-Report_-Final.pdf

Sharma, M. (2021). *Hatchback vs. Sedan.* Spinny Buying Guide. https://www.spinny.com/blog/index.php/hatchback-vs-sedan/

SIAM. (1998-2020). *Annual Reports.* Society of Indian Automobile Manufacturers (SIAM). https://www.siam.in/cpage.aspx?mpgid=42&pgidtrail=89

Simsir, N. B., & Ertan, H. B. (1999). A Comparison of Torque Capabilities of Axial Flux and Radial Flux Type of Brushless DC (BLDC) Drives for Wide Speed Range Applications. *IEEE 1999 International Conference on Power Electronics and Drive Systems, PEDS'99, Hong Kong, China* (pp. 719-724). https://ieeexplore.ieee.org/document/792793

Statista. (2021a). *Sales Volume of Passenger Cars in India from financial year 2016 to 2021, by body type.* Statista. https://www.statista.com/statistics/1265730/india-sales-volume-of-passenger-cars-by-body-type/

Statista. (2021b). *Reserves of rare earth worldwide as of 2021, by country*. Statista. https://www.statista.com/statistics/277268/rare-earth-reserves-by-country

Statista. (2021c). *Distribution of rare earth production worldwide as of 2021, by country*. Statista. https://www.statista.com/statistics/270277/mining-of-rare-earths-by-country/

Sugimoto, S. (2011). Current status and recent topics of rare-earth permanent magnets. *Journal of Physics. D, Applied Physics*, *44*(6), 064001. doi:10.1088/0022-3727/44/6/064001

Tahanian, H., Aliahmadi, M., & Faiz, J. (2020). Ferrite Permanent Magnets in Electrical Machines: Opportunities and Challenges of a Non-Rare-Earth Alternative. *IEEE Transactions on Magnetics*. Advance online publication. doi:10.1109/TMAG.2019.2957468

Tahir, Z., Galadima, A., & Maruf, A. (2019). *A Review of Motors for Electric Vehicles*. International Power Engineering Exhibition & Conference, Nigeria. https://www.researchgate.net/publication/334285900_A_Review_of_Motors_for_Electric_Vehicles

Thakur, D. (2021). *Electric Vehicle Architecture & EV Powertrain Components*. E-Vehicle Info. https://e-vehicleinfo.com/electric-vehicle-architecture-ev-powertrain-components/

Trench, A., & Sykes, J. P. (2020). Rare Earth Permanent Magnets and Their Place in the Future Economy. *Engineering*. doi:10.1016/j.eng.2019.12.007

Un-Noor, F., Padmanaban, S., Mihet-Popa, L., Mohammad Nurunnabi Mollah, M. N., & Hossain, E. (2017). A Comprehensive Study of Key Electric Vehicle (EV) Components, Technologies, Challenges, Impacts, and Future Direction of Development. *Energies*, *10*(8), 1217. doi:10.3390/en10081217

USGS. (2022). *Rare Earths Statistics and Information*. National Minerals Information Center U.S. Geological Survey (USGS). https://www.usgs.gov/centers/national-minerals-information-center/rare-earths-statistics-and-information

Vidhi, R., & Shrivastava, P. (2018). A Review of Electric Vehicle Lifecycle Emissions and Policy Recommendations to Increase EV Penetration in India. *Energies*, *11*(3), 483. doi:10.3390/en11030483

Compilation of References

Aavula, & Deshmukh, & Mane. (2022). Design and Implementation of sensor and IoT based Remembrance system for closed one. *Telematique*, *21*(1), 2769–2778.

Abbas, M. (2018, Dec.12). India has 22 cars per 1,000 individuals: Amitabh Kant. *Economic Times (ET) Auto*. https://auto.economictimes.indiatimes.com/news/passenger-vehicle/cars/india-has-22-cars-per-1000-individuals-amitabh-kant/67059021

Abdullah, A.A., & Hassan, T.M. (2022). Smart Grid (SG) properties and challenges: An overview. *Discover Energy, 2.*

Acciarini, C., Borelli, F., Capo, F., Cappa, F., & Sarrocco, C. (2021). Can digitalization favour the emergence of innovative and sustainable business models? A qualitative exploration in the automotive sector. *Journal of Strategy and Management*, *15*(3), 335–352. doi:10.1108/JSMA-02-2021-0033

Acharya, S., Dvorkin, Y., & Karri, R. (2020). Public plug-in electric vehicles+ grid data: Is a new cyberattack vector viable? *IEEE Transactions on Smart Grid*, *11*(6), 5099–5113. doi:10.1109/TSG.2020.2994177

Afzal, S., Mokhlis, H., Illias, H. A., Mansor, N. N., & Shareef, H. (2020). State-of-the-art review on power system resilience and assessment techniques. *IET Generation, Transmission & Distribution*, *14*(25), 6107–6121. doi:10.1049/iet-gtd.2020.0531

Akansha. (2022). Email Security. *Journal of Image Processing and Intelligent Remote Sensing*, *2*(6).

Alam, M. M., & Lee, Y. (2021). *Cleaner Vehicles and Charging Infrastructure: Greening Passenger Fleets for Sustainable Mobility*. Academic Press.

Alam, M. M., Mekhilef, S., Seyedmahmoudian, M., & Horan, B. (2017). Dynamic charging of electric vehicle with negligible power transfer fluctuation. *Energies*, *10*(5), 701. doi:10.3390/en10050701

Al-Fuqaha, A., Guizani, M., Mohammadi, M., Aledhari, M., & Ayyash, M. (2015). Internet of Things: A Survey on Enabling Technologies, Protocols, and Applications. *IEEE Communications Surveys and Tutorials*, *17*(4), 2347–2376. doi:10.1109/COMST.2015.2444095

Alhayani, B. (2022). 5G standards for the Industry 4.0 enabled communication systems using artificial intelligence: Perspective of smart healthcare system. *Applied Nanoscience*, (0123456789). Advance online publication. doi:10.1007/s13204-021-02152-4 PMID:35096498

Alimisis, V., & Taylor, P. C. (2015). Zoning Evaluation for Improved Coordinated Automatic Voltage Control. *IEEE Transactions on Power Systems*, *30*(5), 2736–2746. doi:10.1109/TPWRS.2014.2369428

Alinejad, M., Rezaei, O., Habibifar, R., & Azimian, M. (2022). A Charge/Discharge Plan for Electric Vehicles in an Intelligent Parking Lot Considering Destructive Random Decisions, and V2G and V2V Energy Transfer Modes. *Sustainability (Basel)*, *14*(19), 12816. doi:10.3390/su141912816

Alizadeh, M., & Jafari-Nokandi, M. (2023). A bi-level resilience-oriented islanding framework for an active distribution network incorporating electric vehicles parking lots. *Electric Power Systems Research*, *218*, 109233. doi:10.1016/j.epsr.2023.109233

Alsmadi, A. A., Alrawashdeh, N., Al-Gasaymeh, A., & Alhawamdeh, L. N., & Al Hazimeh, A. M. D. (2023). Adoption of blockchain technology in supply chain. *SAGE Open*, *13*(1), 21582440231160143. doi:10.1177/21582440231160143

Al-Turjman, F., & Alturjman, S. (2020). 5G/IoT-enabled UAVs for multimedia delivery in industry-oriented applications. *Multimedia Tools and Applications*, *79*(13–14), 8627–8648. doi:10.1007/s11042-018-6288-7

Alzaareer, K., Saad, M., Mehrjerdi, H., Asber, D., & Lefebvre, S. (2020). Development of New Identification Method for Global Group of Controls for Online Coordinated Voltage Control in Active Distribution Networks. *IEEE Transactions on Smart Grid*, *11*(5), 3921–3931. doi:10.1109/TSG.2020.2981145

Amanipoor, A., Golsorkhi, M. S., Bayati, N., & Savaghebi, M. (2023). V-Iq Based Control Scheme for Mitigation of Transient Overvoltage in Distribution Feeders with High PV Penetration. *IEEE Transactions on Sustainable Energy*, *14*(1), 283–296. doi:10.1109/TSTE.2022.3211179

American National Standard for Electric Power Systems and Equipment-Voltage Ratings. (2020). *60 Hertz*. National Electrical Manufacturers Association.

Amin, A., Tareen, W. U. K., Usman, M., Memon, K. A., Horan, B., Mahmood, A., & Mekhilef, S. (2020). An Integrated Approach to Optimal Charging Scheduling of Electric Vehicles Integrated with Improved Medium-Voltage Network Reconfiguration for Power Loss Minimization. *Sustainability (Basel)*, *12*(21), 9211. doi:10.3390/su12219211

Amini, M. H., Kargarian, A., & Karabasoglu, O. (2016). ARIMA-based decoupled time series forecasting of electric vehicle charging demand for stochastic power system operation. *Electric Power Systems Research*, *140*, 378–390. doi:10.1016/j.epsr.2016.06.003

Amirioun, M. H., Aminifar, F., Lesani, H., & Shahidehpour, M. (2019). Metrics and quantitative framework for assessing microgrid resilience against windstorms. *International Journal of Electrical Power & Energy Systems*, *104*, 716–723. doi:10.1016/j.ijepes.2018.07.025

An, Y., Gao, Y., Wu, N., Zhu, J., Li, H., & Yang, J. (2023). Optimal scheduling of electric vehicle charging operations considering real-time traffic condition and travel distance. *Expert Systems with Applications, 213*(B). doi:10.1016/j.eswa.2022.118941

Andersson, Ö., & Börjesson, P. (2021). The greenhouse gas emissions of an electrified vehicle combined with renewable fuels: Life Cycle Assessment and policy implications. *Applied Energy*, *289*, 116621. doi:10.1016/j.apenergy.2021.116621

Andervazh, M., Olamaei, J., & Haghifam, M. (2013). Adaptive multi-objective distribution network reconfiguration using multi-objective discrete particles swarm optimisation algorithm and graph theory. *IET Generation, Transmission & Distribution*, *7*(12), 1367–1382. doi:10.1049/iet-gtd.2012.0712

Antoun, J., Kabir, M. E., Atallah, R., Moussa, B., Ghafouri, M., & Assi, C. (2022). Assisting Residential Distribution Grids in Overcoming Large-Scale EV Preconditioning Load. *IEEE Systems Journal*, *16*(3), 4345–4355. doi:10.1109/JSYST.2021.3104185

Anurag, A., Yang, Y., & Blaabjerg, F. (2015). Thermal Performance and Reliability Analysis of Single-Phase PV Inverters with Reactive Power Injection outside Feed-In Operating Hours. *IEEE Journal of Emerging and Selected Topics in Power Electronics*, *3*(4), 870–880. doi:10.1109/JESTPE.2015.2428432

Arfeen, Z. A., Khairuddin, A. B., Larik, R. M., & Saeed, M. S. (2019). Control of distributed generation systems for microgrid applications: A technological review. *International Transactions on Electrical Energy Systems*, *29*(9), e12072. doi:10.1002/2050-7038.12072

Argueta, R. (2010). *A Technical Research Report: The Electric Vehicle*. University of California. Retrieved from https://www.ehcar.net/library/rapport/rapport044.pdf

Arjomandi-Nezhad, A., Fotuhi-Firuzabad, M., Moeini-Aghtaie, M., Safdarian, A., Dehghanian, P., & Wang, F. (2021). Modeling and Optimizing Recovery Strategies for Power Distribution System Resilience. *IEEE Systems Journal*, *15*(4), 4725–4734. doi:10.1109/JSYST.2020.3020058

Arora, S. (2018). Selection of thermal management system for modular battery packs of electric vehicles: A review of existing and emerging technologies. *Journal of Power Sources*, *400*, 621–640. doi:10.1016/j.jpowsour.2018.08.020

Arpita, M., & Panchal, S. (2023). Smart Health and Cybersecurity in the Era of Artificial Intelligence. In M. S. Kaiser, J. Xie, & V. S. Rathore (Eds.), *Information and Communication Technology for Competitive Strategies (ICTCS 2021). Lecture Notes in Networks and Systems* (Vol. 401). Springer. doi:10.1007/978-981-19-0098-3_5

Arshad, A., Ekström, J., & Lehtonen, M. (2018). Multi-agent based distributed voltage regulation scheme with grid-tied inverters in active distribution networks. *Electric Power Systems Research*, *160*, 180–190. doi:10.1016/j.epsr.2018.02.015

Asif, U., & Schmidt, K. (2021). Fuel cell electric vehicles (Fcev): Policy advances to enhance commercial success. *Sustainability (Basel)*, *13*(9), 5149. doi:10.3390/su13095149

Asrari, A., Lotfifard, S., & Ansari, M. (2016). Reconfiguration of Smart Distribution Systems With Time Varying Loads Using Parallel Computing. *IEEE Transactions on Smart Grid*, *7*(6), 2713–2723. doi:10.1109/TSG.2016.2530713

Audi (2022). *E-Tron Technical Specifications and Data*. Audi Media Centre. https://audimediacenter-a.akamaihd.net/system/production/car_motorizations/984/file_en/d933a6f0a1f3a3bc0915035a491b370621af532d/eTechnische_Daten_Audi_e-tron_50_quattro_230kW.pdf

Aurangzeb, M., Xin, A., Iqbal, S., Afzal, M. Z., Kotb, H., AboRas, K. M., Ghadi, Y. Y., & Ngoussandou, B. P. (2023). A Novel Hybrid Approach for Power Quality Improvement in a Vehicle-to-Grid Setup Using Droop-ANN Model. *International Journal of Energy Research*, *2023*, 2023. doi:10.1155/2023/7786928

Azizi, A., Vahidi, B., & Foroughi Nematollahi, A. (2023). Reconfiguration of Active Distribution Networks Equipped with Soft Open Points Considering Protection Constraints. *Journal of Modern Power Systems and Clean Energy*, *11*(1), 212–222. doi:10.35833/MPCE.2022.000425

Azizivahed, A., Arefi, A., Ghavidel, S., Shafie-khah, M., Li, L., Zhang, J., & Catalao, J. P. S. (2020). Energy Management Strategy in Dynamic Distribution Network Reconfiguration Considering Renewable Energy Resources and Storage. *IEEE Transactions on Sustainable Energy*, *11*(2), 662–673. doi:10.1109/TSTE.2019.2901429

Badran, O., Mekhilef, S., Mokhlis, H., & Dahalan, W. (2017). Optimal reconfiguration of distribution system connected with distributed generations: A review of different methodologies. *Renewable & Sustainable Energy Reviews*, *73*, 854–867. doi:10.1016/j.rser.2017.02.010

Bae, S., Gros, S., & Kulcsár, B. (2022). Can AI Abuse Personal Information in an EV Fast-Charging Market? *IEEE Transactions on Intelligent Transportation Systems*, *23*(7), 8759–8769. doi:10.1109/TITS.2021.3086006

Bagheri Tookanlou, M., Pourmousavi Kani, S. A., & Marzband, M. (2021). An Optimal Day-ahead scheduling framework for E-mobility ecosystem operation with drivers' preferences. *IEEE Transactions on Power Systems*, *36*(6), 5245–5257. doi:10.1109/TPWRS.2021.3068689

Bahramipanah, M., Cherkaoui, R., & Paolone, M. (2016). Decentralized voltage control of clustered active distribution network by means of energy storage systems. *Electric Power Systems Research*, *136*, 370–382. doi:10.1016/j.epsr.2016.03.021

Bai, L., Jiang, T., Li, F., Chen, H., & Li, X. (2018). Distributed energy storage planning in soft open point based active distribution networks incorporating network reconfiguration and DG reactive power capability. *Applied Energy*, *210*, 1082–1091. doi:10.1016/j.apenergy.2017.07.004

Bai, S., & Liu, C. (2021). Overview of energy harvesting and emission reduction technologies in hybrid electric vehicles. *Renewable & Sustainable Energy Reviews*, *147*, 111188. doi:10.1016/j.rser.2021.111188

Bamberg, S. (2013). Applying the Stage Model of Self-Regulated Behavioral Change in Integrated Transportation Choice and Travel Behavior Research and Practice: A Commentary. *Journal of Public Transportation*, *16*(4).

Barman, P., Dutta, L., Bordoloi, S., Kalita, A., Buragohain, P., Bharali, S., & Azzopardi, B. (2023). Renewable energy integration with electric vehicle technology: A review of the existing smart charging approaches. *Renewable & Sustainable Energy Reviews*, *183*, 113518. doi:10.1016/j.rser.2023.113518

Barreto, R., Faria, P., & Vale, Z. (2022). Electric mobility: An overview of the main aspects related to the smart grid. *Electronics (Basel)*, *11*(9), 1311. doi:10.3390/electronics11091311

Bayindir, K. Ç., Gözüküçük, M. A., & Teke, A. (2011). A comprehensive overview of hybrid electric vehicle: Powertrain configurations, powertrain control techniques and electronic control units. *Energy Conversion and Management*, *52*(2), 1305–1313. doi:10.1016/j.enconman.2010.09.028

Bayram, I. S., & Bayhan, S. (2020). Location Analysis of Electric Vehicle Charging Stations for Maximum Capacity and Coverage. *2020 IEEE 14th International Conference on Compatibility, Power Electronics and Power Engineering (CPE-POWERENG)*, *1*, 409-414.

BEE. (2020). *E-Mobility*. Bureau of Energy Efficiency (BEE). https://beeindia.gov.in/content/e-mobility

Begum, H., Alam, A. S. A. F., Mia, M. A., Bhuiyan, F., & Ghani, A. B. (2019). Development of islamic microfinance: A sustainable poverty reduction approach. *Journal of Economic and Administrative Sciences*, *35*(3), 143–157. doi:10.1108/JEAS-01-2018-0007

Behrendt, F. (2017a). Why cycling matters for electric mobility: Towards diverse, active and sustainable e-mobilities. *Mobilities*, *13*(1), 64–80. doi:10.1080/17450101.2017.1335463

Berkeley, N., Jarvis, D., & Jones, A. (2018). Analysing the take up of battery electric vehicles: An investigation of barriers amongst drivers in the UK. *Transportation Research Part D, Transport and Environment*, *63*, 466–481. doi:10.1016/j.trd.2018.06.016

Berssaneti, F. T., Berger, S., Saut, A. M., Vanalle, R. M., & Santana, J. C. (2019). Value generation of remanufactured products: Multi-case study of third-party companies. *Sustainability (Basel)*, *11*(3), 584. doi:10.3390/su11030584

Bharathidasan, M., Indragandhi, V., Suresh, V., Jasiński, M., & Leonowicz, Z. (2022). A review on electric vehicle: Technologies, energy trading, and cyber security. *Energy Reports*, *8*, 9662-9685. doi:10.1016/j.egyr.2022.07.145

Bhusal, N., Abdelmalak, M., Kamruzzaman, M., & Benidris, M. (2020). Power System Resilience: Current Practices, Challenges, and Future Directions. *IEEE Access : Practical Innovations, Open Solutions*, *8*, 18064–18086. doi:10.1109/ACCESS.2020.2968586

Bidgoli, H. S., & Van Cutsem, T. (2018). Combined Local and Centralized Voltage Control in Active Distribution Networks. *IEEE Transactions on Power Systems*, *33*(2), 1374–1384. doi:10.1109/TPWRS.2017.2716407

Biresselioglu, M. E., Kaplan, M. D., & Yilmaz, B. K. (2018). Electric mobility in Europe: A comprehensive review of motivators and barriers in decision making processes. *Transportation Research Part A, Policy and Practice*, *109*, 1–13. doi:10.1016/j.tra.2018.01.017

Blat Belmonte, B., Esser, A., Weyand, S., Franke, G., Schebek, L., & Rinderknecht, S. (2020). *Identification of the Optimal Passenger Car Vehicle Fleet Transition for Mitigating the Cumulative Life-Cycle Greenhouse Gas Emissions until 2050*. Vehicles. doi:10.3390/vehicles2010005

Bleicher, A., & Pehlken, A. (Eds.). (2020). *The material basis of energy transitions*. Academic Press. doi:10.1016/B978-0-12-819534-5.00001-5

Bloomberg, N. E. F. (2021). *Electric vehicles on track to be cheaper than internal combustion engine vehicles by 2027*. Bloomberg NEF.

Bonsu, N. O. (2020). Towards a circular and low-carbon economy: Insights from the transitioning to electric vehicles and net zero economy. *Journal of Cleaner Production*, *256*, 120659. doi:10.1016/j.jclepro.2020.120659

Boulanger, A. G., Chu, A. C., Maxx, S., & Waltz, D. L. (2011). Vehicle electrification: Status and issues. *Proceedings of the IEEE*, *99*(6), 1116–1138. doi:10.1109/JPROC.2011.2112750

Brown, M. A. (2017). *Electric Powertrains: Opportunities and Challenges in the U.S. Light-Duty Vehicle Fleet*. National Academies Press.

Brown, M. A., & Soni, A. (2019). Expert perceptions of enhancing grid resilience with electric vehicles in the United States. *Energy Research & Social Science*, *57*, 101241. doi:10.1016/j.erss.2019.101241

Brown, S., Pyke, D., & Steenhof, P. (2010). Electric vehicles: The role and importance of standards in an emerging market. *Energy Policy*, *38*(7), 3797–3806. doi:10.1016/j.enpol.2010.02.059

Brugger, H., Eichhammer, W., Mikova, N., & Dönitz, E. (2021). Energy efficiency vision 2050: How will new societal trends influence future energy demand in the European countries? *Energy Policy*, *152*, 112216. doi:10.1016/j.enpol.2021.112216

Bunsen, T. (2019). *Global EV Outlook (2018)*. Towards Cross-Modal Electrification.

C40 Cities. (2018). *Electric Buses in Cities: Driving Towards Cleaner Air and Lower* CO_2. Author.

Campbell, P. (2008). *System Cost Analysis for an Interior Permanent Magnet Motor*. AMES Laboratory, US Department of Energy. https://www.osti.gov/servlets/purl/940187

Cannon, P. (2020). How a Lack of Public EV Charging Stations Is Creating Big Business Opportunities. *ABB*, (Jun), 15.

Cao, Y., Cui, J., Liu, S., Li, X., Zhou, Q., Hu, C, & Liu, Z. (2023). A Holistic Review on E-Mobility Service Optimization: Challenges, Recent Progress and Future Directions. *IEEE Transactions on Transportation Electrification*.

Capitanescu, F., Ochoa, L. F., Margossian, H., & Hatziargyriou, N. D. (2015). Assessing the Potential of Network Reconfiguration to Improve Distributed Generation Hosting Capacity in Active Distribution Systems. *IEEE Transactions on Power Systems*, *30*(1), 346–356. doi:10.1109/TPWRS.2014.2320895

Carley, S., Krause, R. M., Lane, B. W., & Graham, J. D. (2013). Intent to purchase a plug-in electric vehicle: A survey of early impressions in large US cities. *Transportation Research Part D, Transport and Environment*, *18*, 39–45. doi:10.1016/j.trd.2012.09.007

Case studies on E-bus development and operation in Chinese cities. (n.d.). https://repository.unescap.org/bitstream/handle/20.500.12870/6285/ESCAP-2023-RP-Case-Studies-E-Bus-Development-China.pdf?sequence=3

Casini, M. (2017, August). Green technology for smart cities. *IOP Conference Series. Earth and Environmental Science*, *83*(1), 012014. doi:10.1088/1755-1315/83/1/012014

Castillo-Calzadilla, T., Alonso-Vicario, A., Borges, C. E., & Martin, C. (2022). E-mobility in positive energy districts. *Buildings*, *12*(3), 264. doi:10.3390/buildings12030264

Cervero, R. (2013). Linking urban transport and land use in developing countries. *Journal of Transport and Land Use*, *6*(1), 7–24. doi:10.5198/jtlu.v6i1.425

Chai, Q., Zhang, C., Xu, Y., & Dong, Z. Y. (2021). PV inverter reliability-constrained volt/var control of distribution networks. *IEEE Transactions on Sustainable Energy*, *12*(3), 1788–1800. doi:10.1109/TSTE.2021.3065451

Chakraborty, R., Chakrabortty, A., Farantatos, E., Patel, M., Hooshyar, H., & Darvishi, A. (2023). Hierarchical frequency and voltage control using prioritized utilization of inverter based resources. *International Journal of Electrical Power & Energy Systems*, *144*, 108527. doi:10.1016/j.ijepes.2022.108527

Chakrapani, B., Chakrapani, K., Kavitha, T., Iqubal, S. M., & Kempanna, M. (2023). Sustainable Smart Energy Systems and Energy Preservation Strategies in Intelligent Transportation Sectors. In Applications of Big Data and Artificial Intelligence in Smart Energy Systems: Volume 2 - Energy Planning, Operations, Control and Market Perspectives. River Publishers.

Chan, C. C. (2016). An overview of electric vehicle technology. *Proceedings of the IEEE, 81*(9), 1202-1213. https://ieeexplore.ieee.org/abstract/document/237530 doi:10.1109/5.237530

Chan, C. C. (2002). The state of the art of electric and hybrid vehicles. *Proceedings of the IEEE*, *90*(2), 247–275. doi:10.1109/5.989873

Charging, M. (n.d.). *Novel, automated charging infrastructure for electric vehicles*. Available at: https://cordis.europa.eu/project/id/830239

Chavhan, S., Gupta, D., Chandana, B. N., Khanna, A., & Rodrigues, J. J. P. C. (2020). IoT-Based Context-Aware Intelligent Public Transport System in a Metropolitan Area. *IEEE Internet of Things Journal*, *7*(7), 6023–6034. doi:10.1109/JIOT.2019.2955102

Chen, P., Wang, Y., Wen, X., Lan, L., Wang, C., Gan, Y., Zhang, L., Zhan, F., & Fu, P., (2010). Summary on the damaging effect of cryogenic freezing rain and snow disaster on power grid in China. *Dianwang Jishu/Power System Technology, 34*, 135–139.

Cheng, S., & Li, Z. (2019). Multi-objective Network Reconfiguration Considering V2G of Electric Vehicles in Distribution System with Renewable Energy. *Energy Procedia*, *158*, 278–283. doi:10.1016/j.egypro.2019.01.089

Chenoy, D., Ghosh, S. M., & Shukla, S. K. (2019). Skill development for accelerating the manufacturing sector: The role of 'new-age' skills for 'make in India.' *International Journal of Training Research, 17*(sup1), 112–130.

Chen, S., Hu, W., & Chen, Z. (2016). Comprehensive Cost Minimization in Distribution Networks Using Segmented-Time Feeder Reconfiguration and Reactive Power Control of Distributed Generators. *IEEE Transactions on Power Systems*, *31*(2), 983–993. doi:10.1109/TPWRS.2015.2419716

Chen, Y., Kang, Y., Zhao, Y., Wang, L., Liu, J., Li, Y., Liang, Z., He, X., Li, X., Tavajohi, N., & Li, B. (2021). A review of lithium-ion battery safety concerns: The issues, strategies, and testing standards. *Journal of Energy Chemistry*, *59*, 83–99. doi:10.1016/j.jechem.2020.10.017

Chien, I., Karthikeyan, P., & Hsiung, P.-A. (2023). Peer to peer energy transaction market prediction in Smart Grids using blockchain and LSTM. *2023 IEEE International Conference on Consumer Electronics (ICCE).* 10.1109/ICCE56470.2023.10043563

Christakou, K., Paolone, M., & Abur, A. (2018). Voltage control in active distribution networks under uncertainty in the system model: A robust optimization approach. *IEEE Transactions on Smart Grid*, *9*(6), 5631–5642. doi:10.1109/TSG.2017.2693212

Clou, S. (2023). *Smart Grid Evolution in 2023: The Future of Energy Management.* Smart Energy International. https://www.smart-energy.com/industry-sectors/smart-grid/smart-grid-evolution-in-2023-the-future-of-energy-management

Constantinides, S. (2012, Sept. 30 – Oct. 3). *The Demand for Rare Earth Materials in Permanent Magnets.* COM 2012, Niagara Falls. https://www.arnoldmagnetics.com/wp-content/uploads/2017/10/Demand-for-rare-earth-materials-in-permanent-magnets-Constantinides-COM-2012-psn-hi-res.pdf

Crow, M. L. (2016). Cost-constrained dynamic optimal electric vehicle charging. *IEEE Transactions on Sustainable Energy*, *8*(2), 716–724.

Csillak, K., & Kuhnke, M. M. (2023). Interoperability Framework for Electromobility (INFRA): The main results from the USER-CHI-framework in a new spotlight. *Transportation Research Procedia*, *72*, 1926–1933. doi:10.1016/j.trpro.2023.11.672

Csonka, B. (2020, June). Centralized charging power distribution method for electric vehicles. In *2020 Smart City Symposium Prague (SCSP)* (pp. 1-6). IEEE. 10.1109/SCSP49987.2020.9134002

Cui, H., Li, F., Fang, X., & Long, R. (2015). *Distribution network reconfiguration with aggregated electric vehicle charging strategy. In 2015 IEEE Power & Energy Society General Meeting*. IEEE. doi:10.1109/PESGM.2015.7285650

Czerwinski, F. (2021). Current trends in automotive lightweighting strategies and materials. *Materials (Basel)*, *14*(21), 6631. doi:10.3390/ma14216631 PMID:34772154

Da Xu, L. (2021). Emerging Enabling Technologies for Industry 4.0 and Beyond. *Information Systems Frontiers*, *2022*. Advance online publication. doi:10.1007/s10796-021-10213-w

Daina, N., Sivakumar, A., & Polak, J. W. (2017). Electric vehicle charging choices: Modelling and implications for smart charging services. *Transportation Research Part C, Emerging Technologies*, *81*, 36–56. doi:10.1016/j.trc.2017.05.006

Danese, A., Torsæter, B. N., Sumper, A., & Garau, M. (2022, March 22). Planning of High-Power Charging Stations for Electric Vehicles: A Review. *Applied Sciences (Basel, Switzerland)*, *12*(7), 3214. doi:10.3390/app12073214

Dargan, J., Gupta, N., & Singh, L. (2021, November). Blockchain Based Energy Management System: A Proposed Model. In *2021 International Conference on Technological Advancements and Innovations (ICTAI)* (pp. 510-514). IEEE. 10.1109/ICTAI53825.2021.9673233

Dar, M. R., & Ganguly, S. (2022.) A review of voltage control strategies in active distribution networks. *2022 IEEE Calcutta Conference (CALCON),* 145-149. 10.1109/CALCON56258.2022.10060319

Das, D. (2006). A Fuzzy Multiobjective Approach for Network Reconfiguration of Distribution Systems. *IEEE Transactions on Power Delivery*, *21*(1), 202–209. doi:10.1109/TPWRD.2005.852335

Das, H. S., Rahman, M. M., Li, S., & Tan, C. W. (2020). Electric vehicles standards, charging infrastructure, and impact on grid integration: A technological review. *Renewable & Sustainable Energy Reviews*, *120*, 109618. doi:10.1016/j.rser.2019.109618

Das, P. K., & Bhat, M. Y. (2022). Global Electric Vehicle Adoption: Implementation and Policy Implications for India. *Environmental Science and Pollution Research International*, *29*(27), 40612–40622. doi:10.1007/s11356-021-18211-w PMID:35083674

Davies, B. E., Mottram, R. S., & Harris, I. R. (2001). Recent developments in the sintering of NdFeB. *Materials Chemistry and Physics*, *67*(1-3), 1–3. doi:10.1016/S0254-0584(00)00450-8

de Melo, V. S., Camargo, R. S., Krebel, F., Brunoro, M., Nunes, W. T., Fernandes, M. C., & Nunes, R. B. (2023). Greenhouse Gas Emissions Estimation with Electric Vehicles Penetration in the Brazilian Power Grid. *2023 15th IEEE International Conference on Industry Applications (INDUSCON),* 254-261. 10.1109/INDUSCON58041.2023.10374611

de Oliveira Neto, G. C. (2022). *How can Industry 4.0 technologies and circular economy help companies and researchers collaborate and accelerate the transition to strong sustainability? A bibliometric review and a systematic literature review, no. 0123456789*. Springer Berlin Heidelberg.

Debnath, B., Biswas, S., & Uddin, M. F. (2020). Optimization of Electric Vehicle Charging to Shave Peak Load for Integration in Smart Grid. *2020 IEEE Region 10 Symposium (TENSYMP),* 483-488. 10.1109/TENSYMP50017.2020.9231029

Deb, S., Tammi, K., Kalita, K., & Mahanta, P. (2018). Impact of Electric Vehicle Charging Station Load on Distribution Network. *Energies*, *11*(1), 178. doi:10.3390/en11010178

Demestichas, K., & Daskalakis, E. (2020). Information and communication technology solutions for the circular economy. *Sustainability (Basel)*, *12*(18), 7272. doi:10.3390/su12187272

Dewan, S., Singh, L., & Gupta, N. (2021). Secure Electronic Voting System based on Mobile-app and Blockchain. *Recent Advances in Computer Science and Communications (Formerly: Recent Patents on Computer Science), 14*(9), 2992-3002.

Dey, S., & Mehta, N. S. (2020). Automobile pollution control using catalysis. *Resource. Environmental Sustainability*, *2*, 100006. Advance online publication. doi:10.1016/j.resenv.2020.100006

Dhar, S., Pathak, M., Shukla, P., & Gupta, A. (2020). Electric vehicles penetration in India for enhanced energy efficiency deployment in the transport sector. *Energy Efficiency in Developing Countries*, 253–270.

Dias, G., Arsenio, E., & Ribeiro, P. (2021). The role of shared e-scooter systems in urban sustainability and resilience during the covid-19 mobility restrictions. *Sustainability (Basel)*, *13*(13), 7084. doi:10.3390/su13137084

Didier, M., Grauvogl, B., Steentoft, A., Ghosh, S., & Stojadinovic, B. (2017). Seismic Resilience of the Nepalese Power Supply System During the 2015 Gorkha Earthquake. *16th World Conference on Earthquake Engineering, 16WCEE 2017.*

Dijk, M., Orsato, R. J., & Kemp, R. (2013). The emergence of an electric mobility trajectory. *Energy Policy*, *52*, 135–145. doi:10.1016/j.enpol.2012.04.024

Dileep, G. J. R. E. (2020). A survey on smart grid technologies and applications. *Renewable Energy*, *146*, 2589–2625. doi:10.1016/j.renene.2019.08.092

Ding, T., Li, C., Yang, Y., Jiang, J., Bie, Z., & Blaabjerg, F. (2017). A Two-Stage Robust Optimization for Centralized-Optimal Dispatch of Photovoltaic Inverters in Active Distribution Networks. *IEEE Transactions on Sustainable Energy*, *8*(2), 744–754. doi:10.1109/TSTE.2016.2605926

Ding, T., Wang, Z., Jia, W., Chen, B., Chen, C., & Shahidehpour, M. (2020). Multiperiod Distribution System Restoration With Routing Repair Crews, Mobile Electric Vehicles, and Soft-Open-Point Networked Microgrids. *IEEE Transactions on Smart Grid, 11*(6), 4795–4808. doi:10.1109/TSG.2020.3001952

Dou, X., Xu, P., Hu, Q., Sheng, W., Quan, X., Wu, Z., & Xu, B. (2018). A distributed voltage control strategy for multi-microgrid active distribution networks considering economy and response speed. *IEEE Access : Practical Innovations, Open Solutions, 6*, 31259–31268. doi:10.1109/ACCESS.2018.2837082

Dubey, A., & Santoso, S. (2015). Electric vehicle charging on residential distribution systems: Impacts and mitigations. *IEEE Access : Practical Innovations, Open Solutions, 3*, 1871–1893. doi:10.1109/ACCESS.2015.2476996

Du, P., & Lu, N. (2014). *Energy Storage for Smart Grids: Planning and Operation for Renewable and Variable Energy Resources (VERs)*. Academic Press.

Dutta, A., Ganguly, S., & Kumar, C. (2022[a]). Coordinated control scheme for EV charging and volt/var devices scheduling to regulate voltages of active distribution networks. *Sustainable Energy, Grids and Networks*, 31.

Dutta, A., Ganguly, S., & Kumar, C. (2020). Model predictive control-based optimal voltage regulation of active distribution networks with OLTC and reactive power capability of PV inverters. *IET Generation, Transmission & Distribution, 14*(22), 5183–5192. doi:10.1049/iet-gtd.2020.0378

Dutta, A., Ganguly, S., & Kumar, C. (2022b). Coordinated Volt/Var Control of PV and EV Interfaced Active Distribution Networks Based on Dual-Stage Model Predictive Control. *IEEE Systems Journal, 16*(3), 4291–4300. doi:10.1109/JSYST.2021.3110509

Dutta, A., Ganguly, S., & Kumar, C. (2022c). MPC-Based Coordinated Voltage Control in Active Distribution Networks Incorporating CVR and DR. *IEEE Transactions on Industry Applications, 58*(4), 4309–4318. doi:10.1109/TIA.2022.3163108

ECFR. (2021). *The Power Atlas – Seven battlegrounds of a networked world.* Published Jointly by the European Council of Foreign Relations (ECFR) and Stiftung Mercator. https://ecfr.eu/wp-content/uploads/power-atlas.pdf

Eisel, M. (2016). *Analyzing the Range Barrier to Electric Vehicle Adoption: The Case of Range Anxiety* (Vol. 86). Cuvillier Verlag.

El-Bayeh, C. Z., Alzaareer, K., Aldaoudeyeh, A.-M. I., Brahmi, B., & Zellagui, M. (2021). Charging and Discharging Strategies of Electric Vehicles: A Survey. *World Electric Vehicle Journal, 12*(1), 11. doi:10.3390/wevj12010011

Eldurssi, A. M., & O'Connell, R. M. (2015). A Fast Nondominated Sorting Guided Genetic Algorithm for Multi-Objective Power Distribution System Reconfiguration Problem. *IEEE Transactions on Power Systems, 30*(2), 593–601. doi:10.1109/TPWRS.2014.2332953

Electrified L-Category Vehicles Integrated Into Transport and Electricity Networks. (n.d.). https://cordis.europa.eu/project/id/769926

Elkhatib, M. E., El-Shatshat, R., & Salama, M. M. A. (2011). Novel coordinated voltage control for smart distribution networks with DG. *IEEE Transactions on Smart Grid, 2*(4), 598–605. doi:10.1109/TSG.2011.2162083

Elwert, T., Goldmann, D., Roemer, F., & Schwarz, S. (2016). Recycling of NdFeB Magnets from Electric Drive Motors of (Hybrid) Electric Vehicles. *Journal of Sustainable Metallurgy, 3*(1), 108–121. Advance online publication. doi:10.1007/s40831-016-0085-1

Enang, W., & Bannister, C. 2017. Modelling and control of hybrid electric vehicles (A comprehensive review). In Renewable and Sustainable Energy Reviews (Vol. 74). Academic Press.

Enang, W., & Bannister, C. (2017). Modelling and control of hybrid electric vehicles (A comprehensive review). *Renewable & Sustainable Energy Reviews*, *74*, 1210–1239. doi:10.1016/j.rser.2017.01.075

Endrenyi, J. (1978). *Reliability modeling in electric power systems*. Wiley.

Energy Efficiency and Renewable Energy, Developing Infrastructure to Charge Plug-In Electric Vehicles. (n.d.). https://afdc.energy.gov/fuels/electricity_infrastructure.html

Erenoglu, A. K., & Erdinc, O. (2023). Real-Time Allocation of Multi-Mobile Resources in Integrated Distribution and Transportation Systems for Resilient Electrical Grid. *IEEE Transactions on Power Delivery*, *38*(2), 1108–1119. doi:10.1109/TPWRD.2022.3206272

Erenoglu, A. K., Sancar, S., Terzi, I. S., Erdinc, O., Shafie-Khah, M., & Catalao, J. P. S. (2022). Resiliency-Driven Multi-Step Critical Load Restoration Strategy Integrating On-Call Electric Vehicle Fleet Management Services. *IEEE Transactions on Smart Grid*, *13*(4), 3118–3132. doi:10.1109/TSG.2022.3155438

Euceda, R., Akundi, A., & Luna, S. (2023). Cybersecurity Challenges in Electric Vehicles: An initial literature review and research agenda. *IEEE International Systems Conference (SysCon)*, 1-8. 10.1109/SysCon53073.2023.10131069

Fang, X., Misra, S., Xue, G., & Yang, D. (2011). Smart grid—The new and improved power grid: A survey. *IEEE Communications Surveys and Tutorials*, *14*(4), 944–980. doi:10.1109/SURV.2011.101911.00087

Farah, R. N., Syahirah, N. A., Misron, N., Azmi, M. S., Karim, N. S., & Husin, N. (2023). A Need Analysis of the Criteria Involved in Determining Suitable Locations for Photovoltaic Electric Vehicle Charging Stations in Malaysia. *Malaysian Journal of Fundamental and Applied Sciences*, *19*(6), 1033–1041. doi:10.11113/mjfas.v19n6.3114

Fears, P. (2021, Jan. 5). Rare Earth Magnets in Electric Vehicle Motors. *Bunting - Berkhamsted.* https://www.bunting-berkhamsted.com/rare-earth-magnets-in-electric-vehicle-motors/

Feng, H., Tavakoli, R., Onar, O. C., & Pantic, Z. (2020). Advances in high-power wireless charging systems: Overview and design considerations. *IEEE Transactions on Transportation Electrification*, *6*(3), 886–919. doi:10.1109/TTE.2020.3012543

Feng, S., & Magee, C. L. (2020). Technological development of key domains in electric vehicles: Improvement rates, technology trajectories and key assignees. *Applied Energy*, *260*, 114264. doi:10.1016/j.apenergy.2019.114264

Fishman, E., Washington, S., & Haworth, N. (2015). Bikeshare's Impact on Active Travel: Evidence from the United States, Great Britain, and Australia. *Journal of Transport & Health*, *2*(2), 135–142. doi:10.1016/j.jth.2015.03.004

Friedel, J. N. (2008). The effect of the Community College Workforce Development Mission on Governance. *New Directions for Community Colleges*, *2008*(141), 45–55. doi:10.1002/cc.314

Furlan, R., & Al-Mohannadi, A. (2020). An urban regeneration planning scheme for the Souq Waqif Heritage Site of Doha. *Sustainability (Basel)*, *12*(19), 7927. doi:10.3390/su12197927

Gallegos, J., Arévalo, P., Montaleza, C., & Jurado, F. (2024). Sustainable Electrification—Advances and Challenges in Electrical-Distribution Networks: A Review. *Sustainability (Basel)*, *16*(2), 698. doi:10.3390/su16020698

Gan, L., Topcu, U., & Low, S. H. (2012). Optimal decentralized protocol for electric vehicle charging. *IEEE Transactions on Power Systems*, *28*(2), 940–951. doi:10.1109/TPWRS.2012.2210288

Gan, W., Shahidehpour, M., Guo, J., Yao, W., Pandey, S., Paaso, A., Vukojevic, A., & Wen, J. (2022). A Tri-Level Planning Approach to Resilient Expansion and Hardening of Coupled Power Distribution and Transportation Systems. *IEEE Transactions on Power Systems*, *37*(2), 1495–1507. doi:10.1109/TPWRS.2021.3107402

Gao, H., Chen, Y., Mei, S., Huang, S., & Xu, Y. (2017). Resilience-Oriented Pre-Hurricane Resource Allocation in Distribution Systems Considering Electric Buses. *Proceedings of the IEEE*, *105*(7), 1214–1233. doi:10.1109/JPROC.2017.2666548

Gao, H., Zhang, Y., & Ji, X. (2020). Scenario clustering based distributionally robust comprehensive optimization of active distribution network. *Dianli Xitong Zidonghua*, *44*(21), 32–41.

Gao, Y., Guo, S., Ren, J., Zhao, Z., Ehsan, A., & Zheng, Y. (2018). An electric bus power consumption model and optimization of charging scheduling concerning multi-external factors. *Energies*, *11*(8), 2060. doi:10.3390/en11082060

Gao, Y., Zhang, X., Cheng, Q., Guo, B., & Yang, J. (2019). Classification and review of the charging strategies for commercial lithium-ion batteries. *IEEE Access : Practical Innovations, Open Solutions*, *7*, 43511–43524. doi:10.1109/ACCESS.2019.2906117

García-Villalobos, J., Zamora, I., San Martín, J. I., Asensio, F. J., & Aperribay, V. (2014). Plug-in electric vehicles in electric distribution networks: A review of smart charging approaches. *Renewable & Sustainable Energy Reviews*, *38*, 717–731. doi:10.1016/j.rser.2014.07.040

Gauß, R., Burkhardt, C., Carencotte, F., Gasparon, M., Gutfleisch, O., Higgins, I., Karajić, M., Klossek, A., Mäkinen, M., Schäfer, B., Schindler, R., & Veluri, B. (2021). *Rare Earth Magnets and Motors: A European Call for Action.* A report by the Rare Earth Magnets and Motors Cluster of the European Raw Materials Alliance. Berlin. https://eitrawmaterials.eu/wp-content/uploads/2021/09/ERMA-Action-Plan-2021-A-European-Call-for-Action.pdf

Gazijahani, F. S., Salehi, J., & Shafie-khah, M. (2022). Benefiting from Energy-Hub Flexibilities to Reinforce Distribution System Resilience: A Pre- and Post-Disaster Management Model. *IEEE Systems Journal*, *16*(2), 3381–3390. doi:10.1109/JSYST.2022.3147075

Gholami, A., Shekari, T., Amirioun, M. H., Aminifar, F., Amini, M. H., & Sargolzaei, A. (2018). Toward a Consensus on the Definition and Taxonomy of Power System Resilience. *IEEE Access : Practical Innovations, Open Solutions*, *6*, 32035–32053. doi:10.1109/ACCESS.2018.2845378

Gholami, K., Islam, M. R., Rahman, M. M., Azizivahed, A., & Fekih, A. (2022). State-of-the-art technologies for volt-var control to support the penetration of renewable energy into the smart distribution grids. *Energy Reports*, *8*, 8630–8651. doi:10.1016/j.egyr.2022.06.080

Ghosh, A. (2020). Possibilities and challenges for the inclusion of the electric vehicle (EV) to reduce the carbon footprint in the transport sector: A review. *Energies*, *13*(10), 2602. doi:10.3390/en13102602

Gilani, M. A., Kazemi, A., & Ghasemi, M. (2020). Distribution system resilience enhancement by microgrid formation considering distributed energy resources. *Energy*, *191*, 116442. doi:10.1016/j.energy.2019.116442

Gilpin, R. G. (2016). *The political economy of international relations*. Princeton University Press.

Gil, S., Zapata-Madrigal, G. D., García-Sierra, R., & Cruz Salazar, L. A. (2022). Converging IoT protocols for the data integration of automation systems in the electrical industry. *Journal of Electrical Systems and Information Technology*, *9*(1), 1–21. doi:10.1186/s43067-022-00043-4

Gomez, J. C., & Morcos, M. M. (2003). Impact of EV battery chargers on the power quality of distribution systems. *IEEE Transactions on Power Delivery*, *18*(3), 975–981. doi:10.1109/TPWRD.2003.813873

Gonsalves, O. (2019, Jan. 17). The Labor Market in India: Structure and Costs. *India Briefing*. https://www.india-briefing.com/news/labor-market-india-structure-costs-18264.html/

Gonzalez Venegas, Petit, & Perez. (2021). Plug-in behavior of electric vehicles users: Insights from a large-scale trial and impacts for grid integration studies. *eTransportation, 10*.

Gouin, V., Alvarez Herault, M., & Raison, B. (2018). Stochastic integration of demand response and reconfiguration in distribution network expansion planning. *IET Generation, Transmission & Distribution*, *12*(20), 4536–4545. doi:10.1049/iet-gtd.2018.5833

Gouse Mohiuddin Kosgiker. (2018). Machine Learning- Based System, Food Quality Inspection and Grading in Food industry. *International Journal of Food and Nutritional Sciences*, *11*(10), 723–730.

Gowda, S. N. (2021). *Assessment and tracking electric vehicle battery degradation cost using blockchain. IEEE Power & Energy Society Innovative Smart Grid Technologies Conference*.

Grace, O., Iqbal, K., & Rabbi, F. (2023). Creating Sustainable Urban Environments: The Vital Link between Development, Health, and Smart Cities. *International Journal of Sustainable Infrastructure for Cities and Societies*, *8*(1), 53–72.

Grijalva, S., & Tariq, M. U. (2011, January). Prosumer-based smart grid architecture enables a flat, sustainable electricity industry. In *ISGT 2011* (pp. 1–6). IEEE. doi:10.1109/ISGT.2011.5759167

Gulia, J., & Jadhav, A. (2021). *E-Cars Market in India- October 2021*. JMK Research & Analytics. https://jmkresearch.com/wp-content/uploads/2021/10/E-Cars-Market-in-India_JMKResearch_Oct-2021.pdf

Gunjal, M., Mathew, L., & Sudha Letha, S. (2023). Electric vehicle controllers for sustainable energy management: Recommendations and challenges. *Environment, Development and Sustainability*, 1–31. doi:10.1007/s10668-023-03561-x

Guo, Y., Wu, Q., Gao, H., Chen, X., Ostergaard, J., & Xin, H. (2019). MPC-Based Coordinated Voltage Regulation for Distribution Networks with Distributed Generation and Energy Storage System. *IEEE Transactions on Sustainable Energy*, *10*(4), 1731–1739. doi:10.1109/TSTE.2018.2869932

Guo, Y., Wu, Q., Gao, H., Huang, S., Zhou, B., & Li, C. (2020a). Double-Time-Scale Coordinated Voltage Control in Active Distribution Networks Based on MPC. *IEEE Transactions on Sustainable Energy*, *11*(1), 294–303. doi:10.1109/TSTE.2018.2890621

Guo, Z., Zhou, Z., & Zhou, Y. (2020). Impacts of Integrating Topology Reconfiguration and Vehicle-to-Grid Technologies on Distribution System Operation. *IEEE Transactions on Sustainable Energy*, *11*(2), 1023–1032. doi:10.1109/TSTE.2019.2916499

Gupta, N., & Saini, I. (2019, April). Renewable Eco friendly Alternate fuels. *International Journal of Technical Innovation in Modern Engineering & Science, 5*, 117-123.

Gupta, N., Mahla, S. K., & Papreja, A. (2017). Experimental Studies of Utilization of Biogas with Biodiesel in Diesel Engine Generator for Electricity Generation. *International Journal of Engineering Research & Science, 3*.

Gupta, N., Mahla, S. K., & Papreja, A. (2022). Experimental Studies of Utilization of Biogas with Biodiesel in Diesel Engine Generator for Electricity Generation. *International Journal of Engineering Research & Science*.

Gupta, N., & Khatri, M. (2017). An experimental analysis of shading on Solar Photovoltaic system. *Int. J. Emerg. Technolog*, *8*(1), 226–228.

Haider, R., & Annaswamy, A. M. (2022). A hybrid architecture for volt-var control in active distribution grids. *Applied Energy*, *312*, 118735. doi:10.1016/j.apenergy.2022.118735

Haines, A., & Frumkin, H. (2021). *Planetary health: Safeguarding human health and the environment in the anthropocene*. Cambridge University Press. doi:10.1017/9781108698054

Hall, D., & Lutsey, N. (2020). Charging infrastructure in cities: Metrics for evaluating future needs. *Europe, 6*.

Halli, U. M. (2022a). Nanotechnology in E-Vehicle Batteries. *International Journal of Nanomaterials and Nanostructures.*, *8*(2), 22–27.

Halli, U. M. (2022b). Voltage Sag Mitigation Using DVR and Ultra Capacitor. *Journal of Semiconductor Devices and Circuits.*, *9*(3), 21–31p.

Hamada, A. T., & Orhan, M. F. (2022). An overview of regenerative braking systems. *Journal of Energy Storage*, *52*, 105033. doi:10.1016/j.est.2022.105033

Hamza, K., Laberteaux, K., & Chu, K. C. (2020). *On modeling the total cost of ownership of electric and plug-in hybrid vehicles.* Academic Press.

Hanemann, P., Behnert, M., & Bruckner, T. (2017). Effects of electric vehicle charging strategies on the German power system. *Applied Energy*, *203*, 608–622. doi:10.1016/j.apenergy.2017.06.039

Hannan, M. A., Azidin, F. A., & Mohamed, A. (2014). Hybrid electric vehicles and their challenges: A review. *Renewable & Sustainable Energy Reviews*, *29*, 135–150. doi:10.1016/j.rser.2013.08.097

Han, Y., An, C., & Wang, J. (2018). A Review of Electric Vehicle Charging Infrastructure Planning. *Renewable & Sustainable Energy Reviews*, *94*, 935–946.

Hapuwatte, B. M., & Jawahir, I. S. (2019). A total life cycle approach for developing predictive design methodologies to optimize product performance. *Procedia Manufacturing*, *33*, 11–18. doi:10.1016/j.promfg.2019.04.003

Haque, A. N. M. M., Xiong, M., & Nguyen, P. H. (2019). March. Consensus algorithm for fair power curtailment of PV systems in LV networks. *2019 IEEE PES GTD Grand International Conference and Exposition Asia (GTD Asia)*, 813-818.

Härer, F., & Fill, H.-G. (2022). Decentralized Attestation and Distribution of Information Using Blockchains and Multi-Protocol Storage. *IEEE Access : Practical Innovations, Open Solutions*, *10*, 18035–18054. doi:10.1109/ACCESS.2022.3150356

Hasankhani, A., Hakimi, S. M., Bisheh-Niasar, M., Shafie-khah, M., & Asadolahi, H. (2021). Blockchain technology in the future smart grids: A comprehensive review and frameworks. *International Journal of Electrical Power & Energy Systems*, *129*, 106811. doi:10.1016/j.ijepes.2021.106811

Hasan, M. M., Hossain, S., Mofijur, M., Kabir, Z., Badruddin, I. A., Yunus Khan, T. M., & Jassim, E. (2023). Harnessing solar power: A review of photovoltaic innovations, solar thermal systems, and the dawn of energy storage solutions. *Energies*, *16*(18), 6456. doi:10.3390/en16186456

Heimerl, A., Weitz, K., Baur, T., & André, E. (2022). Unraveling ML Models of Emotion With NOVA: Multi-Level Explainable AI for Non-Experts. *IEEE Transactions on Affective Computing, 13*(3), 1155-1167. . doi:10.1109/TAFFC.2020.3043603

Hemmatpour, M. H., Rezaeian Koochi, M. H., Dehghanian, P., & Dehghanian, P. (2022). Voltage and energy control in distribution systems in the presence of flexible loads considering coordinated charging of electric vehicles. *Energy*, *239*, 121880. doi:10.1016/j.energy.2021.121880

Hensher, D. A., Mulley, C., Ho, C., Wong, Y., Smith, G., & Nelson, J. D. (2020). *Understanding Mobility as a Service (MaaS): Past, present and future.* Elsevier.

Herold, D. M., & Lee, K. H. (2018). Carbon disclosure strategies in the global logistics industry: Similarities and differences in carbon measurement and reporting. *Pathways to a Sustainable Economy: Bridging the Gap between Paris Climate Change Commitments and Net Zero Emissions*, 87-101.

He, X., Qiu, R. C., Ai, Q., Chu, L., Xu, X., & Ling, Z. (2016). Designing for situation awareness of future power grids: An indicator system based on linear eigenvalue statistics of large random matrices. *IEEE Access : Practical Innovations, Open Solutions*, *4*, 3557–3568. doi:10.1109/ACCESS.2016.2581838

Hidalgo, R., Abbey, C., & Joós, G. (2010). A review of active distribution networks enabling technologies. *IEEE PES General Meeting, PES 2010*. 10.1109/PES.2010.5590051

Hiekata, K., Wanaka, S., Mitsuyuki, T., Ueno, R., Wada, R., & Moser, B. (2021). Systems analysis for deployment of internet of things (IoT) in the maritime industry. *Journal of Marine Science and Technology*, *26*(2), 459–469. doi:10.1007/s00773-020-00750-5

Hoarau, Q., & Perez, Y. (2018). Interactions between electric mobility and photovoltaic generation: A review. *Renewable & Sustainable Energy Reviews*, *94*, 510–522. doi:10.1016/j.rser.2018.06.039

Holmqvist, M., & Johansson, L. (2021). *A self-service design concept for last mile delivery: The solution for retail stores in urban environments*. Academic Press.

Hoogma, R., Kemp, R., Schot, J., & Truffer, B. (2002). *Experimenting for sustainable transport*. Taylor & Francis.

Hoque, M. M., Khorasany, M., Razzaghi, R., Wang, H., & Jalili, M. (2022). Transactive Coordination of Electric Vehicles with Voltage Control in Distribution Networks. *IEEE Transactions on Sustainable Energy*, *13*(1), 391–402. doi:10.1109/TSTE.2021.3113614

Hosmani, S. M. (2022). Implementation of Electric Vehicle system. *Gradiva Review Journal*, *8*(12), 444–449.

Hossain, M. S., Madlool, N. A., Rahim, N. A., Selvaraj, J., Pandey, A. K., & Khan, A. F. (2016). Role of smart grid in renewable energy: An overview. *Renewable & Sustainable Energy Reviews*, *60*, 1168–1184. doi:10.1016/j.rser.2015.09.098

Hosseini, M. M., & Parvania, M. (2020). Quantifying impacts of automation on resilience of distribution systems. *IET Smart Grid*, *3*(2), 144–152. doi:10.1049/iet-stg.2019.0175

Hota, A. R., Juvvanapudi, M., & Bajpai, P. (2014). Issues and solution approaches in PHEV integration to smart grid. *Renewable & Sustainable Energy Reviews*, *30*, 217–229. doi:10.1016/j.rser.2013.10.008

Householder, A. (2022). *Designing Vultron: A Protocol for Multi-Party Coordinated Vulnerability Disclosure (MPCVD) (Version 1)*. Carnegie Mellon University.

Hsieh, I.-Y. L., Pan, M. S., & Green, W. H. (2020). Transition to electric vehicles in China: Implications for private motorization rate and Battery Market. *Energy Policy*, *144*, 111654. doi:10.1016/j.enpol.2020.111654

HT Auto Desk (2021). *Auto Industry in Recovery Mode, Buoyed by Demand for Passenger Vehicles: FADA*. Hindustan Times (HT) Auto Desk. https://auto.hindustantimes.com/auto/news/auto-industry-in-recovery-mode-buoyed-by-demand-for-passenger-vehicles-fada-41625722389929.html

Huang, S., Wu, Q., Cheng, L., & Liu, Z. (2016). Optimal Reconfiguration-Based Dynamic Tariff for Congestion Management and Line Loss Reduction in Distribution Networks. *IEEE Transactions on Smart Grid*, *7*(3), 1295–1303. doi:10.1109/TSG.2015.2419080

Huang, X., & Ge, J. (2019). Electric Vehicle Development in Beijing: An analysis of consumer purchase intention. *Journal of Cleaner Production*, *216*, 361–372. doi:10.1016/j.jclepro.2019.01.231

Hua, W., Chen, Y., Qadrdan, M., Jiang, J., Sun, H., & Wu, J. (2022). Applications of blockchain and artificial intelligence technologies for enabling prosumers in smart grids: A review. *Renewable & Sustainable Energy Reviews*, *161*, 112308. doi:10.1016/j.rser.2022.112308

Hu, D., Ye, Z., Gao, Y., Ye, Z., Peng, Y., & Yu, N. (2022). Multi-Agent Deep Reinforcement Learning for Voltage Control with Coordinated Active and Reactive Power Optimization. *IEEE Transactions on Smart Grid*, *13*(6), 4873–4886. doi:10.1109/TSG.2022.3185975

Hudda, V., & Kumar, A. (2014). *A Summary of Electric Vehicle Propulsion Technologies.* GRIN Verlag, Munich. Document no. 277317. https://www.grin.com/document/277317

Hu, J., Ye, C., Ding, Y., Tang, J., & Liu, S. (2022). A Distributed MPC to Exploit Reactive Power V2G for Real-Time Voltage Regulation in Distribution Networks. *IEEE Transactions on Smart Grid*, *13*(1), 576–588. doi:10.1109/TSG.2021.3109453

Hu, R., Wang, W., Wu, X., Chen, Z., Jing, L., Ma, W., & Zeng, G. (2022). Coordinated active and reactive power control for distribution networks with high penetrations of photovoltaic systems. *Solar Energy*, *231*, 809–827. doi:10.1016/j.solener.2021.12.025

Hussain, A., & Musilek, P. (2022). Resilience Enhancement Strategies For and Through Electric Vehicles. *Sustainable Cities and Society*, *80*, 103788. doi:10.1016/j.scs.2022.103788

Hyundai. (2022). *Kona Electric Specifications.* Hyundai. https://www.hyundai.com/in/en/find-a-car/kona-electric/specification

IEA. (2017). *Technical report.* https://www.iea.org/reports/global-ev-outlook-2017

IEEE Power & Energy Society. (2019). *Big shift in power & energy.* IEEE PES GTD 2019 Grand International Conference & Exposition Asia. Bangkok International Trade & Exhibition Centre (BITEC), Bangkok, Thailand.

Indian Bureau of Mines. (2020). *Indian Minerals Yearbook 2019 (Part- III: Mineral Reviews), 58th Edition, Rare Earths (Advance Release).* https://ibm.gov.in/writereaddata/files/10012020172151RareEarth_2019_AR.pdf

International Council on Clean Transportation. (2021). *Global Battery Demand and Supply Update 2021.* ICCT.

International Energy Agency. (2020). *Global EV Outlook 2020.* IEA.

Invest India (2022). *Sector Automobile – National Investment Promotion and Facilitation Agency.* https://www.investindia.gov.in/sector/automobile

Iqbal, S., Sarfraz, M., Ayyub, M., Tariq, M., Chakrabortty, R. K., Ryan, M. J., & Alamri, B. (2021). A comprehensive review on residential demand side management strategies in smart grid environment. *Sustainability (Basel)*, *13*(13), 7170. doi:10.3390/su13137170

Irfan, M., Deilami, S., Huang, S., & Veettil, B. P. (2023). Rooftop Solar and Electric Vehicle Integration for Smart, Sustainable Homes: A Comprehensive Review. *Energies*, *16*(21), 7248. doi:10.3390/en16217248

Islam, S., Badsha, S., Sengupta, S., Khalil, I., & Atiquzzaman, M. (2023). An Intelligent Privacy Preservation Scheme for EV Charging Infrastructure. *IEEE Transactions on Industrial Informatics*, *19*(2), 1238-1247.. doi:10.1109/TII.2022.3203707

Islam, F. R., Lallu, A., Mamun, K. A., Prakash, K., & Roy, N. K. (2021). Power Quality Improvement of Distribution Network Using BESS and Capacitor Bank. *Journal of Modern Power Systems and Clean Energy*, *9*(3), 625–632. doi:10.35833/MPCE.2019.000304

Islam, M. R., Lu, H., Hossain, M. J., & Li, L. (2019). Mitigating unbalance using distributed network reconfiguration techniques in distributed power generation grids with services for electric vehicles: A review. *Journal of Cleaner Production*, *239*, 117932. doi:10.1016/j.jclepro.2019.117932

Ismael, S. M., Abdel Aleem, S. H. E., Abdelaziz, A. Y., & Zobaa, A. F. (2019). State-of-the-art of hosting capacity in modern power systems with distributed generation. *Renewable Energy*, *130*, 1002–1020. doi:10.1016/j.renene.2018.07.008

Itani, K., & De Bernardinis, A. (2023). Review on New-Generation Batteries Technologies: Trends and Future Directions. *Energies*, *16*(22), 7530. doi:10.3390/en16227530

Jabr, R. A., Dzafic, I., & Huseinagic, I. (2018). Real Time Optimal Reconfiguration of Multiphase Active Distribution Networks. *IEEE Transactions on Smart Grid*, *9*(6), 6829–6839. doi:10.1109/TSG.2017.2724766

Jaguar. (2022). *E – Pace Specifications.* Jaguar. https://www.jaguar.com/jaguar-range/e-pace/specifications/index.html

Jaguemont, J., Boulon, L., & Dubé, Y. (2016). A comprehensive review of lithium-ion batteries used in hybrid and electric vehicles at cold temperatures. *Applied Energy*, *164*, 99–114. doi:10.1016/j.apenergy.2015.11.034

Jangdoost, A., Keypour, R., & Golmohamadi, H. (2021). Optimization of distribution network reconfiguration by a novel RCA integrated with genetic algorithm. *Energy Systems*, *12*(3), 801–833. doi:10.1007/s12667-020-00398-5

Jang, Y. J. (2018). Survey of the operation and system study on wireless charging electric vehicle systems. *Transportation Research Part C, Emerging Technologies*, *95*, 844–866. doi:10.1016/j.trc.2018.04.006

Jerew, B. (2021). *EV Lifespan: Do They Last as Long as Gasoline Cars?* Lifewire – Tech for Humans. https://www.lifewire.com/do-evs-last-as-long-as-gasoline-cars-5202392

Jiajie, Z., Xinru, J., Jia, F., Wenbo, Z., Penghui, L., Chen, L., & Zheng, L. (2022). Online Reactive Power Optimization of Distribution Network with Soft Open Point Based on Hierarchical Reinforcement Learning. *Proceedings - 2022 4th International Conference on Electrical Engineering and Control Technologies, CEECT*, 66–72. 10.1109/CEECT55960.2022.10030151

Jiang, T., Fang, H., & Wang, H. (2018). Blockchain-Based Internet of Vehicles: Distributed Network Architecture and Performance Analysis. *IEEE Internet of Things Journal.*

Ji, H., Yu, H., Song, G., Li, P., Wang, C., & Wu, J. (2019). A Decentralized Voltage Control Strategy of Soft Open Points in Active Distribution Networks. *Energy Procedia*, *159*, 412–417. doi:10.1016/j.egypro.2018.12.067

Johnson, J., Anderson, B., Wright, B., Quiroz, J., Berg, T., Graves, R., Daley, J., Phan, K., & Kunz, K. (2022). *Cybersecurity for Electric Vehicle Charging Infrastructure.* Sandia Report. https://www.osti.gov/servlets/purl/1877784

Juan, L. P. (2017). *Regenerative Braking.* Stanford University. http://large.stanford.edu/courses/2017/ph240/leis-pretto1/

Kabir, F., Yu, N., Gao, Y., & Wang, W. (2023). Deep reinforcement learning-based two-timescale Volt-VAR control with degradation-aware smart inverters in power distribution systems. *Applied Energy*, *335*, 120629. doi:10.1016/j.apenergy.2022.120629

Kabir, M. N., Mishra, Y., Ledwich, G., Dong, Z. Y., & Wong, K. P. (2014). Coordinated control of grid-connected photovoltaic reactive power and battery energy storage systems to improve the voltage profile of a residential distribution feeder. *IEEE Transactions on Industrial Informatics*, *10*(2), 967–977. doi:10.1109/TII.2014.2299336

Kalambe, S., & Agnihotri, G. (2014). Loss minimization techniques used in distribution network: Bibliographical survey. *Renewable & Sustainable Energy Reviews*, *29*, 184–200. doi:10.1016/j.rser.2013.08.075

Kalapsazova, M., Kukeva, R., Harizanova, S., Markov, P., Nihtianova, D., Zhecheva, E., & Stoyanova, R. (2023). High-performance layered oxides for sodium-ion batteries achieved through combined aluminum substitution and surface treatment. *Batteries*, *9*(2), 144. doi:10.3390/batteries9020144

Kamath, H. G., Majumdar, R., Krishnan, A. V., & Srikanth, R. (2022). Cost and environmental benefits of coal-concentrated solar power (CSP) hybridization in India. *Energy*, *240*, 122805. Advance online publication. doi:10.1016/j.energy.2021.122805

Kamrani, A. S., & Dagdougui, H. (2023). A Two-Stage Optimization Framework for Electric Vehicle Fleet Day-ahead Charging Management. *2023 IEEE 11th International Conference on Smart Energy Grid Engineering (SEGE),* 16-21.

Kandaperumal, G., & Srivastava, A. K. (2020). Resilience of the electric distribution systems: Concepts, classification, assessment, challenges, and research needs. *IET Smart Grid*, *3*(2), 133–143. doi:10.1049/iet-stg.2019.0176

Kang, E. S., Pee, S. J., Song, J. G., & Jang, J. W. (2018, April). A blockchain-based energy trading platform for smart homes in a microgrid. In *2018 3rd international conference on computer and communication systems (ICCCS)* (pp. 472-476). IEEE. 10.1109/CCOMS.2018.8463317

Kapassa, E., & Themistocleous, M. (2022). Blockchain technology applied in IoV demand response management: A systematic literature review. *Future Internet*, *14*(5), 136. doi:10.3390/fi14050136

Kaur, K., Kaddoum, G., & Zeadally, S. (2021, August). Blockchain-Based Cyber-Physical Security for Electrical Vehicle Aided Smart Grid Ecosystem. *IEEE Transactions on Intelligent Transportation Systems*, *22*(8), 5178–5189. doi:10.1109/TITS.2021.3068092

Kavousi-Fard, A., Abbasi, A., Rostami, M.-A., & Khosravi, A. (2015a). Optimal distribution feeder reconfiguration for increasing the penetration of plug-in electric vehicles and minimizing network costs. *Energy*, *93*, 1693–1703. doi:10.1016/j.energy.2015.10.055

Kavousi-Fard, A., & Khodaei, A. (2016). Efficient integration of plug-in electric vehicles via reconfigurable microgrids. *Energy*, *111*, 653–663. doi:10.1016/j.energy.2016.06.018

Kavousi-Fard, A., Niknam, T., & Fotuhi-Firuzabad, M. (2015b). Stochastic Reconfiguration and Optimal Coordination of V2G Plug-in Electric Vehicles Considering Correlated Wind Power Generation. *IEEE Transactions on Sustainable Energy*, *6*(3), 822–830. doi:10.1109/TSTE.2015.2409814

Kavousi-Fard, A., Rostami, M. A., & Niknam, T. (2015c). Reliability-Oriented Reconfiguration of Vehicle-to-Grid Networks. *IEEE Transactions on Industrial Informatics*, *11*(3), 682–691. doi:10.1109/TII.2015.2423093

Kazi, K. (2022a). Hybrid optimum model development to determine the Break. *Journal of Multimedia Technology & Recent Advancements*, *9*(2), 24–32.

Kazi, K. (2022b). Big data and HR Analytics in Talent Management: A Study. *Recent Trends in Parallel Computing*, *9*(3), 16–26.

Kazi, K. S. (2022c). IoT-Based Healthcare Monitoring for COVID-19 Home Quarantined Patients. *Recent Trends in Sensor Research & Technology*, *9*(3), 26–32.

Kazimierczuk, M. K., & Stando, T. (2019). Electric Vehicles Integration with Smart Grid: Opportunities, Challenges, and Solutions. *Energies*, *12*(10), 1942.

Kazi, S. S. L. (2023). Integrating IoT and Mechanical Systems in Mechanical Engineering Applications. *Journal of Mechanisms and Robotics*, *8*(3), 1–6.

Kazi, V. (2023). Deep Learning, YOLO and RFID based smart Billing Handcart. *Journal of Communication Engineering & Systems*, *13*(1), 1–8.

Kekatos, V., Wang, G., Conejo, A. J., & Giannakis, G. B. (2014). Stochastic reactive power management in microgrids with renewables. *IEEE Transactions on Power Systems*, *30*(6), 3386–3395. doi:10.1109/TPWRS.2014.2369452

Kemp, R., Schot, J., & Hoogma, R. (1998). Regime shifts to sustainability through processes of niche formation: The approach of strategic niche management. *Technology Analysis and Strategic Management*, *10*(2), 175–196. doi:10.1080/09537329808524310

Kene & Olwal. (2023). Energy Management and Optimization of Large-Scale Electric Vehicle Charging on the Grid. *World Electric Vehicle Journal, 14*(4), 95.

Kern, D., & Krauß, C. (2021, November). Analysis of e-mobility-based threats to power grid resilience. In *Proceedings of the 5th ACM Computer Science in Cars Symposium* (pp. 1-12). 10.1145/3488904.3493385

Khan, M. I., Shahrestani, M., Hayat, T., Shakoor, A., & Vahdati, M. (2019). Life cycle (well-to-wheel) energy and environmental assessment of natural gas as transportation fuel in Pakistan. *Applied Energy*, *242*, 1738–1752. doi:10.1016/j.apenergy.2019.03.196

Khan, N., Shahid, Z., Alam, M. M., Bakar Sajak, A. A., Mazliham, M. S., Khan, T. A., & Ali Rizvi, S. S. (2022). Energy management systems using SMART GRIDS: An exhaustive parametric comprehensive analysis of existing trends, significance, opportunities, and challenges. *International Transactions on Electrical Energy Systems*, *2022*, 1–38. doi:10.1155/2022/3358795

Khan, S., Ahmad, A., Ahmad, F., Shafaati Shemami, M., Saad Alam, M., & Khateeb, S. (2018). A comprehensive review on solar powered electric vehicle charging system. *Smart Science*, *6*(1), 54–79. doi:10.1080/23080477.2017.1419054

Khanum, S., & Mustafa, K. (2023). A systematic literature review on sensitive data protection in blockchain applications. *Concurrency and Computation: Practice and Experience, 35*(1), e7422.422.

Khawaja, Y., Shankar, N., Qiqieh, I., Alzubi, J., Alzubi, O., Nallakaruppan, M. K., & Padmanaban, S. (2023). Battery management solutions for li-ion batteries based on artificial intelligence. *Ain Shams Engineering Journal*, *14*(12), 102213. doi:10.1016/j.asej.2023.102213

Khetarpal, P., Nagpal, N., Al-Numay, M. S., Siano, P., Arya, Y., & Kassarwani, N. (2023). Power Quality Disturbances Detection and Classification Based on Deep Convolution Auto-Encoder Networks. *IEEE Access : Practical Innovations, Open Solutions*, *11*, 46026–46038. doi:10.1109/ACCESS.2023.3274732

Khor. (n.d.). *Battery Electric Vehicle (BEV)*. https://medium.com/@ackhor/battery-electric-vehicle-bev-d69f8683b141

Kianmehr, E., Nikkhah, S., Vahidinasab, V., Giaouris, D., & Taylor, P. C. (2019). A Resilience-Based Architecture for Joint Distributed Energy Resources Allocation and Hourly Network Reconfiguration. *IEEE Transactions on Industrial Informatics*, *15*(10), 5444–5455. doi:10.1109/TII.2019.2901538

Kiasari, M. (2023). *Comparative Analysis of Machine Learning Techniques for an Hour-Ahead Forecasting of Electric Vehicle States*. Academic Press.

Kim, J., & Dvorkin, Y. (2019). Enhancing Distribution System Resilience With Mobile Energy Storage and Microgrids. *IEEE Transactions on Smart Grid*, *10*(5), 4996–5006. doi:10.1109/TSG.2018.2872521

Kim, S. K., & Huh, J. H. (2018). A study on the improvement of smart grid security performance and blockchain smart grid perspective. *Energies*, *11*(8), 1973. doi:10.3390/en11081973

Kley, F., Lerch, C., & Dallinger, D. (2011). New business models for electric cars—A holistic approach. *Energy Policy*, *39*(6), 3392–3403. doi:10.1016/j.enpol.2011.03.036

Knieke, C., Lawrenz, S., Fröhling, M., Goldmann, D., & Rausch, A. (2019). Predictive and flexible circular economy approaches for highly integrated products and their materials as given in E-mobility and ICT. *Materials Science Forum*, *959*, 22–31. doi:10.4028/www.scientific.net/MSF.959.22

Koh, L., Dolgui, A., & Sarkis, J. (2020). Blockchain in transport and Logistics – Paradigms and transitions. *International Journal of Production Research*, *58*(7), 2054–2062. doi:10.1080/00207543.2020.1736428

Köhler, J., Turnheim, B., & Hodson, M. (2020). Low carbon transitions pathways in mobility: Applying the MLP in a combined case study and simulation bridging analysis of Passenger Transport in the Netherlands. *Technological Forecasting and Social Change*, *151*, 119314. doi:10.1016/j.techfore.2018.06.003

Koirala, B. P., Koliou, E., Friege, J., Hakvoort, R. A., & Herder, P. M. (2016). Energetic communities for community energy: A review of key issues and trends shaping integrated community energy systems. *Renewable & Sustainable Energy Reviews*, *56*, 722–744. doi:10.1016/j.rser.2015.11.080

König, A., Mayer, S., Nicoletti, L., Tumphart, S., & Lienkamp, M. (2022). The Impact of HVAC on the Development of Autonomous and Electric Vehicle Concepts. *Energies*, *15*(2), 441. doi:10.3390/en15020441

Kothona, D., & Bouhouras, A. (2022). A Two-Stage EV Charging Planning and Network Reconfiguration Methodology towards Power Loss Minimization in Low and Medium Voltage Distribution Networks. *Energies*, *15*(10), 3808. doi:10.3390/en15103808

Krause, T., Ernst, R., Klaer, B., Hacker, I., & Henze, M. (2021). Cybersecurity in power grids: Challenges and opportunities. *Sensors (Basel)*, *21*(18), 6225. doi:10.3390/s21186225 PMID:34577432

Kryonidis, G. C., Kontis, E. O., Chrysochos, A. I., Demoulias, C. S., & Papagiannis, G. K. (2018). A Coordinated Droop Control Strategy for Overvoltage Mitigation in Active Distribution Networks. *IEEE Transactions on Smart Grid*, *9*(5), 5260–5270. doi:10.1109/TSG.2017.2685686

Kulmala, A., Mutanen, A., Koto, A., Repo, S., & Järventausta, P. (2010.) RTDS verification of a coordinated voltage control implementation for distribution networks with distributed generation. *2010 IEEE PES Innovative Smart Grid Technologies Conference Europe (ISGT Europe)*, 1-8.

Kulmala, A., Repo, S., & Järventausta, P. (2014). Coordinated voltage control in distribution networks including several distributed energy resources. *IEEE Transactions on Smart Grid*, *5*(4), 2010–2020. doi:10.1109/TSG.2014.2297971

Kumar, M., Panda, K. P., Naayagi, R. T., Thakur, R., & Panda, G. (2023). Comprehensive Review of Electric Vehicle Technology and Its Impacts: Detailed Investigation of Charging Infrastructure, Power Management, and Control Techniques. *Applied Sciences (Basel, Switzerland)*, *13*(15), 8919. doi:10.3390/app13158919

Kumar, R. R., & Alok, K. (2020). Adoption of electric vehicle: A literature review and prospects for sustainability. *Journal of Cleaner Production*, *253*, 119911. doi:10.1016/j.jclepro.2019.119911

Kumar, R., Jain, V., Leong, W. Y., & Teyarachakul, S. (2023). *Convergence of IoT*. Blockchain, and Computational Intelligence in Smart Cities.

Kumar, S., Roy, D. N., & Dey, V. (2021). A comprehensive review on techniques to create the anti-microbial surface of biomaterials to intervene in biofouling. *Colloid and Interface Science Communications*, *43*, 100464. doi:10.1016/j.colcom.2021.100464

Kumar, V., Jain, V., Sharma, B., Chatterjee, J. M., & Shrestha, R. (Eds.). (2022). *Smart City Infrastructure: The Blockchain Perspective*. John Wiley & Sons. doi:10.1002/9781119785569

Kutubuddin, K. (2022). Detection of Malicious Nodes in IoT Networks based on packet loss using ML. *Journal of Mobile Computing, Communication & Mobile Networks*, *9*(3), 9–16.

Kutubuddin, K. (2022). Smart Grid energy saving technique using Machine Learning. *Journal of Instrumentation Technology and Innovations*, *12*(3), 1–10.

Kuzey, C., Fritz, M. M. C., Uyar, A., & Karaman, A. S. (2022). Board gender diversity, CSR Strategy, and eco-friendly initiatives in the transportation and logistics sector. *International Journal of Production Economics, 247*, 108436.

Lakshmi, D. N., Sandhyaa, R., Kanwar, P. D., & Priya, L. S. (2017). Energy Efficient Electric Vehicle Using Regenerative Braking System. *International Journal of Advance Research, Ideas, and Innovations in Technology, 2*(3). http://large.stanford.edu/courses/2017/ph240/leis-pretto1/docs/lakshmi.pdf

Lakshmi, D., Nagpal, N., Chandrasekaran, S., & Hemanth, D. J. (2023). A quantum-based approach for offensive security against cyber-attacks in electrical infrastructure. *Applied Soft Computing, 136*. doi:10.1016/j.asoc.2023.110071

Lakshmi, D., Tiwari, R. S., Nagpal, N., Kassarwani, N., Vishnuvarthanan, G., & Srivastava, A. (2023). Comparative Study on Forecasting of Schedule Generation in Delhi Region for the Resilient Power Grid Using Machine Learning. *IEEE Transactions on Industry Applications*. Advance online publication. doi:10.1109/TIA.2023.3316646

Larminie, J., & Lowry, J. (2017). *Electric Vehicle Technology Explained*. Wiley.

Lebrouhi, B. E., Khattari, Y., Lamrani, B., Maaroufi, M., Zeraouli, Y., & Kousksou, T. (2021). Key challenges for a large-scale development of battery electric vehicles: A comprehensive review. *Journal of Energy Storage*, *44*, 103273. doi:10.1016/j.est.2021.103273

Lee, C. (n.d.). *What is EV, BEV, HEV, PHEV? Here's your guide to types of electric cars*. www.marketwatch.com/story/what-is-ev-bev-hev-phev-heres-your-guide-to-types-of-electric-cars-11617986782tch

Lee, J., & Kim, G. H. (2017). Comparison analysis of the voltage variation ranges for distribution networks. *2017 IEEE International Conference on Environment and Electrical Engineering and 2017 IEEE Industrial and Commercial Power Systems Europe (EEEIC/I&CPS Europe),* 1-3. 10.1109/EEEIC.2017.7977500

Lei, S., Chen, C., Zhou, H., & Hou, Y. (2019). Routing and Scheduling of Mobile Power Sources for Distribution System Resilience Enhancement. *IEEE Transactions on Smart Grid, 10*(5), 5650–5662. doi:10.1109/TSG.2018.2889347

Lei, Z., Chun, H., & Haoming, Y. (2014). Study of charging station short-term load forecast based on wavelet neural networks for electric buses. *Proc. Chin. Conf. Pattern Recognit.,* 555-564. 10.1007/978-3-662-45643-9_59

Lemian, D., & Bode, F. (2022). Battery-supercapacitor energy storage systems for electrical vehicles: A review. *Energies*, *15*(15), 5683. doi:10.3390/en15155683

Lewis, A. (2015). Blockchain technology explained. *Blockchain Technol*, 1-27.

Li, Y., Gerling, D., Ma, J., Liu, J., & Yu, Q. (2010). The Comparison of Control Strategies for the Interior PMSM Drive used in the Electric Vehicle. *World Electric Vehicle Journal*. https://www.researchgate.net/publication/279899500_The_Comparison_of_Control_Strategies_for_the_Interior_PMSM_Drive_used_in_the_Electric_Vehicle

Liang, K. T., Afzal, S., Mokhlis, H., Illias, H. A., Mansor, N. N., & Khairuddin, A. S. M. (2022). Assessing the Resilience of a Power Distribution System Considering Different Restoration Strategies. *2022 IEEE International Conference on Power Systems Technology (POWERCON),* 1–5. 10.1109/POWERCON53406.2022.9929870

Liang, Y., Saner, C. B., Kenneth Koh, T. J., Dou, H., Huang, Z., Wolfram, F. A., & Tan, J. (2022). *Energy-economic-environmental feasibility of building-integrated photovoltaic systems in Singapore. In 2022 IEEE PES Innovative Smart Grid Technologies - Asia.* ISGT Asia.

Liao, F., & Correia, G. (2020). Electric carsharing and micromobility: A literature review on their usage pattern, demand, and potential impacts. *International Journal of Sustainable Transportation, 16*(3), 269–286. doi:10.1080/15568318.2020.1861394

Li, B., Chen, Y., Wei, W., Huang, S., & Mei, S. (2021). Resilient Restoration of Distribution Systems in Coordination With Electric Bus Scheduling. *IEEE Transactions on Smart Grid, 12*(4), 3314–3325. doi:10.1109/TSG.2021.3060801

Li, B., Chen, Y., Wei, W., Mei, S., Hou, Y., & Shi, S. (2022). Preallocation of Electric Buses for Resilient Restoration of Distribution Network: A Data-Driven Robust Stochastic Optimization Method. *IEEE Systems Journal, 16*(2), 2753–2764. doi:10.1109/JSYST.2021.3123623

Li, H., Yang, J., & Zhu, J. (2021). An Y. Electric vehicle charging scheduling algorithm based on traffic condition. *Computer System & Applications, 30*(7), 178–183.

Li, J., Xu, X., Yan, Z., Wang, H., Shahidehpour, M., & Chen, Y. (2023). Coordinated Optimization of Emergency Response Resources in Transportation-Power Distribution Networks under Extreme Events. *IEEE Transactions on Smart Grid, 14*(6), 1–1. doi:10.1109/TSG.2023.3257040

Lin, Y., Bie, Z., & Qiu, A. (2018). A review of key strategies in realizing power system resilience. *Global Energy Interconnection, 1*, 70–78. https://doi.org/https://doi.org/10.14171/j.2096-5117.gei.2018.01.009

Li, P., Ji, H., Wang, C., Zhao, J., Song, G., Ding, F., & Wu, J. (2017). Coordinated Control Method of Voltage and Reactive Power for Active Distribution Networks Based on Soft Open Point. *IEEE Transactions on Sustainable Energy, 8*(4), 1430–1442. doi:10.1109/TSTE.2017.2686009

Li, P., Ji, J., Ji, H., Jian, J., Ding, F., Wu, J., & Wang, C. (2020). MPC-Based Local Voltage Control Strategy of DGs in Active Distribution Networks. *IEEE Transactions on Sustainable Energy, 11*(4), 2911–2921. doi:10.1109/TSTE.2020.2981486

Li, P., Wei, M., Ji, H., Xi, W., Yu, H., Wu, J., Yao, H., & Chen, J. (2022). Deep Reinforcement Learning-Based Adaptive Voltage Control of Active Distribution Networks with Multi-terminal Soft Open Point. *International Journal of Electrical Power & Energy Systems, 141*, 108138. doi:10.1016/j.ijepes.2022.108138

Litman, T. (2022). *Well Measured: Developing Indicators for Sustainable and Livable Transport Planning.* Victoria Transport Policy Institute.

Liu, Y., Gao, H., Liu, J., Ma, Z., Chen, J., & Yang, Y. (2015). Multi-agent based hierarchical power scheduling strategy for active distribution network. *Proceedings - 2015 International Symposium on Smart Electric Distribution Systems and Technologies, EDST 2015, 1*, 151–158.

Liu, C., Chai, K. K., Zhang, X., Lau, E. T., & Chen, Y. (2018). Adaptive blockchain-based electric vehicle participation scheme in smart grid platform. *IEEE Access : Practical Innovations, Open Solutions, 6*, 25657–25665. doi:10.1109/ACCESS.2018.2835309

Liu, C., Guo, H., Xu, M., Wang, S., Yu, D., Yu, J., & Cheng, X. (2022). Extending On-Chain Trust to Off-Chain – Trustworthy Blockchain Data Collection Using Trusted Execution Environment (TEE). *IEEE Transactions on Computers, 71*(12), 3268–3280. doi:10.1109/TC.2022.3148379

Liu, H. J., Shi, W., & Zhu, H. (2018). Hybrid voltage control in distribution networks under limited communication rates. *IEEE Transactions on Smart Grid, 10*(3), 2416–2427. doi:10.1109/TSG.2018.2797692

Liu, M. (2018). *Deep Reinforcement Learning Based Performance Optimization in Blockchain-Enabled Internet of Vehicle, ICC 2019-2019 IEEE Int'l. Conf. Commun.* ICC.

Liu, W., & Yang, Q. (2021). Electric Vehicle Charging Infrastructure Planning: A Review. *Energies, 14*(2), 226.

Liu, Y., Ning, P., & Reiter, M. K. (2011). False data injection attacks against state estimation in electric power grids. *ACM Transactions on Information and System Security, 14*(1), 1–33. doi:10.1145/1952982.1952995

Liu, Z., & Mi, C. (2019). *Electric Vehicle Integration into Modern Power Networks*. Springer.

Liyakat, K. K. S. (2023). Machine Learning Approach Using Artificial Neural Networks to Detect Malicious Nodes in IoT Networks. In P. K. Shukla, H. Mittal, & A. Engelbrecht (Eds.), *Computer Vision and Robotics. CVR 2023. Algorithms for Intelligent Systems*. Springer. doi:10.1007/978-981-99-4577-1_3

Liyakat, K. K. S. (2024). Machine Learning Approach Using Artificial Neural Networks to Detect Malicious Nodes in IoT Networks. In S. K. Udgata, S. Sethi, & X. Z. Gao (Eds.), *Intelligent Systems. ICMIB 2023. Lecture Notes in Networks and Systems* (Vol. 728). Springer. doi:10.1007/978-981-99-3932-9_12

Liyakat, S. (2022). *Multiple object detection and classification based on Pruning using YOLO. Lambart Publications.*

Li, Z., Khajepour, A., & Song, J. (2019). A comprehensive review of the key technologies for pure electric vehicles. *Energy, 182*, 824–839. doi:10.1016/j.energy.2019.06.077

Lombardi, F., Aniello, L., De Angelis, S., Margheri, A., & Sassone, V. (2018). *A blockchain-based infrastructure for reliable and cost-effective IoT-aided smart grids*. Academic Press.

Lopes, J. A. P., Hatziargyriou, N., Mutale, J., Djapic, P., & Jenkins, N. (2007). Integrating distributed generation into electric power systems: A review of drivers, challenges and opportunities. *Electric Power Systems Research, 77*(9), 1189–1203. doi:10.1016/j.epsr.2006.08.016

Lopez, K. L., Gagne, C., & Gardner, M. A. (2019). Demand-side management using deep learning for smart charging of electric vehicles. *IEEE Transactions on Smart Grid, 10*(3), 2683–2691. doi:10.1109/TSG.2018.2808247

Lu, J., & Hossain, J. (2015). Introduction to the smart grid using pevs. *Vehicle-to-Grid: Linking electric vehicles to the smart grid*, 1–10.

Lu, S., & Shi, W. (2023). Vehicle as a Mobile Computing Platform: Opportunities and Challenges. *IEEE Network.*

Luo, C., Wu, H., Zhou, Y., Qiao, Y., & Cai, M. (2021). Network partition-based hierarchical decentralised voltage control for distribution networks with distributed PV systems. *International Journal of Electrical Power & Energy Systems, 130*(5), 4087–4098. doi:10.1016/j.ijepes.2021.106929

Lutsey, N., & Nicholas, M. (2019). *Electric vehicle costs and consumer benefits in Colorado in the 2020–2030 time frame*. Academic Press.

Lu, Z., Lu, X., Wang, W., & Wang, C. (2010). *Review and evaluation of security threats on the communication networks in the smart grid. In 2010 - MILCOM 2010 Military Communications Conference*. IEEE. doi:10.1109/MILCOM.2010.5679551

Lv, Z., & Shang, W. (2023). Impacts of intelligent transportation systems on energy conservation and emission reduction of transport systems: A comprehensive review. *Green Technologies and Sustainability, 1*(1), 100002. doi:10.1016/j.grets.2022.100002

Ma, T., & Zhang, T. (2019). Research on Reactive Power and Voltage Characteristics Analysis and Reactive Power Configuration Strategy of Active Distribution Network. *2019 12th International Conference on Intelligent Computation Technology and Automation (ICICTA)*, 108-111.

Maharjan, S., Khambadkone, A. M., & Peng, J. C. H. (2021). Robust constrained model predictive voltage control in active distribution networks. *IEEE Transactions on Sustainable Energy*, *12*(1), 400–411. doi:10.1109/TSTE.2020.3001115

Maheriya, A., & Panchal, S. (2023). A Neoteric Approach to Improvise Privacy Shielding of Sensitive Healthcare Information & Prediction of Disease. *International Journal of Computing and Digital Systems*, *13*(1), 1–12.

Maheswari, C., Priyanka, E. B., Thangavel, S., Vignesh, S. V. R., & Poongodi, C. (2020). Multiple regression analysis for the prediction of extraction efficiency in mining industry with industrial IoT. *Production Engineering*, *14*(4), 457–471. doi:10.1007/s11740-020-00970-z

Mahiban, L. N., & Emimal, M. (2023). *Longevity of Electric Vehicle Operations*. Qeios.

Mahindra. (2022). *E-Verito Electric Car. Features and Specifications*. Mahindra Electric. https://www.mahindraelectric.com/vehicles/everito/

Mai, T.T., Haque, A.N.M.M., & Nguyen, P.H. (2019, June). Consensus-based distributed control for overvoltage mitigation in LV microgrids. *2019 IEEE Milan Power Tech,* 1-6.

Ma, L., Wang, L., & Liu, Z. (2023). Soft Open Points-Assisted Resilience Enhancement of Power Distribution Networks Against Cyber Risks. *IEEE Transactions on Power Systems*, *38*(1), 31–41. doi:10.1109/TPWRS.2022.3165169

Manduchi, E., Le, T.T., Fu, W., & Moore, J.H. (2022). Genetic Analysis of Coronary Artery Disease Using Tree-Based Automated Machine Learning Informed By Biology-Based Feature Selection. *IEEE/ACM Transactions on Computational Biology and Bioinformatics, 19*(3), 1379-1386. . doi:10.1109/TCBB.2021.3099068

Manisalidis, I., Stavropoulou, E., Stavropoulos, A., & Bezirtzoglou, E. (2020). Environmental and health impacts of air pollution: A review. *Frontiers in Public Health*, *8*, 14. doi:10.3389/fpubh.2020.00014 PMID:32154200

Mannaro, K., Pinna, A., & Marchesi, M. (2017, September). Crypto-trading: Blockchain-oriented energy market. In *2017 AEIT International Annual Conference* (pp. 1-5). IEEE.

Maqbool, R., Saiba, M. R., & Ashfaq, S. (2022). Emerging industry 4.0 and Internet of Things (IoT) technologies in the Ghanaian construction industry: Sustainability, implementation challenges, and benefits. *Environmental Science and Pollution Research International*, *30*(0123456789), 37076–37091. Advance online publication. doi:10.1007/s11356-022-24764-1 PMID:36565423

Martinenas, S., Knezovic, K., & Marinelli, M. (2017). Management of Power Quality Issues in Low Voltage Networks Using Electric Vehicles: Experimental Validation. *IEEE Transactions on Power Delivery*, *32*(2), 971–979. doi:10.1109/TPWRD.2016.2614582

Martinez, C. M., Hu, X., Cao, D., Velenis, E., Gao, B., & Wellers, M. (2017). Energy Management in Plug-in Hybrid Electric Vehicles: Recent Progress and a Connected Vehicles Perspective. *IEEE Transactions on Vehicular Technology*, *66*(6), 4534–4549. doi:10.1109/TVT.2016.2582721

Ma, S., Su, L., Wang, Z., Qiu, F., & Guo, G. (2018). Resilience Enhancement of Distribution Grids Against Extreme Weather Events. *IEEE Transactions on Power Systems*, *33*(5), 4842–4853. doi:10.1109/TPWRS.2018.2822295

Masias, A. (2018). Lithium-Ion Battery Design for Transportation. *Behaviour of Lithium-Ion Batteries in Electric Vehicles: Battery Health, Performance, Safety, and Cost*, 1-33.

Mastoi, M. S., Zhuang, S., Munir, S., Malik, H., Hassan, M., Alqarni, M., & Alamri, B. (2023). A study of charging-dispatch strategies and vehicle-to-grid technologies for electric vehicles in distribution networks. *Energy Reports, 9*, 1777-1806. doi:10.1016/j.egyr.2022.12.139

Ma, Z., Yang, N., Zou, S., & Shao, Y. (2018). Charging coordination of Plug-In electric vehicles in distribution networks with capacity constrained feeder lines. *IEEE Transactions on Control Systems Technology*, *26*(5), 1917–1924. doi:10.1109/TCST.2017.2730825

Mazhar, T. (2023). Electric Vehicle Charging System in the Smart Grid Using Different Machine Learning Methods. *Sustainability, 15*(3), 2603.

Mazhar, T., Asif, R. N., Malik, M. A., Nadeem, M. A., Haq, I., Iqbal, M., & Ashraf, S. (2023). Electric Vehicle Charging System in the Smart Grid Using Different Machine Learning Methods. *Sustainability (Basel)*, *15*(3), 2603. doi:10.3390/su15032603

McQueen, M., Abou-Zeid, G., MacArthur, J., & Clifton, K. (2020). Transportation transformation: Is micromobility making a macro impact on sustainability? *Journal of Planning Literature*, *36*(1), 46–61. doi:10.1177/0885412220972696

Mehrjerdi, H. (2021). Resilience-robustness improvement by adaptable operating pattern for electric vehicles in complementary solar-vehicle management. *Journal of Energy Storage*, *37*, 102454. doi:10.1016/j.est.2021.102454

Meliani, M., Barkany, A. E., Abbassi, I. E., Darcherif, A. M., & Mahmoudi, M. (2021). Energy Management in the smart grid: State-of-the-art and future trends. *International Journal of Engineering Business Management*, *13*, 184797902110329. doi:10.1177/18479790211032920

Mengelkamp, E., Notheisen, B., Beer, C., Dauer, D., & Weinhardt, C. (2018). A blockchain-based smart grid: Towards sustainable local energy markets. *Computer Science (Berlin, Germany)*, *33*(1-2), 207–214. doi:10.1007/s00450-017-0360-9

Mercedes – Benz. (2022). *Mercedes – Benz EQC Technical Specification*. https://www.mercedes-benz.co.in/passenger-cars/mercedes-benz-cars/models/eqc/specifications/aerodynamics.module.html

Merlin, A., & Back, H. (1975). Search for a minimal-loss operating spanning tree configuration in an urban power distribution system. *5th Power System Computation Conf.*, 1–18.

Meskers, C., Caffarey, M., & Van Camp, M. (2019). Circular cities, e-mobility and the metals industry—a world in transition. *The Minerals, Metals & Materials Series*, 313–318.

Metke, A. R., & Ekl, R. L. (2010, January). Smart grid security technology. In *2010 Innovative Smart Grid Technologies (ISGT)* (pp. 1-7). IEEE.

MG Motors (2022). *MG ZS EV Specifications and Features.* MG Motors. https://s7ap1.scene7.com/is/content/mgmotor/mgmotor/documents/mg-dc-pdf-0186.pdf

Miglani, S. (2019). Innovation, Economic Development, and Intellectual Property in India and China. ARCIALA Series on Intellectual Assets and Law in Asia. In *The Growth of the Indian Automobile Industry: Analysis of the Roles of Government Policy and Other Enabling Factors* (pp. 439-463). https://link.springer.com/chapter/10.1007/978-981-13-8102-7_19

Mihai, S. (2022). Digital Twins: A Survey on Enabling Technologies, Challenges, Trends and Future Prospects. *IEEE Communications Surveys & Tutorials, 24*(4), 2255-2291. Advance online publication. doi:10.1109/COMST.2022.3208773

Miloslavich, P., Seeyave, S., Muller-Karger, F., Bax, N., Ali, E., Delgado, C., Urban, E. (2018). Challenges for global ocean observation: The need for increased human capacity. *Journal of Operational Oceanography, 12*(sup2).

Ming, T., Derichter, R., Shen, S., & Caillol, S. (2016). Fighting global warming by greenhouse gas removal: Destroying atmospheric nitrous oxide thanks to synergies between two breakthrough technologies. *Environmental Science and Pollution Research International*, *23*, 6119–6138. doi:10.1007/s11356-016-6103-9 PMID:26805926

Mishra, S., Das, D., & Paul, S. (2017). A comprehensive review on power distribution network reconfiguration. *Energy Systems*, *8*(2), 227–284. doi:10.1007/s12667-016-0195-7

Mobarak, M. H., Kleiman, R. N., & Bauman, J. (2020). Solar-charged electric vehicles: A comprehensive analysis of grid, driver, and environmental benefits. *IEEE Transactions on Transportation Electrification*, *7*(2), 579–603. doi:10.1109/TTE.2020.2996363

Mohammadi, F., & Saif, M. (2023). A comprehensive overview of electric vehicle batteries market. *e-Prime-Advances in Electrical Engineering, Electronics and Energy*, 100127.

Mohammadi, M., & Thornburg, J. (2023). Strategizing EV Charging and Renewable Integration in Texas. *arXiv preprint arXiv:2310.17056.*

Mohammed, A., Saif, O., Abo-Adma, M., Fahmy, A., & Elazab, R. (2024). Strategies and sustainability in fast charging station deployment for electric vehicles. *Scientific Reports*, *14*(1), 283. doi:10.1038/s41598-023-50825-7 PMID:38168937

Mohanty, S., Panda, S., Parida, S. M., Rout, P. K., Sahu, B. K., Bajaj, M., Zawbaa, H. M., Kumar, N. M., & Kamel, S. (2022). Demand side management of electric vehicles in Smart Grids: A survey on strategies, challenges, modeling, and Optimization. *Energy Reports*, *8*, 12466–12490. doi:10.1016/j.egyr.2022.09.023

Mojumder, M., Ahmed Antara, F., Hasanuzzaman, M., Alamri, B., & Alsharef, M. (2022). Electric vehicle-to-grid (V2G) technologies: Impact on the Power Grid and battery. *Sustainability (Basel)*, *14*(21), 13856. doi:10.3390/su142113856

Mololoth, V. K., Saguna, S., & Åhlund, C. (2023). Blockchain and machine° learning for future smart grids: A review. *Energies*, *16*(1), 528. doi:10.3390/en16010528

Momen, H., Abessi, A., & Jadid, S. (2020). Using EVs as distributed energy resources for critical load restoration in resilient power distribution systems. *IET Generation, Transmission & Distribution*, *14*(18), 3750–3761. doi:10.1049/iet-gtd.2019.1561

Montes, F., Meinen, R., Dell, C., Rotz, A., Hristov, A. N., Oh, J., & Dijkstra, J. (2013). SPECIAL TOPICS—Mitigation of methane and nitrous oxide emissions from animal operations: II. A review of manure management mitigation options. *Journal of Animal Science*, *91*(11), 5070–5094. doi:10.2527/jas.2013-6584 PMID:24045493

Morfeldt, J., Kurland, S. D., & Johansson, D. J. (2021). Carbon footprint impacts of banning cars with internal combustion engines. *Transportation Research Part D, Transport and Environment*, *95*, 102807. doi:10.1016/j.trd.2021.102807

Mossali, E., Gentilini, L., Merati, G., & Colledani, M. (2020). Methodology and application of Electric Vehicles Battery packs redesign for Circular Economy. *Procedia CIRP*, *91*, 747–751. doi:10.1016/j.procir.2020.01.139

Mousavian, S., Erol, K., Melike, & Ortmeyer, T. (2015). *Cyber Attack Protection for a Resilient Electric Vehicle Infrastructure.* . doi:10.1109/GLOCOMW.2015.7414174

Movahednia, M., Kargarian, A., Ozdemir, C. E., & Hagen, S. C. (2022). Power Grid Resilience Enhancement via Protecting Electrical Substations Against Flood Hazards: A Stochastic Framework. *IEEE Transactions on Industrial Informatics*, *18*(3), 2132–2143. doi:10.1109/TII.2021.3100079

Mozaffari, M., Askarian Abyaneh, H., Jooshaki, M., & Moeini-Aghtaie, M. (2020). Joint Expansion Planning Studies of EV parking lots placement and distribution network. *IEEE Transactions on Industrial Informatics*, *16*(10), 6455–6465. doi:10.1109/TII.2020.2964049

Muhammad, Z., Anwar, Z., Saleem, B., & Shahid, J. (2023). Emerging Cybersecurity and Privacy Threats to Electric Vehicles and Their Impact on Human and Environmental Sustainability. *Energies*, *16*(3), 1113. doi:10.3390/en16031113

Mukherjee, A., Bagchi, J., Roy, K., & Mondal, K. (2016). Catalytic Converter in Automobile Exhaust Emission. *Journal for Research., 02*(10), 29–33. http://www.journal4research.org/articles/J4RV2I10020.pdf

Mulani, A. O. (2019). Effect of Rotation and Projection on Real time Hand Gesture Recognition system for Human Computer Interaction. *Journal of The Gujrat Research Society, 21*(16), 3710–3718.

Muneeb, M., Raza, Z., Haq, I. U., & Shafiq, O. (2022). SmartCon: A Blockchain-Based Framework for Smart Contracts and Transaction Management. *IEEE Access, 10*, 23687-23699. . doi:10.1109/ACCESS.2021.3135562

Muratori, M., Alexander, M., Arent, D., Bazilian, M., Cazzola, P., Dede, E. M., & Ward, J. (2021). The rise of electric vehicles—2020 status and future expectations. *Progress in Energy, 3*(2), 022002. doi:10.1088/2516-1083/abe0ad

Mustafi, N. N. (2022). An overview of hybrid electric vehicle technology. *Engines and Fuels for Future Transport*, 73-102.

Muzir, N. A. Q., Mojumder, M. R. H., Hasanuzzaman, M., & Selvaraj, J. (2022). Challenges of electric vehicles and their prospects in Malaysia: A comprehensive review. *Sustainability (Basel), 14*(14), 8320. doi:10.3390/su14148320

Mwasilu, F., Justo, J. J., Kim, E. K., Do, T. D., & Jung, J. W. (2014). Electric vehicles and smart grid interaction: A review on vehicle to grid and renewable energy sources integration. *Renewable & Sustainable Energy Reviews, 34*, 501–516. doi:10.1016/j.rser.2014.03.031

Naguib, M. A., Omran, W. E. A., & Talaat, H. (2022). Performance Enhancement of Distribution Systems via Distribution Network Reconfiguration and Distributed Generator Allocation Considering Uncertain Environment. *Journal of Modern Power Systems and Clean Energy, 10*, 647–655. doi:10.35833/MPCE.2020.000333

Nair, A., Kapoor, C., NB, S. S., & Balamurugan, S. (2021, June). Exploring blockchain enabled smart community with electric vehicles. In *2021 10th IEEE International Conference on Communication Systems and Network Technologies (CSNT)* (pp. 745-750). IEEE.

Najafi Tari, A., Sepasian, M. S., & Tourandaz Kenari, M. (2021). Resilience assessment and improvement of distribution networks against extreme weather events. *International Journal of Electrical Power & Energy Systems, 125*, 106414. doi:10.1016/j.ijepes.2020.106414

Naseri, F., Farjah, E., & Ghanbari, T. (2016). An efficient regenerative braking system based on battery/supercapacitor for electric, hybrid, and plug-in hybrid electric vehicles with BLDC motor. *IEEE Transactions on Vehicular Technology, 66*(5), 3724–3738. doi:10.1109/TVT.2016.2611655

Navesi, R. B., Nazarpour, D., Ghanizadeh, R., & Alemi, P. (2022). Switchable Capacitor Bank Coordination and Dynamic Network Reconfiguration for Improving Operation of Distribution Network Integrated with Renewable Energy Resources. *Journal of Modern Power Systems and Clean Energy, 10*(3), 637–646. doi:10.35833/MPCE.2020.000067

Nazemi, M., Moeini-Aghtaie, M., Fotuhi-Firuzabad, M., & Dehghanian, P. (2020). Energy Storage Planning for Enhanced Resilience of Power Distribution Networks Against Earthquakes. *IEEE Transactions on Sustainable Energy, 11*(2), 795–806. doi:10.1109/TSTE.2019.2907613

Nerkar, P. M., & Shinde, S. S. (2023). Monitoring Fresh Fruit and Food Using Iot and Machine Learning to Improve Food Safety and Quality. *Tuijin Jishu/Journal of Propulsion Technology, 44*(3), 2927 – 2931. https://propulsiontechjournal.com/index.php/journal/article/view/914

Newman, P., Beatley, T., & Boyer, H. (2009). *Resilient cities: responding to peak oil and climate change*. Island Press.

Newman, P., & Kenworthy, J. (1999). *Sustainability and cities: overcoming automobile dependence*. Island press.

Nguyen, H. M., Torres, J. L. R., Leki, A., & Pham, H. V. (2021). MPC Based Centralized Voltage and Reactive Power Control for Active Distribution Networks. *IEEE Transactions on Energy Conversion*, *36*(2), 1537–1547. doi:10.1109/TEC.2021.3054844

Nikita, K., & Supriya, J. (2020). Design of Vehicle system using CAN Protocol. *International Journal for Research in Applied Science and Engineering Technology*, *8*(V), 1978–1983. doi:10.22214/ijraset.2020.5321

Nikitas, A., Kougias, I., Alyavina, E., & Njoya Tchouamou, E. (2017). How can autonomous and connected vehicles, electromobility, BRT, hyperloop, shared use mobility and mobility-as-a-service shape transport futures for the context of smart cities? *Urban Science (Basel, Switzerland)*, *1*(4), 36. doi:10.3390/urbansci1040036

Nimalsiri, N. I., Mediwaththe, C. P., Ratnam, E. L., Shaw, M., Smith, D. B., & Halgamuge, S. K. (2019). A survey of algorithms for distributed charging control of electric vehicles in smart grid. *IEEE Transactions on Intelligent Transportation Systems*, *21*(11), 4497–4515. doi:10.1109/TITS.2019.2943620

NITI Aayog, & RMI India (2022). *Banking on Electric Vehicles in India – A Blueprint for inclusion of EVs in Priority Sector Lending Guidelines*. A Joint Study by NITI Aayog, and RMI India. https://www.niti.gov.in/sites/default/files/2022-01/Banking-on-EV_web_2.0a.pdf

Nordelöf, A., Messagie, M., Tillman, A. M., Ljunggren Söderman, M., & Van Mierlo, J. (2014). Environmental impacts of hybrid, plug-in hybrid, and battery electric vehicles—What can we learn from life cycle assessment? *The International Journal of Life Cycle Assessment*, *19*(11), 1866–1890. doi:10.1007/s11367-014-0788-0

Novak, A., & Ivanov, A. (2023). Network Security Vulnerabilities in Smart Vehicle-to-Grid Systems Identifying Threats and Proposing Robust Countermeasures. *Journal of Artificial Intelligence and Machine Learning in Management*, *7*(1), 48–80.

Novakova, L. (2020). The impact of technology development on the future of the labour market in the Slovak Republic. *Technology in Society*, *62*, 101256. doi:10.1016/j.techsoc.2020.101256

Novel, automated charging infrastructure for electric vehicles. (n.d.). https://cordis.europa.eu/project/id/781780

Ntombela, M., Musasa, K., & Moloi, K. (2023). A Comprehensive Review for Battery Electric Vehicles (BEV) Drive Circuits Technology, Operations, and Challenges. *World Electric Vehicle Journal*, *14*(7), 195. doi:10.3390/wevj14070195

Obama, B. (2013). Presidential Policy Directive 21: Critical Infrastructure Security and Resilience. Academic Press.

Obebe, S. B., Kolo, A., & Yusuf, P. P. (2021). Automotive emissions: Sources, environmental effects, and control measures. *Int. J. Innov. Sci. Res. Technol.*, *6*(4), 288–293.

Oboudi, M. H., Mohammadi, M., Trakas, D. N., & Hatziargyriou, N. D. (2021). A Systematic Method for Power System Hardening to Increase Resilience Against Earthquakes. *IEEE Systems Journal*, *15*(4), 4970–4979. doi:10.1109/JSYST.2020.3032783

Olival, P. C., Madureira, A. G., & Matos, M. (2017). Advanced voltage control for smart microgrids using distributed energy resources. *Electric Power Systems Research*, *146*, 132–140. doi:10.1016/j.epsr.2017.01.027

Omer, A. M. (2008). Energy, environment and sustainable development. *Renewable & Sustainable Energy Reviews*, *12*(9), 2265–2300. doi:10.1016/j.rser.2007.05.001

Ons, S., & Ammar, O. (2017). Electric vehicle scheduling and optimal charging problem: Complexity, exact and heuristic approaches. *International Journal of Production Research*, *5*(2), 519–535. doi:10.1080/00207543.2016.1192695

Operator, G. (n.d.). *gridX*. www.gridx.ai/knowledge/what-is-a-grid-operatorned – gridX

Ouyang, M., & Duenas-Osorio, L. (2014). Multi-dimensional hurricane resilience assessment of electric power systems. *Structural Safety*, *48*, 15–24. doi:10.1016/j.strusafe.2014.01.001

Paalosmaa, T., & Shafie-khah, M. (2021). Feasibility of innovative smart mobility solutions: A case study for Vaasa. *World Electric Vehicle Journal*, *12*(4), 188. doi:10.3390/wevj12040188

Pahlavanhoseini, A., & Sepasian, M. S. (2019). Scenario-based planning of fast charging stations considering network reconfiguration using cooperative coevolutionary approach. *Journal of Energy Storage*, *23*, 544–557. doi:10.1016/j.est.2019.04.024

Panday, A., & Bansal, H. O. (2014). A review of optimal energy management strategies for hybrid electric vehicle. *International Journal of Vehicular Technology*, *2014*. doi:10.1155/2014/160510

Pandey, A., & Majumdar, R. (2023). Anthropogenic Environmental Hazards: Compensation and Mitigation. In P. Pathak, R. R. Srivastava, & S. Ilyas (Eds.), *Assessing End-of-Life Room Air Conditioner Recycling Potential for Sustainable Resource Utilization in India: A Case Study for Reducing Environmental Burden'* (pp. 41–69). Springer Nature Switzerland. Retrieved from https://link.springer.com/chapter/10.1007/978-3-031-41013-0_3

Pandiyan, P., Saravanan, S., Usha, K., Kannadasan, R., Alsharif, M. H., & Kim, M. K. (2023). Technological advancements toward smart energy management in smart cities. *Energy Reports*, *10*, 648–677. doi:10.1016/j.egyr.2023.07.021

Pan, J.-S., Wang, H.-J., Nguyen, T.-T., Zou, F.-M., & Chu, S.-C. (2022). Dynamic reconfiguration of distribution network based on dynamic optimal period division and multi-group flight slime mould algorithm. *Electric Power Systems Research*, *208*, 107925. doi:10.1016/j.epsr.2022.107925

Pankaj, & Kulkarni. (2019). Implementation of Low Power and area efficient carry select Adder. *International Journal of Research in Engineering, Science and Management*, *2*(4), 183–184.

Panteli, M., & Mancarella, P. (2015a). Influence of extreme weather and climate change on the resilience of power systems: Impacts and possible mitigation strategies. *Electric Power Systems Research*, *127*, 259–270. doi:10.1016/j.epsr.2015.06.012

Panteli, M., & Mancarella, P. (2015b). The Grid: Stronger, Bigger, Smarter?: Presenting a Conceptual Framework of Power System Resilience. *IEEE Power & Energy Magazine*, *13*(3), 58–66. doi:10.1109/MPE.2015.2397334

Panteli, M., Trakas, D. N., Mancarella, P., & Hatziargyriou, N. D. (2017). Power Systems Resilience Assessment: Hardening and Smart Operational Enhancement Strategies. *Proceedings of the IEEE*, *105*(7), 1202–1213. doi:10.1109/JPROC.2017.2691357

Paudel, A., Supingklad, W., & Ongsakul, W. (2016). Distribution network reconfiguration to support electric vehicles integration. *2016 International Conference on Cogeneration, Small Power Plants and District Energy (ICUE),* 1–6. 10.1109/COGEN.2016.7728975

Peng, Y., & Wu, I. C. (2021). A cloud-based monitoring system for performance analysis in IoT industry. *The Journal of Supercomputing*, *77*(8), 9266–9289. doi:10.1007/s11227-021-03640-8

Peters, J. F., Baumann, M., Zimmermann, B., Braun, J., & Weil, M. (2017). The environmental impact of Li-Ion batteries and the role of key parameters–A review. *Renewable & Sustainable Energy Reviews*, *67*, 491–506. doi:10.1016/j.rser.2016.08.039

Photovoltaics, D. G., & Storage, E. (2018). IEEE standard for interconnection and interoperability of distributed energy resources with associated electric power systems interfaces. *IEEE Std.*, *1547*, 1547–2018.

Pilo, F., Pisano, G., & Soma, G. G. (2011). Optimal Coordination of Energy Resources With a Two-Stage Online Active Management. *IEEE Transactions on Industrial Electronics*, *58*(10), 4526–4537. doi:10.1109/TIE.2011.2107717

Poal, V., Panchare, D., & Mehta, B. (2021). *Balancing of HVAC System Energy Consumption in Electric Vehicles*. SAE Technical Paper 2021-28-0124. doi:10.4271/2021-28-0124

Pokhriyal, A., Rueda-García, D., & Gómez-Romero, P. (2023). To flow or not to flow. A perspective on large-scale stationary electrochemical energy storage. *Sustainable Energy & Fuels*, *7*(23), 5473–5482. doi:10.1039/D3SE00955F

Poudel, S. (2020). *Quantifying and Improving Resilience in Power Distribution Systems. Doctoral dissertation.* Washington State University.

Poullikkas, A. (2015). Sustainable options for electric vehicle technologies. *Renewable & Sustainable Energy Reviews*, *41*, 1277–1287. doi:10.1016/j.rser.2014.09.016

Pourmirza, Z., & Walker, S. (2021). Electric Vehicle Charging Station: Cyber Security Challenges and Perspective. *IEEE 9th International Conference on Smart Energy Grid Engineering (SEGE),* 111-116. 10.1109/SEGE52446.2021.9535052

Prabawa, P., & Choi, D. (2021). Hierarchical Volt-VAR Optimization Framework Considering Voltage Control of Smart Electric Vehicle Charging Stations Under Uncertainty. *IEEE Access : Practical Innovations, Open Solutions*, *9*, 123398–123413. doi:10.1109/ACCESS.2021.3109621

Pradeepa, M. (2022). Student Health Detection using a Machine Learning Approach and IoT. *2022 IEEE 2nd Mysore sub section International Conference (MysuruCon).*

Prado-Lorenzo, J. M., Rodríguez-Domínguez, L., Gallego-Álvarez, I., & García-Sánchez, I. M. (2009). Factors influencing the disclosure of greenhouse gas emissions in companies world-wide. *Management Decision*, *47*(7), 1133–1157. doi:10.1108/00251740910978340

Pramuanjaroenkij, A., & Kakaç, S. (2023). The fuel cell electric vehicles: The highlight review. *International Journal of Hydrogen Energy*, *48*(25), 9401–9425. doi:10.1016/j.ijhydene.2022.11.103

Press Trust of India. (2021). '*Government aims to raise auto sector contribution to GDP, job creation: Gadkari*'. Business Standard. https://www.business-standard.com/article/automobile/govt-aims-to-raise-auto-sector-contribution-to-gdp-job-creation-gadkari-121082501375_1.html

Pujari, A. S., Majumdar, R., Saha, S. K., Saha, S. K., & Subramaniam, C. (2023). Annular vertical cylindrical thermochemical storage system with innovative flow arrangements for improved heat dispatch towards space heating requirements. *Renewable Energy*, *217*, 119168. Advance online publication. doi:10.1016/j.renene.2023.119168

Pullins, S. (2019). Why microgrids are becoming an important part of the energy infrastructure. *The Electricity Journal*, *32*(5), 17–21. doi:10.1016/j.tej.2019.05.003

Putrus, G., Suwanapingkarl, P., Johnston, D., Bentley, E., & Narayana, M. (2009). Impact of electric vehicles on power distribution networks. *Proceedings of the 2009 IEEE Vehicle Power and Propulsion Conference*, 827–831. 10.1109/VPPC.2009.5289760

Radanliev, P., De Roure, D., Nicolescu, R., Huth, M., & Santos, O. (2022). Digital twins: Artificial intelligence and the IoT cyber-physical systems in Industry 4.0. *International Journal of Intelligent Robotics and Applications*, *6*(1), 171–185. doi:10.1007/s41315-021-00180-5

Ragavendran, S., & Kumar, P. (2023). Exploring Barriers and Challenges of Electric Vehicles in India and Vehicle-to-Grid Optimization: A Comprehensive Review. *International Journal Of Novel Research In Engineering Sciences,* 16-19.

Rahimi, K., & Davoudi, M. (2018). Electric vehicles for improving resilience of distribution systems. *Sustainable Cities and Society*, *36*, 246–256. doi:10.1016/j.scs.2017.10.006

Rahman, A., Chakraborty, C., Anwar, A., Karim, M. R., Islam, M. J., Kundu, D., Rahman, Z., & Band, S. S. (2022). SDN–IoT empowered intelligent framework for industry 4.0 applications during COVID-19 pandemic. *Cluster Computing*, *25*(4), 2351–2368. doi:10.1007/s10586-021-03367-4 PMID:34341656

Rahman, F. A., Aziz, M. M. A., Saidur, R., Bakar, W. A. W. A., Hainin, M. R., Putrajaya, R., & Hassan, N. A. (2017). Pollution to solution: Capture and sequestration of carbon dioxide (CO2) and its utilization as a renewable energy source for a sustainable future. *Renewable & Sustainable Energy Reviews*, *71*, 112–126. doi:10.1016/j.rser.2017.01.011

Rahmani-Andebili, M., & Fotuhi-Firuzabad, M. (2018). An Adaptive Approach for PEVs Charging Management and Reconfiguration of Electrical Distribution System Penetrated by Renewables. *IEEE Transactions on Industrial Informatics*, *14*(5), 2001–2010. doi:10.1109/TII.2017.2761336

Rainieri, G., Buizza, C., & Ghilardi, A. (2023). The psychological, human factors and socio-technical contribution: A systematic review towards range anxiety of battery electric vehicles' drivers. *Transportation Research Part F: Traffic Psychology and Behaviour*, *99*, 52–70. doi:10.1016/j.trf.2023.10.001

Ramasamy, S., Ganesan, K., & Koodalsamy, B. (2023). Realistic choice-based decision-making mechanism in electric vehicles and secure energy transaction using ethereum blockchain. *Electrical Engineering*, 1–21. doi:10.1007/s00202-023-02074-0

Ramraj, R., Pashajavid, E., Alahakoon, S., & Jayasinghe, S. (2023). Quality of Service and Associated Communication Infrastructure for Electric Vehicles. *Energies*, *16*(20), 7170. doi:10.3390/en16207170

Ranamuka, D., Agalgaonkar, A. P., & Muttaqi, K. M. (2016). Online Coordinated Voltage Control in Distribution Systems Subjected to Structural Changes and DG Availability. *IEEE Transactions on Smart Grid*, *7*(2), 580–591.

Rane, S. B., & Narvel, Y. A. M. (2022). Data-driven decision making with Blockchain-IoT integrated architecture: A project resource management agility perspective of industry 4.0. *International Journal of System Assurance Engineering and Management*, *13*(2), 1005–1023. doi:10.1007/s13198-021-01377-4

Rathee, G., Balasaraswathi, M., Chandran, K. P., Gupta, S. D., & Boopathi, C. S. (2021). A secure IoT sensors communication in industry 4.0 using blockchain technology. *Journal of Ambient Intelligence and Humanized Computing*, *12*(1), 533–545. doi:10.1007/s12652-020-02017-8

Raval, A. S., & Borisagar, K. R. (n.d.). *Document Recommendation: An anatomization of neoteric advancements, problems and potential scope*. Academic Press.

Raval, A., & Borisagar, K. (2022, March). A Survey on Techniques and Methods of Recommender System. In *International Conference on Computational Intelligence in Data Science* (pp. 97-114). Cham: Springer International Publishing. 10.1007/978-3-031-16364-7_8

Raval, A., Lohia, A., & Bhatt, U. Y. AI Business Model is Emerging Energy Market and Smart Grid. In *Applications of Big Data and Artificial Intelligence in Smart Energy Systems* (pp. 169–192). River Publishers. doi:10.1201/9781003440864-8

Raven, R. P., Schot, J., & Berkhout, F. (2019). Space and scale in socio-technical transitions. *Environmental Innovation and Societal Transitions*.

Razavi, S.-M., Momeni, H.-R., Haghifam, M.-R., & Bolouki, S. (2022). Multi-Objective Optimization of Distribution Networks via Daily Reconfiguration. *IEEE Transactions on Power Delivery*, *37*(2), 775–785. doi:10.1109/TPWRD.2021.3070796

RazeghiG.LeeJ.SamuelsenS. (2021). *Resiliency Impacts of Plug-in Electric Vehicles in a Smart Grid.* doi:10.7922/G2CV4G13

Relan, D., Khatter, K., & Nagpal, N. (2023)/ Blockchain based Solution for Electricity Supply Chain in Smart Grids. In *Applications of Big Data and Artificial Intelligence in Smart Energy Systems: Volume 2 - Energy Planning, Operations, Control and Market Perspectives.* River Publishers.

Revesz, A., Dunham, C., Jones, P., Bond, C., Fenner, R., Mody, S., Nijjhar, R., Marques, C., & Maidment, G. (2022). A holistic design approach for 5th generation smart local energy systems: Project greenscies. *Energy*, *242*, 122885. doi:10.1016/j.energy.2021.122885

Ribeiro, P., & Fontes, T. (2019). Electric Vehicles and Renewable Energy: A Review. *Renewable & Sustainable Energy Reviews*, *102*, 54–64.

Ritchie, H., Rosado, P., & Roser, M. (2022). *Natural Disasters.* https://ourworldindata.org/natural-disasters

Romvary, J. J., Ferro, G., Haider, R., & Annaswamy, A. M. (2021). A proximal atomic coordination algorithm for distributed optimization. *IEEE Transactions on Automatic Control*, *67*(2), 646–661. doi:10.1109/TAC.2021.3053907

Roomi, M. M., Hussain, S. M., Mashima, D., Chang, E.-C., & Ustun, T. S. (2023). Analysis of false data injection attacks against automated control for parallel generators in IEC 61850-based Smart Grid Systems. *IEEE Systems Journal*, *17*(3), 1–12. doi:10.1109/JSYST.2023.3236951

Rosati, R., Romeo, L., Cecchini, G., Tonetto, F., Viti, P., Mancini, A., & Frontoni, E. (2022). From knowledge-based to big data analytic model: A novel IoT and machine learning based decision support system for predictive maintenance in Industry 4.0. *Journal of Intelligent Manufacturing*, *34*(1), 107–121. doi:10.1007/s10845-022-01960-x

Rostami, M.-A., Kavousi-Fard, A., & Niknam, T. (2015). Expected Cost Minimization of Smart Grids With Plug-In Hybrid Electric Vehicles Using Optimal Distribution Feeder Reconfiguration. *IEEE Transactions on Industrial Informatics*, *11*(2), 388–397. doi:10.1109/TII.2015.2395957

Roudbari, A., Nateghi, A., Yousefi-khanghah, B., Asgharpour-Alamdari, H., & Zare, H. (2021). Resilience-oriented operation of smart grids by rescheduling of energy resources and electric vehicles management during extreme weather condition. Sustainable Energy. *Grids and Networks*, *28*, 100547. doi:10.1016/j.segan.2021.100547

Roytelman, I., Melnik, V., Lee, S. S. H., & Lugtu, R. L. (1996). Multi-objective feeder reconfiguration by distribution management system. *IEEE Transactions on Power Systems*, *11*(2), 661–667. doi:10.1109/59.496136

Ruan, H., Gao, H., Liu, Y., Wang, L., & Liu, J. (2020). Distributed voltage control in active distribution network considering renewable energy: A novel network partitioning method. *IEEE Transactions on Power Systems*, *35*(6), 4220–4231. doi:10.1109/TPWRS.2020.3000984

Ruggieri, R., Ruggeri, M., Vinci, G., & Poponi, S. (2021). Electric mobility in a smart city: European overview. *Energies*, *14*(2), 315. doi:10.3390/en14020315

Sabounchi, M., & Wei, J. (2017, November). Towards resilient networked microgrids: Blockchain-enabled peer-to-peer electricity trading mechanism. In *2017 IEEE conference on energy internet and energy system integration (EI2)* (pp. 1-5). IEEE.

Sabri, M., Danapalasingam, K. A., & Rahmat, M. F. (2016). A review on hybrid electric vehicles architecture and energy management strategies. *Renewable & Sustainable Energy Reviews*, *53*, 1433–1442. doi:10.1016/j.rser.2015.09.036

Sabzi, Asadi, & Moghbeli. (2019). Regenerative energy management of electric drive based on Lya-punov stability theorem. *J. Mod. Power Syst. Clean Energy*, *7*(2), 321–328. doi:10.1007/s40565-018-0497-y

Safayet, A., Fajri, P., & Husain, I. (2017). Reactive Power Management for Overvoltage Prevention at High PV Penetration in a Low-Voltage Distribution System. *IEEE Transactions on Industry Applications*, *53*(6), 5786–5794. doi:10.1109/TIA.2017.2741925

Salunke. (2022). Announcement system in Bus. *Journal of Image Processing and Intelligent Remote Sensing, 2*(6).

Sanguesa, J. A., Torres-Sanz, V., Garrido, P., Martinez, F. J., & Marquez-Barja, J. M. (2021). A review on electric vehicles: Technologies and challenges. *Smart Cities*, *4*(1), 372–404. doi:10.3390/smartcities4010022

Santinelli, M., & Sapountzakis, E. J. (2020). The Role of Electric Vehicles in Sustainable Urban Mobility: Challenges and Opportunities. *Sustainability*, *12*(17), 6729.

Saurabh, K., & Majumdar, R. (2022). Clean Fuels for Mobility. Energy, Environment, and Sustainability. In G. Di Blasio, A. K. Agarwal, G. Belgiorno, & P. C. Shukla (Eds.), *Fuels for Sustainable Transport in India* (pp. 27–55). Springer. doi:10.1007/978-981-16-8747-1_3

Saurabh, K., & Majumdar, R. (2023a). Renewable Fuels for Sustainable Mobility. Energy, Environment, and Sustainability. In P. C. Shukla, G. Belgiorno, G. D. Blasio, & A. K. Agarwal (Eds.), *Functional Use-Based Positioning of Conventional Vehicles in Conjunction with Alternate Low-Emission Fuels* (pp. 97–159). Springer. doi:10.1007/978-981-99-1392-3_5

Saurabh, K., & Majumdar, R. (2023b, December). Sub-segment-based approach for assessing growth and energy transition trajectories of Indian two-wheelers. *Journal of Cleaner Production*, *430*(10), 139559. doi:10.1016/j.jclepro.2023.139559

Schiller, P. L., & Kenworthy, J. (2017). *An introduction to sustainable transportation: Policy, planning and implementation*. Routledge. doi:10.4324/9781315644486

Schot, J., & Geels, F. W. (2008). Strategic niche management and sustainable innovation journeys: Theory, findings, research agenda, and policy. *Technology Analysis and Strategic Management*, *20*(5), 537–554. doi:10.1080/09537320802292651

Schot, J., & Kanger, L. (2018). Deep transitions: Emergence, acceleration, stabilization and directionality. *Research Policy*, *47*(6), 1045–1059. doi:10.1016/j.respol.2018.03.009

Schwartfeger, L., Santos-Martin, D., Wood, A., Watson, N., & Miller, A., (2014, JUNE). Review of Distributed Generation Interconnection Standards. *Proceedings of Electricity Engineers' Association Conference*.

Sedighizadeh, M., Shaghaghi-shahr, G., Esmaili, M., & Aghamohammadi, M. R. (2019). Optimal distribution feeder reconfiguration and generation scheduling for microgrid day-ahead operation in the presence of electric vehicles considering uncertainties. *Journal of Energy Storage*, *21*, 58–71. doi:10.1016/j.est.2018.11.009

Sekhavatmanesh, H., & Cherkaoui, R. (2020). A Multi-Step Reconfiguration Model for Active Distribution Network Restoration Integrating DG Start-Up Sequences. *IEEE Transactions on Sustainable Energy*, *11*(4), 2879–2888. doi:10.1109/TSTE.2020.2980890

Serohi, A. (2021). E-mobility ecosystem innovation – impact on downstream supply chain management processes. is India ready for inevitable change in auto sector? *Supply Chain Management*, *27*(2), 232–249. doi:10.1108/SCM-11-2020-0588

Serpell, O., Paren, B., & Chu, W.-Y. (2021). *Rare Earth Elements – A Resource Constraint of the Energy Transition.* Kleinman Center for Energy Policy. https://kleinmanenergy.upenn.edu/wp-content/uploads/2021/05/KCEP-Rare-Earth-Elements.pdf

SESEI. (2018). *Indian Automobile Industry*. Seconded European Standardization Expert in India (SESEI). https://www.sesei.eu/wp-content/uploads/2018/12/Automotive-Sector-Report_-Final.pdf

Shafik, M. B., Chen, H., Rashed, G. I., El-Sehiemy, R. A., Elkadeem, M. R., & Wang, S. (2019). Adequate Topology for Efficient Energy Resources Utilization of Active Distribution Networks Equipped With Soft Open Points. *IEEE Access : Practical Innovations, Open Solutions*, *7*, 99003–99016. doi:10.1109/ACCESS.2019.2930631

Shaheen, S., Martin, E., & Bansal, A. (2019). Peer-to-Peer (P2P) Carsharing: Understanding Early Markets, Social Dynamics, and Behavioral Impacts. *Transport Research Board 98th Annual Meeting.*

Shaheen, S. A., & Chan, N. D. (2015). Mobility and the sharing economy: Potential to facilitate the first- and last-mile public transit connections. *Built Environment*, *41*(4), 573–588. doi:10.2148/benv.42.4.573

Sha, L., Xiao, F., Chen, W., & Sun, J. (2018). IIoT-SIDefender: Detecting and defense against the sensitive information leakage in industry IoT. *World Wide Web (Bussum)*, *21*(1), 59–88. doi:10.1007/s11280-017-0459-8

Shapsough, S., Qatan, F., Aburukba, R., Aloul, F., & Al Ali, A. R. (2015, October). Smart grid cyber security: Challenges and solutions. In *2015 international conference on smart grid and clean energy technologies (ICSGCE)* (pp. 170-175). IEEE.

Sharma, M. (2021). *Hatchback vs. Sedan.* Spinny Buying Guide. https://www.spinny.com/blog/index.php/hatchback-vs-sedan/

Sharma, A., & Sharma, S. (2019). Review of power electronics in vehicle-to-grid systems. *Journal of Energy Storage*, *21*, 337–361. doi:10.1016/j.est.2018.11.022

Shen, C., Shan, P., & Gao, T. (2011). A comprehensive overview of hybrid electric vehicles. *International Journal of Vehicular Technology*, *2011*, 2011. doi:10.1155/2011/571683

Shi, L., Lv, T., & Wang, Y. (2019). Vehicle-to-grid service development logic and management formulation. *Journal of Modern Power Systems and Clean Energy*, *7*(4), 935–947. doi:10.1007/s40565-018-0464-7

Shin, H., & Baldick, R. (2017). Plug-In Electric Vehicle to Home (V2H) Operation Under a Grid Outage. *IEEE Transactions on Smart Grid*, *8*(4), 2032–2041. doi:10.1109/TSG.2016.2603502

Shiramagond, T., & Lee, W. J. (2018, September). Integration of renewable energy into electric vehicle charging infrastructure. In *2018 IEEE International Smart Cities Conference (ISC2)* (pp. 1-7). IEEE. 10.1109/ISC2.2018.8656981

Short, T. A. (2014). *Electric power distribution handbook.* CRC Press.

Showers, S. O., & Raji, A. K. (2020). Benefits of electric vehicle as Mobile Energy Storage System. *2020 IEEE PES/IAS PowerAfrica.*

Shreya & Afrin. (2022). 3D E-Commers using AR. *International Journal of Information Technology and Computer Engineering, 2*(6), 18-27.

Shuvra, M. A., & Chowdhury, B. H. (2018). Distributed voltage control of active MV distribution networks in the presence of high PV penetration. *2018 IEEE Power and Energy Society Innovative Smart Grid Technologies Conference*, 1–5.

Shweta Nagare. (2014). Different Segmentation Techniques for brain tumor detection: A Survey. *MM- International Society for Green, Sustainable Engineering and Management*, *1*(14), 29–35.

SIAM. (1998-2020). *Annual Reports.* Society of Indian Automobile Manufacturers (SIAM). https://www.siam.in/cpage.aspx?mpgid=42&pgidtrail=89

Siddique, A. B., & Gabbar, H. A. (2023). Adaptive Mixed-Integer Linear Programming-Based Energy Management System of Fast Charging Station with Nuclear–Renewable Hybrid Energy System. *Energies*, *16*(2), 685. doi:10.3390/en16020685

Silvestri, L., Forcina, A., Silvestri, C., & Traverso, M. (2021). Circularity potential of rare earths for sustainable mobility: Recent developments, challenges and future prospects. *Journal of Cleaner Production*, *292*, 126089. doi:10.1016/j.jclepro.2021.126089

Simsir, N. B., & Ertan, H. B. (1999). A Comparison of Torque Capabilities of Axial Flux and Radial Flux Type of Brushless DC (BLDC) Drives for Wide Speed Range Applications. *IEEE 1999 International Conference on Power Electronics and Drive Systems, PEDS'99, Hong Kong, China* (pp. 719-724). https://ieeexplore.ieee.org/document/792793

Singh, H., Ambikapathy, A., Logavani, K., Arun Prasad, G., & Thangavel, S. (2021). Plug-in hybrid electric vehicles (PHEVs). *Electric Vehicles: Modern Technologies and Trends*, 53-72.

Singh, J., & Tiwari, R. (2019). Real power loss minimisation of smart grid with electric vehicles using distribution feeder reconfiguration. *IET Generation, Transmission & Distribution*, *13*(18), 4249–4261. doi:10.1049/iet-gtd.2018.6330

Singh, J., & Tiwari, R. (2020). Cost Benefit Analysis for V2G Implementation of Electric Vehicles in Distribution System. *IEEE Transactions on Industry Applications*, *56*(5), 5963–5973. doi:10.1109/TIA.2020.2986185

Singh, K. V., Bansal, H. O., & Singh, D. (2019). A comprehensive review on hybrid electric vehicles: Architectures and components. *Journal of Modern Transportation*, *27*(2), 77–107. doi:10.1007/s40534-019-0184-3

Singh, P. P., Wen, F., Palu, I., Sachan, S., & Deb, S. (2022). Electric vehicles charging infrastructure demand and deployment: Challenges and solutions. *Energies*, *16*(1), 7. doi:10.3390/en16010007

Singh, S. P., Pamshetti, V. B., & Singh, S. P. (2019). Time Horizon-Based Model Predictive Volt/VAR Optimization for Smart Grid Enabled CVR in the Presence of Electric Vehicle Charging Loads. *IEEE Transactions on Industry Applications*, *55*(6), 5502–5513. doi:10.1109/TIA.2019.2928490

Skeete, J.-P., Wells, P., Dong, X., Heidrich, O., & Harper, G. (2020). Beyond the event horizon: Battery waste, recycling, and sustainability in the United Kingdom Electric Vehicle Transition. Energy Research &amp. *Social Science*, *69*, 101581.

Soares, N., Martins, A. G., Carvalho, A. L., Caldeira, C., Du, C., Castanheira, É., & Garcia, R. (2018). The challenging paradigm of interrelated energy systems towards a more sustainable future. *Renewable & Sustainable Energy Reviews*, *95*, 171–193. doi:10.1016/j.rser.2018.07.023

Sobiech-Grabka, K., Stankowska, A., & Jerzak, K. (2022). Determinants of electric cars purchase intention in Poland: Personal attitudes v. economic arguments. *Energies*, *15*(9), 3078. doi:10.3390/en15093078

Solanki, A., Jain, V., & Gaur, L. (Eds.). (2022). *Applications of Blockchain and Big IoT Systems: Digital Solutions for Diverse Industries*. CRC Press. doi:10.1201/9781003231332

Souto, L., Yip, J., Wu, W.-Y., Austgen, B., Kutanoglu, E., Hasenbein, J., Yang, Z.-L., King, C. W., & Santoso, S. (2022). Power system resilience to floods: Modeling, impact assessment, and mid-term mitigation strategies. *International Journal of Electrical Power & Energy Systems*, *135*, 107545. doi:10.1016/j.ijepes.2021.107545

Sovacool, B. K., Kester, J., Noel, L., & de Rubens, G. Z. (2019). Energy injustice and Nordic electric mobility: Inequality, elitism, and externalities in the electrification of vehicle-to-grid (V2G) transport. *Ecological Economics*, *157*, 205–217. doi:10.1016/j.ecolecon.2018.11.013

Spickermann, A., Grienitz, V., & Von der Gracht, H. A. (2014). Heading towards a multimodal city of the future?: Multi-stakeholder scenarios for urban mobility. *Technological Forecasting and Social Change*, *89*, 201–221. doi:10.1016/j.techfore.2013.08.036

Srinivasan, M., Stonier, A. A., Visalaxi, G., Revanth, M., Sanjeevkumar, K., Sinduja, B., & Subiksha, S. (2021). Design of bidirectional battery charger for electric vehicle. *IOP Conference Series. Materials Science and Engineering*, *1055*(1), 012141. doi:10.1088/1757-899X/1055/1/012141

Statista. (2021a). *Sales Volume of Passenger Cars in India from financial year 2016 to 2021, by body type*. Statista. https://www.statista.com/statistics/1265730/india-sales-volume-of-passenger-cars-by-body-type/

Statista. (2021b). *Reserves of rare earth worldwide as of 2021, by country*. Statista. https://www.statista.com/statistics/277268/rare-earth-reserves-by-country

Statista. (2021c). *Distribution of rare earth production worldwide as of 2021, by country*. Statista. https://www.statista.com/statistics/270277/mining-of-rare-earths-by-country/

Steinschaden, S., & Baptista, J. (2020, January). Development of an Efficient Tool for Solar Charging Station Management for Electric Vehicles. *Energies*, *13*(11), 11. Advance online publication. doi:10.3390/en13112979

Sugimoto, S. (2011). Current status and recent topics of rare-earth permanent magnets. *Journal of Physics. D, Applied Physics*, *44*(6), 064001. doi:10.1088/0022-3727/44/6/064001

Sultana, B., Mustafa, M. W., Sultana, U., & Bhatti, A. R. (2016). Review on reliability improvement and power loss reduction in distribution system via network reconfiguration. *Renewable & Sustainable Energy Reviews*, *66*, 297–310. doi:10.1016/j.rser.2016.08.011

Sultan, V., Aryal, A., Chang, H., & Kral, J. (2022). Integration of EVs into the smart grid: A systematic literature review. *Energy Informatics*, *5*(1), 1–28. doi:10.1186/s42162-022-00251-2 PMID:35252758

Sun, Y., Hu, Y., Zhang, H., Chen, H., & Wang, F.Y. (2023). A Parallel Emission Regulatory Framework for Intelligent Transportation Systems and Smart Cities. *IEEE Transactions on Intelligent Vehicles, 8*(2), 1017-1020. . doi:10.1109/TIV.2023.3246045

Sun, H., Guo, Q., Qi, J., Ajjarapu, V., Bravo, R., Chow, J., Li, Z., Moghe, R., Nasr-Azadani, E., Tamrakar, U., Taranto, G. N., Tonkoski, R., Valverde, G., Wu, Q., & Yang, G. (2019). Review of Challenges and Research Opportunities for Voltage Control in Smart Grids. *IEEE Transactions on Power Systems*, *34*(4), 2790–2801. doi:10.1109/TPWRS.2019.2897948

Sverdrup, H. U. (2016). Modelling global extraction, supply, price and depletion of the extractable geological resources with the LITHIUM model. *Resources, Conservation and Recycling*, *114*, 112–129. doi:10.1016/j.resconrec.2016.07.002

Swami. (2022). Sending notification to someone missing you through smart watch. *International Journal of Information Technology and Computer Engineering, 2*(8), 19 – 24.

Sze, K. Y., & Xu, W. (2020). *Electric Vehicles and Power Grid Integration*. CRC Press.

Tahanian, H., Aliahmadi, M., & Faiz, J. (2020). Ferrite Permanent Magnets in Electrical Machines: Opportunities and Challenges of a Non-Rare-Earth Alternative. *IEEE Transactions on Magnetics*. Advance online publication. doi:10.1109/TMAG.2019.2957468

Tahir, Z., Galadima, A., & Maruf, A. (2019). *A Review of Motors for Electric Vehicles*. International Power Engineering Exhibition & Conference, Nigeria. https://www.researchgate.net/publication/334285900_A_Review_of_Motors_for_Electric_Vehicles

Tamanni, L., & Haji Besar, M. H. (2019). Profitability vs poverty alleviation: Has banking logic influences Islamic microfinance institutions? *Asian Journal of Accounting Research*, *4*(2), 260–279. doi:10.1108/AJAR-05-2019-0039

Tanaka, K., Nagakubo, K., & Abe, R. (2017, June). Blockchain-based electricity trading with Digitalgrid router. In *2017 IEEE international conference on consumer electronics-Taiwan (ICCE-TW)* (pp. 201-202). IEEE.

Tanaka, K. (2009). Decentralized voltage control in distribution systems by distributed generators. *IEEE International Symposium on Industrial Electronics*, 554-559. 10.1109/ISIE.2009.5213299

Tang, Z., Hill, D. J., & Liu, T. (2021). Distributed Coordinated Reactive Power Control for Voltage Regulation in Distribution Networks. *IEEE Transactions on Smart Grid*, *12*(1), 312–323. doi:10.1109/TSG.2020.3018633

Tan, K. M., Babu, T. S., Ramachandaramurthy, V. K., Kasinathan, P., Solanki, S. G., & Raveendran, S. K. (2021). Empowering smart grid: A comprehensive review of energy storage technology and application with renewable energy integration. *Journal of Energy Storage*, *39*, 102591. doi:10.1016/j.est.2021.102591

Tan, K. M., Ramachandaramurthy, V. K., & Yong, J. Y. (2016). Integration of electric vehicles in smart grid: A review on vehicle to grid technologies and optimization techniques. *Renewable & Sustainable Energy Reviews*, *53*, 720–732. doi:10.1016/j.rser.2015.09.012

Tata Motors (2022). *Specifications and Features of TATA Nexon EV.* TATA Motors. https://nexonev.tatamotors.com/features/

Thakur, D. (2021). *Electric Vehicle Architecture & EV Powertrain Components*. E-Vehicle Info. https://e-vehicleinfo.com/electric-vehicle-architecture-ev-powertrain-components/

Themelis, N. J., & Ulloa, P. A. (2007). Methane generation in landfills. *Renewable Energy*, *32*(7), 1243–1257. doi:10.1016/j.renene.2006.04.020

Thimm, T. (2020). E-destination: The future of E-mobility in the Lake Constance Region, Germany. *Journal of Tourism Futures*, *8*(1), 24–36. doi:10.1108/JTF-10-2019-0116

Tian, M.-W., & Talebizadehsardari, P. (2021). Energy cost and efficiency analysis of building resilience against power outage by shared parking station for electric vehicles and demand response program. *Energy*, *215*, 119058. doi:10.1016/j.energy.2020.119058

Tian, Y., Zeng, G., Rutt, A., Shi, T., Kim, H., Wang, J., Koettgen, J., Sun, Y., Ouyang, B., Chen, T., Lun, Z., Rong, Z., Persson, K., & Ceder, G. (2020). Promises and challenges of next-generation "beyond Li-ion" batteries for electric vehicles and grid decarbonization. *Chemical Reviews*, *121*(3), 1623–1669. doi:10.1021/acs.chemrev.0c00767 PMID:33356176

Torres, B. S., Borges da Silva, L. E., Salomon, C. P., & de Moraes, C. H. (2022). Integrating smart grid devices into the traditional protection of distribution networks. *Energies*, *15*(7), 2518. doi:10.3390/en15072518

Trakas, D. N., & Hatziargyriou, N. D. (2018). Optimal Distribution System Operation for Enhancing Resilience Against Wildfires. *IEEE Transactions on Power Systems*, *33*(2), 2260–2271. doi:10.1109/TPWRS.2017.2733224

Trench, A., & Sykes, J. P. (2020). Rare Earth Permanent Magnets and Their Place in the Future Economy. *Engineering*. doi:10.1016/j.eng.2019.12.007

Troester, E. (2009). New German grid codes for connecting PV systems to the medium voltage power grid. *2nd International workshop on concentrating photovoltaic power plants: optical design, production, grid connection*, 9-10.

Umunnakwe, A., Huang, H., Oikonomou, K., & Davis, K. R. (2021). Quantitative analysis of power systems resilience: Standardization, categorizations, and challenges. *Renewable & Sustainable Energy Reviews*, *149*, 111252. doi:10.1016/j.rser.2021.111252

Unnerstall, T. (2022). *Factfulness Sustainability: What you should know about Ecological Crises and Resource Consumption.* Springer Nature.

Un-Noor, F., Padmanaban, S., Mihet-Popa, L., Mollah, M. N., & Hossain, E. (2017). A comprehensive study of key electric vehicle (EV) components, technologies, challenges, impacts, and future direction of development. *Energies*, *10*(8), 1217. doi:10.3390/en10081217

ur Rehman. (2022). A robust vehicle to grid aggregation framework for electric vehicles charging cost minimization and for smart grid regulation. *International Journal of Electrical Power & Energy Systems, 140*. doi:10.1016/j.ijepes.2022.108090

USGS. (2022). *Rare Earths Statistics and Information*. National Minerals Information Center U.S. Geological Survey (USGS). https://www.usgs.gov/centers/national-minerals-information-center/rare-earths-statistics-and-information

Ushashree, P., & Kumar, K. S. (2023). Power System Reconfiguration in Distribution System for Loss Minimization Using Optimization Techniques: A Review. *Wireless Personal Communications*, *128*(3), 1907–1940. doi:10.1007/s11277-022-10026-3

Vaijnath, S. P., & Prajakta, M. (2022). Smart safety device for women. *International Journal of Aquatic Science*, *13*(1), 556–560.

Valverde, G., & Van Cutsem, T. (2013). Model predictive control of voltages in active distribution networks. *IEEE Transactions on Smart Grid*, *4*(4), 2152–2161. doi:10.1109/TSG.2013.2246199

Vargas, A., & Samper, M. E. (2012). Real-Time Monitoring and Economic Dispatch of Smart Distribution Grids: High Performance Algorithms for DMS Applications. *IEEE Transactions on Smart Grid*, *3*(2), 866–877. doi:10.1109/TSG.2012.2187078

VDE. (2011). *Power generation systems connected to the low-voltage distribution network*. VDE Verlag.

Venugopal, D. S. (2022). *Building an electric mobility ecosystem in India*. Deccan Herald.

Verma, S., Mishra, S., Gaur, A., Chowdhury, S., Mohapatra, S., Dwivedi, G., & Verma, P. (2021). A comprehensive review on energy storage in Hybrid Electric Vehicle. *Journal of Traffic and Transportation Engineering*, *8*(5), 621–637. doi:10.1016/j.jtte.2021.09.001

Verzijlbergh, R. A., Grond, M. O., Lukszo, Z., Slootweg, J. G., & Ilic, M. D. (2012). Network impacts and cost savings of controlled EV charging. *IEEE Transactions on Smart Grid*, *3*(3), 1203–1212. doi:10.1109/TSG.2012.2190307

Vidhi, R., & Shrivastava, P. (2018). A Review of Electric Vehicle Lifecycle Emissions and Policy Recommendations to Increase EV Penetration in India. *Energies*, *11*(3), 483. doi:10.3390/en11030483

Vu, V.-B., Gonzalez-Gonzalez, J. M., Pickert, V., Dahidah, M., & Trivino, A. (2021). A Hybrid Charger of Conductive and Inductive Modes for Electric Vehicles. IEEE Trans. Ind. Electron., 68(12), 12021–12033. doi:10.1109/TIE.2020.3042162

Waghmare. (2022). Smart watch system. *International Journal of Information Technology and Computer Engineering, 2*(6), 1 - 9.

Waghmode. (2022). *Voltage Sag mitigation in DVR based on Ultra capacitor. Lambart Publications.*

Wale, A. D., & Dipali, R. (2019). Smart Agriculture System using IoT. *International Journal of Innovative Research In Technology*, *5*(10), 493–497.

Walling, R. A., Saint, R., Dugan, R. C., Burke, J., & Kojovic, L. A. (2008). Summary of distributed resources impact on power delivery systems. *IEEE Transactions on Power Delivery*, *23*(3), 1636–1644. doi:10.1109/TPWRD.2007.909115

Wang, T., Zhao, C., Yang, Q., Zhang, S., & Liew, S.C. (2021). Ethna: Analyzing the Underlying Peer-to-Peer Network of Ethereum Blockchain. *IEEE Transactions on Network Science and Engineering, 8*(3), 2131-2146. . doi:10.1109/TNSE.2021.3078181

Wang, B., Gou, Y., & Zhang, S. (2023). V2G Blind Signature Privacy Protection Scheme Based on Blockchain. *Journal of Electronics and Information Science*, *8*(1), 36–41. doi:10.1016/j.ins.2023.01.021

Wang, J., Wang, W., Wang, H., & Zuo, H. (2020a). Dynamic Reconfiguration of Multiobjective Distribution Networks Considering DG and EVs Based on a Novel LDBAS Algorithm. *IEEE Access : Practical Innovations, Open Solutions*, *8*, 216873–216893. doi:10.1109/ACCESS.2020.3041398

Wang, J., Wang, W., Yuan, Z., Wang, H., & Wu, J. (2020b). A Chaos Disturbed Beetle Antennae Search Algorithm for a Multiobjective Distribution Network Reconfiguration Considering the Variation of Load and DG. *IEEE Access : Practical Innovations, Open Solutions*, *8*, 97392–97407. doi:10.1109/ACCESS.2020.2997378

Wang, L., Yan, R., Bai, F., Saha, T., & Wang, K. (2020). A Distributed Inter-Phase Coordination Algorithm for Voltage Control with Unbalanced PV Integration in LV Systems. *IEEE Transactions on Sustainable Energy*, *11*(4), 2687–2697. doi:10.1109/TSTE.2020.2970214

Wang, Q., & Su, M. (2020). Integrating blockchain technology into the energy sector—From theory of blockchain to research and application of energy blockchain. *Computer Science Review*, *37*, 100275. doi:10.1016/j.cosrev.2020.100275

Wang, S., Ouyang, L., Yuan, Y., Ni, X., Han, X., & Wang, F.-Y. (2019). Blockchain-Enabled Smart Contracts: Architecture, Applications, and Future Trends. *IEEE Transactions on Systems, Man, and Cybernetics. Systems*, *49*(11), 2266–2277. doi:10.1109/TSMC.2019.2895123

Wang, Y., John, T., & Xiong, B. (2019). A two-level coordinated voltage control scheme of electric vehicle chargers in low-voltage distribution networks. *Electric Power Systems Research*, *168*, 218–227. doi:10.1016/j.epsr.2018.12.005

Wang, Y., Qiu, D., Strbac, G., & Gao, Z. (2023). Coordinated Electric Vehicle Active and Reactive Power Control for Active Distribution Networks. *IEEE Transactions on Industrial Informatics*, *19*(2), 1611–1622. doi:10.1109/TII.2022.3169975

Wang, Y., Zhao, Z., & Baležentis, T. (2023). Benefit distribution in shared private charging pile projects based on modified Shapley value. *Energy*, *263*, 125720. doi:10.1016/j.energy.2022.125720

Willyanto, H. R., & Dhafi Iskandar, M. (2023). *Implementation Marketing Business Plan for Electric Vehicle Charging Station (EV-Station) in Indonesia*. Syntax Idea. doi:10.46799/syntax-idea.v5i11.2594

Woo, J., & Magee, C. L. (2020). Forecasting the value of battery electric vehicles compared to internal combustion engine vehicles: The influence of driving range and battery technology. *International Journal of Energy Research*, *44*(8), 6483–6501. doi:10.1002/er.5382

World Health Organization. (2012). *Health in the green economy: health co-benefits of climate change mitigation-transport sector*. World Health Organization.

Wu, C., Chen, C., Ma, Y., Li, F., Sui, Q., Lin, X., Wei, F., & Li, Z. (2022). Enhancing resilient restoration of distribution systems utilizing electric vehicles and supporting incentive mechanism. *Applied Energy*, *322*, 119452. doi:10.1016/j.apenergy.2022.119452

Wu, G., Yi, C., Xiao, H., Wu, Q., Zeng, L., Yan, Q., & Zhang, M. (2023). Multi-objective optimization of Integrated Energy Systems considering renewable energy uncertainty and Electric Vehicles. *IEEE Transactions on Smart Grid*, *14*(6), 1–1. doi:10.1109/TSG.2023.3250722

Wu, J., Wu, Z., Wu, F., & Mao, X. (2017). A power balancing method of distributed generation and electric vehicle charging for minimizing operation cost of distribution systems with uncertainties. *Energy Science & Engineering*, *5*(3), 167–179. doi:10.1002/ese3.157

Wu, R., & Liu, S. (2023). Multi-Objective Optimization for Distribution Network Reconfiguration With Reactive Power Optimization of New Energy and EVs. *IEEE Access : Practical Innovations, Open Solutions, 11*, 10664–10674. doi:10.1109/ACCESS.2023.3241228

Wurster, S., Heß, P., Nauruschat, M., & Jütting, M. (2020). Sustainable circular mobility: User-integrated innovation and specifics of electric vehicle owners. *Sustainability (Basel), 12*(19), 7900. doi:10.3390/su12197900

Wu, W., Cheung, C., Lo, S. Y., Zhong, R. Y., & Huang, G. Q. (2020). An IOT-enabled real-time logistics system for a third party company: A case study. *Procedia Manufacturing, 49*, 16–23. doi:10.1016/j.promfg.2020.06.005

Wu, W., & Lin, B. (2021). Benefits of electric vehicles integrating into power grid. *Energy, 224*, 120108. doi:10.1016/j.energy.2021.120108

Wu, Y., Wang, Z., Huangfu, Y., Ravey, A., Chrenko, D., & Gao, F. (2022). Hierarchical Operation of Electric Vehicle Charging Station in Smart Grid Integration Applications — An Overview. *International Journal of Electrical Power & Energy Systems, 139*, 108005. doi:10.1016/j.ijepes.2022.108005

X-enginner. (n.d.). *What is a battery electric vehicle (BEV).* https://x-engineer.org/battery-electric-vehicle-bev/

Xia, Y., Zhang, L., & Liu, Y. (2016). Special issue on big data driven Intelligent Transportation Systems. *Neurocomputing, 181*, 1-3. doi:10.1016/j.neucom.2015.10.121

Xiang, Y., Hu, S., Liu, Y., Zhang, X., & Liu, J. (2019). Electric vehicles in smart grid: A survey on charging load modelling. *IET Smart Grid, 2*(1), 25–33. doi:10.1049/iet-stg.2018.0053

Xiao, H., Huimei, Y., Chen, W., & Hongjun, L. (2014, June). A survey of influence of electrics vehicle charging on power grid. In *2014 9th IEEE conference on Industrial Electronics and Applications* (pp. 121-126). IEEE. 10.1109/ICIEA.2014.6931143

Xie, W., Zhang, C., Kumar, V., & Chen, T. (2023). A Novel Stochastic Hierarchical Framework for Smart Management of Electric Vehicle Aggregation with Demand Response Mechanism. *2023 IEEE Sustainable Power and Energy Conference (iSPEC),* 1-6.

Xie, F., Liu, C., Li, S., Lin, Z., & Huang, Y. (2018). Long-term strategic planning of inter-city fast charging infrastructure for battery electric vehicles. *Transportation Research Part E, Logistics and Transportation Review, 109*, 261–276. doi:10.1016/j.tre.2017.11.014

Xue, Q., Zhang, X., Teng, T., Zhang, J., Feng, Z., & Lv, Q. (2020). A comprehensive review on classification, energy management strategy, and control algorithm for hybrid electric vehicles. *Energies, 13*(20), 5355. doi:10.3390/en13205355

Xu, S., Chen, X., & He, Y. (2021). EVchain: An Anonymous Blockchain-Based System for Charging-Connected Electric Vehicles. *Tsinghua Science and Technology, 26*(6), 845–856. doi:10.26599/TST.2020.9010043

Xu, Y., Wang, Y., He, J., Su, M., & Ni, P. (2019). Resilience-Oriented Distribution System Restoration Considering Mobile Emergency Resource Dispatch in Transportation System. *IEEE Access : Practical Innovations, Open Solutions, 7*, 73899–73912. doi:10.1109/ACCESS.2019.2921017

Yang, S., Cheng, W., Hsu, Y., Gan, C., & Lin, Y. (2013). Charge scheduling of electric vehicles in highways. *Mathematical and Computer Modelling, 57*(11–12), 2873-2882. doi:10.1016/j.mcm.2011.11.054

Yang, C., Zha, M., Wang, W., Liu, K., & Xiang, C. (2020). Efficient energy management strategy for hybrid electric vehicles/plug-in hybrid electric vehicles: Review and recent advances under intelligent transportation system. *IET Intelligent Transport Systems, 14*(7), 702–711. doi:10.1049/iet-its.2019.0606

Yang, Q., Wang, G., Sadeghi, A., Giannakis, G. B., & Sun, J. (2020). Two-Timescale Voltage Control in Distribution Grids Using Deep Reinforcement Learning. *IEEE Transactions on Smart Grid*, *11*(3), 2313–2323. doi:10.1109/TSG.2019.2951769

Yang, W., Joo, M., Kim, Y., Kim, S. H., & Chung, J.-M. (2020). Hybrid Machine Learning Scheme for Classification of BECTS and TLE Patients Using EEG Brain Signals. *IEEE Access : Practical Innovations, Open Solutions*, *8*, 218924–218935. doi:10.1109/ACCESS.2020.3038948

Yang, Z., Dehghanian, P., & Nazemi, M. (2020). Seismic-Resilient Electric Power Distribution Systems: Harnessing the Mobility of Power Sources. *IEEE Transactions on Industry Applications*, *56*(3), 2304–2313. doi:10.1109/TIA.2020.2972854

Yan, R., Marais, B., & Saha, T. K. (2014). Impacts of residential photovoltaic power fluctuation on on-load tap changer operation and a solution using DSTATCOM. *Electric Power Systems Research*, *111*, 185–193. doi:10.1016/j.epsr.2014.02.020

Yan, R., & Saha, T. K. (2012). Investigation of voltage stability for residential customers due to high photovoltaic penetrations. *IEEE Transactions on Power Systems*, *27*(2), 651–662. doi:10.1109/TPWRS.2011.2180741

Yan, X., & Sun, S. (2021). Impact of electric vehicle development on China's energy consumption and greenhouse gas emissions. *Clean Technologies and Environmental Policy*, *23*(10), 2909–2925. doi:10.1007/s10098-021-02209-6

Yin, S.-A., & Lu, C.-N. (2009). Distribution Feeder Scheduling Considering Variable Load Profile and Outage Costs. *IEEE Transactions on Power Systems*, *24*(2), 652–660. doi:10.1109/TPWRS.2009.2016300

YoCharge. (n.d.). *The Different Types of Electric Vehicles*. https://yocharge.com/ev/types/

Yogita Shirdale,. (2014). Analysis and design of Capacitive coupled wideband Microstrip antenna in C and X band: A Survey. *Journal GSD-International Society for Green, Sustainable Engineering and Management*, *1*(15), 1–7.

Yong, C., Li, Y., Zeng, Z., Zhang, Z., Zhang, Z., & Liu, Y. (2020). Coordinated active and reactive power optimization considering load characteristics for active distribution network. *Chinese Journal of Electrical Engineering*, *6*(4), 97–105. doi:10.23919/CJEE.2020.000034

Yoro, K. O., & Daramola, M. O. (2020). CO2 emission sources, greenhouse gases, and the global warming effect. In *Advances in carbon capture* (pp. 3–28). Woodhead Publishing. doi:10.1016/B978-0-12-819657-1.00001-3

Yuan, W., Wang, J., Qiu, F., Chen, C., Kang, C., & Zeng, B. (2016). Robust Optimization-Based Resilient Distribution Network Planning Against Natural Disasters. *IEEE Transactions on Smart Grid*, *7*(6), 2817–2826. doi:10.1109/TSG.2015.2513048

Yu, D., Adhikari, M. P., Guiral, A., Fung, A. S., Mohammadi, F., & Raahemifar, K. (2019, May). The impact of charging battery electric vehicles on the load profile in the presence of renewable energy. In *2019 IEEE Canadian Conference of Electrical and Computer Engineering (CCECE)* (pp. 1-4). IEEE. 10.1109/CCECE.2019.8861730

Yu, L., Czarkowski, D., & De León, F. (2012). Optimal distributed voltage regulation for secondary networks with DGs. *IEEE Transactions on Smart Grid*, *3*(2), 959–967. doi:10.1109/TSG.2012.2190308

Zaidi, I., Oulamara, A., Idoumghar, L., & Basset, M. (2023). Electric Vehicle Charging Scheduling Problem: Heuristics and Metaheuristic Approaches. *SN Computer Science*, *4*(3), 283. Advance online publication. doi:10.1007/s42979-023-01708-1

ZAPINAMO. (n.d.). *Affordable, Future-proof Rapid Charging Infrastructure For Electric Vehicles From stored [clean and economical] energy*. https://cordis.europa.eu/project/id/790471

Zeng, X., Li, M., Abd El-Hady, D., Alshitari, W., Al-Bogami, A. S., Lu, J., & Amine, K. (2019). Commercialization of lithium battery technologies for electric vehicles. *Advanced Energy Materials*, *9*(27), 1900161. doi:10.1002/aenm.201900161

Zeraati, M., Golshan, M. E. H., & Guerrero, J. M. (2017). A consensus-based cooperative control of PEV battery and PV active power curtailment for voltage regulation in distribution networks. *IEEE Transactions on Smart Grid*, *10*(1), 670–680. doi:10.1109/TSG.2017.2749623

Zhang, B., Zhang, L., Tang, W., Wang, Z., & Wang, C. (2021). A coordinated restoration method of electric buses and network reconfiguration in distribution systems under extreme events. *CSEE Journal of Power and Energy Systems*, 1–9. doi:10.17775/CSEEJPES.2020.04320

Zhang, C., & Xu, Y. (2020). Hierarchically-Coordinated Voltage/VAR Control of Distribution Networks Using PV Inverters. *IEEE Transactions on Smart Grid*, *11*(4), 2942–2953. doi:10.1109/TSG.2020.2968394

Zhang, W., Yang, W., Chen, C., Li, N., Bao, Z., & Luo, M. (2022). Toward Privacy-Preserving Blockchain-Based Electricity Auction for V2G Networks in the Smart Grid. *Security and Communication Networks*, *2022*. doi:10.1155/2022/6911463

Zhang, X., Chan, K. W., Li, H., Wang, H., Qiu, J., & Wang, G. (2021, June). Deep-Learning-Based Probabilistic Forecasting of Electric Vehicle Charging Load With a Novel Queuing Model. *IEEE Transactions on Cybernetics*, *51*(6), 3157–3170. doi:10.1109/TCYB.2020.2975134 PMID:32248136

Zhang, X., Xie, J., Rao, R., & Liang, Y. (2014). Policy incentives for the adoption of electric vehicles across countries. *Sustainability (Basel)*, *6*(11), 8056–8078. doi:10.3390/su6118056

Zhang, Z., da Silva, F. F., Guo, Y., Bak, C. L., & Chen, Z. (2021). Double-layer stochastic model predictive voltage control in active distribution networks with high penetration of renewables. *Applied Energy*, *302*, 117530. doi:10.1016/j.apenergy.2021.117530

Zhang, Z., Du, W., Lu, Y., Xu, G., & Zhao, Y. (2022). Two-Stage Robust Distribution Network Reconfiguration Against Failures of Lines and Renewable Generations. *IEEE Access : Practical Innovations, Open Solutions*, *10*, 108614–108624. doi:10.1109/ACCESS.2022.3213803

Zhan, J., Liu, W., Chung, C. Y., & Yang, J. (2020). Switch Opening and Exchange Method for Stochastic Distribution Network Reconfiguration. *IEEE Transactions on Smart Grid*, *11*(4), 2995–3007. doi:10.1109/TSG.2020.2974922

Zhao, W., Liu, D., Li, Q., & Nong, J. 2015. A novel flexible load coordination control based on distributed generation consumption in the active distribution network. *IET Conference Publications, 2015*(CP679), 7–11.

Zhao, Y., Feng, C., Lin, Z., Wen, F., He, C., & Lin, Z. (2018). Development of Optimal Bidding Strategy for an Electric Vehicle Aggregator in a Real-Time Electricity Market. *2018 IEEE Innovative Smart Grid Technologies - Asia (ISGT Asia)*, 288-293.

Zhao, B., Xu, Z., Xu, C., Wang, C., & Lin, F. (2018). Network Partition-Based Zonal Voltage Control for Distribution Networks with Distributed PV Systems. *IEEE Transactions on Smart Grid*, *9*(5), 4087–4098. doi:10.1109/TSG.2017.2648779

Zhao, J., Mili, L., & Wang, M. (2018). A Generalized False Data Injection Attacks Against Power System Nonlinear State Estimator and Countermeasures. *IEEE Transactions on Power Systems*, *33*(5), 4868–4877. doi:10.1109/TPWRS.2018.2794468

Zhao, J., Yao, M., Yu, H., Song, G., Ji, H., & Li, P. (2020). Decentralized voltage control strategy of soft open points in active distribution networks based on sensitivity analysis. *Electronics (Basel)*, *9*(2), 295. doi:10.3390/electronics9020295

About the Contributors

Lakshmi D. has 26 years of teaching experience. She has addressed innumerable guest lectures, acted as a session chair, and was invited as a keynote speaker at several international conferences. She has conducted FDPs that cover approximately ~1,00,000 members. She has 20 international conference presentations and 2 papers that got best paper awards, 30 international journal papers inclusive of SCOPUS & SCI (cumulative impact factor 38.33). A total of 20 SCOPUS-indexed book chapters. A total of 24 patents are in various states at both national and international levels. She was appointed as an external examiner for several Ph.D candidates. Under her supervision, one scholar has completed a Ph.D. She received two educational awards in the year 2022. She has 3 books under self-publication. She acted as a SPOC to SIH-2022, SIH-2023, and KAVACH-2023. She has created 15 hours of the high-quality course "Python Essentials" at VIDyarthi Platform. She has 7 SCOPUS Indexed book edits.

Neelu Nagpal (Senior Member, IEEE) received the bachelor's degree in electrical engineering from the Delhi College of Engineering, the master' degree (Hons.) in control and instrumentation from Delhi University, and the Ph.D. degree in electrical engineering from Delhi Technological University, Delhi, India. She is presently working as an Associate Professor in EEE Department of Maharaja Agrasen Institute of Technology, Delhi (affiliated to GGSIP University, Delhi). She has a grant of one Australian patent. Her research interests include stochastic and nonlinear control, state estimation, smart grid technologies, renewable energy integration, and artificial intelligence. She was a recipient of the Commendable Research Award from Delhi Technological University, during her Ph.D. course. She is the Vice-Chair of the 2023 IEEE Smart Cities Ambassador Program. She is Associate Editor of IEEE Journal of Emerging and Selected Topics in Power Electronics (Special issue) and Editor of 2 books.

Neelam Kassarwani received her B. E. degree in electrical engineering from Gorakhpur University, Gorakhpur, her M.E. degree in electrical engineering (Power Apparatus and Systems) from Delhi University, Delhi, India, and her Ph.D. degree in electrical engineering from the National Institute of Technology (NIT), Kurukshetra, Haryana, India. She did her Ph.D. on the topic "Performance Study of Dynamic Voltage Restorer for Mitigation of Voltage Sags". Currently, she is working as an Associate Professor in the Electrical and Electronics Engineering Department at the Maharaja Agrasen Institute of Technology, Delhi, India. She received the award of national scholarship during her graduation. She has vast experience in teaching various subjects, power systems, control systems, materials in electrical systems, and power electronics for more than eighteen years in various Technical Institutes. She has also worked with a Govt. Consultancy as an Electrical Engineer for eight years. She has published various papers in international journals and conferences. Her research findings are related to artificial intelligence, machine learning applications in power systems, power systems, dynamic voltage restorer, load frequency control, load forecasting, prediction, and optimization techniques. She has authored and co authored various research papers, and chapters published.

Vishnuvarthanan Govindaraj is an Associate Professor at VIT-Bhopal University, Bhopal, Madhya Pradesh, in the Division of Data Science of the School of Computing Science and Education. He earned a PhD in Electronics and Communication Engineering for his thesis "Tumor Detection and Tissue Segmentation in Multimodal MR Brain Images Using Fuzzy and Optimization Techniques" in 2015. He taught at three of Tamil Nadu's top engineering schools for fourteen years. He has published more than 100 journal papers in 31 international journals included in the Science Citation Index Database, with the greatest impact factor being 17.560 and the average being 4.923. His publications in IEEE Transactions on Fuzzy Systems (12.253), Industrial Informatics (11.648), and Information Fusion (Elsevier) are notable. He has written or co-written more than 70 SCOPUS-indexed conference papers. He has two Scopus-indexed book chapters and has reviewed fourteen journals, including IEEE Transactions on Fuzzy Systems, Neural Networks and Learning Systems, Current Medical Imaging and Reviews, and Neural Processing Letters. Three journals—Computers in Biology and Medicine, Applied Soft Computing, and OPTIK—have accredited him as a reviewer. Medical Image Processing, AI, Machine Learning, and Pattern Recognition are his specialties.

Pierluigi Siano (M'09–SM'14) received the M.Sc. degree in electronic engineering and the Ph.D. degree in information and electrical engineering from the University of Salerno, Salerno, Italy, in 2001 and 2006, respectively. He is a Full Professor of Electrical Power Systems and Scientific Director of the Smart Grids and Smart Cities Laboratory with the Department of Management & Innovation Systems, University of Salerno. In 2013 he received the Italian National Scientific Qualification as Full Professor in the competition sector electrical energy engineering. His research activities are centered on the integration of distributed energy resources in smart distribution systems and on planning and management of power systems. In these research fields, he has co-authored more than 760 articles including more than 460 international journals that received in Scopus more than 20200 citations with an H-index equal to 68. Since 2019 he has been awarded as a Highly Cited Researcher in Engineering by Web of Science Group. He has been the Chair of the IES TC on Smart Grids. He is an Editor for the Power & Energy Society Section of IEEE Access, IEEE TRANSACTIONS ON POWER SYSTEMS, IEEE TRANSACTIONS ON INDUSTRIAL INFORMATICS, IEEE TRANSACTIONS ON INDUSTRIAL ELECTRONICS, IEEE SYSTEMS.

* * *

Fairosebanu Abdul received her M.Sc., (CS) degree in Government Arts College (Autonomous), Kumbakonam, India, in 2012. She also received her M.Phil. (CS) degree in the Jamal Mohammed College (Autonomous), Trichy, India, in 2013. Now she is employed as an Assistant Professor in PG & Research Department of Computer Science, Idhaya College for Women, Kumbakonam, India. In addition, she is pursuing a PhD (Computer Science) in Government Arts and Science College, Kumulur, Lalgudi, Trichy, India.

Mohammed Hyder Ali A. H. is pursuing B.E. program specializing in Electronics and Communication at Sahyadri College of Engineering & Management, Mangalore, Karnataka, India. His area of interest includes Digital System Design.

Jafar A. Alzubi is an associate professor at Al-Balqa Applied University, School of Engineering, Jordan. Received Ph.D. degree in Advanced Telecommunications from Swansea University, Swansea – UK (2012). Master of Science degree (Hons.) in electrical and computer engineering from New York Institute of Technology, New York - USA (2005). And Bachelor of Science degree (Hons.) in Electrical Engineering, majoring in Electronics and Communications, from the University of Engineering and Technology, Lahore – Pakistan (2001). Jafar works and researches in multi and interdisciplinary environment involving Machine Learning, classifications and detection of Web scams, Internet of Things, Wireless Sensor Networks and its security, and using Algebraic– Geometric theory in channel coding for wireless networks. Managed and directed few projects funded by the European Union. A cumulative research experience for over ten years resulted in publishing more than sixty papers in highly impacted journals.

Manjunatha Badiger has been working as an Assistant Professor in the Department of Electronics and Communication Engineering at Sahyadri College of Engineering & Management. He has 12 years of experience in Academics, Research, and Administration. He is currently pursuing his Ph.D. in the field of Deep Learning at Visvesvaraya Technological University. He has obtained an M.Tech in VLSI & Embedded System Design from Reva Institute of Technology and Management. He has obtained a B.E. in Electronics & Communication Engineering from the NMAM Institute of Technology, Nitte. His areas of interest are Machine Learning, Deep Learning & Image processing, and VLSI. He has contributed to National and International journals, and book chapters to various reputed journals and Conferences.

Sanjib Ganguly was born in West Bengal, India. He was awarded with the Ph.D degree from the Department of Electrical Engineering, Indian Institute of Technology Kharagpur in 2011. He worked in the Tata Power Company Ltd., as a Sr. officer, electrical maintenance of 2×500 MW thermal power units in Trombay Thermal Power Station, Mumbai from 2006-2007. He worked as the Assistant Professor in NIT Rourkela from 2011-2015, before joining in IIT Guwahati in 2015. He is presently working as Associate Professor in the Department of Electronics and Electrical Engineering, IIT Guwahati. His research interest includes distribution system planning and optimization, voltage control and stability, multi-objective optimization, custom power devices, hybrid energy systems, and evolutionary algorithms. He has authored 45 papers in international journals, which include IEEE Transactions, IET, and different Elsevier, Springer, and Wiley journals with more than 2265 citations according to the Google scholar, till date. He has published sole-authored papers in IEEE Transactions Power Systems, IEEE Transactions Power Delivery, and IET Generation, Transmissions, and Distribution. He is enlisted to the World's top 2% scientist list in the years of 2019, 2020, 2021, 2022 according to the list published by authors of Stanford University, USA.

Neha Gupta is an experienced educator with over 22 years of teaching and research experience in artificial neural network, Smart Grids, Blockchain, Power system, Renewable Energy Fuels. Her research has been published in many reputable international conferences, book chapters, and major international journals She is an author of 4 books and registered 7 Patents on her research work with the Indian Ministry of Patents. She has acted as a keynote speaker and delivered several professional talks on various national and international platforms. She has also been invited by ABP news channel as a subject matter expert. She has her MOOC courses live on Udemy. She has chaired various special sessions at international conferences and also served as a reviewer in several reputable international journals and conferences. She is a lifetime member of the The Institution of Electronics and Telecommunication Engineers (IETE).

Nidhi Gupta has done her Ph.D from Delhi Technological University, Delhi. She worked as Lecturer (2002-2007), as Senior Lecturer (2007-2011) in G.P.M.C.E., GGSIPU, Delhi and as an Asst. Prof. in MSIT (2011- till date). Her broad area of interest includes power system operation and control, FACTS, Renewable Energy and electric vehicle Battery Energy Sources. She uses various artificial techniques to design controllers for frequency control of interconnected power systems with different sources like thermal, hydro, wind, gas etc. She build various projects with Internet of things. She is particularly interested in problems that can only be tackled by a combination of renewable power system simulations and energy sources with soft computing approaches.

Kutubuddin Kazi has completed his B.E., M.E., and Ph.D. in E&TC Engineering and is nowadays working as a Professor in the Electrical Engineering Department and also as Dean R&D. He is Post Doctoral Fellow working on IoT in Healthcare. He has published more than 90+ papers in various Journals. Also published 11 books in the field of Engineering. He has published 13 Indian Patents, 1 South African Grant Patent, and 6 UK Grant Patent. All patents are in the field of IoT in Healthcare. He worked as a Reviewer for Scopus Conferences and Journal. Also work as Editorial Board Member for various Journals.

Arnel Ralf Lobo is pursuing B.E. program specializing in Electronics and Communication at Sahyadri College of Engineering & Management, Mangalore, Karnataka, India. His area of interest includes automotive electronics.

Arpita Maheriya is a full-time Ph.D. scholar at Gujarat Technological University – Graduate School of Engineering and Technology. She is pursuing a Ph.D. under the AICTE Doctoral Fellowship scheme. https://scholar.google.com/citations?user=MNIebogAAAAJ&hl=en.83201b6f-afa4-4272-aef0-23f6899d6124

Rudrodip Majumdar has been working as an Assistant Professor in the Energy, Environment, and Climate Change Programme at NIAS since August 2019. He has a doctorate in Nuclear Engineering from North Carolina State University in the USA and an MTech in Nuclear Engineering from IIT Kanpur. Prior to joining NIAS, Rudrodip worked as an Institute Postdoctoral Fellow at IIT Bombay and as a Postdoctoral Research Assistant at North Carolina State University. Rudrodip's professional research interests revolve around Sustainability, encompassing Renewable and Clean Energy, Nuclear and Plasma Fusion Energy, Industrial Applications of Plasma, and Analyses of Energy-efficient Sustainable Systems.

Ashish Makanadar is a Civil Engineer, Designer, Urban Planner and researcher at the forefront of developing integrative methodological frameworks for addressing entangled socioenvironmental challenges facing global urbanization. Through an interdisciplinary body of work spanning urban simulation modelling, spatial econometrics, mixed-methods ethnography and machine learning applications in transportation and land-use planning, Ashish advances our understanding of the feedback dynamics between the built environment, human decision-making and planetary boundaries. His recent project in Environment and Urbanization, "Operationalizing The Nexus: A Neuro-adaptive Systems Approach to Planning for Interdependent Shocks" proposes a novel agent-based operationalization of the food-energy-water nexus to strategically appraise mitigation and adaptation portfolios amid deep uncertainty. Currently, Ashish is leading research applying cognitive mapping and neural semantic representation learning to analyze discursive dimensions of informal settlement upgrade negotiation processes. The work evaluates how planners can leverage conceptual metaphor and cognitive biases to craft more ethically grounded and politically feasible solutions.

Sanketh C. Naik is pursuing B.E. program specializing in Electronics and Communication at Sahyadri College of Engineering & Management, Mangalore, Karnataka, India. His area of interest includes Digital System Design.

Arpita Pandey is a Research Associate at The Energy and Resources Institute (TERI). She has worked on projects related to transport sector decarbonisation, energy transition, just transition, energy efficiency and sustainable agriculture. She holds a Master's degree in Economics. Her research interest lies in understanding the economic implications of energy transition and the role of critical minerals.

Madhumala R. B. is an experienced Associate Professor with a demonstrated history of working in the education management industry with 15 years. International Academic MoU, Faculty Exchange and Research Exchange from different parts of the world. Senior IEEE Member, International conference and Project Competition organizer. Skilled in Enterprise IT Strategy, IT Automation, Enterprise IT Infrastructure, IT Service Management, and Interaction Design. Organized International Seminars from different countries online/offline. Cyber Safe Campus, Research-Oriented technical Sessions. Session chair, Technical Program Committee member for International conferences. Reviewer for Elsevier science direct Inderscience and Quality Journals.

Yadav Krishna Kumar Rajnath was born in Ghazipur in Uttar Pradesh (India) on 10th July 1988. He migrated to Maharashtra (India) during the childhood and received his schooling from Maharashtra State Board. He obtained Bachelor's in Engineering (B.E.) in Mechanical Engineering from K.J. Somaiya College of Engineering, Vidyavihar (Mumbai) in 2010 with First Class with Distinction and completed the Masters of Technology (M.Tech.) in Mechanical Engineering with specialization on Refrigeration, Air-conditioning and Heat Transfer from the National Institute of Technology (NIT) Patna, Bihar (India) in 2014 with First Class with Distinction. In the following year, Dr. Yadav joined as a Junior Research Fellow (JRF) in a Govt. of India sponsored research project on "Flow control in complex duct using synthetic jets" in the Department of Applied Mechanics at the Motilal Nehru National Institute of Technology (MNNIT) Allahabad, Prayagraj (India) and continued in the project till June 2017. He also joined in the Ph.D. programme in the same department in July 2015 and completed Doctor of Philosophy (Ph.D.) in February 2021.

Sona S S is pursuing B.E. program specializing in Electronics and Communication at Sahyadri College of Engineering & Management, Mangalore, Karnataka, India. Her area of interest includes Digital System Design.

Index

O

P

R

S

T

U

V

W

Z

www.ingramcontent.com/pod-product-compliance
Lightning Source LLC
LaVergne TN
LVHW080630011026
849160LV00014B/1166